"We are fortunate to now have a new full translation of Nikolai Aleksandrovich Bernstein's book *On the Construction of Movements* (1947). Few if any researchers have had such long-lasting impact as Bernstein on our understanding of motor control in intact and neurologically impaired subjects. Bernstein began his work in Russia early in the last century and his publications extend almost to his death in 1966. His ideas about muscle synergy, hierarchical control and feedback mechanisms in the nervous system have seen a major resurgence in recent years, in spite of the fact that the technologies he used were quite limited in their precision. We now have a crisply written and highly readable translation of this pivotal volume written by a distinguished international leader in human motor control research, who continues current exploration of Bernstein's ideas using new technologies, and relying heavily on modern instrumentation, modern physics and mathematics."

– *William Zev Rymer, Northwestern University, USA*

BERNSTEIN'S CONSTRUCTION OF MOVEMENTS

Nikolai Aleksandrovich Bernstein was one of the great neuroscientists of the twentieth century and highly respected by Western scientists even though most have never read his most important book entitled *On the Construction of Movements. Bernstein's Construction of Movements: The Original Text and Commentaries* is the first English translation. It supplements the translated text with a series of commentaries by scientists who knew Bernstein personally, as well as leaders in related fields including physics, motor control, and biomechanics.

While written in 1947, Bernstein's book is anything but obsolete, making this English translation and accompanying commentaries an invaluable text. The translated original text presents in detail Bernstein's views on the evolutionary history of biological movement and his multi-level hierarchical scheme of the construction of movements in higher animals, including humans. The following commentaries address Bernstein's personality, the history of the book, and current views on different aspects of neuroscience covered in Bernstein's text. Ultimately, they present "a book within the book" to showcase how Bernstein's heritage has developed over the past years.

This classic, available for the first time to an English-speaking audience, will prove beneficial to students, instructors, and experts of neuroscience, physics, neurophysiology, motor control, motor rehabilitation, biomechanics, dynamical systems, and related fields.

Mark L. Latash was trained in physics of living systems and neurophysiology. He has been working in the fields of motor control and movement disorders, published over 400 papers, six books, and ten edited volumes. He is a recipient of the Bernstein Prize from the International Society of Motor Control.

BERNSTEIN'S CONSTRUCTION OF MOVEMENTS

The Original Text and Commentaries

Edited by Mark L. Latash

Routledge
Taylor & Francis Group
NEW YORK AND LONDON

First published 2021
by Routledge
52 Vanderbilt Avenue, New York, NY 10017

and by Routledge
2 Park Square, Milton Park, Abingdon, Oxon, OX14 4RN

Routledge is an imprint of the Taylor & Francis Group, an informa business

Library of Congress Cataloging-in-Publication Data
Names: Bernshteĭn, N. A. (Nikolaĭ Aleksandrovich), 1896–1966 |
Latash, Mark L., 1953– editor, translator.
Title: Bernstein's construction of movements: the original text and commentaries / edited by Mark L. Latash.
Other titles: O postroenii dvizheniĭ. English
Description: New York, NY: Routledge, 2021. |
Includes bibliographical references and index. |
Identifiers: LCCN 2020022403 (print) | LCCN 2020022404 (ebook) |
ISBN 9780367418908 (hardback) | ISBN 9780367418922 (paperback) |
ISBN 9780367816797 (ebook)
Subjects: LCSH: Bernshteĭn, N. A. (Nikolaĭ Aleksandrovich), 1896–1966.
O postroenii dvizheniĭ. | Human mechanics. | Kinesiology.
Classification: LCC QP303.B4713 2020 (print) | LCC QP303 (ebook) |
DDC 612.7/6–dc23
LC record available at https://lccn.loc.gov/2020022403
LC ebook record available at https://lccn.loc.gov/2020022404

ISBN: 9780367418908 (hbk)
ISBN: 9780367418922 (pbk)
ISBN: 9780367816797 (ebk)

Typeset in Bembo
by Newgen Publishing UK

"To the eternal bright memory of comrades who gave their lives fighting for the Soviet Motherland."

CONTENTS

TRANSLATING BERNSTEIN'S CLASSIC

The name of Nikolai Aleksandrovich Bernstein (1896–1996) is known to all educated persons working in areas related to human movement, physiology, neuroscience, and psychology. Bernstein's works are frequently cited as groundbreaking in discussions of such issues as synergies, the problem of redundancy (also known as the Bernstein problem, Turvey 1990), hierarchical control, anticipation, skill acquisition, physiology of activity, and many others. Surprisingly, until now, his main book related to the control and coordination of movements has not been available for the English-speaking audience. The fate of this book may be viewed as typical for other Bernstein's works. His very first major book, *Contemporary Studies on the Physiology of the Neural Process*, written in the mid-1930s, remained unpublished until the next millennium (Bernstein 2003). His book *On Dexterity and Its Development*, written in the late 1940s, waited for over 40 years to be published in Russian and a few more years to become available in English (Bernstein 1991, 1996). *On the Construction of Movements* waited for over 70 years to be translated into English.

I envy the reader who opens this book for the first time. It is a true intellectual treat to follow the line of thinking, including both impeccable logical constructions and wild guesses, of one of the greatest minds of the twentieth century. When presented with a data set reflecting behavior of an animal, a mathematician sees an equation or a statistical distribution, a physicist sees a law of nature, and Bernstein saw a product of the evolution. Of course, he knew mathematics and mechanics (this is clear from his movement analyses presented in the book), but, first and foremost, in this book he presents himself as an evolutionary biologist looking for processes of the emergence of brain structures and their interactions driven by the fight for survival in the presence of new motor tasks.

The history of the book and the situation in the USSR is reflected in one of the commentaries (Talis 2020). So, I will limit myself to just a few comments. The book was written immediately after the end of World War II. The European part of the USSR was in ruins, population decimated, and economy struggling trying to provide bare necessities for the citizens. The effects of the war were exacerbated by the inhumane communist regime. Food and other basic goods were rationed. To publish a scientific book at that time looked next to impossible. Nevertheless, Bernstein managed to persuade those in power that such a book would be very much needed to facilitate recovery of soldiers and civilians suffering from various war-related injuries. The paperback book was published on poor-quality yellowish paper and became highly successful. In particular, it was awarded the

Stalin Prize, which was the highest award of those times in the USSR. The award could be seen as unexpected given that the book contained no references to Lenin and Stalin (expected from any major publication at those times) and paid only lip service to other official themes such as those of "stakhanovist" labor, etc.

It is feasible that the Stalin Prize protected Bernstein during the Pavlovian Session of the USSR Academy of Sciences and Academy of Medical Sciences (1951), which declared Pavlov's theory the ultimate truth and started persecution of a number of prominent physiologists who were declared "anti-Pavlovian." In fact, by the time of the Session, Bernstein was arguably the only scientist who openly criticized Pavlov's theory (some of those criticisms are contained in the current book), but he was spared during and immediately after the Session and fired a couple of years later (see Feigenberg 2014).

This book is not easy reading. I would compare it to hard candy: It is a pleasure to consume but takes time and effort.

Bernstein's style is anything but reader-friendly. He wrote with very long compound sentences that look like riddles – fun to decipher, but only if one has unlimited time. As the translator, I tried to be true to Bernstein's style and kept such complex sentences unbroken. On the other hand, as the editor, I sometimes took the liberty of splitting some of the particularly long and convoluted sentences into two or even three to facilitate understanding.

Bernstein had high expectations with respect to the level of education of future readers. He had been trained as a medical doctor before the 1917 revolution in Russia and knew Latin and Greek well. So, he did not hesitate using words with Latin and Greek roots, sometimes very rarely used ones and sometimes even words invented by Bernstein himself. Among my favorites are "ecphoria" and "effrenation." Since Bernstein did not consider it necessary to explain the meaning of those words, the reader of the translation will have to find out their meanings with no extra help.

The text contains a number of terms that are no longer used (or very rarely used) in scientific literature. For example, Bernstein addresses motor units as "mions" and, sometimes, "motons." He uses "cathode electrotone" and "anode electrotone" to refer to phenomena that are currently addressed as excitatory and inhibitory post-synaptic potentials (EPSPs and IPSPs) and local currents. At the time of writing the book, Bernstein was not convinced that synapses actually existed and even made in one of his (numerous!) footnotes a statement that the chances of seeing a synapse in the microscope could be equal to the chances of stumbling over a meridian while walking in one's backyard.

Another frequently used pair of ambiguous notions is "paleokinetic" and "neokinetic." Driven by the evolutionary views, Bernstein used the former concept to address a variety of relatively slow processes such as humoral, diffusion (synaptic), and slow electrical potentials. The latter term related to fast processes such as the generation of action potentials. Some of the other terms may be viewed as misleading given their current usage. For example, Bernstein used "innervation" and "denervation" in situations when researchers of the twenty-first century would expect "excitation" and "inhibition."

The book contains a number of views and concepts that can be currently addressed as obsolete. When faced with insufficient knowledge, Bernstein frequently tried to guess and, in a surprisingly large number of cases, he guessed right! But sometimes he guessed wrong thus misleading generations of future researchers. This relates, in particular, to some of his views on interactions among brain structures, on muscle tonus (tone), on the issue of motor redundancy, and on the nature of neural control signals for individual muscles. There are also high-school level inaccuracies (e.g., expressing force in kg and direction in mm), probably related to the fact that the book was written in a hurry. Let us not judge those occurrences too harshly: Only boring people are always right, a genius has the right to be wrong, at least occasionally.

I would like to finish these introductory comments with expression of gratitude toward people who introduced me many years ago to Bernstein's ideas and writings. These include, first and foremost, my late father – Lev Latash, my first mentor – Victor Gurfinkel, and my senior friends-colleagues – Michael Berkinblit, Iosif Feigenberg, Anatol Feldman, Israel Gelfand, and Vladimir Zatsiorsky. I am also grateful to those who encouraged me to translate this wonderful classic – Alexei Chernavsky, Micha Mirsky, Mihai Nadin, and Vera Talis, and to those who made earlier attempts at translating "On the Construction of Movements" but, for various reasons, did not finish this truly enormous undertaking – David Adams and Micha Mirsky.

And now – enjoy!

Mark L. Latash (Translator/Editor)

References

Bernstein NA (1991) *О ловкости и ее развитии. (On dexterity and its development.)* Fizkultura i Sport: Moscow, Russia. (In Russian)

Bernstein NA (1996) On dexterity and its development. In: Latash ML, Turvey MT (Eds.) *Dexterity and Its Development*, pp. 1–244, Erlbaum Publ.: Mahwah, NJ.

Bernstein NA (2003) *Современные искания в физиологии нервного процесса. (Contemporary studies on the physiology of the neural process.)* Smysl: Moscow, Russia. (In Russian)

Feigenberg IM (2014) *Nikolai Bernstein: From Reflex to the Model of the Future*. Studien zur Geschichte des Sports (Book 17), LIT Verlag: Berlin-Münster-Wien-Zürich-London.

Talis VL (2020) *The Current Volume*.

Turvey MT (1990) Coordination. *American Psychologist* 45: 938–953.

ABBREVIATIONS

CES	Central excitatory state
CIS	Central inhibitory state
CIT	Central Institute of Traumatology
CNIIFK	Central Scientific Research Institute of Physical Culture
GCOLIFK	State Central Lenin Order Institute of Physical Culture
NKSO	People's Commissary of Social Security
NTK NKPS	Scientific-Technical Committee of the People's Commissary of Transportation Railways
USSR	Union of Soviet Socialist Republics
VIEM	All-Union Institute of Experimental Medicine
VNIIFK	All-Union Scientific Research Institute of Physical Culture

PREFACE

Work on this book started in 1939, and already in 1940, detailed theses on its first parts were published in the journal *Theory and Practice of Physical Culture*. The thunderstorm of the Great Patriotic War (World War II) interrupted the preparation of this book for publication. During the final year of the war, the author re-wrote the whole text, did a major revision, refined it and supplemented it with new materials. Excerpts from the central part of the book – the theory of movement construction levels – were published in an abbreviated version in 1945 in the volume *Movement and Activity*. At the same time, the author had the opportunity to contribute a lay-reader description of the basics of the theory of movement coordination to the *Physiology Textbook for Physical Culture Colleges* edited by Professor M. Marshak and published in 1946; the author is very much grateful to the editor of the volume for this opportunity.

In a sense, this book summarizes the results of the work by the author and his scientific group over 25 years in the field of human movement physiology. As such, the book is also an attempt to review and summarize the data and concepts related to issues of motor coordination in health and pathology that have accumulated in the world literature. A substantial portion of both ideas and concepts in this book are not novel; in particular, the idea of multi-level structure and control of motor actions can be found for the first time in works by Hughlings Jackson, whereas the term "level" in the context of "levels of switching from afferent pathways to the motor system" was used by academician L.A. Orbeli in 1934 in his *Lectures on the Physiology of the Nervous System*. The author of this book uses this term in a different meaning compared to its interpretation by L.A. Orbeli: (a) I emphasize that the material substrate of each of the coordination levels of the central nervous system contains both effector and specialized receptor formations; (b) the main reasons for a particular level to play the role of the leading level for a given movement (not discussed by L.A. Orbeli) are the content and structure of the motor task to be solved; and (c) the book provides a specific list of coordination levels in the human body and their detailed description.

In addition to the two mentioned classics of neurophysiology, fragmented statements resembling those in this book can be found in works by a number of other authors such as Sherrington, Graham Brown, Goldstein, O. Foerster, Lindhard, Wachholder, Monakow, M.O. Gurevich, and others; as it is commonly said, these ideas are indeed "in the air."

Without doubt, it is time to generalize and summarize the progress in the physiology of coordination, and it would be wrong to postpone doing this. Nevertheless, the author appreciates, better than anybody else, all the drawbacks of his work: the insufficiently complete factual material and the speculative nature of some of the constructs. In a sense, this book is, to a larger degree, a program for future urgently needed studies rather than a dogmatic summary of firmly established facts. The author is encouraged by the awareness that, at some stages of scientific work, a successfully designed program of research may be helpful. At least, as of now, a number of views expressed in the book have already attracted attention and even helped in practical work of colleagues working in such diverse fields as physical education, neuropathological semiotics and diagnostics, psychophysiology of labor and defense programs, pedagogy of artistic performance, etc.

The author is ready to welcome most severe criticisms as long as they help general success of natural science. Even if this critique were destructive for this book, as long as it promotes thoughts and concepts moving us closer to objective truth, the author would be satisfied by appreciation that this book played the role of one of the modest bricks in the foundation of *future physiology*.

Moscow, January 1947

Nikolai Aleksandrovich Bernstein.

PART 1

Movements

1

ON THE ORIGIN OF THE MOTOR FUNCTION

Of all the domains of general physiology, none is more specifically human as the field of physiology of motor function in spite of its obvious, continuous evolvement from phylogenetic ancestors. The point is that no other physiological function has shown such an intense and, moreover, accelerating phylogenetic progression. It would be hard to point out essential changes during evolutionary progression in any other functional domain between humans and other warm-blooded animals, for example, in the domain of autonomic functions such as breathing, circulation, metabolism, etc.[1] There is a single obvious exception showing even more dramatic progress, which has been singled-out into a separate domain based on related methodology and has become a separate science – the domain of psychic life or, as we will address it here for convenience, the domain of central processing systems of the neural machinery. The methodological issue that contributed to turning it into an independent branch of science also created an insurmountable additional difficulty. If we drew this process as a graph for the functions of breathing or metabolism – within the phylogenesis of warm-blooded animals – the line would be nearly parallel to the abscissa axis. The development of psychic functions would look in such a graph as a steeply rising curve; unfortunately, we would have the right to draw only its rightmost part related to humans. The rest of the curve would reflect only hypotheses because of the total absence of objective material related to animals despite the united heroic efforts by researchers in the fields of animal psychology, behaviorism, and conditioned reflexes. Only for motor function can we draw a fully objective evolutionary curve, steeply increasing toward the right part of the graph and leaving far behind its morphological central nervous substrate, which, by itself, displays evolution that is anything but slow. Even this fact, by itself, makes movement physiology interesting for psychologists and neurologists, independently of its importance in the role of a necessary ornament for the much better developed field of psychophysiology of receptor functions.

Besides the mentioned factors, the overwhelming evolutionary importance of motor function has been emphasized by the total time when it dominated in the phylogenesis of the overall somatic apparatus. The highest rate of evolutionary growth of the central processing system is due to the fact that these systems, over the same time interval, had to develop from the effector apparatuses to higher functions. This system gained the leading role in both its position and leading importance in the phylogenetic evolution relatively recently, whereas earlier it played (and still plays in less developed organisms) much more modest, auxiliary roles related to integration of receptor and effector signals. The current picture of the animal world – the living book of phylogenetic history – keeps memories on the early biography of this organ, barely starting (in coelenterates and echinoderms) its future mind-boggling career with the modest role of a switchboard operator, just introducing into the physiological repertoire the new, bioelectrical, (telegraph) transmission method instead of the more ancient and substance-based, humoral (kind of postal) signal transmissions. However, the crucial step in the history of the central processing system was the emergence of elongated animal shapes instead of the ancient circular-symmetrical (ray-like) shapes. This defined the predominance of the front, mouth-bearing end of the body, which becomes the first to encounter both food and danger, thus confronting the biological necessity to signal all other metamers, to lead and unite their movements and initiatives to produce movements. The *head* end became the *main* end of the body. This is the embryo of centralized nervous systems instead of the ancient diffuse ones (Reflex-Republics, Üexkull). Further, the head metamers acquired all the prerequisites for the emergence and development of *telereceptors*, which transformed in the process of refinement and improvement from one of the ancient contact categories (olfaction from taste chemoreceptors, hearing from vibration receptors, and vision from skin photochemical sensitivity). Telereceptors turned out to be a powerful centralizing factor by giving the animal possibilities of reacting to stimuli from infinitely far away, compared to the body size. This brought about, as actions of primary importance, locomotor movements of the whole body in space, which moved aside, into the second-rate category, local metamer reactions, which had dominated in the era of tangoreceptors. The biological necessity of locomotion led to the development of powerful, integrative, synergic apparatuses of the central nervous system – the most ancient in the phylogenesis of actions by vertebrates of a central neural origin – which have not been exceeded, as we will see later, up to humans, with respect to their ability of vast motor integrations and muscle synergies: we speak now of the so-called thalamo-pallidar motor system or *level*, as we are going to address it in future (Chapter 4).

As correctly pointed out by Sherrington, "telereceptors created the brain" – or, more exactly, structures that we have so far addressed as central processing systems (at the same time, telereceptors contributed to the centralizing properties of the spinal cord, which used to be purely metamer-specific and acquired during later phylogenesis obvious properties of a central formation). The point is that receptors, namely telereceptors, are overwhelmingly secondary, derivative instruments, and here we have to continue and develop the line of thinking of Sherrington.

In the process of evolution of the somatic system (maybe with the exception of the latest phylogenetic period), the role of the dominating factor has been played by effector functions. The fate of an individual in the struggle for existence is defined by its actions – their being more or less adequate within the ever-complicating process of adaptation. In this process, the receptor function plays a secondary, service role. Nowhere in the phylogenesis, watching the world is the ultimate, something self-sufficient, goal. Receptor systems happen to be either signal systems – we have already discussed this role – and then any degree of their sophistication cannot by itself ensure biological advantage for the animal if the effector apparatus is defective, or they are part of the process ensuring adequate, coordinated work of the effectors – we will discuss this role later – and here the secondary character of their activity follows directly from the semantic content of performed tasks. Hence, both the signal and corrective roles of receptors is ensured only by effectors of the animal or species. In this aspect, central processing systems historically play the role of servant systems for servants. We will show further how the emergence and development of telereceptors themselves and of the even more important for coordination *sensory syntheses*, which are based on central processing systems, are defined by the increasing and complicating demands from the effector systems.

The complication of motor tasks faced by an organism and the resulting enrichment of coordination resources of the individual proceed along two paths. On the one hand, motor tasks become more and more complicated. Diversity of reactions required from the organism increases. These reactions have to become more varied. The reactions have to meet higher and higher demands with respect to their differentiation and accuracy. Ultimately, requirements to the meaning of movements, actions, and behaviors of an animal also increase. It is sufficient to recall, for example, that the aerodynamics of the bird's flight is more complex compared to the almost completely hydrostatic swimming of a fish, and the hunting strategy of a team of mammal predators is more sophisticated compared to hunting by sharks. The young generation of warm-blooded mammals won over the Jurassic "zaurs" because of their more sophisticated motor abilities.[2] On the other hand, the overall composition of emerging motor problems contains an ever-increasing percentage of problems that are unique, unexpected, and extemporaneous at the expense of more ancient, stereotypical situations. All the numerous studies of "plasticity of the nervous system" show, in addition to the evolutionary increase in adaptability of the central nervous system to unusual changes in external conditions, immediate, nearly instantaneous adjustments to the most fanciful experimental conditions. But even if we forget about exquisite experimental anastomoses of nerves and muscles, the much more mundane fact of the phylogenetic increase in the ability to accumulate individual experience and to create new conditional connections, i.e., once again to go beyond the species-specific stereotypes, supports the aforementioned thesis.

In a somewhat schematic way, one can say that the first of the aforementioned routes of the development of motor coordination is ensured and accompanied primarily by the evolution of receptor functions, whereas the second one – by the evolution of central processing systems. First, as far as receptor function is concerned, there is a systematic qualitative improvement of receptor devices, which originates from the most ancient phylogenesis: overlaying of the ancient (paleokinetic, see Chapter 3) protopathic tactile sensitivity and more novel and sophisticated epicritic sensitivity realized by the neokinetic process; the emergence of the younger (also neokinetic) forms of proprioceptors – geometrical that perceive postures and velocities and led by the neo-labyrinth of the semicircular canals – on the background of proprioceptors reflecting tropisms and led by the otolithic apparatus (the paleo-labyrinth) and adapted for measuring pressures, tensions, efforts, orientation in the gravity field, etc. Second, there is ongoing development of the system of telereceptors, which assume the leading role and contribute to the evolution of central processing systems and the brain and to the whole qualitative revolution, which has been described before and which follows the gradual dominance of receptors of this type. In particular, the main accompanying structural phenomena include: (1) The transition from the single-neuron thalamic scheme of ascending neuronal pathways to the schemes of cortical afferentation involving two or more neurons, which signifies not only the emergence of a couple of extra synaptic connections in the pathway of sensory signals but the profound qualitative signal processing in the intermediate ganglia; (2) The transition from a system of isolated sensory neural nuclei to a two-dimensional elaborate, layered system typical of the cortex of the large hemispheres; the importance of these transitions for the evolution of coordinative function will be clarified later; and (3) The adaptive evolution of sensory function develops in parallel to the formation of progressively more and more elaborate synthetic sensory fields, which will be discussed in Chapters 4 and 5. These sensory syntheses, where raw receptor signals from different sensory organs are combined with stored in memory components related to the individual experience of the animal related to the generation of deeply processed and generalized directives for future movements and actions, stimulate and direct growth and development of central processing systems at least as much as telereceptors do. The phylogenetic formation of these gradually elaborating fields is related to the continuous increase in the role of memory-related components – in other words, of individual memory.[3]

Following a similar schematic interpretation, the second line of development of effector function corresponding to the growing importance of singular reactions based on the accumulated and organized individual experience of the animal, is primarily related to the evolution of central processing systems located in the cortex of the large hemispheres. The development of the cortex provides a prerequisite for the ongoing elaboration of the purposive structure of the organism's actions and an increase in its memory capabilities. This is how central processing systems, at a certain stage of the evolution, transform from a subordinate position to a leading position, thus directing further development of the entire neuro-somatic system.

The phylogenetic evolution of the central nervous system (Figure 1.1), unlike all other organs and systems within the body, is not only (and even, not so much) limited to *quantitative growth* but rather includes qualitative *accretion* with new formations that have no analogues at preceding stages of the phylogenesis and represent primarily superstructures by one (or more) neuronal level within the overall neural process. This principle leads to inevitable leaps in the development of the central nervous system, in particular due to the discrete nature of the neural scheme: The elaboration of a reflex arc or any other pathway for a neural impulse within the central nervous system is possible only if a discrete number of relay neurons have been added. The ways

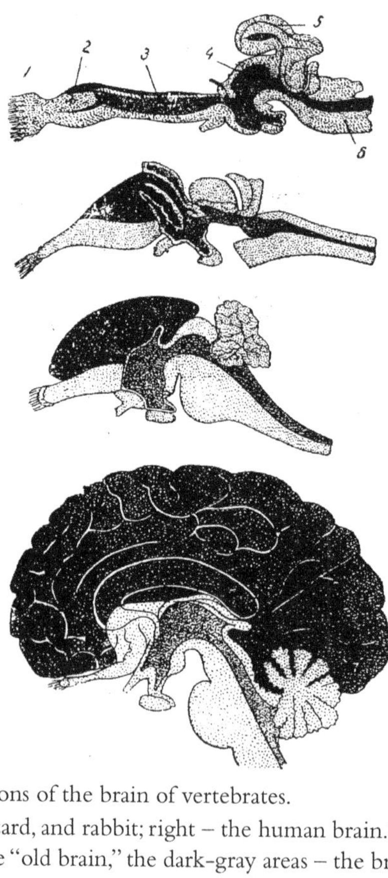

FIGURE 1.1 Longitudinal sections of the brain of vertebrates.
Left – the brain of the shark, lizard, and rabbit; right – the human brain. The black areas show the "new brain," the light-gray areas – the "old brain," the dark-gray areas – the brain ventricles. 1 – the olfactory bulbs; 2 – the pallidum; 3 – the diencephalon; 4 – the thalamus; 5 – the cerebellum; 6 – the medulla.

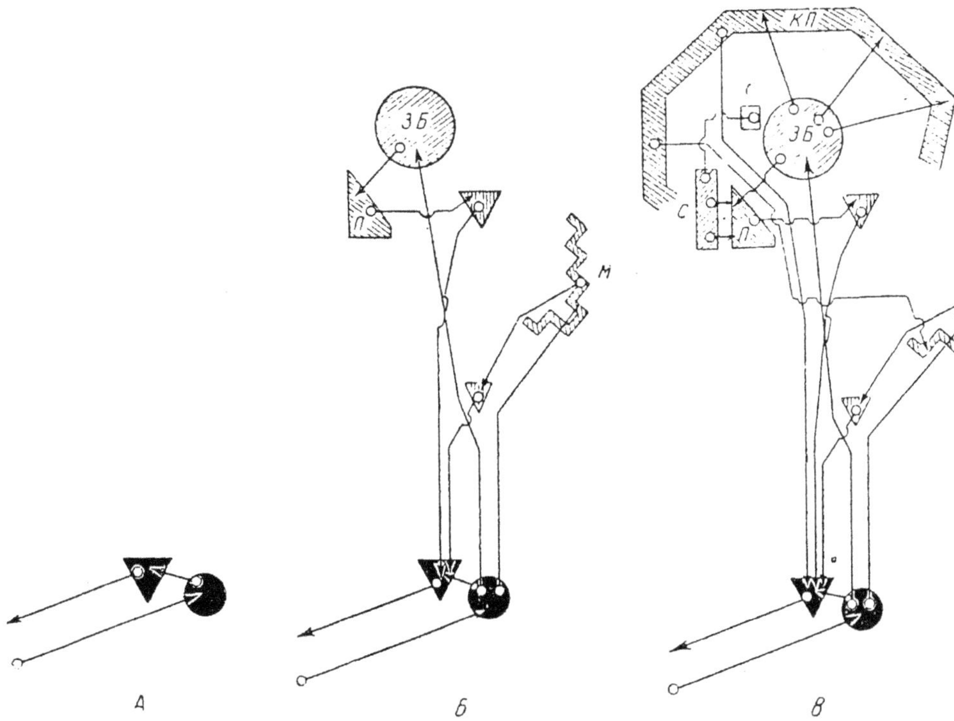

FIGURE 1.2 A scheme of gradual accretion of the brain effector systems.
Sensory nuclei are shown with circular forms, effector nuclei – with angular forms. A – the spinal level: a peripheral sensory neuron making a synaptic connection with a motoneuron; Б – the thalamo-pallidar level; ЗБ – the thalamus; П – the pallidum; M – the cerebellar cortex; B – the emergence of the motor cortex (КП) and pyramidal tract; C – phylogenetically most recent and dominating effector nucleus (the striatum) of the extrapyramidal system.

to overcome such leap-like development are illustrated by episodes similar, for example, to the development of the visual system from amphibia to birds, with the transition from a single-neuron scheme "retina – tectum" to a two-neuron scheme "retina – lateral geniculate – visual areas of the cortex." At some stage of the development, both apparatuses, old and new, work side-by-side, and then the old one shows involution, as it happened with the mesencephalic visual centers, or, more frequently, is modified to form, together with the new system, a more complex functional synthesis. This happened, for example, with the gradual development of the brain effector systems (Figure 1.2). This process leads gradually to the formation of structures including several collaborating neuronal levels.

It is necessary to emphasize that the neuronal principle of the construction of the central nervous system did not emerge in a single step and was not present over the whole process of evolution of the central nervous system. This corresponds to the general biological principle of gradual replacement of leading levels, which we have already witnessed in the alternation of effector, sensory, and central processing systems as leading factors in brain evolution. The nervous systems in praevertebrata are not neuron-based; in vertebrates, up to the highest mammals, the autonomous systems in their post-ganglionic part are much closer to neuropil than to neuron-based schemes. It is most peculiar that even the most central neural structures in higher vertebrates work with respect to some functions based on the neuronal principle, whereas, at the same time, they act with respect to other classes of functions like an unarguably continuous, diffuse neuropil. It is possible that the first layer of the cortex is built morphologically based on the neuropil principle; this is also more than likely with respect to a whole range of cortical intercellular clusters.[4]

Many of the aforementioned neuronal superstructures that emerged in the process of evolutions of the central nervous system played the leading role with respect to the whole central nervous system at certain stages of the phylogenetic process; later, they were overridden by younger formations, which took the leading role. Here, we have to mention one more complicating factor, which is due to shifts and re-weighting of the main functions of the central nervous system. This is the progressive "encephalization" of functions emphasized by many researchers of brain history starting from Monakow and Economo. This term implies two facts or, maybe, two facets of phenomena: (1) The progressive loss of independence and functional impoverishment of caudal parts of the central nervous system – the spinal cord; and (2) The gradual transition of centers of various physiological brain functions toward more rostral nuclei. This relentless process may be directly related to the described role switching and domination by the brain. Starting from a particular moment in the evolution, head ganglia transform from auxiliary, integrating tools serving telereceptors into a dominant organ, solely responsible for further evolutionary process. The domination of the central nervous system at the current stage of the evolution and its formative influence not only on narrowly defined behavior, but also on the autonomic, trophic, metabolic, immune, and other functions, is beyond doubt.

The increasing complexity of motor tasks facing an animal does not proceed smoothly and gradually; in contrast, changes in the lifestyle, zoological environment, ecology, etc. lead to the accumulation of numerous qualitatively novel coordination problems characterized by new features of their purposeful structure, involved motor subsystems, and necessary sensory corrections, which have not had a chance to be incorporated into the existing repertoire. For some time, animals can cope with such problems using existing resources; however, sooner or later, the contradiction between new purposes and sensory demands associated with the novel tasks and inadequate means of coordination leads, by the process of selection, to domination of animals able to handle the new problems; these animals suddenly acquire a whole class of new movements homogenous with respect to their type, complexity, and requisite sensory control. If evolution followed Lamarck's scheme, with gradual exercise of effectors, one could possibly expect certain hypertrophic, slowly developing adaptive brain changes. However, the principle of selection does not leave room for development of the central nervous system in response to new types of motor tasks along a route different from accumulating dominance of individuals with the qualitatively different, mutated brain. The emergence of a *new brain adjustment* in the process of phylogenesis signifies a biological reaction to a new quality or type of *motor task*. As it will be shown later, this necessarily leads to the emergence of a *new synthetic sensory field* affording the realization of a *new class or group of movements* with qualitatively different construction and control as compared to those available earlier. We are going to refer to the totality of morphological and functional aspects typical of such a new class of movements as the next level of the construction of movements and motor coordinations.

Everything said so far about the increase in the complexity of motor tasks encountered by the organism allows estimating directions of development of the emerging, one after another, coordination levels of the movement construction.

Phylogenetically newer levels, which are also higher levels, become:

1) More and more tightly linked to telereceptors and related superstructures representing integrative systems in the brain cortex;

2) More and more extemporaneous, i.e. adequate to find singular coordinative solutions and perform plastic switching;
3) More and more synthetic, i.e. based on complex synthetic sensory fields organized based on psychological factors; and
4) More and more enriched with memory elements based on the accumulated individual experience.

Patterns of movements and actions led by the corresponding levels develop in the same directions.

Each new level brings a complex of new movements, which used to be impossible for the organism. We should reject upfront the false old view that phylogenetically younger superstructures primarily ensure new features of coordination and, therefore, each of the various morphological brain levels is equivalent to a particular facet of the coordination picture of any given *whole* movement. Each new morphological brain level, each functional level of movement construction contains and contributes *not new features of movements, but new complete movements.* The nervous system of a highly developed vertebrate containing N structural levels and available N construction levels can control not N groups of movement features but N special lists or contingents of movements, which are complete and biologically ready to solve certain corresponding motor tasks. It would be very hard to understand what biological purpose and what justification of their existence could offer "preterm movements" deprived over the centuries of phylogenetic evolution of any coordinative features or, vice versa, representing sets of secondary, auxiliary features without the main defining purpose – a background without a picture. In the history of its development, each construction level available in humans was, at some point, the leading level (of course, taking into consideration the evolution of movement contingents, Chapter 3), defined the ceiling of the coordination abilities of the organism, and curtailed the list of available at the time movements. However, at each such stage, the movements were complete and showed well-formed coordination corresponding to the modest motor problems, which they were required to solve.

The thesis of *movement contingents at each specific level* is most vividly illustrated by clinical facts of movement *losses* during precisely localized loci of impairment or precise systemic diseases within the central nervous system. Typically, as already noted in the clinical neuropathological literature, such cases demonstrate losses not of *features of all movements but whole lists or classes of movements* or their background components. Particularly striking for the researcher in such cases is the clearly delineated selective pattern of losses with perfectly intact other movements, sometimes very similar in their appearance to the lost ones, but dramatically different in their purposes. One patient cannot lift an arm when commanded "lift your arm" but lifts it without any difficulty following the command "take the peak-cap off." Another patient shows no mimics to a degree suggesting complete paralysis of the facial musculature while being able to perform effortlessly and precisely any voluntary movements of the lips, nose, brows, eyelids, forehead, etc. in an attempt to reproduce purposefully demonstrated movements or following a verbal command. Still another patient (with hemiplegia) is unable to perform voluntary shoulder movements in the affected arm but can, in particular when dosing off, in a state of affect, or under light narcosis perform those very movements as components of synergic involuntary actions. The fourth patient cannot draw on a piece of paper a circle or a tilted cross following a command but has no problems with drawing letters "O" and "X." The fifth patient cannot make a step over the smooth floor, but drawing on the floor regular stripes, as if by a miracle, returns to him the ability to walk. There are an infinite number of such examples, and they are quite variable. In all such cases, it is frequently sufficient to switch skillfully the lost movement to a different, intact level by modifying formulation of the motor task to reach at once nearly complete restitution.

The aforementioned principle of the morphogenesis of the central nervous system based on its accretion leads to the central nervous system of highly developed vertebrates, for example anthropoids or humans, representing a kind of a geological section reflecting the whole history of the development of nervous systems, starting from the diffuse neuropils of lower invertebrates and the simplest spinal reflex loops of ancient chordate. All this is reproduced in the highly organized nervous system, in its sequential layers, levels, and superstructures with the accuracy comparable to that of the individual tree history reflected in its wood rings.

It seems also very interesting and highly meaningful that the human coordinative movement contingents form exactly the same simultaneous recapitulation of the whole history of animal movements, starting from such ancient movements as the peristaltic-like movements of annelids or the swallowing-vomiting movements of the sea cucumber. This recapitulation reveals unquestionable advantages compared to neuromorphology, because the former reproduces phylogenetics not in statics or the symbolic histological images of neural ganglia, which fail to inform us on their functional significance, but in dynamics, in movements themselves, which can be accurately compared with respect to their contents, structure, and patterns to movements of the contemporary representatives of the whole phylogenetic ladder. At the very bottom of the deep mine dug into the thickness of human motor coordination, we can see the most ancient paleokinetic coordinations that have been delegated to the autonomous function in higher vertebrates: the peristalsis of the intestines, the constricting-dilating movements of the blood vessels, the stomach sphincters, the gallbladder, the colon, etc. At higher levels, we encounter primary, elementary in their structure and inherent afferentation, neokinetic coordinations – spinal reflexes – studied in detail by Sherrington and his school. One step higher, we enter the realm of movements with more complex biological motivation and afferentation, which synthesizes telereceptors and individual memory-related components – the domain of true psychophysiology. Even higher, there are the youngest and most phylogenetically specific to humans coordinations, which rely even stronger on functioning of the cortex of large hemispheres. These coordinations cannot be reduced to purely biological factors: primarily, this group involves the coordination of speech, writing, and labor movements with objects, characterized by their social-psychological determination. Each of these sequential layers is related to a new morphological substrate, and, as it will be shown further in the book, each of the layers does not negate the lower, more ancient coordination layers but merge with them in a very unique and multifaceted synthesis.

In the following chapters, starting with Chapter 3, we will present general characteristics of these sequentially formed construction levels that form a hierarchy. We will also describe the basic facts related to the theory of coordination. However, first, a few general explanations have to be made.

Notes

1 We do not mention here the physiology of labor, which represents an indisputably and uniquely human area of human physiology. Physiology of labor – primarily of physical labor – either explores the process of labor itself, i.e., labor movements, and then it becomes physiology of movement, or it addresses autonomic functions, which have not changed in humans compared to animals but are being performed in conditions not encountered by animals. Movements, labor and innumerable other ones, in humans changed themselves compared to what is possessed by animals.

2 We will see later (Chapter 7) that this was primarily the victory of cortical adaptability of movements over the ancient extrapyramidal motor stereotypes.

3 An increase in the leading role of telereceptors in the evolution of motor functions is also due to the fact that telereceptors brought about complex integrative movement forms (locomotion etc.), whereas the latter require sensory corrections (see Chapter 2). Therefore, movements started "volens-nolens" to rely on sensory function, whereas, in the most ancient forms, vice versa, sensory signals followed movements (touching in worms, caterpillars, etc.)

4 It is hard to foresee how the morphological argument between "neuronists" and "antineuronists" will resolve. Maybe, chances of seeing a synapse under the microscope do not exceed the probability of stumbling over a meridian during walking in the field. However, it is already undoubtfully clear that: (1) Functionally, synapses are points of interrupted continuity between individual neural elements; and (2) The neuropil structures exist together with neurons in highly organized neural systems, the neuropil intimately linked to ancient structures, and neurons – to modern structures.

2

ON THE CONSTRUCTION OF MOVEMENTS

The motor system of vertebrates includes: (a) A passive part – the rigid, articulated skeleton; and (b) An active part – the striatal muscles with all their equipment. The passive motor apparatus consists of bone links, oriented primarily along the main axes of the effectors (axially), which prevents them from contributing to stability of the system in the absence of continuous active participation of the muscles.[1] These links are flexibly connected to each other forming the so-called *kinematic chains*. Muscle arrays are subdivided anatomically into separate muscles, which has primarily morphological significance, without major importance for biodynamics. These muscle arrays shroud the kinematic axial chains from the outside, following rules of also primarily morphological origin. Indeed (and this theorem is easy to prove), the location and direction of the end-segments of muscle tendons is of biodynamic, crucial importance, whereas the location of muscle bellies is of no importance whatsoever. Further, I will imply under skeletal kinematic chains not only bones and joints but whole moving organs.

Mutual mobility of two links of a kinematic chain is measured in mechanics using so-called *degrees of freedom of mobility and deformation*. Every degree of freedom of mobility more or less coincides with a separate, independent direction of mobility in a joint. Single-axis, e.g., pulley-like, joints have one degree of freedom. Egg-like and saddle-like joints (examples include the wrist joint and the carpal-metacarpal joint of the thumb) have two degrees of freedom each, spherical joints – three degrees of freedom of mobility. Degrees of freedom of mobility describe not the range or another quantitative measure of mobility (for example, flexion over more or fewer degrees in a joint) but a qualitative measure of the manifold of directions and forms of the mobility, which can happen to be very large even if the motion amplitudes are small. Examples are the mobility of the ulna with respect to the humerus, which has a single degree of freedom, and the deformability of the thoracic part of the spine, which theoretically has 66 degrees of freedom.

The number of degrees of freedom of mutual mobility of links in a kinematic chain (or, in other words, degrees of freedom of deformability of the kinematic chain) is nothing else than the necessary and sufficient number of *mutually independent coordinates* that have to be set to define the organ posture unambiguously. For example, to identify position of the shoulder with respect to the scapula (given the three degrees of freedom of the scapula-humeral joint), it is necessary and sufficient to state three coordinates (for example, coordinates of flexion-extension, abduction-adduction, and axial rotation). It is important to note that the number of degrees of freedom of a chain does not depend on choosing a system of coordinates or notation, i.e., it represents an objective characteristic of the chain. Note also that the number of degrees of freedom of deformation of a multi-link chain is either equal to the sum of the degrees of freedom of all its joints (so-called open chains) or is somewhat smaller (for so-called closed chains).

The mobility of the kinematic chains in the human body is enormous and reaches tens of degrees of freedom. The mobility of the wrist with respect to the scapula and of the tarsus with respect to the pelvis adds up to seven degrees of freedom, the mobility of a fingertip with respect to the thorax is 16 degrees of freedom. The presence of mobile fingers enriches the mobility and deformability of the hand compared, for example, with the forelimb of hoofed quadrupeds by 22 additional degrees of freedom. For comparison, let us mention that most machines that work without continuous human supervision, with all the seeming complexity of lever and rack-and-pinion kinematic chains, have only a *single degree of freedom*, i.e., they show what is called *forced motion*. Examples include the multi-valve diesel engine and the rotator newspaper printing machine. Two degrees of freedom can be encountered very rarely (for example, in centrifuge regulators), and three degrees of freedom are never used. This is a consequence of the enormous increase in the complexity of control with addition of new degrees of freedom. Theoretically, a flying projectile (a cannonball, a bullet, or a mine) has six degrees of freedom – an object of study of external ballistics. Here, we have to note: (a) the major inaccuracy of the control of a projectile and of hitting the target; and (b) The necessity to perform aiming and correction, to be discussed later.

We have to emphasize, in the strongest possible way, the aforementioned first sharp difference between the kinematic chains of the living body and artificial machines.

In artificial machines, the absence of kinematic chains with many degrees of freedom is due to the significant difficulties in the control of movements of such chains. The main difficulty is as follows. For a multi-link kinematic chain of any complexity, a single degree of freedom characterizes motion of a *forced type*. This means that, in such a system, each of its moving points is strictly constrained to a single trajectory. This trajectory may have any shape, simple or complex; the point can move along the trajectory forward or backwards, fast or slowly, etc., but the type of motion for the point is pre-defined. If the system gets at least one additional degree of freedom, this signifies that each of its points shifts from a single trajectory not to a few or even many trajectories but to an area of a surface over which a point with two degrees of freedom can move in any arbitrary way along an infinite number of equally accessible trajectories. For example, the endpoint of a pen, as long as it does not lose contact with paper, has two degrees of freedom; clearly, the diversity of its possible trajectories coincides with the diversity of everything that has ever been written or drawn on a piece of paper.

Therefore, transition from a single degree of freedom, i.e., from the forced type of mobility, to two or several degrees of freedom signifies the emergence of necessity to make *choice* of a movement trajectory or its *tracing*. A living organism always has an opportunity to *justify* its choice and planning of a trajectory. For a machine, in a similar situation, it is necessary to design a special *device*, which can make such choice in a meaningful way; otherwise, the movement will be doomed to be chaotic. An example of such a device is an automatic gyro-pilot. Mobility of a vessel (viewed here as a material point) on the sea surface has exactly two degrees of freedom; the gyro-pilot ensures selection of a trajectory corresponding to the set direction with respect to the compass from the infinite variety of equally possible trajectories.

Everything we have discussed so far points at the very important qualitative leap between one and few degrees of freedom. The extreme paucity of non-forced mobile systems among artificial machines is primarily due to the difficulties with devices for making *automatic, continuous, meaningful choice*. Besides, with more degrees of freedom in a system, errors contributed by each degree of freedom sum up. With an increase in the number of degrees of freedom, the overall error may reach magnitudes that would exceed the advantages afforded in principle by the rich variety of the mobility of a complex chain. For example, if each of the degrees of freedom of the arm and hand of a pianist gives an error of one degree, these errors may sum up to produce 5–6 cm deviations of the fingertip leading to a third or quart error (although in various individual links, e.g., finger phalanges, the contributing errors may be under 0.05 cm). It is also necessary to consider the unavoidable cumulative time effects on errors, which cannot be compensated for by any phenomenal accuracy of the initial calibration of the moving parts; these effects have to be prevented from reaching very high magnitudes in kinematic chains of the living body, e.g., of the spine.

Even more significant are dynamic complications. In a complex kinematic chain, every link has a specific gravitational and inertial mass. As a result, any force emerging in one of the links immediately brings about a large system of *reactive* or reflective forces, which are transmitted to all other links. This mutual influence of all the links on each other in all feasible combinations creates altogether a huge number of force interactions, mathematically boundless and posing unsolvable analytical problems. These reactive forces are superposed on the forces available for the organism to produce movements and on the external forces, which are always only partly controlled by the organism. This makes the overall dynamic pattern of movement of the kinematic chain extremely complicated and, most importantly, practically *unpredictable* because of its extreme mechanical entanglement. Nevertheless, it is possible to make movement of a kinematic chain *accurate*, at least theoretically, which requires producing incredibly accurate mutual adjustment of its parts. Making such a multi-link chain *controllable* is impossible in principle, because no theory can handle the violently increasing abundance and complexity of the reactive forces and interactions among the links of the chain. For a system such as the human arm, it is possible to define mathematically only the very initial moment of its motion under the action of a specific muscle; tracking down the progression of the movement is an unsolvable problem.

In order to *statically* fix a posture of a complex kinematic chain, one has to fix each of the available degrees of freedom using constraints independent of each other, one constraint per degree of freedom. The role of such constraints in the body of a vertebrate is typically played by muscles, less commonly and only in specific situations – by external forces. An analogous situation emerges in movement *dynamics*.

No matter how complex a kinematic chain is, its motion always happens to be unpredictable in advance but defined and in principle possible to analyze dynamically with any accuracy post factum. This means that, for any movement of any kinematic chain, the resultants of all the applied forces and moments constrain all the degrees of freedom of its elements, except one degree of freedom per element – the one along which the element motion has been observed. Hence, if one considers, in addition to static forces, all the dynamic forces, any movement of any chain may be viewed as *dynamically forced*. In this case, the role of missing constraints for the fixed degrees of freedom is played by dynamic forces, both external and internal. They are of little use, however. There is no argument that the totality of all the acting forces, internal, reactive, and inertial, constrains all the excessive degrees of freedom of the links and moves the links along some well-defined trajectories; the problem is that these trajectories are very likely to differ from the ones we need.

Obviously, we have the right to call a kinematic chain *controlled* only if we can *prescribe* certain, desired trajectories (and velocities) for each element of the chain and *force* the elements to move along the prescribed routes. To accomplish this goal, we have to possess means of binding the redundant degrees of freedom of the chain. This implies having methods of making obedient all the forces that emerge and develop during movement of the chain. The main problem of movement *coordination* is indeed in *overcoming the redundant degrees of freedom of a moving organ, i.e. in turning it into a controllable system.*

There are major difficulties associated with the fact that the organism always has control over only a small fraction of all the forces that produce the resultant, which defines movement of the chain. These difficulties are amplified by the generosity of the organism in endowing its kinematic chains with degrees of freedom. This "nonchalant attitude" to the number of degrees of freedom should by itself suggest that the method of control inherent to the organism is in principle different from those known to us from artificial devices. In spite of this situation, over the long tens of years of the development of neurophysiology, a view has dominated (and continues to dominate the textbooks in our time) that the dependence between muscle force and movement is as simple, direct, and unambiguous as, for example, the dependence between motion of the piston in a steam engine cylinder and rotation of the powered wheel. Unfortunately, the available facts in biodynamics present plenty of cases when *only flexor muscles* are activated along a kinematic chain, and, at the same time, all the links in the chain experience *only extensor angular accelerations*, or vice versa. Cases when a muscle crossing joint A causes angular acceleration in all other joints, B, C, D, etc., of a kinematic chain outnumber by far cases when this does not happen. Later, we will analyze a few typical examples. Ultimately, as if for the purpose of attracting our attention and forcing us to focus on the coordination process, nature piles a new difficulty on top of the problems caused by the enormous mobility of the skeletal kinematic chains, the difficulty that complicates much more the problem of the central control of movement. This problem is in using elastic cords connecting the links – *the skeletal muscles.*

The problem is that the striated muscles represent elastic bodies, which do not show direct proportionality between changes in their length and tension but, nevertheless, can be characterized for any of their physiological states by a well-defined non-linear curve reflecting a dependence between these two variables. In other words, muscle tension (or, equivalently, force generated by the muscle) is a function of two variables, its physiological state and its length. The complete picture of the dependence between the effector process or physiological muscle state and muscle tension can be represented only as a whole family of curves (Figure 2.1). Each curve within this family reflects a particular physiological state of the muscle;[2] each point on a curve shows the amount of tension as a function of length corresponding to this physiological state. When the central nervous system sends to a muscle a certain set of impulses, it sets one of the curves of the aforementioned family but, clearly, this is very far from defining a specific point on the curve, i.e., a specific magnitude of muscle force. We arrive at a conclusion that, from the whole totality of forces that define motion of a complex kinematic chain – internal, reactive, and external forces – the organism has at least a degree of control over the former category of forces only; but, as we see now, even for the internal forces, there cannot be an unambiguous relationship between an efferent process and forces produced by this process. The same level of impulses observed 20 times in a row can cause different forces depending on posture (and velocity of deformation) of the kinematic chain – variables that are largely dependent on forces outside the control of the organism, in particular external and reactive forces.

In fact, the situation is even more complicated than it has seemed to be until now. Muscle tension, in some way, is a contributor to a system of forces that causes displacement and deformation of the kinematic chain. Chain deformation leads to displacements of the points of muscle attachment to the bones leading to a secondary change in its length in one of the directions.[3] As a result, a change in muscle tension changes its length, whereas this length change causes a change in muscle tension. We deal here with a circular mutual interdependence of causes and consequences,

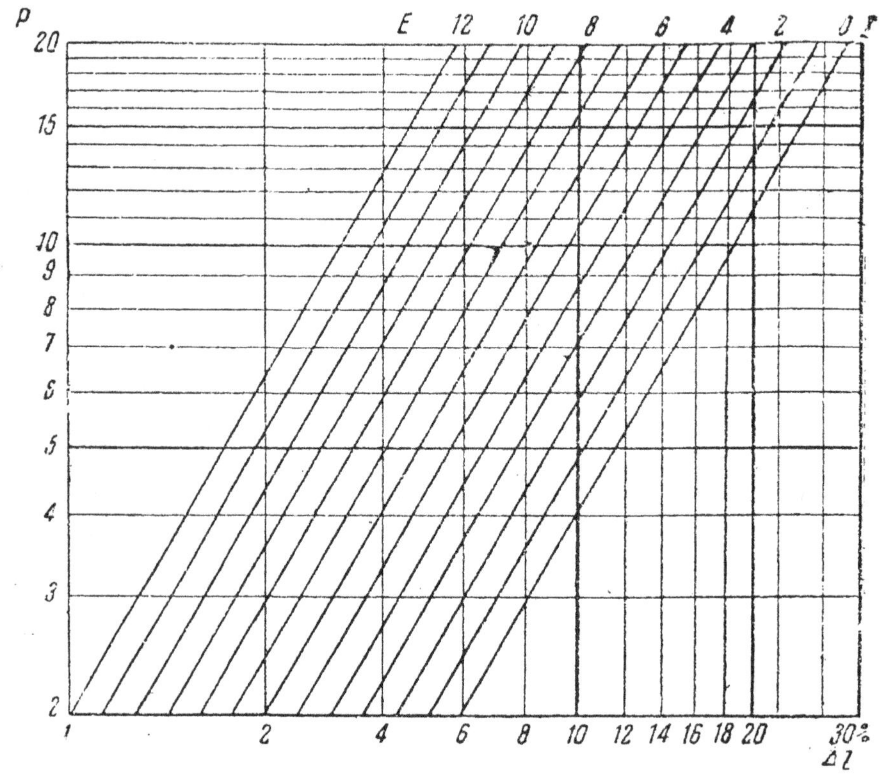

FIGURE 2.1 A family of lines showing the dependence between a measure of muscle excitation, its length and tension (schematic).
Lines 0–12 correspond to a gradual increase in the mechanical measure of muscle excitation, from complete denervation (I) to the maximal amount of excitation (E). The abscissa shows (on a logarithmic scale) percent change in muscle length with respect to its maximal contraction taken as unity (1). The ordinate (also on a logarithmic scale) shows an increase in tension P. Details are in the text.

which is expressed in mathematical language with second-order differential equations.[4] We are going to address this interdependence as *peripheral cycle of interactions.*

Hence, there is no and there cannot be an unambiguous dependence between muscle tension and resulting movement; we deal here with *inherent indeterminacy.*[5] This fact is the second crucial difference between the mechanics of the living organism of a vertebrate and the mechanics of artificial devices.

It might seem that a system of links connected not by a single elastic cord, as in the described example, but with two antagonist cords (Figures 2.2 and 2.3) would be free of the mentioned uncertainty. In fact, the difference is only seeming. A system with two elastic antagonists can be brought into any angular posture by applying appropriate external forces, independently of the elastic connections within the system. Given a *certain set of constant external forces*, the organism can select a combination of tensions in the two antagonists to ensure a desired joint angle. However, if the external forces change, maintenance of the same joint angle requires a completely different combination of the two tensions. Since in this example external forces do not depend on the central nervous system, the thesis on inherent indeterminacy remains fully valid.

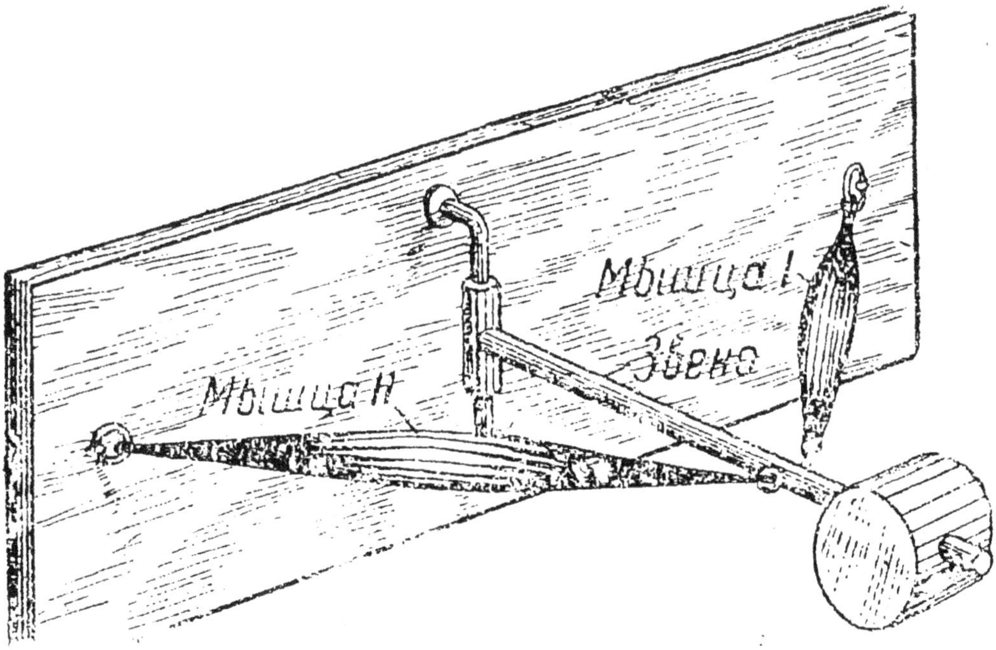

FIGURE 2.2 A moving link controlled by two muscles-antagonists. Мышца = muscle, звено = link.

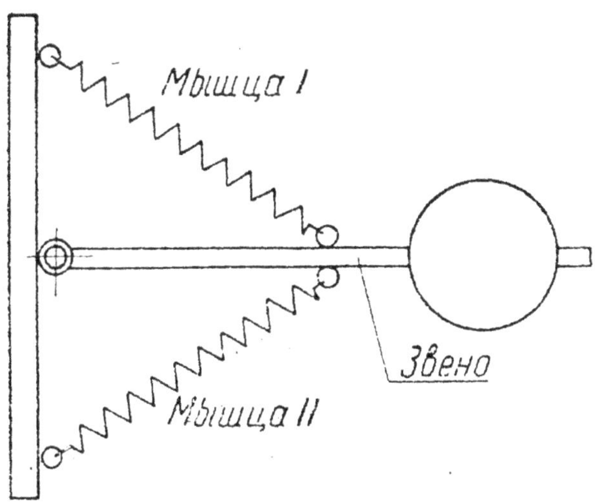

FIGURE 2.3 A schematic of a system shown in the previous figure. Details are in the text. Мышца = muscle, звено = link.

One has to note also that, for a given muscle, external forces include not only exactly external forces, such as the gravity force, the force of an attacking opponent, etc., but also the forces produced by muscles crossing other, remote joints of the same organism in their direct and reactive action. Whereas exactly external forces are absolutely impossible to predict, with a few exceptions, the second type of forces, which can be addressed as conditionally external forces, in principle could be predicted in advance and taken into consideration during motor coordination because the organism itself defines what impulses to send to specific muscle groups. However, it is sufficient to recall the intractable complexity of reactive interactions in multi-link chains to understand that it is practically impossible to predict all those reactive forces and their effects on remote mobile parts of the body. To ensure a specific angular position or angular velocity of a joint it is insufficient to establish a certain balance between the elasticities of the two antagonist muscles, but it is also necessary to modify this balance within a wide range depending on the position, loads, and accelerations in all the neighboring joints. This means that there is no time-invariant unambiguous dependence between states of muscles crossing a joint and its motion. Consequently, everything said about external forces is equally applicable to reactive forces. Moreover, since reactive forces in multi-link chains almost always are more complex and variable than exactly external forces, the complicating and distorting effects on the former forces on the movement dynamics are much stronger than the effects of the latter forces.

Later, we will consider how the central nervous system finds a solution to all the mentioned problems with movement coordination. At this time, it is only necessary to point out that difficulties introduced by the interference of external, reactive, and inertial forces and by the ambiguous relations between muscle excitation, tension, and motion, are much more frequent and important than commonly considered. Our overall extensive experience with studying human movements has shown that cases when, during a specific movement, very different muscle groups are activated at different times and in different patterns, as compared to what could be expected from elementary anatomical analysis, are much more common that cases when muscle action is classical and fully understandable. There are many movement elements, where we are still unable to interpret the behavior of each muscle group; sometimes, analysis of external and reactive forces can be performed and it clearly reveals the logic of those unexpected at the first glance muscle actions, but this logic is very much different from the elementary logic of high-school textbooks.

Consider several examples.

EXAMPLE 2.1 EFFECT OF ACTION OF INERTIAL FORCES

During chisel cutting, about half of the swing motion (Figure 2.4) is performed during active contraction of all the muscles that pull the hand *forward and downward*, while the motion itself at that phase is directed *upward and backwards*. This is explained by the fact that the arm with the hammer has a substantial moment of inertia and, consequently, a rather large backwards-directed inertia is generated by the reactive and muscle forces during the first phase of the swing, which has to be stopped and redirected to stretch the extensor muscles as needed to produce the strike. There are many examples of movements against the direction of muscle forces during rhythmical actions (walking, running, playing piano, etc.) This example is easy to analyze because the mentioned seeming contradiction reveals itself in the decelerating nature of the movement.

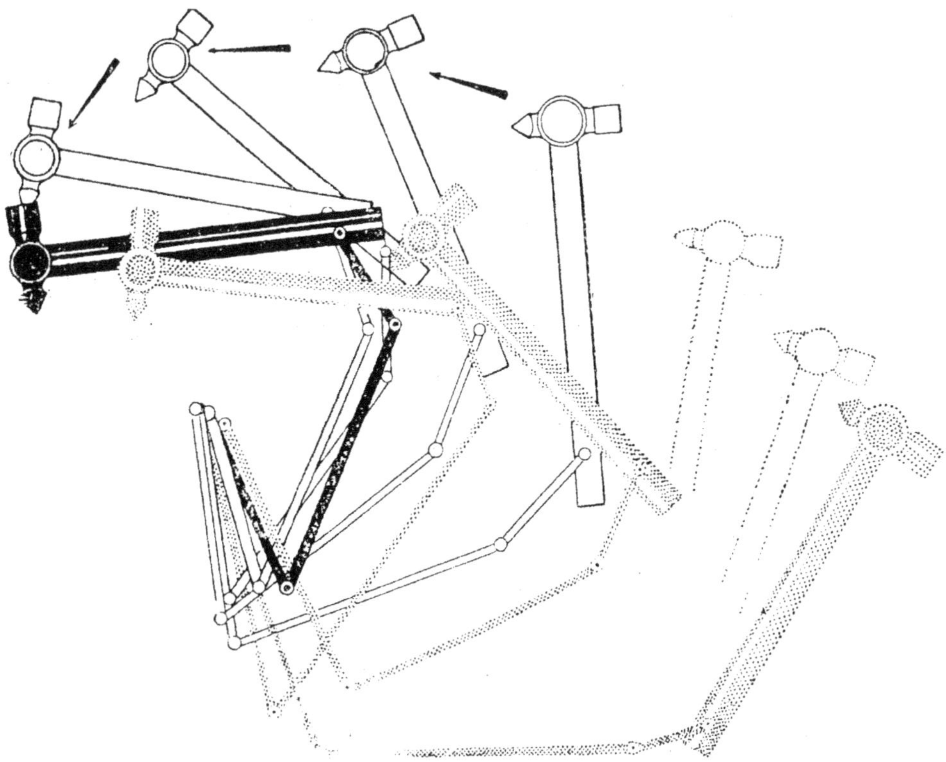

FIGURE 2.4 Sequential positions of the arm and hammer during cutting with the chisel. Top – swing, bottom – strike. The four postures connected with arrows are phases of the swing motion, during which efforts are directed forward.
(Source: from the author's work, 1923).

EXAMPLE 2.2 EFFECT OF ACTION OF THE EXTERNAL GRAVITY FORCE

The terminal phase of the hand-stand motion (Figure 2.5) consists of a slow lifting of the whole body by slow extension of the elbow joints. Due to its slowness, the movement is not complicated by inertial and reactive forces. However, it turns out that elbow extension over 90°, which is so powerful that it lifts the whole body, is performed not by elbow extensors that are nearly quiescent, but by muscles that perform shoulder anterior flexion (m. deltoideus, m. pectoralis major, and m. serratus anterior.)

It is relatively easy to explain the unusual action of the strongly loaded joint against an external load exclusively by muscles that do not cross that joint. Figure 2.5 shows that, during the described action, the center of mass of the body is within a single vertical plane with the two points of support – the hands (otherwise, during a slow movement, the gymnast would lose equilibrium), the two forearm axis, and the elbow joints. The moment generated by the gravity force with respect to these joints is close to zero. Hence, due to the rule of equality

FIGURE 2.5 Three sequential phases of the lifting motion and hand stance. The motion of elbow extension is produced by the action of shoulder flexor muscles.
(Source: study by M. Ukran, GCOLIFK – the movement study laboratory ofVNIIFK).

of action and reaction, moments of force generated by elbow muscles are also close to zero. Activation of elbow extensors in the postures shown in Figure 2.5 would lead not to lifting the body but to its fall counterclockwise.

Across all the arm joints, the gravitational moment of force is large only for the shoulder joints; indeed, those muscles work and increase the angle between the shoulder and the trunk thus lifting the body.

EXAMPLE 2.3 EFFECT OF ACTION OF REACTIVE AND INERTIAL FORCES

This example is more complex compared to the previous ones. During running, soon after the toe-off of the swing leg, powerful accelerating knee flexion begins (Figure 2.6) reaching very large angular speed, up to 3.6–3.7 revolutions per second in sprinters, which is similar to the speed of wheel rotation of an express steam engine at full speed. This motion, moving the heel very close to the gluteus muscle over 0.1–0.15 s, is performed over nearly its full duration on the background of a substantial excess of forces by *knee extensors*. During running of the world record holder Ladumegue, this flexion with the total duration of 0.273 s

FIGURE 2.6 One of the phases during running by the world record-holder J. Ladumegue. The phase when experienced runners typically show accelerating knee flexion of the trailing leg proceeds during continuous domination of extensor forces in the knee muscles.
(Source: from the author's work, CNIIFK, 1930–1939).

is accelerating during the first 0.198 s and decelerates only during the remaining 0.075 s. This happens exactly at the end of the interval when the work by the extensors reveals itself in the decelerating flexion, which is seen 0.011 s before the balance shifts toward flexor muscles.

In general terms, this phenomenon is explained by the foot motion upward primarily due to reactive effects from the contralateral leg. These effects are so strong that the runner not only does not have to push the foot upward with active knee flexion but, in contrast, has to slow it down to avoid hitting one's own buttocks.

EXAMPLE 2.4 EFFECT OF ACTION OF REACTIVE AND INERTIAL FORCES

The next example is similar to the previous one and is also taken from the author's study of running. After the swing leg is moved by the support leg (Figure 2.7), the former shows: (a) a drop in the forward velocity of the knee, i.e., braking of the thigh; and (b) acceleration

FIGURE 2.7 Phases during running by the world record-holder J. Ladumegue demonstrating the striking contradiction between the direction of muscle forces and angular accelerations in the swing leg joints.
(Source: from the author's work, CNIIFK, 1930–1939).

TABLE 2.1

Joint	Runner's posture (Fig. 9)	Observed motion acceleration	"Logical" expected force	Actual force
Hip	Strongly extended (forward)	Slows down its extension (under flexion acceleration)	Flexion (?)	Drop of the earlier flexion effort to zero (wave ζ)
Knee	Maximally flexed	Maximum of extending acceleration	Extension	Large flexion wave (ζ)

of the foot motion, i.e., forward extending angular acceleration in the knee joint. Comparison of what one could expect from muscle action with what is actually observed is shown in Table 2.1 following Figure 2.7.

So, all the actions are in fact opposite, in particular in the knee joint. In example 2.1, the direction of *acceleration* coincided with the direction of muscle efforts, although the direction of *motion* was opposite to the muscle efforts. In the current example, there is a contradiction between the directions of muscle efforts and resultant accelerations. This would be unimaginable during dynamics of a material *point*; dynamics of a linked kinematic system allows for such contradictions that can be due to collisions of reactive and inertial forces. The general explanation of this example, as well as of a number of similar examples, is that powerful reactive forces act in the direction of the actually observed acceleration. The local muscular efforts acting in the opposite direction are insufficient to counteract the reaction force, which ultimately accelerates the segment motion in its direction of action.

Difficulties created by the ambiguity and circular dependence for the planned motor coordination are, by themselves, so deep and basic that, on their background, the aforementioned problems associated with disobedience of chains with multiple degrees of freedom look secondary. This explains our earlier metaphorical expression of the nature "carelessness" with respect to the abundance of allowed degrees of freedom of mobility: Finding a way to overcome the basic problem of ambiguity allows nature to solve the less difficult and basic problem of multiple degrees of freedom completely, and then, as we will see soon, it actually does not matter whether we deal with a chain consisting of 5 or 75 degrees of freedom.

The solution discovered by nature to solve the aforementioned difficulties is directly suggested by the fact of two-factor conditionality of muscle forces, interpreted earlier with a family of curves (see Figure 2.1). Since for a given physiological state of a muscle its tension depends on its current length (for the time being, we ignore the complicating role of muscle viscosity, which does not change the situation qualitatively), the central nervous system would be able to ensure a specific magnitude of muscle tension only if it had information on the current muscle length and its ongoing change. Solving the problem of ambiguity is based on using, for the control of the effector process, sensory signals reflecting the posture of the kinematic chain and the degree of stretch of each of the muscles that affect its motion. It is easy to imagine that the presence of such continuous flow of information from the periphery of the central nervous system allows, in principle, dealing with any profligacy with respect to degrees of freedom of mobility. In fact, as soon as an organ under the action of external and reactive forces, with an addition of intrinsic muscle forces, deviates in its resultant motion from what has been planned by the central nervous system, it receives exhaustive information about this deviation sufficient to introduce adequate corrections into the effector process. The whole described method of coordination, therefore, deserves to be addressed as *the principle of sensory corrections*.[6]

The previous text explains why disorders of effector components of the central nervous system usually do not lead to pronounced disorders of coordination, leading only to syndromes of paralysis, paresis, contractures, etc., and why problems in afferent systems cause obligatory ataxic movement disorders, i.e., problems with coordination. Later we will show that afferent systems, except for the secondary corrective ones, have another highly important role for the motor process role in initiating, setting, and triggering action. Therefore, it is of no surprise that purely afferent disorders frequently lead, in addition to discoordination, to disorders with pronounced paralysis, paresis, etc., which show good motor recovery after some vicarious substitutions for the lost afferentation.

All the known clinical forms of organic disorders of coordination are always linked to dysfunction of receptor systems and their conduction pathways: vestibular (labyrinth and vestibular ataxia), receptor systems of the cerebellum (cerebellar ataxia), dorsal columns of the spinal cord that conduct proprioceptive and tactile afferentation (tabetic ataxia), etc. Animal experiments with cutting of the motor (ventral) roots for one of the extremities lead to paralysis of this extremity, whereas cutting the dorsal roots (deafferentation) leads to dramatic disorders of coordination. In the frog, cutting the dorsal roots to a hindlimb may lead to no noticeable disorders of coordination, but it is sufficient to paralyze or amputate the ipsilateral forelimb to produce an immediate dramatic disorder of motor coordination in the earlier deafferented hindlimb. Clearly, the presence of the normal mobility in the forelimb produces some bypassing (collateral) compensation for the sensory deprived hindlimb; as a result, hindlimb coordination barely balances on this tether. But it is sufficient to cut this remaining thread, without touching the hindlimb itself, to reveal the full-blown discoordination.

Compensations are possible in humans as well, which allow overcoming to some extent the organic ataxia; they always involve recruitment of new types of afferentation into the motor process. It is well known that a person with tabes dorsalis shows a dramatic deterioration of movements after closing the eyes, revealing the very important role of vision for compensation. These patients could show gait recovery, to a larger or smaller extent, with the help of bandages, which produced oscillating pressure on the skin over the belly during hip movements thus causing compensatory tactile sensations.

As it will be shown later, all types of *afferentation in the organism* participate to different degrees and in different situations in *implementing sensory corrections*. In other words, each type of sensation, happens, in alternation with its main exteroceptive (sometimes, also interoceptive) function, to perform the function of a guardian of movements of the body and to signal on the movements to facilitate sensory corrections. To use the terminology of Sherrington, we are going to address the whole broad range of such receptor functions – *proprioceptive function in a broad or functional sense*. However, the basic fact that makes such corrections obligatory – the fact of muscle force dependence on *muscle length* – suggests that the primary and most direct role in realization of such corrections is played by the *proprioceptive system in the narrow sense* – the system of sensory signals on postures, joint angular velocities, and muscle stretch and force. A muscle, which by its nature causes changes in movement of the kinematic chain, excites by its action the sensory endings of proprioceptors sensu stricto ("*peripheral closure*"), and these proprioceptive signals connect within the central nervous system with effector pathways and induce changes in the effector volleys, i.e., in the physiological muscle state ("*central closure*"). Hence, we encounter here not a reflex arc but a different relationship between the afferent and efferent processes characteristic of all coordinative processes – *the reflex loop* (Figure 2.8). Therefore, here we once again reveal the picture of circular interdependence very much reminiscent of the one described earlier during the analysis of interrelations between muscle force and movement. However, here it emerges using different mechanisms, not purely biomechanical, but with participation of the central nervous system. This case of interdependence could also be theoretically represented in the form of a differential equation, although we are currently very far from formulating it.

The reflex loop, which represents a fundamental form of the movement-related neural process, can be, for didactical reasons, represented in the form of the following rectangle (Figure 2.9).

It is very important to emphasize from the very beginning that coordination is not some kind of extreme accuracy or sophistication of *effector* neural impulses but rather a special group of physiological mechanisms, which create a continuous *organized cyclical interaction between receptor and effector processes*. No sophisticated analysis would be able to detect in the effector impulse signs or elements of "coordination": they are not there. Coordination is always beyond the effector impulse, in a sense above it, no matter whether it prepares the peripheral motor apparatus to accept an effector impulse or forms and regulates the effector impulse to match the current peripheral state.

Let us summarize the main lessons so far. There are two crucial factors: (1) abundance of *kinematic* degrees of freedom depending on the design of the joints; and (2) elastic muscle connections between links of mobile chains, which cause uncertain, ambiguous dependences between muscle activity and movement and may be considered as an equivalent of several additional *dynamic* degrees of freedom. Taken together, these two factors make movement organs in principle uncontrollable by systems with any, even infinitely accurate, forms of purely effector sequences of impulses. Forces

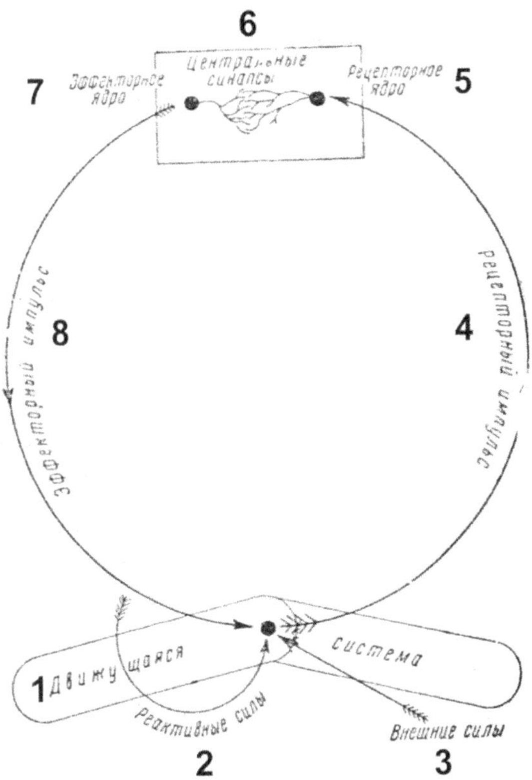

FIGURE 2.8 Scheme of a proprioceptive reflex loop.
1 – motor system, 2 – reactive forces, 3 – external forces, 4 – receptor impulse, 5 – receptor nucleus, 6 – central synapses, 7 – effector nucleus, 8 – effector impulse.

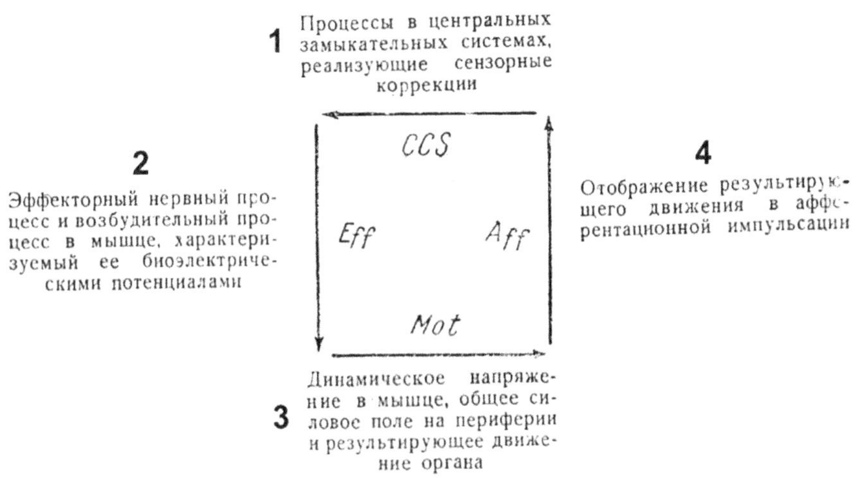

FIGURE 2.9 1 – processes in central closure systems that realize sensory corrections; 2 – effector neural and muscle excitation process, characterized by muscle biopotentials; 3 – dynamic muscle tension, general peripheral force field, and the resulting motion of the organ; 4 – reflection of the resultant movement in afferent impulses.

that define actual movement of each link of a kinematic chain can be represented as the geometric sum of three components: (1) Forces produced by the active motor of the system – muscle in our case; typically, these forces are internal forces of the system; (2) external forces (gravity, resistance of the environment, etc.); and (3) reactive forces – their number and variety increases dramatically with an increase in the number of degrees of freedom. Sensory corrections of effector impulses controlling muscle activity are organized in such a way that the resultant of all the mentioned forces contributing to the movement – internal, external, and reactive – leads the moving system in the desired direction with the desired force and speed. Effector mechanisms contribute only one summand to each of such resultant forces, consisting of three summands (active, external, and reactive). It is clear that the structure of this single summand differs from the resultant more for movements with larger contributions of reactive and external forces and for movements that are designed more economically with respect to active muscle force expenditure. This contradiction between the first category of forces – the only one directly controllable – and the resultant kinetics of the chain makes the control of kinematic chains with multiple kinematic and dynamic degrees of freedom very difficult. In addition, even with minor allowed errors and design variations, which are always possible even in very accurately built machines and can be rather large in biological kinematic chains, there is a sharp increase in variability and unpredictability of reactive forces over multiple cycles of the same movement. This makes reactive force contributions practically unpredictable in addition to their inherent complexity.

On the other hand, it is beyond doubt (as it will be analyzed in detail based on facts in Chapter 8) that movements are more economical and, therefore, more rational when the organism uses reactive and external force to implement them and, hence, it can add less active muscle forces. It is obvious that, for situations with lower contributions from muscle forces, there is less similarity between patterns of these forces and of the resultant of all three types of forces that factually defines movements produced by the organism. In the most perfect in their biodynamics movements (dynamically stable ones, see Chapters 4 and 8), this phenomenon reaches its pinnacle, and similarity between the muscle contribution pattern and movement is not more than, for example, between functioning of an auxiliary ship engine, which is turned on every now and again, and the course of a sail boat moved by the strong tailwind. Everybody is aware of the art of soaring by sea birds that can cover long distances with negligible wing work, only by the masterful usage of the vibrations of air streams. This art is better and better copied by humans in glider sports, but it is less known that, even in everyday movements of walking, running, labor movements, etc., the dependences between curves of muscle activation and curves of resultant forces and movements do not differ much from the aforementioned kinetics of the albatross.

These are the reasons for the complex process of deciphering from the language of spatial-kinematic representations used to create a psychological, primary movement plan to the language of actual muscle dynamics. This deciphering is more complex and intricate for more elaborate movements, i.e. for better developed motor skills. One has to add that, because of the aforementioned ambiguity, these codes change across repetitive executions of a movement. As a result, we are left with very little of those outdated views on new skill or conditioned response elaboration, according to which this process involves "beating the trail" as a consequence of identical repetitions. For the sake of the following discussion, we have to emphasize also that, during complex motor acts realized by higher cortical systems, there always have to be several superimposed sequential encodings of different nature and meaning.

Let us consider an example, which provides an impressive illustration to these statements.

To integrate a differential second-order equation, i.e. to find one of the infinite number of possible solutions, it is necessary to substitute into the general solution at least two initial conditions, which are independent of the equation itself. In case of an equation describing movement of a kinetic chain with elastic connections, such initial conditions can be, for example, the initial positions and velocities of the elements of the chain. These are indeed data of a type signaled to the central nervous system by proprioceptive organs in the course of executing sensory corrections. Obviously, if, for some reasons, proprioceptive afferentation (in general sense) is turned off, the central nervous system would have no data, of the mentioned kind or equivalent ones, to select one or another solution of the differential equation. As a result, instead of facilitating adaptively variable sequences of impulses, which are sent in normal conditions and produce strikingly similar cycles during cyclical movements, the central nervous system would sent to peripheral organs each time stereotypical, identical sequences of impulses with no knowledge of what conditions these impulses will encounter. Obviously, as a result, we will see a pattern drastically different from normal: identical sequences of impulses will lead to strikingly different movement cycles.

Indeed, this happens in patients with tabes dorsalis, which causes a typical pathological shutoff of pathways carrying proprioceptive signals along the spinal cord. Figures 2.10 and 2.11 confirm this with comparisons of two groups of curves. The right panel of Figure 2.10 shows the curves of vertical acceleration and dynamical *forces* in the center of mass of the foot of a patient with advanced tabes dorsalis and the left panels of this figure show two curves of the resulting *displacements* of the front and rear ends of the foot of the same patient in two corresponding consecutive steps. For comparison, Figure 2.11 shows similar curves during healthy, normal gait. It is clear from these figures that inter-step variability of the *force curves* is similar between the patient and the healthy person. At the same time, whereas the *movement curves* are nearly indistinguishable during the two successive steps in the healthy person, the patient shows a striking difference

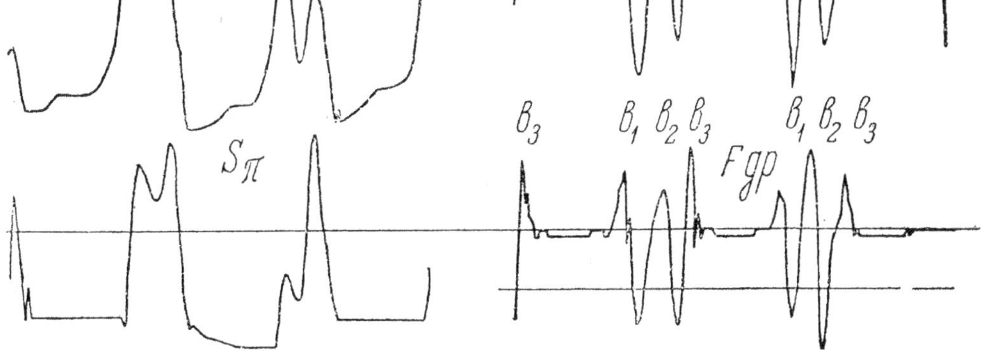

FIGURE 2.10 Left – displacement curves (along the vertical direction) of the heel (p) and toes (π) in two successive steps by a patient with tabes dorsalis. Right – the curves of the vertical accelerations of the heel (top) and vertical forces in the center of mass of the foot (bottom) during the same two steps. The drawing clearly shows that the two strikingly different steps by the patient can be accompanied by very similar cycles of accelerations and forces.
(Source: work by V. Farfel and the author, VIEM, 1935).

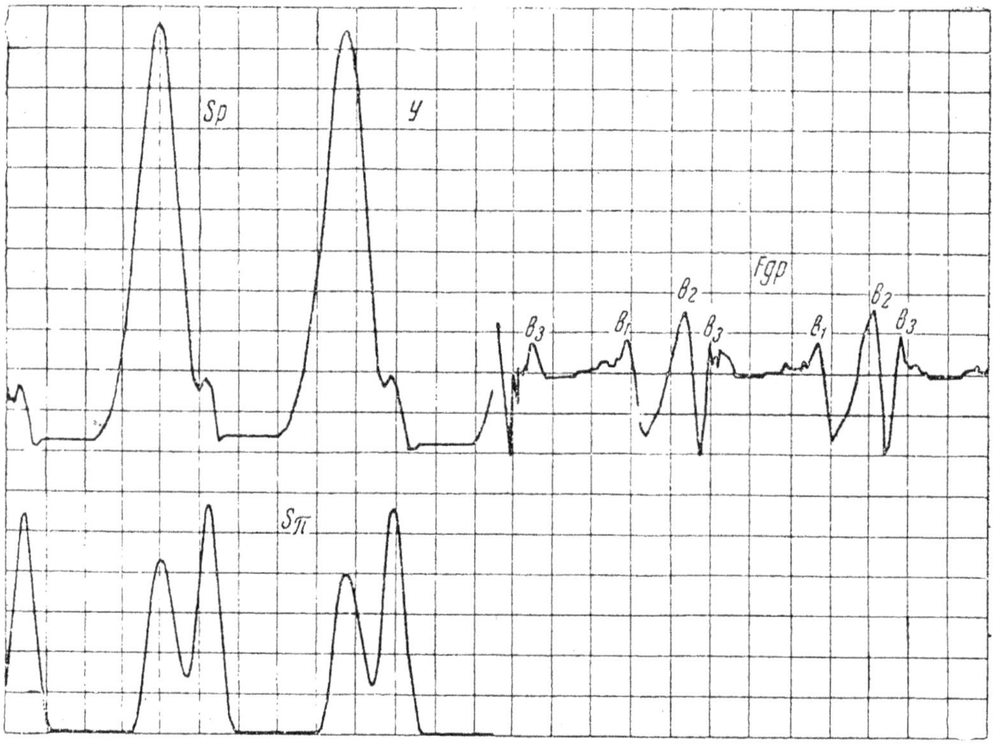

FIGURE 2.11 Left – displacement curves (along the vertical direction) of the heel (p) and toes (π) in two successive steps by a healthy subject. Right – the curves of vertical forces in the center of mass of the foot during the same two steps.
The displacements are perfectly identical in the two steps; inter-step variability of forces is not much different from the one shown in the previous figure.
(Source: author's work, VIEM, 1935).

between the cycles. The organism pays for ignoring data on initial conditions in each step and for the stereotypical similarity of series of impulses in the absence of proper decoding, in the best-case scenario, by a major lack of automaticity of gait and, in the worst-case scenario, – by complete loss of stability.

So, in the most exact definition, *movement coordination is overcoming redundant degrees of freedom of the moving organ, or, in other words, turning it into a controllable system.* The problem mentioned in this definition is solved using the principle of sensory corrections, executed together by various systems of afferentation and proceeding according to the main structural formula of the reflex loop.

The composition of afferent ensembles participating in the coordination of a given movement by executing required corrections and ensuring adequate decoding for effector impulses, and also the totality of system interactions among them all, are defined by us as *the construction of a given movement.*

It is necessary to emphasize that, even though all the types of receptor apparatuses available in the organism participate in ensuring sensory corrections and executing requisite recoding processes at different levels and for different movement aspects, however, there is not a single case (except, maybe, the simplest ancient reflexes) when such acts would be mediated by raw receptor signals isolated according to features of particular afferent systems. In contrast, sensory corrections are always driven by *whole syntheses* that become more and more complex as we move from peripheral to central levels built on the deep integrative processing of sensory signals of different modalities. These syntheses, addressed also as sensory fields, define what we call levels of the construction of various movements. *Each motor task discovers for itself, depending on its content and semantic structure, one of the levels or, in other words, one of the sensory syntheses, which is most adequate by its qualitative features, the contents of its afferent components, and their synthetic unification for the desired solution.* This level is defined as the leading level for a given movement with respect to the execution of the main, crucial sensory corrections and realization of the required recoding processes.

The notion of different *leading levels of the construction of movement* can be understood best by comparing examples of movements similar in their appearance but very different in their level structure.

For example, a man can perform a *circular arm motion* in a number of very dissimilar situations. *A*. During the very fast piano vibrato, i.e. during repetition of a note or octave with the frequency of 6–8 per second; frequently, points on the hand and forearm in exceptional virtuosi move along small circles (or ovals). *B*. One can make a circular arm movement during the performance of a gymnastic exercise or choreographic motion. *C*. A man can trace with the pencil a circle drawn on a piece of paper (*C1*) or copy a circle he sees in front of himself (*C2*). *D*. He can perform a circular arm movement while making a stitch with the needle or untying a knot. *E*. While proving a geometric theorem, he can draw a circle on the blackboard, which is part of the drawing used to prove the theorem. In all these cases, the person draws a circle or something that looks like a circle; however, in all the mentioned examples, the central neural foundations, the *levels of construction* would be essentially different (as we will show later). In the mentioned examples, we also encounter differences in the movement mechanics, in its spatial-dynamic pattern and, more importantly, profound differences in the coordination mechanism that defines these movements.

First of all, we have to mention that all these circular movements are linked in different examples to *different afferent systems*. Circles in example *A* (proofs will be presented in Chapters 3–4) are produced involuntarily, as a result of a subconscious *proprioceptive reflex*. The dancing or gymnastic circle (*B*) is also performed under the dominance of *proprioceptive corrections*, but already not an elementary-reflex one, but consciously perceived and with predominance of joint-level spatial proprioceptive afferentation rather than the muscle-force-related one. The traced (*C1*) and copied (*C2*) circles are drawn primarily under *visual control*; in the former example, by the most direct and primitive one, and in the latter – by a very complex synthetic afferent system of the "visual-spatial field." In example *D*, the leading afferent system is representation of the *object*, apperception of the object, understanding its shape and meaning, which produces an active outcome in the form of *action* or sequence of actions directed at goal-specific manipulation with the object. Finally, in example *E*, – the circle drawn on the blackboard by a mathematics teacher, the main point is not so much representing the geometric shape of the circle (as it would have been if the circle were drawn by a teacher of art, not mathematics), but a semi-conventional representation of relationships between the drawn circle and other elements of the mathematical scheme. Distorting the exact shape of the circle would not hurt the idea of the teacher and would not cause any corrective impulses in his or her motor system, which could be expected in the art teacher.

All the mentioned movements (from *A* to *E*) would be circles according to their muscle-joint schemes, but their realization, their *construction*, performed by the central nervous system, would proceed at different levels for each of the aforementioned varieties of circles.

A bright example of practical use of these facts for rehabilitation movement therapy is given by a series of studies performed by A.N. Leontjev and his colleagues (VIEM – Institute of Psychology) during the recent war. According to their observations, even in cases of major peripheral movement disorders caused by ankylosis or severe contractures, the amplitude of possible voluntary arm movements can change within a very broad range following changes in the formulation of the motor task, i.e., switching the executed movement to one or another level. For example, when instructed "to lift the arm as high as possible," the patient lifts it to a certain mark on the measuring stick (which he cannot see). During the next task, to touch a visible point drawn on a piece of paper, the patient lifts the arm 10–12 cm higher. If the task is formulated as "take an object from the hook," this ensures an additional increase in the amplitude of lifting the arm by about 10 cm more. A control test with lifting the arm without a target object (as in the beginning of the experiment) shows that the gained tens of centimeters can be seen only in trials associated with the aforementioned tasks. It is easy to see that the three sequential tasks of Leontjev are related to the mentioned levels *B*, *C*, and *D*. This example shows the differences between the innervation and muscle formulas involved in seemingly same movements performed at different levels.

Characteristics of different levels of the construction of movements, as they can be identified at this time, are presented in Chapters 3–6; the dynamics of their emergence and of the development of motor coordination in the phylo- and ontogenesis is described in Chapters 7 and 8. Let us make one more important point.

No movement (maybe with very rare exceptions) is served in all its coordination details by exclusively the leading level of its construction. We will see later that, at the beginning of the formation of a new individual motor skill, indeed, nearly all corrections are performed in a surrogate way by the leading level – the initiator of the skill. However, soon the situation changes. Each of the technical aspects and details of the complex movement sooner or later finds for itself, among the *lower levels*, the one with afferent signals that are most adequate in features of their sensory corrections for this particular detail. Gradually, after a chain of sequential switches and leaps, a complex multi-level structure is formed led by the *leading level*, adequate for the *conceptual structure* of the motor act, which performs only the main corrections, crucial for the meaning of the action. Further, under its leadership, movement execution gets contributions from a number of *background levels*, which provide background or technical components of the action: tonus, innervation and denervation, reciprocal inhibition, complex synergies, etc. *The process of switching technical movement components to lower, background levels represents what is commonly addressed as movement automation. In every movement*, independently of its overall level of construction, *only the leading level and only those corrections that are executed at that level are consciously perceived*. For example, if a required motor action is tying a knot, performed at level *D*, its technical components at the level of spatial field *C*, as a rule do not reach threshold of consciousness. If the planned movement is, for example, stretching or smiling, performed at level *B*, this level is perceived consciously, even though it is lower than level *C*. Of course, this does not mean that degree of consciousness is the same for each leading level; quite the opposite, both degrees of consciousness and voluntariness grow with transition from lower to higher levels.

According to the previous text, delegation of a technical component from the leading level to one of the lower, background levels leads to elimination of this component from the sphere of consciousness, and this phenomenon has in fact deserved the name of automation. Advantages of automation are clear: It leads to relieving consciousness from secondary, technical aspects and allowing it to concentrate on most essential and responsible aspects of movement, typically complicated by numerous unexpected events, which require quick and resourceful transitions. The opposite process of transient or complete destruction of automation is termed *deautomation*. Both these processes are described in more detail in Chapters 8 and 9.

Let me finish this chapter with the list of the *levels of construction*, which are characterized in the second part of this book. *A* – the level of paleokinetic regulations, known also as the rubro-spinal level within the central nervous system. *B* – the level of synergies, also known as the thalamo-pallidar level. *C* – the level of spatial field. It subdivides into two sublevels: *C1* – the striatal level that belongs to the extrapyramidal system, and *C2* – the pyramidal level related to the group of cortical levels. *D* – the level of actions (object actions, actions with meaningful serial order, etc.), also known as the parieto-premotor level. *E* – the group of the highest cortical levels of symbolic coordinations (writing, speech, etc.)

In the following description of the levels of construction, I will follow, as much as possible, a single plan: localization and substrate; leading afferentation; typical properties of movements; individual movements controlled by the level under discussion; background role of the level in actions led by higher levels; and dysfunction and pathological syndromes.

Notes

1 Neokinetic motor systems (see Chapter 3) can be found in the phylogenesis of arthropods and vertebrates. In both classes of animals, these systems brought about fast and powerful mobility, which distinguishes them from more ancient, soft-body classes. However, the problem of stability (the statokinetic problem) is solved in arthropods and vertebrates in qualitatively different ways. In the former, the skeletons of body segments cover the segments from outside, as an armor, which does not require muscle activation to keep stable posture. This can be proven, for example, by the fact that a carefully killed insect (under narcosis) does not fall, as a vertebrate does. As a result, the muscle tissue of arthropods does not carry static load; it is poor in sarcoplasma, striated irregularly, etc. Rigid skeletons are necessary equipment for the transmission of dynamic efforts of the fast and powerful striated musculature. The arguably only exception is the skeletonless striated heart muscle, which acts against the hydrodynamic resistance of the incompressible liquid of blood instead of a rigid external skeleton.

2 Which is a function of the percent of actively working mions (motor units), their properties, parameters of their excitability, etc.

3 An attentive reader can already conclude that this fact makes muscle shortening not a cause of a movement but its consequence. Whereas this conclusion looks paradoxical, it is correct, and the actual sequence of causes and consequences is: (1) A change in muscle tension; (2) Motion of bones with the points of muscle attachment; and (3) A change in muscle length. Similarly, gas expansion in a steam cylinder is not a cause but a consequence of the piston motion.

4 The mentioned circular interdependence is additionally complicated by the fact that joint motion changes the angle between the muscle axis and joint skeletal links, i.e. the length of the lever arm changes, which is a multiplier in the expression for the joint moment of force produced by the muscle. This complicates the equation expressing the dependence between the muscle moment of force and motion, and it cannot be reduced to a simple second-order differential equation, which would have been possible without this latest complicating factor.

5 *Indeterminacy does not mean indefinitiveness (undefinability).* The latter word would imply rejection of causality; the former means only *lack of unambiguousness* (compare, for example, with the term "indeterminate equation").

6 The principle of sensory corrections does not play a major role in movements of animals produced by smooth muscles, which has a characteristic effect on their movements: (a) primarily metameric; and (b) chaotically feeling.

PART 2

Levels of the Construction of Movements

3

SUBCORTICAL CONSTRUCTION LEVELS

Level A: The Rubro-spinal Level of Paleokinetic Regulations

We will now begin an orderly psycho-physiological characterization of the levels of construction of movement, as we understand them now, starting from the lowest ones to the highest, most novel according to their genesis levels, which are characterized by complex structure. Because the formulation of the problem is so novel, further discourse will be unable to avoid unclear and questionable points.

Before we characterize the lowest level, which displays functional independence in humans – the rubro-spinal level of paleokinetic regulations – it is necessary to describe some data related to specific features of microphysiology of the neuro-muscular process, which plays the leading role at that level.

The most ancient motor apparatuses, which are still retained by human from the times of the most primitive multi-cellular organisms, are the mobile systems of internal organs equipped with smooth muscles and innervated by the autonomic nervous system. Taken together, all the neural and muscular elements of this group deserve the name of the paleokinetic system in contrast to the neokinetic system of the somatic skeletal-articular-muscular apparatus linked by the network of peripheral, myelinated neural fibers to the central nervous system.

Movement of smooth muscles of the paleokinetic system is slow and diffuse; these are not exactly movements but rather leisurely changes in stationary length magnitudes of muscle cells, able to dwell for an unlimited time at each of the states. Under certain conditions, muscles of the paleokinetic system can develop substantial forces, but only slowly, at low power (e.g., the closing force of the shell leaves produced by the mollusc).

Muscles of the paleokinetic system tend to form continuous grids (syncytia), the nerves also form uninterrupted networks with numerous anastomoses (neuropil). Because of this factor, the paleokinetic neural process tends to display irradiation. A major role in complex synergies of the paleokinetic apparatus is played by stato-kinetic (adaptive to shape) processes of smooth changes in the forms and contour of the organ. This is facilitated by the skeleton-less construction of the soma of animals, in which the paleokinetic system is the only one available, and by internal, also skeleton-less, organs in vertebrates.

The slow maturation of the need for fast and powerful movements in the phylogenesis led, at one of the stages, to the emergence and parallel development of: (a) rigid, skeleto-articular kinematic chain; and (b) striated musculature with its neural equipment; taken together, they form structures that we address with the term "neokinetic system." The passive part (a) of this system is an obligatory satellite of the active part (b), because the full employment of the speed and power of skeletal muscles requires rigid lever systems to transmit the large and quickly changing efforts developed by such a muscle and to resist the sometimes enormous inertial forces emerging during its work (Figure 3.1).[1]

Both neural and striated muscle elements of the *neokinetic process* are characterized by a quick and brief discharge, which takes in warm-blooded animals just a few milliseconds and is related to the comparably quick development on the fiber of a short-lasting negative potential (spike). This burst is standard in its characteristic processes and the accompanying development of the short-lasting transient non-excitability (absolute refractoriness); it spreads along the fiber at a high speed, which is a function of the caliber of the neural or muscle fiber (the speed is higher for thicker fibers)[2] and represents the so-called *phase velocity*. In other words, this is not the actual speed of actual displacement of something – matter or energy – but the *speed of successive emergence* of local stationary processes, which provoke each other, from one point to the next. The phase character of the process defines its lack of decrement. The neokinetic system is fully governed by the *law of isolated transmission* of the phase excitation wave.

The emergence of a burst of excitation at some point along a fiber leads to two sequences of processes representing two aspects of a single physiological phenomenon: (1) a change in the trace bioelectric *potentials*; and (2) shifts in the levels of *excitability* and all amplitude and speed characteristics of the *ongoing excitation* (Figure 3.2). Following the excitation burst, during the spike declining phase, there is gradual recovery of excitability (relative refractoriness). After a few milliseconds from the beginning of the spike (these times differ within a wide range across different fibers depending on their caliber and related general speed characteristics), there is the phase of negative after-potential with the amplitude of only a small fraction of the spike amplitude. This phase matches exactly the supernormal phase of elevated excitability by both its beginning and end. When this phase is over, two parallel processes develop with the duration that is 10–20 times longer and an order of magnitude smaller by amplitude compared to the negative after-potential: the positive after-potential and a decline in the excitability (subnormal). In fibers of different type, both the duration and subtle qualitative features of this sequence of phases vary somewhat, but there is always strict correspondence between the sign and time interval of the potential and the phases of changes of its physiological characteristics.

It is possible to describe the sequence of events that accompany a burst of neurokinetic excitation in such detail only because this sequence is *absolutely standard* for any given neokinetic element and shows minimal variation across different elements. The inventory of the phenomena and resources able to show high-speed phase transmission and to induce physical contraction of the striated muscle is *fully described* by the spike with its after-potential, which always show the same composition.

In addition, at a given physiological state of a neokinetic element, no change in the nature and magnitude of excitation can produce changes in the qualitative and even quantitative pattern of the described phenomena. The last fact is known as the "all or none" law. It means that neokinetic excitability has *qualitatively contrasting properties*: It either emerges or not without a quantitative dependence on the excitatory stimulus. As such, this law presupposes existence of exactly defined *thresholds of excitability* inherent to all aspects of the neokinetic process.

In the anisotropic disks of the *striated fiber*, a burst of excitation causes similarly short-lasting *change* in all its mechanical characteristics: zero length, modulus of elasticity, and coefficient of viscosity in the direction of a decrease in the former and an increase in the latter two;[3] then, all these characteristics return nearly instantaneously to their original states. Neither neural, nor muscular elements of the neokinetic system can stay in the state of activation, according to their indices of

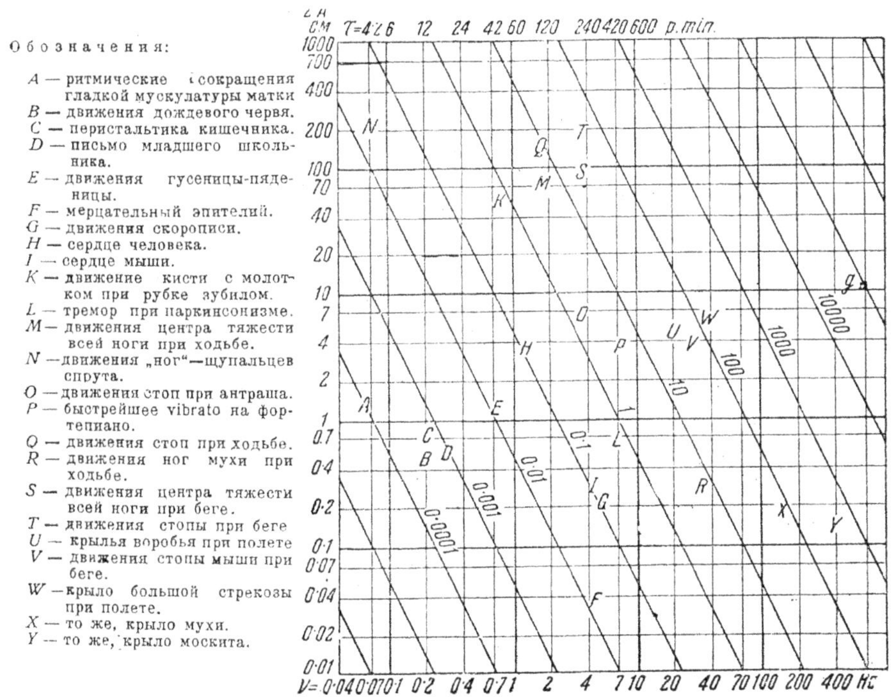

FIGURE 3.1 The amplitude of accelerations during different sine-like cyclical movement in humans and animals.

Abscissa (on the logarithmic scale): below – frequencies V in Hz; top – tempi T per minute. Ordinate: peak-to-peak movements (twice the amplitude), 2A in centimeters. The slanted lines correspond to different amplitudes of acceleration in units of gravity acceleration, from 0.0001 to 10000 g. The letters from A to V are shown in the points corresponding to different cyclical movements; the nearby slanted lines characterize the corresponding acceleration amplitudes.

A – rhythmical contractions of the smooth muscle of the uterus;
B – movement of the worm;
C – peristalsis of the intestines;
D – writing by an elementary school student;
E – movement of the caterpillar;
F – ciliary epithelium;
G – movement of the scorpion;
H – human heart;
I – mouse heart;
K – movement of the arm with the hammer during cutting with the chisel;
L – parkinsonian tremor;
M – motion of the leg center of mass during walking;
N – motion of the squid tentacles;
O – foot motion during entre-chat;
R – leg motion during walking by the fly;
S – motion of the leg center of mass during running;
T – foot motion during running;
U – wing motion of the sparrow during flight;
V – motion of mouse foot during running;
W – wing motion of a large dragon-fly during flight;
X – same for the fly wing;
Y – same for the mosquito wing.

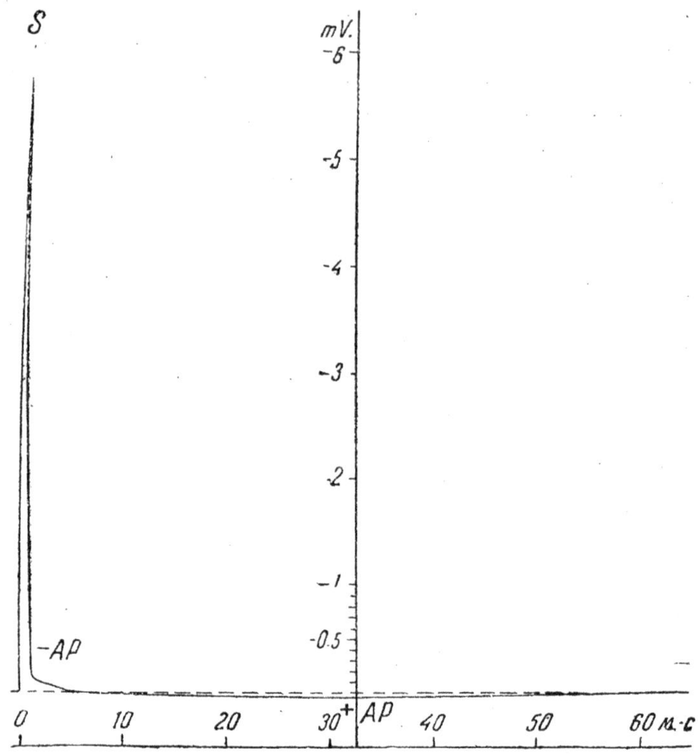

FIGURE 3.2 Half-schematic changes in the bioelectric potentials in an excitable element during a discharge of neokinetic excitation. The graph is *drawn to scale*. The potentials are in millivolts, time – in millisecond.

excitation and mechanical muscle characteristics, longer than for a few tens of their chronaxies. Hence, *the state of activity of a neo-element is a very much unstable state*, whereas its relaxed state is stable, able to last for an unlimited time. Let us also note that the state of activity of a neo-element is always accompanied by a burst of electro-negativity, never showing signs of something like "spike-positrons."

Mechanical events in the striated muscle induced by an excitation burst are complex, primarily due to a whole range of external factors of both mechanical and structural nature. At the time of excitation, anisotropic disks of the muscle fiber show nearly instantaneous, enormous contracting forces. These forces stretch the elastic passive isotropic disks, which play the role of buffers-accumulators of the elastic energy and shorten much more slowly and smoothly, ultimately transmitting this energy, via the tendons and skeletal levers, to the external world. The excitation discharge emerging originally on the muscle fiber close to the neuro-muscular plate spreads as a phase wave in both directions along the fiber. Parallel to this process, an elastic wave of mechanical tension spreads along the muscle fiber with the speed of about the same order of magnitude but of a completely different nature. The non-instantaneous spread of both waves along the muscle fiber leads, for example, to the fact that, at the physiological frequency of tetanus – about 100 Hz – in a long muscle, such as sartorius, the length of both elastic and excitation waves is 4–5 times shorter than the muscle itself. It also leads to asynchronous involvement in the excitation process of all the mions within a muscle and to the aforementioned buffer function of the isotropic disks, resulting in substantial softening and fusion of the rough, lightning-like contractile explosions turning their sequences into the familiar smooth and accurately graded work of skeletal muscles. However, a different process, described later, plays the leading role in the regulation of the stereotypical neokinetic discharges.

The paleokinetic process in the non-myelinated neural fiber and smooth muscle cell differ dramatically in all its features – up to being totally opposite – from the aforementioned neokinetic stereotype. First, the neural process in the paleokinetic system does not display a standard discharge; it does not show a spike with its large negative potential and accompanying refractoriness. Instead, it is characterized by a slow, long-lasting shift of the potential with highly variable patterns and possibility of shifts to both negative and positive values. In contrast to the typical for the neokinetic process antithesis, "excitation" (unstable) – "relaxation" (stable), the paleokinetic process proceeds based on another antithesis, "state of excitation – state of inhibition." The former is accompanied by a shift of the potential to negative values, the latter – toward positive values, and both are characterized by comparable stability. In the paleokinetic process, there is no special zero or some other resting state that would be qualitatively different from other states.

Further, the paleokinetic process is not constrained by the law "all or none." Its potential shifts and changes in muscle length and force proceed with any gradations of activity and force. Being free from constraints of the law "all or none," the paleokinetic process does not include its satellite – the phenomenon of stable thresholds: Measures of excitability of a paleokinetic element can vary within a much broader range, compared to a neokinetic element, but even the slightest excitation causes weak changes in the degree of excitability of a paleokinetic element. *Neokinetic* elements are characterized by *minimal magnitudes of excitation*, and they do not respond to smaller stimuli; *paleokinetic* elements are characterized by *coefficients of proportionality*, which define the dependence between change in the magnitude of excitation and magnitude of the induced response.

In the absence of burst-like discharges emerging on the homogenous rest state background, the paleokinetic process does not lead to a phase wave: Its long-lasting changes spread along the fiber slowly with substantial decrement. Finally, in contrast to the responses to suprathreshold stimuli in a "strike for strike" style, typical for excitation of neokinetic elements, paleokinetic elements are not excitable without repetitive excitation and consistent buildup (so-called iterative type of excitability, Lapicque). After the termination of a series of excitatory stimuli, these elements frequently display the other side of the same inertia: for some time, they show lasting residual discharge.

The mechanism of excitation spread, in particular transmission of the neural process from one element to another, is also qualitatively different in the two systems. In the neokinetic apparatus, a highly visible role, if not the dominating role, is played by the bioelectrical fuse process, whereas in the paleokinetic system, the dominating process is the phylogenetically ancient mechanism of humoral transmission.[4]

The ultimate inflexibility, uniformity, and burst-like nature of the neokinetic process, the only one available to the somatic nervous system, would be too high price to pay for the advantages of speed and power, if not for a fundamentally important group of facts, which are necessary at this stage of the discourse.

First, note that the range of physiological functions available for a neokinetic element is limited by the standard neokinetic discharge only for phenomena characterized by fast conduction of phase waves. Beyond that range, there is a whole broad spectrum of phenomena, which were described for these elements much later due to difficulties in their observation. These phenomena, which also belong to the group of excitation processes, differ dramatically from the neokinetic process in a whole range of characteristics. We will address this group of phenomena as *alterations*, although this pushes the accepted limits of this term in neurophysiology, and stimuli that bring them about – as *alteration agents*.

It is convenient to consider three groups of such agents: (1) electric field with subthreshold magnitude; (2) other adequate stimuli of the neokinetic process with subthreshold magnitude; and (3) a number of pharmaco-chemical agents, which usually are adequate stimuli for the paleokinetic system, but are not adequate for neokinetic excitation. Typical alteration agents can be best exemplified by the first group. Their action can be seen consistently in a number of groups of phenomena.

First, they lead to changes in all the *thresholds* that characterize *excitability* of an element to the neokinetic process. Second, at the same time, there are changes in *all* the amplitude and speed characteristics of the neokinetic excitation discharge *process*: amplitude of the spike and bioelectrical after-effects, duration of all the consecutive phases of the excitation process, speed of the phase wave, etc. Third, specific to the muscle fiber, parallel to the mentioned effects, there are shifts in *all the mechanical parameters*: the resting length, the elasticity modulus, and the viscosity coefficient. As a result, this leads to displacement of the curve describing the dependence between the length of a muscle element and its force, i.e., the characteristic force-length diagram (see Figure 2.1).

If an external electric field plays the role of an alteration agent, all the aforementioned shifts representing nothing else than *Pflüger's electrotonus*. All the available data on electrotonus in neokinetic elements speak in favor of a universal *rule of parallelism* for all the mentioned processes. An increase in the positive potential (*anode electrotonus*) leads to a slow shift toward overall inhibition, which can progress into complete paralysis of excitability and biological function. An increase in the negative potential demonstrates two phases, as demonstrated and explored by N.E. Vvedensky, who coined the term *parabiosis* for this group of phenomena. Moderate negative potentials (*cathode electrotonus*) are accompanied by shifts toward lower thresholds, higher amplitude and speed characteristic of excitation, and shifts of muscle parameters that favor contraction. Progressive increase of the cathode electrotonus leads to a maximum in all these excitatory shifts followed by their involution toward increasing inhibition able to lead, as in the case of anode electrotonus, to complete paralysis. This cathode electrotonic or parabiotic suppression *following* hyperexcitation represents, most likely, inhibition *caused by* hyperexcitation and, indeed, some experiments by Vvedensky create an impression of the neuro-muscular substrate being stunned by the excessive excitatory load.

Currently, the rule of parallel shifts with respect to all alternation agents of all groups may be viewed as firmly established. Although the same agents, in different doses and environmental conditions, can lead to cathode or anode electrotonic shifts, but in either case, the sign of the shift defines the whole list of changing characteristics.

It is particularly important to emphasize that, independently of the causes leading to alteration shifts in a neokinetic element, these shifts are always accompanied by the emergence of *endogenous electric field* of a specific sign, i.e. by shifts of the bioelectrical potential. The amplitude of such potential shifts is of the same order of magnitude (about 1 mV or lower) as the after-potentials of the neokinetic discharge and the "slow potentials" of the paleokinetic neural process, i.e. they are much lower than potential shifts during the neokinetic spike. The sign of the endogenous field brought about by alteration, i.e., the emergence of a cathode or anode electrotonic shift is tightly linked to excitatory or inhibitory character of alteration changes. As a result, all types of alterations, i.e. all processes that obey the rule of parallel shifts of indices of excitability and excitation of a neokinetic element, are intimately linked to the phenomenon of electrotonus.

Arguably, the most remarkable feature of the alteration phenomena is as follows. As mentioned earlier, alteration shifts are accompanied by relatively slow, long-lasting, low-amplitude shifts of the bioelectrical potential, which can be directed toward positive or negative voltages. They do not have more or less stable states; they proceed in an antithetic fashion to the "excitation state – inhibition state," where both states show identical stability. They have no special state, different from all other states, such as zero or resting state potential.

Alteration shifts of characteristics are not constrained by the law "all or none" and proceed with most variable grades of intensity. Not being constrained by the law "all or none," alteration does not have specific, reliable thresholds; even the weakest alteration stimulus produces in a neokinetic element at least weak shifts of its characteristic.

Further, the process of alteration shift spreads along an element not in the phase wave form: its long-lasting shifts spread along the fiber slowly and with a substantial decrement. Finally, electrotonic shifts of characteristics of excitability and excitation, caused by an adequate but subthreshold stimulus, may not be seen at the beginning of the stimulation but require repetitive stimulation and buildup (so-called summation of subthreshold stimuli), but, after a series of stimuli is gone, it displays frequently a phenomenon strongly resembling the after-discharge.

Given that currently the humoral, mediator-based nature of alteration shifts is beyond doubt, the described pattern of *changes of all the characteristics of excitability of neokinetic elements and of the time course of their excitation discharge* is nothing else than a *paleokinetic process* transplanted to a neokinetic substrate with all its features, including even those (for example, the ability to show irradiation) that seem to be in stark contrast with the substrate's morphological structure. So, all the features and characteristics of the neokinetic process, including the excitability threshold, the amplitude and speed of the spike, the speed of the phase wave, and mechanical characteristics of the striated muscle *possess regular variation, tightly linked and parallel to features of the paleokinetic process developing in the same substrate*. Clearly, these links and parallelism reflect *direct causation* pointing at an ability of the paleokinetic process to function as a *regulator of the neokinetic process*, which ensures its flexibility and adjustability, which are lacking in the isolated neokinetic process.

In particular, with respect to the skeletal muscle, the established identity between the whole set of electrotonic shifts and the paleokinetic process leads to a very important conclusion. Both the described above electrotonic shifts of muscle characteristics – the resting length and the coefficients of elasticity and viscosity – are fully analogous to the main mechanical process of the paleokinetic smooth muscle; this represents *physiological behavior of the striatal muscle mimicking that of the smooth muscle*. Summarizing all the known relevant facts, we are able to claim that this is the very circle of phenomena, which are united under the term of *physiological tonus* of the skeletal muscle. All this allows formulating an exact definition: *muscle tonus is the paleokinetic mode of skeletal muscle work* or, in other words, its work mimicking functioning of the smooth muscle. This definition agrees with all the available factual material and opens new wide horizons for physiological generalizations of the accumulated facts.

First of all, everything said so far shows that physiological tonus of the skeletal muscle is a *complex and comprehensive set* of phenomena that are not limited to mechanical events. The rule of parallelism makes the cathode-electrotonic shift of muscle tonus display itself in the form of a *decrease in the fiber resting length* (i.e., its shortening) accompanied by an increase in its *coefficients of stiffness and viscosity*. On the other hand, the same shift of muscle tonus leads to an increase in its excitability or, in other words, to *a decrease of all its thresholds* and *an increase in the amplitude* of all the characteristics of the neokinetic physical contraction, its magnitude, power, speed, spike amplitude, etc. To sacrifice exactness of the definition for didactic purposes, the cathode-electrotonic shift of the tonus reveals itself in the mechanical domain in *two ways*: as an increase in the tonic muscle contraction and as ensuring favorable conditions for a *phasic contraction and force production* (neokinetic discharge) developing on the background of the tonic contraction. Anode-electrotonic shift (inhibition) of the tonus reveals itself exactly in the opposite way, up to complete blockage of the neokinetic process.

We would also like to note that the emphasized strict parallelism between post-excitation phases of alternating excitability and the accompanying oscillations of the after-potentials may now be naturally interpreted as phenomena of the same electrotonic nature. Hence, the sequence of post-excitatory shifts, describing the whole set of properties of excitability and excitation according to the rule of parallelism, is nothing else but an extremely fast *sequence of alterations* with changing signs. This conclusion, taken together with the presented definition of physiological tonus of the skeletal muscle, allows tonus shifts to be not only slow in their nature; the sequence of post-excitation shifts includes fast, consistent alterations of tonus mechanical characteristics. As shown by exact analysis, during optimal frequency of tetanic stimulation, these fast post-contraction tonus shifts contribute directly to an increase of the tetanus amplitude and coalescence.

The definition of *tonic activity* of the striated muscle as the *paleokinetic mode of its functioning* is supported by the observations of low-amplitude, slow shifts of bioelectrical potentials during tonic contractions and by the facts that these contractions can show any duration and any gradation of their magnitude, i.e. their immunity with respect to the law "all of none," the lack of a phase wave, and finally, by the equal probability of potential changes of either sign. In fact, tonic activity of the striated muscle can show both active shortening and active lengthening, which is alien to the neokinetic process. Such active lengthening can be seen, for example, during reciprocal relaxation of muscles-antagonists, which play an important role in the following discussion.

The preceding detailed analysis of alterational, alias – electrotonical, alias – paleokinetic phenomena in the neokinetic neuro-muscular system and the control possibilities afforded by the paleokinetic process, as compared to the neokinetic burst-like stereotypy, is needed in the current context (as it has been alluded earlier) because all these phenomena are definitely not part of exogenously initiated artificial processes. As shown by a broad range of various observations, all the phenomena, united here under the name of paleokinetic process, take place continuously in neokinetic substrates of an intact organism under physiological conditions and play a very important coordinating role. In higher vertebrates, this process is a tool of *central regulation* emerging in the *brain stem mesencephalic ganglia* and transmitted via descending spinal pathways to synapses in the anterior horns and further – via the motoneurons – to skeletal muscles. This centrally regulated paleokinetic control process has revealed itself in studies of many authors who did not suspect that they observed one and the same cycle of processes. The spinal "slow potentials" described by American authors, Sherrington's alternating states of excitation and inhibition (CES and CIS), the intermittent factor of Barton and Matthews, Lapicque's subordination, Sechenov's inhibition, and many other phenomena may, without doubt, all be viewed as reflections of the described regulation cycle. As it will be show later, this cycle represents the physiological meaning of the lowest of human coordination levels of movement construction – the rubro-spinal level of paleokinetic regulations *A*, which is going to be described next.

To summarize, the dull stereotypy of the blind energy discharge, which is the essence of the explosive neokinetic process, undergo essential corrections by the central nervous system physiology. The *paleokinetic, tonic* (electrotonic and humoral) *activity of the brain stem apparatuses* in a sense takes the responsibility for *playing the leading role and regulating* the phylogenetically younger, but characterized by many drawbacks, *neokinetic physical* (bioelectrical) *process*. This regulation does not interfere with the neokinetic spike processes, does not add anything to them and does not take anything away, but it creates *presetting* states including values of constants and parameters, which will define further time course of the neokinetic process. In a metaphorical way, this can be compared to pressing a violin string with a finger, which does not produce sound by itself but defines what sound will be produced by motion of the bow. It can also be compared to setting a multi-digit number on a mechanical calculator, which then will be revealed in its readings after the handle has been turned. It is no coincidence that Lapicque compared one of the observed by him phenomena of this preparatory-regulatory process with moving a railroad switch (aiguillage) – we will return to this comparison later.

The level of paleokinetic regulations, which is the rubro-spinal level, designated in the list as level *A* is the lowest and phylogenetically oldest level according to its function but not to its substrates in the human body. The anatomical nuclei of the red nucleus group formed only in mammals, and a whole range of functions migrated to those nuclei in the process of encephalization; these functions were performed and are still performed in lower vertebrates by systems within the spinal cord. This process of phylogenetic development will be discussed in detail in Chapter 7.

By itself, the spinal cord (mentally separated from the brain stem below the red nucleus level) involuted so much in humans in the course of encephalization that it fails to define not only any independent movement but even any clear background component. Everything it is capable of doing in isolation is, at best, to produce one or two artificially induced clinical reflexes. It would be wrong to conclude, based on those observations, that we assign to the spinal cord the transmission role only. As we are going to show now, the active role of the spinal cord in humans is still rather varied, but there is not a single natural physiological process where the spinal cord would act alone, without forming an inseparable whole with the stem (mesencephalic) part of the brain. That is why, we address this system as a rubro-spinal one.

The anatomic substrate of the rubro-spinal level (i.e., all the crucially essential organs for the functions at this level) is composed of the spinal cord with its cellular formations and at least some of the transmission pathways, a group of cellular formations in the brain stem, which we address for brevity as the *red nucleus group*, involving the red nucleus itself with its two parts (paleorubrum and neorubrum), substantia nigra, the Darkszevic nucleus, and maybe the Lewes body, an area of the hypothalamus, the Deiters nucleus, the paleocerebellum, and finally, in some poorly understood functions, the central part of the autonomic (parasympathetic and sympathetic) neural apparatus.

Afferent systems, defining the functioning of the level built on this substrate, are represented primarily by: (a) the most ancient components of proprioceptive sensitivity – those that can be defined as proprioception of tropisms – originating from the endings sensitive to muscle tensions and forces and from the otolith apparatus of the inner ear (paleo-labyrinth); and (b) The most ancient components of "tango-receptors," those united under the term "protopathic sensitivity." The clear paleokinetic character of the process typical of the latter group is particularly obvious in the so-called hyperpathies, i.e. syndromes of lacking epicritic sensitivity during injuries to certain brain areas: the iterative slowness of the buildup, the after-discharge, and the clearly seen irradiation. The micro-physiological nature of the ancient-proprioceptive process is not clear as of now. All the mentioned afferent systems, united into a rather complex synthesis, signal the animal on the location and orientation of its body in the field of gravity and, maybe more importantly for motor coordination, on the magnitude of stretch (along the muscle axis) and tension (force) of skeletal muscles. This level is primarily responsible for the circular correction process of adjusting the effector activity of each muscle to its current length: this scheme was described in Chapter 2 under the name of reflex loop.

At this point, we need to clarify a very important general principle. Based on the aforementioned evolution of the central nervous system based on its qualitative overgrowth (not quantitative growth), one could imagine a central nervous system functioning at some level *P* in such a way as if it were truncated rostrally corresponding to the nuclei of that particular level. This would be, however, altogether wrong. If we had a light filter allowing to see the central nervous system of a higher mammal in a strictly "monochromatic light" corresponding to a single level (this is a metaphorical wish corresponding to possibility of observing the surface of the Sun in the monochromatic light of a specific part of the spectrum available to contemporary astronomers), in the light filter of level *P*, the central nervous system would be still seen in its completeness, with all the active formations up to the most modern ones, i.e., working as an unbreakable whole. However, two basic facts will be revealed, following indirectly some of the known facts. First, such a monochromatic view would change dramatically the proportions, functional relative weights, and even sizes of structural brain parts compared to their actual characteristics; at the same time, some of the synapses and conductions pathways would be emphasized whereas others would be diminished to complete lack of identity. Second, the physiological process isolated during such an observation from all the processes within the central nervous system would display clear peculiarities and characteristics, and some of these would be similar to those typical of this process at lower phylogenetic levels in species where the level *P* and its physiological substrates define the highest capabilities and morphogenetic achievements of those species. Therefore, the central nervous system of a higher mammal leads a physiological process, ancient by all its characteristics typical of level *P*, using a modified and enriched substrate, which inevitably modifies it somewhat and makes it both similar and dissimilar to its ancient form when level *P* was dominating. If a motor act realized by the central nervous system represents a complex structure with a number of background levels controlled by a supreme leading level, one and the same indivisible substrate within the central nervous system of a higher mammal displays simultaneous, parallel, highly dissimilar physiological processes, which not only show lack of negative interference in a healthy organism but in fact interact and complement each other.

This is exactly the case clearly seen with respect to functions of the rubro-spinal level *A*. Starting with the substrate, the monochromatic lens of type "*A*" would show the brain as a very small, indigent, and primarily small-cell structure; there is sparse, diffuse inflow of sensory impulses of *A* type to the cortex, which does not reach consciousness. Most importantly, this "head end" is not the "main end" but rather the "spinal cord of the brain" serving the brain metamers in the same way as the coronary artery serves the heart muscle. Overall, the brain stem related to level *A* is clearly metamer-based and tightly linked to the autonomous nervous system (other monochromatic lenses barely show this connection at all).

The process developing in this substrate is exactly the paleokinetic excitatory-inhibitory process described earlier in this chapter. This process, assigned to this substrate, facilitates and controls actions in internal organs, actions that have little importance for psychomotorics but play a role in the control of somatic coordination. As it will become clear from the following material, physiological process at each level of movement construction has its own method of performing sensory corrections, its own specific features of patterns and functioning, which are typical of its function as much as the qualitative patterns of its afferent systems. The method of control typical of the rubro-spinal level *A* is the regulating action of the neokinetic, tetanic effector process via paleokinetic, electrotonic changes of its characteristics – the cycle of phenomena described in the first part of this chapter; this regulating action is produced by the central nervous system and develops on the neokinetic neural and muscular network. It is rather peculiar that the paleokinetic process born during the phylogenesis together with its specific substrate – the non-myelinated, neuropil-like neural network and the system of smooth muscles – had continued at a limited scale even in the higher mammals within the autonomic apparatus, and could, at the same time, adjust to the neokinetic substrate, which seems alien to it in both the time of its emergence within the evolution and its structural characteristics. This substrate, with the myelin sheath ensuring all the necessary conditions for the "law of isolated transmission" of the phase wave of neokinetic excitation, somehow contrives to irradiate as successfully as in the paleokinetic neuropil permeated with anastomoses. When the burst-like neokinetic process encounters strict, hardly changeable, thresholds, paleokinetics flows in its typical non-threshold manner causing broad, iterative shifts of characteristics of excitability. We have already discussed the significance of this process for the organization of the neokinetic chains of excitability; now, we have to discuss what phenomena reveal this regulation and what aspects of the coordination process are served by it.

First of all, the rubro-spinal level controls the so-called spinal reflexes, which were studied in detail by the school of Sherrington. These reflexes represent a transition phase from the purely metamer mode of the spinal cord function to some kind of integration. At this level, integration is realized by multi-segmental innervation, slow, irradiation involvement of an increasing number of metamers, and, finally, representing a qualitatively higher level, simultaneous and sequential distribution of excitation and inhibition not only over metamer muscles but also over antagonist muscle pairs of the *extremities* – phylogenetically young structures compared to metamers. The most typical reflexes of this group include the reflex of *reciprocal innervation and denervation* of the antagonists, which forms the basis of any movement of an extremity. This reflex ensures denervation of the antagonist muscle when a muscle of an extremity or the trunk shows activation tetanus. From the material discussed earlier in this chapter, it follows that *denervation is an anode-electrotonic decrease in the tonus*, which is of course accompanied by parallel anode-electrotonic shifts of all the characteristics of excitability and parameters of excitation. Reciprocal denervation is accompanied by spontaneous muscle lengthening, which is not associated with any stretching force, because the same lengthening can be observed in a muscle with the cut and detached distal tendon. This muscle lengthening proves clearly the paleokinetic nature of the intrinsic muscle processes because the neokinetic spike always shows negative voltage and always leads to muscle shortening. Only a shift of the potential toward positive values, which is available within the bipolar paleokinetic process, is able to induce muscle lengthening.

Since we are discussing spinal reflexes, it is impossible to ignore an opinion, which has only recently had numerous champions, that any complex coordinated movement may be viewed as a mosaic (or synthesis, the word does not change the essence of the position) of reflexes. The contemporary neurological concept resolutely rejects such an atomism. To get simple reflexes, one does not have to cut a whole movement into small pieces, but to arrange all possible human movements in a row according to their increasing complexity, and then, at the end containing the simplest movements, we will find all the reflexes.

Many arguments supporting this view can be supplemented by another one, arguably the strongest, originating from the theory of coordination. Each level of the construction of movement differs from other levels by the composition and quality of the inherent afferent signals. To build a single movement at a higher level from numerous movements at lower levels, e.g., from reflexes at level A, is as impossible as forming a single visual sensation from thousands of tactile or proprioceptive ones.

Participation of elementary movements or elements of movements, controlled by a low level, in the construction of a complex coordinated motor action, controlled by one of the higher levels, is a very different issue. We have to be perfectly clear. There are no *movements* controlled by a certain level, which could be used as bricks to build a movement at a higher level. However, processes of switching from afferent to efferent signals within a functional-proprioceptive reflex loop, which is the process of encoding of impulses related to coordination, necessarily proceed at lower levels during movements controlled at higher levels; these processes create what can be called a "rough draft" of the movement using necessary afferent signals, which do not reach higher levels, and relieving the higher levels from taking care of all the auxiliary ingredients necessary for any movement. This is the reason that, during the definition of the notion of levels, we invoked the notion of *leading levels*, rather than the only or monopolizing levels for any given movement. Later, we will see a few examples revealing the multi-layer structure of any movement: at this time, let us only mention that the leading level clearly controls all the lower, background levels participating in any given movement such that each of the background levels performs its part in the movement orchestra, while, to a degree, losing its individuality and only following the baton of the conductor.

The participation of the red nucleus has been proven for the next group of processes performed by the rubro-spinal level; these processes unite what is known as the *subordination of chronaxies*, i.e. the central control of characteristics of the rate of excitability of motor units (motons).[5] This phenomenon was originally discovered by L. and M. Lapicque during the studies of chronaxie and is commonly related to this characteristic only. But the rule of parallelism of shifts has no exceptions, and the subordinate process shifts all the characteristics of excitability and of the phase process in the neural and muscle elements, as well as muscle mechanical parameters that characterize its tonus. The subordination is typically inhibitory, i.e., anode-electrotonic, although there have been clear proofs of subordination shifts of the opposite sign. Note that, whereas the process of excitation with the accompanying negative shift of the potential is undoubtfully universal in the whole nervous system of higher vertebrates, phenomena of suppression and inhibition are many and varied in their physiological nature.

The coordination importance of subordination was already emphasized by the original authors who had discovered this phenomenon. Since, according to Lapicque, necessary condition for synaptic transmission is isochrony (at least, approximate one) of both excitable structures separated by the synapse: two neurons or a motoneuron and a muscle fiber, because consistent shifts of chronaxies due to the subordination, which are able to convert isochrony into heterochrony and vice versa in synapses of the ventral horns of the spinal cord, can, according to the expression coined by Lapicque, "switch the tracks" for the effector impulse and prevent it from exciting simultaneously both muscles within an antagonist pair. Clearly, such a "track switch" ("aiguillage") is identical to the act of reciprocal innervation and denervation of antagonists described by Sherrington, but interpreted based on a different viewpoint. Clearly, compared to Sherrington, Lapicque managed to get deeper into the physiological essence of this phenomenon.

The next group of phenomena originating from the red nucleus are those originally described by Sherrington and studied in detail by Magnus, De Kleijn, and Rademaker; they include the group of tonic reflexes, which do not require much explanation to establish their identity with all the aforementioned phenomena. Currently, it has been firmly established that muscle tonus (at least, in its most common representations), the subordination, and a whole range of phenomena to be described later represent not only different facets of the same basic phenomenon, but are different methods to look at the same phenomenon and induce it in experiments.

The term "muscle tonus" currently unites so many phenomena that it is no surprise that this concept has become very hard to define. The group of phenomena under the umbrella of this term includes the purely humoral "prae-tonus," studied by Uexkull in lower invertebrates, which flows like a liquid along their rays and extremities. It also refers to the, presumably main, complex paleokinetic synthesis of humoral-electrotonic *shifts of excitatory and mechanical characteristics* of the striated muscle, which has already been described and looks, in terms of mechanics, as *work* (contraction and relaxation) of the *striated muscle similar to that of smooth muscles*. The two version of the paleokinetic tonus are so unlike each other that it is hard to see them as phenomena of the same genesis: (a) *elastic* tonus elevated in cases of cortical hemiplegias; and (b) *viscous* tonus increasing during injuries to the mid-structures of the extrapyramidal system. The mentioned forms of tonus are followed by younger ones, which are also clearly tonic formations: tonic contractions observed in muscles of the decerebrate cat by Briscoe (followed by similar observations by Gorshkov and

Gusev from A. Ukhtomsky's laboratory) in response to indirect electrical stimuli at subthreshold frequency and amplitude. According to the frequencies used by those experimenters, this kind of tonus is reflected in electromyographic signals at the frequencies between 15 and 30 Hz. Finally, we have to assign to this group the transitory forms between tonus and tetanus, which are sometimes addressed as "*tetanic tonus*" and can be observed in neck-trunk static-kinetic reflexes. To use the same term "tonus" for this huge group of dissimilar phenomena is not better than addressing all processes that take place in the digestive tract, from its top to bottom, with one word "digestion" without further subdivision of this notion.

For clarity, it is necessary, first, to extract from this large set a narrow, homogenous subset of phenomena, which should retain the name "tonus," and, second, to define this group as exactly as possible. Further, we will understand under muscle tonus the *paleokinetic mode of work of the striatal muscles taken as a whole*, i.e. including not only shifts in muscle mechanical parameters but all other changes linked to those shifts according to the law of parallelism. From this point of view, muscle tonus is not only the current state of muscle tissue stiffness and viscosity and changes in this state, but the whole set of *flexible and plastic changes of excitability* of muscles during their functioning in an intact organism. *Tonus is the flowing state of preparation of the neuro-muscular periphery to selective acceptance of an effector process and its realization*. Therefore, tonus involves both independent tonic contractions and the overall mechanical background, described with coefficients of stiffness and viscosity, for active neokinetic tetanic phenomena, an finally, the whole preparatory setting of the neuro-muscular periphery to expected effector impulses.

In particular, the unity of tonic and subordinate phenomena lead to the same methods of stimulation of the vestibular apparatus inducing both typical Magnus shifts of the neck and trunk tonus and subordinate shifts of chronaxies. Just like the subordinate "track switch" in spinal synapses is accompanied by a decrease in the antagonist tonus, in compliance with the rule of parallelism, i.e., reciprocal relaxation, a redistribution of tonus seen during head rotation and various stimulations of the vestibular (otolith) apparatus is accompanied by redistribution of excitability in the muscle groups and their readiness for tetanus: Muscle groups showing an increase in tonus also demonstrate and increase of excitability (of various types, including electrical).

Paleokinetic phenomena in different parts and levels of the central nervous system, which have clear regulatory significance and frequently display their relatedness to the control of excitability and conduction of neural structures with respect to neokinetic volleys, should also be considered as parts of the analyzed level. Here we have to mention the broad range of phenomena united by Sherrington under the label of central states of excitation and inhibition (central excitatory state and central inhibitory state, CES and CIS), which characterize general state of the spinal cord or one of its reflex fields; interestingly, both the author of these terms and, more recently, Fulton emphasize their unquestionably humoral nature. This feature of humoral phenomena makes these states related to the paleo-process and allow viewing them as arguably the most general manifestations of the central paleokinetic regulation of rubro-spinal level mechanisms. The characteristic slow potentials recorded in the spinal cord show patterns that very much similar to those of paleokinetic processes recorded by an oscillograph from their inherent substrates in internal organs.

One of the convincing proofs that the central nervous system works for the rubro-spinal paleokinetics with all its structures is the phenomenon of slow bioelectrical curves, similar to the aforementioned ones, but recorded from the cortex of the large hemispheres. Cortex encephalogram recorded synchronously with bioelectrical potentials from a working muscle shows two superimposed groups of rhythms. One of them covers the frequency range from 40 to 80 Hz and is frequently synchronized, spike by spike, with the simultaneous tetanic activity of the muscle; this is the so-called beta-rhythm (Berger). Analysis of the second group reveals a whole spectrum of different frequencies from 6–8 to 15 Hz; this group is called alpha-rhythm. The paleokinetic nature of this slow group is beyond any doubt, although these phenomena are much more complex, as expected from the brain compared to the periphery.

Alpha-waves typically emerge in the small-cell fields and layers of the cortex, which belong to phylogenetically ancient structures (for example, the visual field arca striata), whereas the spike-like beta-waves are generated, as a rule, by phylogenetically younger large-cell formations. Most ganglionic formations, which are the sources of long effector axons, are composed of large cells; e.g., the fifth layer of the cortex pyramidal field, the pallidum, effector cells of the corporis striati, red nucleus formations, which are the origins of Monakow's rubro-spinal tract, etc. In contrast, small granular structures are typical of non-projectional areas of the cortex, and it is possible that most of the small cells of these areas do not send their axons to the white matter; rather, they are part of the ancient neuropil formations.

The next group of arguments in favor of the paleokinetic nature of the alpha-rhythms has been expressed in various ways by many established researchers (Rohracher, Kornmüller, Adrian, Walsh, etc.) and seems to be broadly accepted. These arguments, speaking in favor of tight links of alpha-rhythms with humoral, metabolic processes in ganglionic formations (compare to the metabolic

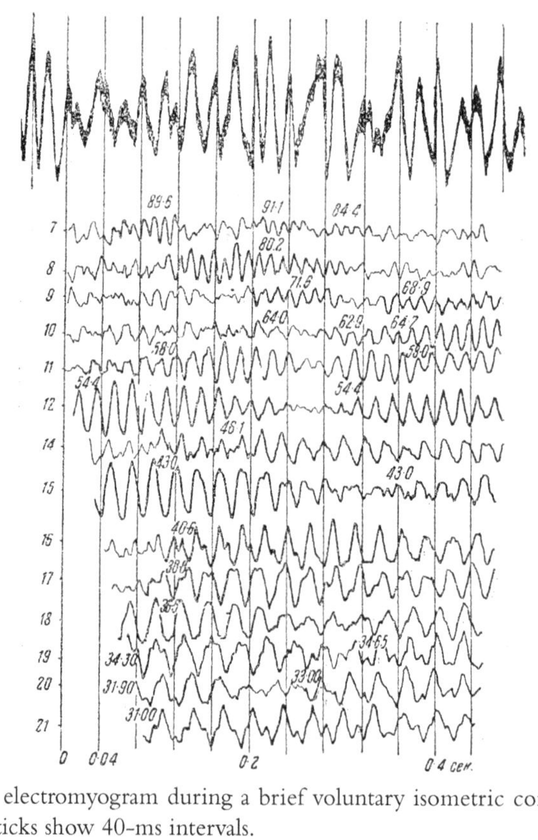

FIGURE 3.3 Top – the electromyogram during a brief voluntary isometric contraction of the human biceps. Ordinate: time ticks show 40-ms intervals.
Under the electromyogram – its decomposition into elementary oscillations according to the method developed by the author. The frequency in Hz is shown over each of the oscillations. Numbers on the left – the durations of the used in the analysis "test periods" in 1/650 of a second.
(Source: work of the author and L. Osipova, CNIIFK, 1941).

interpretation of the weak after-potentials in axons!) and, in particular, with processes of assimilation and re-synthesis. This is reflected in beta-rhythms as "resting currents," which is most clearly seen during full emotional and physical relaxation of the subject, isolated from external stimuli, and disappearing immediately with initiation of brain activity. At the same time, the alpha-waves do not represent pure basal metabolism because they, as a rule, disappear during sleep. Their association with states of rest and inaction, and the fact that during wakefulness they cannot be suppressed over more than a few tens of second and then they emerge again inevitably (as in voluntary breathing delay) both speak in favor of their assimilation role. They are absent in sleep, exactly because in the absence of dissimilation, there is no need in assimilation.[6]

There is one more phenomenon frequently seen in the electroencephalogram, which speaks clearly in favor of the control and electrotonic character of the influence exerted by the paleokinetic alpha-process on tetanic beta-spikes. Many electroencephalographic recordings show that, every time the slow wave of the alpha-rhythms rises toward negative (cathode-electrotonic) potentials, the amplitude of the accompanying beta-spikes increases; and vice versa, during the troughs of the slow alpha-wave, i.e., in its anode-electrotonic phases, beta-spikes become smaller and even disappear altogether (M.N. Livanov). It would be hard to imagine a more impressive and didactic illustration for the coexistence of paleo- and neo-processes and the electrotonic regulation effects of the former on the latter. Anode-electrotonic troughs may reach such depths that they cause complete inhibitory blocking of the "spike" process over their duration. This phenomenon was observed in the spinal cord in 1935 by Barron and Matthews and interpreted by the authors as action of some "intermittent factor"; this gave the paleo-regulatory process one more, (n + 1)-th name, something it could live without.

The functional process that defines the coordination function of the rubro-spinal level is, of course, the process of ensuring sensory corrections assigned to this level. As shown in Chapter 2, muscle force is defined by an equation with two unknowns; it depends on: (1) the physiological muscle state (which can be called mechanical measure of its excitability); and (2) its current length (and also speed of its deformation). Instantaneous magnitude of the second variable is not under the control of the central nervous system; it is signaled via proprioceptive afferents, and the goal

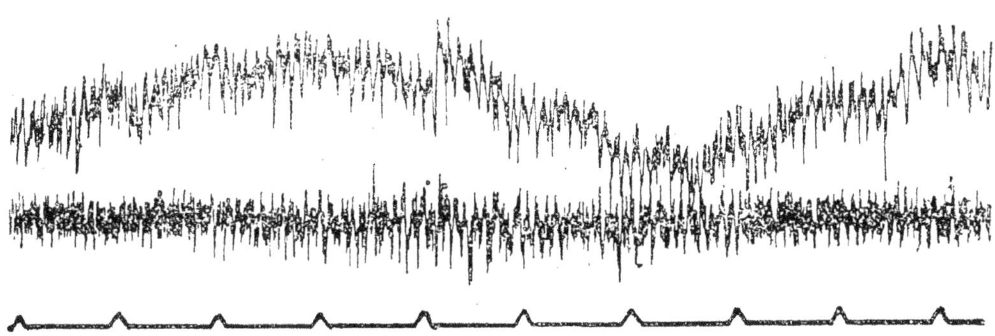

FIGURE 3.4 Simultaneous recordings of the oscillations of bioelectric potentials from the motor cortex of the exposed rabbit cortex (top) and from a muscle of a hindlimb during its spontaneous movements. One can clearly see: (a) characteristic features of the two patterns: slow, large-amplitude changes in the potential on the brain curve, and the lack of such changes in the muscle curve; (b) close similarity bordering with total identity of the beta-rhythms on the order of 50–60 Hz in both curves. Below: The 1/5–s time ticks.

(Source: The electromyogram was generously given to the author by Prof. M.N. Livanov (Moscow Brain Institute).

of the central nervous system is to select a value of the first variable, in other words, to form the excitatory process in the muscle, in such a way that the resultant solution of the equation would produce a magnitude of force required by the movement specification. This is exactly what level *A* does at all times, from a millisecond to the next millisecond, and now we know how it is performing this task. It does not add to the neokinetic process flowing through its effector substrate a single extra spike but always modulates with electrotonic means these tetanic combs, which can be done in three ways. First, it can influence the *amplitude of the spikes* directly by changing their electrotonic background. Second, it can create, in a certain percentage of motor units, *transient synaptic blockade* – either by heterochronism or by turning on the "intermittent factor." Third, it can modulate directly and flexibly the mechanical characteristics of *muscle tonus.*[7]

It is very important to note that the red nucleus is the terminal effector – so-called final cascade – of the extrapyramidal effector system; the red nucleus is the origin of the final common path of this system to the ventral horn cells – the Monakow rubro-spinal tract. The neural process emerging in formations of the extrapyramidal system, such as the striatum and pallidum, and descending along this tract is, of course, a neokinetic, tetanic process, and the red nucleus should possess necessary properties to transmit it, even if its normal independent function were limited to paleokinetic actions. A very likely contributor to this process in the magnocellular caudal nucleus, which is the origin of most axons forming the Monakow tract: its parallel nucleus, the substantia nigra, also has a large-cell part architectonically close to the pallidum and linked to centripetal projection pathways. Besides, level *A* itself, at least in pathological cases, is able to generate neokinetic processes, at least as revealed in the phenomenon of "rest tremor" discussed later.

One can hardly find in a healthy human a single *independent movement* controlled by the level of paleokinetic regulations as the leading level. Most frequent cases of independent movements produced by this level when, despite its inherently background role, it somehow becomes a leader are, first, involuntary trembling actions such as shivering in cold environment, chattering teeth because of being scared, shuddering, etc. and, in the area of voluntary movements, *quick rhythmical vibrating movements* following the "reflex loop" mechanism; second, these are movements associated with *occupying and maintaining a certain posture.* The first group almost certainly includes the fastest piano "vibrato" with the speed of 7–8 strikes per second in cases when they proceed as forced vibrations (Figure 3.5), finger vibrations of the left hand of violinists and cellists, maybe quick,

FIGURE 3.5 The kimocyclographic recording of the hand motion during the fastest octave vibrato on the piano.

The subject was a professional, Europe-wide renowned pianist. The movement proceeds as a forced elastic oscillation and shows typical features of the leading level *A* (which disappear completely when the movement is slowed down). The hand movement frequency is 7.2 Hz; the frequency of recording is 300 points per second.

(Source: the work by T. Popova and the author, 1928).

FIGURE 3.6 Extension dismount by the gymnast A.K. during the competition organized by the journal *Gymnastics* (1937).

automated pronation-supination movements of the hand swaying the fan, etc. The second group is most clearly seen when level *A* happens to play the leading role within a phase of a complex chain action, for example during the aerial phase of the dismount or dive into water (Figures 3.6 and 3.7). Among typical pathological cases of the first group of movements, when level *A* plays the role of the leading level, we have to mention the "rest tremor" of patients with Parkinson's disease, involuntary, precisely rhythmical, following perfect sine waves (Figure 3.10).

In contrast, the technical backgrounds executed by this level are so many and varied that it is impossible to imagine a single motor act controlled by higher levels that would not be soaked in them as either simultaneous or successive components. The main background, making any movement possible, is the *background of the flexible, reactive tonus* of all the muscles of the body – tonus disturbances during local injuries to level *A* produce the so-called amyostatic (violating muscle statics) complex of symptoms. Many authors have emphasized that tonus of skeletal muscles is very flexible, reactive, and adaptive; it represents the most basic and deep background for the whole movement picture, which is also most ancient phylogenetically. Interestingly, in addition to the role of "background for all backgrounds," the paleokinetic process at the rubro-spinal level brings about one more paleokinetic feature, clearly seen in the motorics of lower invertebrates: the statokinetic, shape-adaptive function. It is most peculiar that, if a higher mammal resembles in its movements skeleton-less invertebrates (this is most vividly seen, for example, in feline carnivores), this is seen in neck and trunk tonic rubro-spinal movements of the spine.

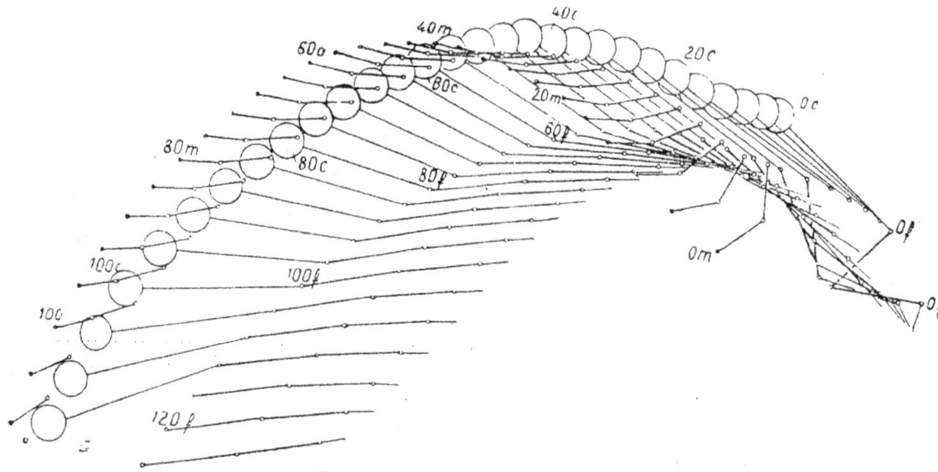

FIGURE 3.7 The stick figures of the sequence of body positions during swimming race start. Drawn after a cyclogram. A likely example of a more or less clean leading role of level *A*. (Source: work by S. Spitsyn, CNIIFK, 1939).

FIGURE 3.8 The characteristic posture of the trunk and arms during parkinsonism (after Seiffer).

Static components of the rubro-spinal level reveal themselves also in gripping postures. We mean here not the grasping movement (performed by higher levels) but gripping, both simple and skilled holding of an instrument, a handle, a processed object, etc. These tonically maintained postures illustrate vividly the adaptive to shape, plastic in the kinematic meaning of the word, function of this level; examples include the principle of plastic correspondence (Prinzip der gleitenden Kopplüng) in the fingers described by Bethe, which he tried to copy in his hand prostheses (Figures 6.6 and 6.7).

The kinetic background components of this level can be seen in a whole range of rhythmical voluntary movements, controlled by higher levels, which show elements of forced oscillations. They can be clearly seen during movement of the arm with the hammer during rhythmical hitting of the chisel, and movements of the bird's wing during flight (as it follows from analysis of Marey's recordings), arm and leg movements during locomotion, etc.[8] The fast convergence of Fourier series, which can be used to interpret biological rhythmical movements, is likely a very reliable reflection of the presence in the movement of kinetic components of level *A*. If a fast rhythmical movement involves a simple kinematic chain, for example only the hand (pill-rolling in patients with Parkinson's disease) or only the pronation-supination system (moving the fan), movements controlled by this level represent nearly perfect sine waves.

Pathological violations of the rubro-spinal level function reveal themselves primarily as disorders of the tonus – dystonias.[9] The well-known phenomenon of decerebrate rigidity, observed in mammals following transection of the brain stem below the red nucleus level and leading to disappearance of functional subordination and fixation of muscle tonus at some stationary level (different in different animals), is due to the loss of functions of this level. As mentioned earlier, in humans, dysfunction of level *A* leads to the amyostatic complex of symptoms; in severe cases, this can lead to cases of overall severe hypertonus, catalepsy, and flexibilitas cerea. The tremulous paralysis in Parkinson's disease represents a superposition of many phenomena of disorders of level *A*, primarily leading to *hyperkinesis* due to the loss of descending control. Disorders of the distribution and adaptive activity of muscle tonus frequently accompany problems with other, higher levels, but they always signify involvement of the rubro-spinal level into the disordered process – the level that monopolizes the control of tonus throughout the central nervous system. There may be both hypotonic and hypertonic syndromes, most frequently syndromes of not only quantitative tonus shifts in one or the other direction, but primarily violations of its reflex, adaptive regulation. Ataxia during tabes dorsalis involves the higher level *C* (described later), but it is always accompanied by disorders of tonus – hypotonia originating from level *A*.

The next characteristic manifestation of level *A* dysfunction is *tremors*, typically the two most typical types of tremors described in the semiotics of neural disorders. Effector hyperfunction of level *A* is manifested by the mentioned *rest tremor*, which is a cardinal component of parkinsonism – the modestly fast (8–10 Hz, interestingly, it overlaps with the Berger alpha-rhythm) rhythmical, monotonous, and presenting in cyclographic recordings as a perfect sine wave (Figure 3.10). This tremor can involve both the head and distal segments of all extremities. During voluntary movements, it either becomes less visible on the movement background or disappears altogether.[10]

Hypofunction of the rubro-spinal level reveals itself as a much more interesting phenomenon of *intentional tremor*. At rest, unlike the previous type of tremor, this tremor is absent, but if the patient initiates a movement or even intends to initiate it, immediately the affected limb shows irregular, fidgeting, uncontrolled oscillations. The more the patient tries to suppress the tremor, the more it develops (Figure 3.11).

The essence of intentional tremor can be easily understood in the light of the described theory of coordination; it is in the disordered or totally lost function of reciprocal innervation and

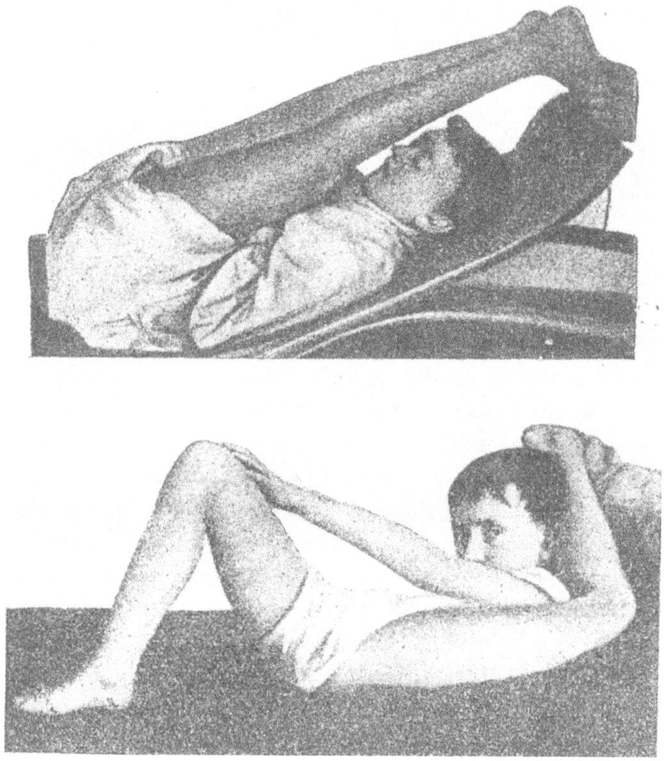

FIGURE 3.9 Two photographs of the hypotonic syndrome in tabes dorsalis (Seiffert, from the neurological clinic in Galle).

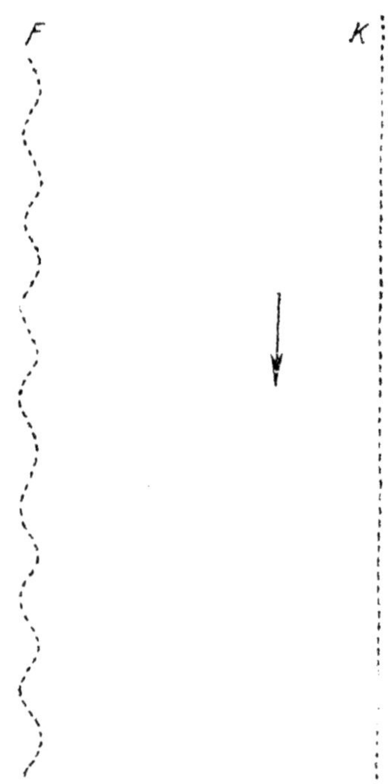

FIGURE 3.10 The kimocyclogram of the parkinsonian hand tremor as an independent movement at level *A*. Note the purely sinusoidal pattern of the curve.
(Source: patient of Prof. M. Gurevich, the work by the author, 1924).

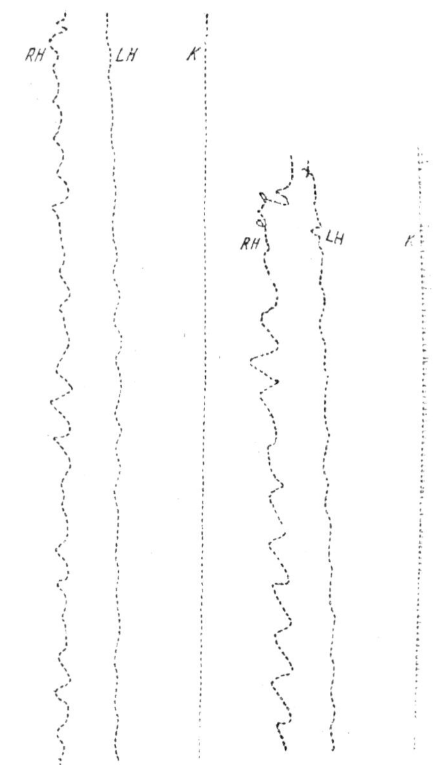

FIGURE 3.11 The kimocyclogram of the intentional tremor in both hands of a patient with a complex extrapyramidal syndrome.
(Source: Little disease, work by the author, 1926).

denervation of antagonists. In normal conditions, an effector impulse to muscle *P* automatically turns-off its antagonist *Q* by creating heterochronous conditions in synapses on its motoneurons. When the mechanism of reciprocity is lost, the effector process flows into both antagonists simultaneously in a disordered way, without any balanced dosage leading to a fight between the two muscles, which is revealed as oscillating throws of the extremity in different directions. Similarly to the mentioned symptom of level *A* hypofunction – hypotonia, intentional tremor usually accompanies ataxia linked to disorders of level *C*. Every neurologist knows how hard it is to perform the finger-to-nose or heel-to-knee test in this syndrome.

Notes

1 In insects, accelerations experienced by their wings can reach magnitudes a hundred times higher than the acceleration of gravity.
2 Within the neokinetic system, the fiber caliber of both neural and muscle fibers, besides its effects on conduction speed, affects other parameters of excitability and excitation itself. In higher mammals, both neural and striatal muscle fibers form whole classes ("sortments") of caliber ranges and parameters, which can be classified onto groups (Gasser's types A, B, and C, and Krause's white and red muscle fibers) only with a grain of salt. Definitely, these "sortments" represent auxiliary biological means to compensate for the lack of variation and flexibility of the neokinetic excitation process (see later).
3 Currently, there is no agreement on the direction of change in the modulus of elasticity during the burst of excitation.

4 Whereas synapses within the paleokinetic system do not show major functional differences from the axons and muscle cells, the neokinetic synapses show a number of deep functional distinctions from the conducting and contracting neokinetic elements (more so in the spinal system, less so in neuromuscular synapses). Due to so far unknown reasons, synapses of the neokinetic system are kind of paleokinetic islands, anachronisms, in the neo-motor system, which affect all the functions within a living organism.

The spinal synapses do not obey the law "all or none." They possess an ability to show iterative behavior, i.e., they either require lengthy buildup or convergence of the effects from multiple afferent neurons. The same inertia is reflected in their inherent phenomenon of after-discharge. The spinal synapses show very large amplitudes of changes of electrotonic states. These states may be conditioned by impulses from brain ganglia as well as by summation of excitations coming from sensory neurons or interneurons. These electrotonic states can be of either sign (these are the so-called Sherrington's central excitatory state and central inhibitory state), i.e. able to show an increase or a decrease in their excitability and conduction. The first of these changes is sometimes addressed as "facilitation"; the second is known as subordinate inhibition, Sechenov's inhibition, and by other names, when it is of a central origin.

Synapses display most vividly the role of humoral mechanisms in the transmission of excitation in neokinetic systems. The whole process of transmission of bursts of excitation through "spinal synapses" proceeds with an important role of bioelectrical factors; this is convincingly proven by their ability to transmit sequences of spikes and produce exact numbers of output spikes. At the same time, the presence of a humoral component in the process of synaptic transmission of excitation has been proven beyond doubt, although there is still no unanimous agreement on mechanisms of the combined action of the two factors. It is very likely that the synaptic delay in the transmission of excitation is due to transformation of an excitation process in the area of the synapse due to the bioelectric wave into a more complex electrochemical phenomenon.

Hence, one has to admit that synapses of the spinal neokinetic system demonstrate a whole range of properties typical of paleokinetic systems, which are absolutely alien to the neokinetic systems of the neural and muscle fibers.

5 We use the term "moton" to address the system involving a motoneuron and the muscle fibers connected to it – the so-called mion. The term "moton" is a synonym of an English term "motor unit."

6 For a discussion on the high amplitude of the alpha-waves in encephalograms in contrast to low voltage typical of peripheral paleo-processes – see my paper "Urgent problems of the contemporary neurophysiology" (*Physiological Journal of USSR*, 1945, issues 5–6.)

7 There is a seeming fourth possibility: To affect the amplitude of tetanic spikes by inhibiting synaptic transmission. This possibility, however, is removed by the "all or none" law and non-decrement conduction of neo-excitation: If synaptic transmission to cells in the ventral root is weakened, this would not be perceived a few millimeters downstream.

8 See our experimental studies of the mentioned movements and their analyses.

9 Let us note for future discussion that it is convenient to address typical disorders of the next level of synergies *B* as asynergies or dis-synergies, disorders of level *C* – ataxias and dystaxias, and finally, disorders at level *D* – apraxias and dyspraxias.

10 The similarity between the parkinsonian "Ruhetremor" and the alpha-rhythm of brain potentials is not limited to the similarities between the two frequencies, which could easily be coincidental. Both phenomena: (a) disappear during sleep; (b) are seen best during complete rest; (c) disappear with the initiation of activity; and (d) if activity goes on, both phenomena can resume. This clear similarity provides one more proof of the paleokinetic origin of the alpha-rhythm phenomenon. The electromyographic patterns during the rest tremor – intermittent rhythmical bursts of spikes – speak directly in favor of the aforementioned central blockage.

4

SUBCORTICAL CONSTRUCTION LEVELS

Level B: The Level of Synergies and Patterns, or the Thalamo-pallidar Level

The next level within the hierarchy of movement construction is the level of synergies and patterns,[1] alias the thalamo-pallidar level, which we will address as level *B*. It is not exactly the next level after level *A* in the chronology of phylogenesis – this is proven, for example, by the fact that the level of paleokinetic regulations from the very beginning had something to regulate, namely the neokinetic stream passing through its substrates and delivered in higher vertebrates by the pallidum. Hence, on the one hand, the *substrates* of level *A* in higher vertebrates are phylogenetically younger than the substrates of the level of synergies, because the most ancient functions of level *A* have relatively recently acquired a new anatomical dwelling in the red nucleus and hypothalamus, moving there from the depth of the spinal cord. On the other hand, the *functions* of the level of synergies *B* are phylogenetically younger. This is revealed, first, by the fact that this level reaches the pinnacles of its development in phylogenesis after level *A*; second, by the fact that level *B* is, in its phylogenesis, the first truly neokinetic level, which realizes its effector outcomes with the help of tetanic, spike-based processes; third, the principles of functioning of level *B* are younger than those characterizing level *A*. Whereas level *A* is primarily the level of movements of the trunk and its segments, level *B* is the level of the body as a *locomotor muscle*, equipped with the four extremities-actuators.

In phylogenesis, the level of synergies is the first truly controlling, leading level, which, at some level of development, rules all the somatic movements. In the morphogenesis of vertebrates, its substrate is the most ancient brain, not only a head appendage of the spinal cord, as in the case of level *A* substrate. The level of synergies integrates its processes that include very large groups of muscles: it was born as a result of the necessity to perform locomotor actions, although the process of progressive encephalization moved the control of locomotor movements to higher brain levels. As far as its ability to conduct huge synergic choirs is concerned, this level reaches a pinnacle, after which the evolution, in a sense, made a step backward: The next level *C* (see Chapter 5), a clearly much younger level, is much more frugal and even stingy with respect to simultaneous involvement of limbs and muscle groups into movements.

The anatomical substrate of the level of synergies in higher mammals and humans – the two pairs of the largest in the brain subcortical nuclei: thalami optici as afferent centers and globi pallidi as efferent nuclei (Figures 4.1a and 4.1b). Thalamic nuclei are the loci where the second-order neurons, counting from the body periphery, converge from all the proprioceptors,

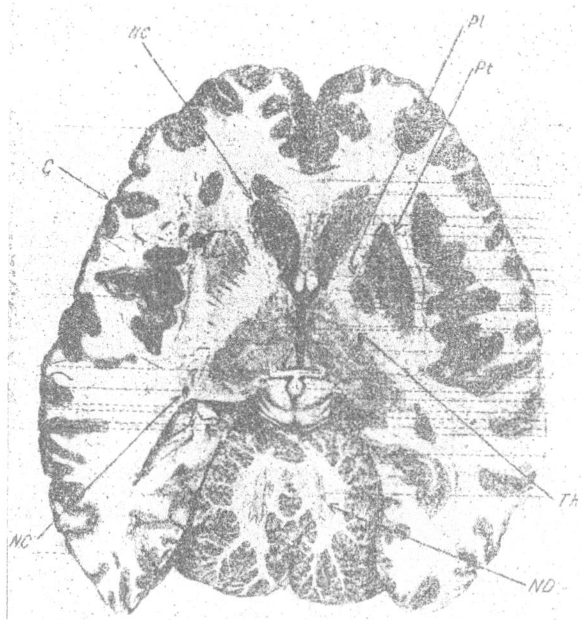

FIGURE 4.1A A horizontal slice of the brain not far from its base.
NC, NC – the rostral and caudal part of the caudate nucleus; *Pt* – the putamen; *Pl* – the globus pallidus; *Th* – the thalamus opticus; *Nd* – the dentate nucleus of the cerebellum; *C* – the brain motor cortex (from the brain atlas by E. Flatau).

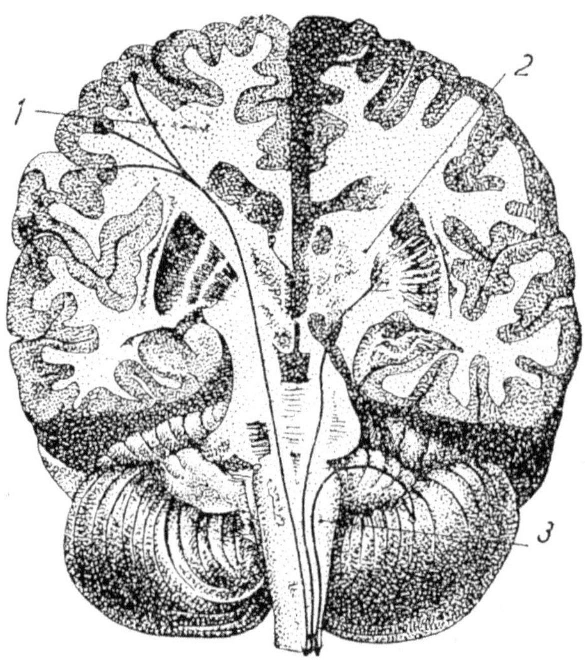

FIGURE 4.1B A section of the human brain from the top toward caudal ventral-to-rostral parts.
1 – the scheme of the pyramidal pathway; 2 – the pallido-rubro-spinal pathway; 3 – the cerebellar (dentate-vestibulo-spinal) pathway; the crossover of pathways 2 and 3 are not shown.

exteroceptors, and tango-receptors. The sensory pathways from the telereceptors of olfaction, hearing, and vision have no relation to this brain formation either in the ancient or in the recent phylogenesis.

Thalamus opticus is the most ancient and, before the cortex of the large hemispheres had developed, was the most important center of all the tango-receptor function, which is "the central processing shop of sensory percepts" (the expression coined by Monakow). Between the thalamus and pallidum, there are well-developed two-way pathways. Pallidum is a pure effector with a homogenous, microscopic structure (see Chapter 7), which is an intermediate level within the extrapyramidal effector system. It controls in a hierarchical way the lower level of this system, which is the group of the red nucleus discussed earlier; the pallidum itself is subordinate to another major subcortical effector – the corpus striatum, a formation composed of nucleus caudatus and putamen. *All* the centrifugal pathways from the pallidum terminate in the red nucleus group and have no direct projections to cells in the ventral roots of the spinal cord.

Similar to the previous level, typical leading afferents of the thalamo-pallidar level are primarily proprioceptors, but they contain completely different components and demonstrate a different style compared to proprioceptive afferents of level *A*. First, based on typical movements of level *B*, the dominating afferents are those reflecting new, joint-angle-based, geometric proprioception of positions and velocities, which is also supplemented by the broad range of exteroceptive sensitivity, both protopathic (receptors of pressure and deep touch) and epicritic (receptors of differentiating touch, pinprick, friction, pain, vibration, and temperature with typical of these sensations precise "local signs"). Second, all these receptor reactions, both purely proprioceptive and the supplementary exteroceptive ones, are responsible at level *B* for encoding and correcting movements, after being already subjected to preliminary central processing and synthesis at a much more complex and differentiated level, compared to the primitive synthesis of the rubro-spinal level. The described levels seem to have no perceivable links to the vestibular system.

If any of the sensory syntheses controlling the coordination of different levels within the human central nervous system deserves to be addressed as proprioceptive, this is the described synthesis of the thalamo-pallidar level. To summarize all typical for this level afferents, this would be *afferents from one's own body*, proprioceptors par excellence. At this level, the body is the original reference system for both perception and motion, and the ultimate goal of these sensations and movements. All the innumerable tactile and proprioceptive signals unite, produce local signs, and develop some king of a common for all signals coordinate system linked to the body; this generates complete and detailed information on the motor apparatus of the body. These processes are also reflected in movements, performed by this level; typically, these are movements of the whole body, linked to the body alone with no reference to outside objects, movements that are goals of themselves, which are corrected only along the lines of their intrinsic harmony and organization. To use the analogy of their correcting sensory synthesis, these movements deserve the name of *propriomotor* actions and reactions. Under the aforementioned limitations, these propriomotor movements reach very high levels of development and differentiation. They have been addressed with a variety of names: motor formula, synergy, patterns, lower automatisms, etc.

There are three main coordination features characteristic of the relatively well-studied thalamo-pallidar system, which make it different from other human kinetic systems.

The first is the adjustment of level *B* to *broad muscle synergies*, i.e. its ability to lead highly coordinated movements of the whole body involving many dozens of muscles in a highly coordinated fashion. Local injuries to this system, e.g. those of the thalamus or pallidum, result in typical dyssynergias, i.e. loss of such large-ensemble movements with surrogate substitution by deliberate, rigid, clumsy movements seen, for example, in patients with Parkinson's disease.

From the classical point of view, it was natural to assign responsibility for synergies in normal conditions and for their loss in pathological cases to the pallidum as the main effector of the system. There were attempts to find an explanation of its mysterious ability, unavailable for all other effectors of the central nervous system, in its cytoarchitecture, conduction pathways, and even peculiarities of biochemistry. Only the theory of coordination, described in this book as a process fully based on the principle of sensory corrections, has turned the question 180° and focused crucial attention not on the effector of the system, the pallidum, but on its afferent systems thus clarifying many issues. Obviously, the pallidum is able to lead complex synergies of kinematic chains smoothly and easily, primarily because no other effector of the body, neither from higher nor from lower brain levels, possesses such a full and detailed first-hand information on positions and movements of the body compared to what the pallidum gets from the thalamus. Its coordination is "propriomotor" in its character and meaning exactly because it is led and corrected by the main proprioceptive nucleus of the central nervous system, which is completely detached in higher mammals and humans from distant receptors but represents the center of all processes related to the body and its contact sensations.

We have already mentioned in Chapter 2 that the disobedience and difficult control of kinematic chains grow rapidly with an increase in the number of degrees of freedom, i.e., in particular, with an increase in the number of involved joints, because it is accompanied by intensive increase and complication of reactive forces perturbing movement of the chain. Having at its disposal the quick (with a minimal number of synaptic delays) and comprehensive afferent system signaling on all the dynamic phenomena on body periphery, the thalamo-pallidar level has all the tools to counteract those forces in a timely manner and to make the kinematic chain a controllable system. The huge problems introduced by the reaction forces into any movement allow making a general statement that the problem is not in controlling 30 muscles at the same time but in controlling three joints of a chain. It is clear that the level of synergies, always ready to solve in real time the most complex part of the problem – overcoming the reactive forces and, at the same time, controlling the multi-joint pendulums of the extremities, is also solving easily the simpler part of the same problem – controlling dozens of protagonists, antagonists, and synergists.

The second coordination feature characterizing the style of functioning of level *B* is the ability to *control movement in time* gracefully, for example, to ensure appropriate alternations of all the extremities during locomotion, to unite in a common rhythm, with a millisecond precision, movements of the multi-link pendulums of the extremities, which have their own variable and complex spectra of natural oscillations, etc. This feature is obviously closely related to the first one. One has only to add that the meaning of movements at level *B* is focused on one's own body, independently of the external world, and naturally, due to the design of our extremities with their periodic, reproducible form, it leads to the visible inclination of the level of synergies toward rhythmical, rocking, repetitive movements with highly variable and complex, but perfectly reproducible, rhythmical patterns (Figures 4.5, 4.6, and 4.7). Naturally, among movements directed at an external goal (reaching a point in space, touching an external object, etc.), there is a much larger percentage of those where reaching the ultimate goal, playing the role of the end-switch of the central nervous system, works to terminate the movement compared to self-sustaining movements linked to one's own body. In contrast, here, the absence of a meaningful external ultimate goal, given the presence of numerous kinematic and elastic factors that favor return movements, may easily produce end-switches favoring return movements. In the realm of rhythmical construction, the described level possesses the highly structured, exquisite technique, which beats in this aspect all the other construction levels.

We believe that everything said above about end-switches should clarify the problem of the nature of rhythmical and cyclical movements in general. Clearly, no single construction level can monopolize rhythmical and cyclical actions. And no action (possibly with the exception of most elementary tremors and forced elastic oscillations at level *A*) possesses obligatory, immanent cyclicity inherent to its motor essence. Cyclicity can be seen, and is actually seen, at each of the levels, the same is true for rhythmicity, and decision whether the movement will proceed with returns and repetitions or will be limited to a single excursion *depends not on its motor structure but only on its meaning*: Some movements tend to be singular, while others – repeated. It is a different matter that, at different levels, switches involved in repetitive actions work differently leading to different rhythmical-cyclical features. Each level forms movements in its own mode and tonality.

When providing background components for movements controlled by higher levels, the level of synergies commonly takes the responsibility for rhythmical sequences. Moreover, it somehow, in a manner not completely understood, creates an intrinsic propriomotor rhythm and brings it into the action of the higher level, which has no reasons for a rhythm or repetition in its task and signaling or triggering afferents: for example, during locomotion as transition of the whole body to a more or less remote point in space. Some forms of disorders at the level of synergies lead to disruptions of this mechanism of recoding of the problem of non-rhythmical transition into rhythmical stepping; this may be accompanied by peculiar cases when the patient in unable to make a single step over the smooth floor, but can walk easily and briskly upstairs and downstairs, with the visual perception of the steps substituting for the lost intrinsic mechanism, which normally brings rhythm into the aperiodic space.

Finally, the third feature of level *B*, which has to be mentioned, is its striking propensity toward stereotypes, exact repetition of movements, both rhythmical and singular, similar to each other as two coins. There were attempts to explain this phenomenon, well-known to neurologists, with some special (still hypothetical) features of the pallidum as an effector, but the situation is in fact quite different.

First of all, it is necessary to recall that external, particularly reactive, forces during cyclical actions do not show stereotypical repetition. Quite the opposite, because a single rhythm of a forced repetitive movement of a multi-link extremity-pendulum cannot coincide with the various natural frequencies of all its elements, reactive forces create very complex dynamical interferences with some exogenous frequency transmitted to all the links via the muscles. This means that movements of such a multi-link pendulum could be the same across cycles only if active muscle dynamical contributions would vary adaptively over successive cycles. The level of synergies needs all the rich resources and perfection of the thalamic afferents to maneuver flexibly among all the reactive forces turning their dynamical mess into a stereotypical pattern. Not being familiar with mechanics, neurologists do not even suspect about the huge background work performed during such movements by level *B* patiently in the subcortical bilges of the brain. Hence, the pallidum should not be viewed as some kind of a warehouse of stereotypies, which it recruits on demand as a phonograph disk. Movements it produces can be so stereotypical only because it itself is not stereotypical at all.

Analysis of physiological and biodynamical regularities of movements controlled by the level of synergies reveals the actual causes of its propensity for stereotypical actions. We will perform more detailed analysis of this issue in Chapter 8. At this time, let us mention that, in the process of skill acquisition, one can typically observe significant qualitative changes in the principles of control of the practiced movement. For most large-scale, whole-body movements, which are subjects of level *B*, as well as for rhythmical and singular movements accompanied by large-amplitude changes of inertial forces, one can find forms when any deviation from the proper trajectory causes reactive forces that tend to return the moving link to the original path. If an organism discovers movement structures that satisfy this description, so-called *dynamically stable movements*,[2] then not only does it not have to correct perturbing reactive forces with phasic tetanic volleys (as in early stages of skill acquisition), but it gains an ability to use them to its advantage. As shown by exact cyclogrammetric analyses, this produces two outcomes. First, removing active forces that used to be spent on counteracting reactive forces, by itself, yields both energetic and inertial economy, because the movement starts to flow by itself without the necessity for steadfast control and corrections. In addition, such dynamically stable forms, obviously, represent forms that correspond most closely to the design of the kinematic chain and, as such, are most economical. In fact, achieving such a stage in the process of skill development is accompanied by both a drop in the number of force waves in the cyclogrammetric dynamical curves and a decrease in their amplitude, i.e. by a substantial increase in the coefficient of efficiency.

Clearly, for each motor task, there cannot be many such solutions. Indeed, there is either a single solution or, in the best-case scenario, a few for each given task. It is easy to understand that movement forms that are close to dynamically stable ones, whereas not being stable themselves, will be forced by the reactive forces into trajectories of stable movements. This makes dynamically stable movements discrete, separated from each other by unstable, self-destructive forms, which are hard to learn because of the difficulties with their repetitions. This creates divergence of stable forms or movement styles during both conscious practice by grown-ups and movement development in children, for example for walking or running, which may be at first indistinguishable. The elaboration and fixation of such dynamically stable movement forms, which automatically lead to a small number of discrete stereotypies characterized by stable patterns or movement equations, is characteristic of the thalamo-pallidar level. The heart of the matter is not in effector stereotypies, which have been viewed as answers to the puzzle of pallidar patterns and intrinsic controversies analyzed earlier. The actual reason is that, within the group of huge, complex, and harmonious synergies, such as locomotions, it is possible to create and perform only a few motor solutions for the problem of combining the actual external movement goal, the kinetic structure of the motor apparatus, and the reactive dynamics emerging as a result of the fight between the two.

Despite the exquisite perfection of the efferent mechanisms and richness of coordinative possibilities of the level of synergies, the organism retains only a few *independent movements led by that level*. First, we have to mention the triad of movements related to *expressive mimics, pantomime, and plastics*, i.e. the totality of not symbolic, but directly emotional movements of the face, limbs, and the whole body. To a large degree, this group also involves choreographic movements – not of the western, locomotor style, but of the eastern, plastic one. The act of dance, as a whole, is built above the discussed level, similar to all other movements with an exogenous rhythm. The level of synergies, with its poor links to telereceptors, is not well fit to use either visual-based or hearing-based control.

Many movements of the *floor gymnastic exercise* barely move beyond the level of synergies: trunk bends, twists, and various plastic-rhythmical movements. Finally, this group involves semi-involuntary movements – stretching the whole body, stretching the extremities, caressing movements (embrace, kiss, etc.), routine monotonously machinelike movements, etc. All such movements are smooth, harmonious, display gracefulness and repetitiveness, even in non-gracious people. If these movements are rhythmical and repetitive, they do not follow the primitive sine pattern (as in level *A*), but show more intricate and variable patterns (Figures 4.5, 4.6, and 4.7).

The percentage of independent movements at this level is much higher in the motor repertoire of infants, which will be discussed in Chapter 7. Here, we have to mention the pre-walking bilateral alternating reflex of stepping – the rhythmical leg jerks in the supine position, which point at the synergic nature of the main backgrounds of walking, which are located not higher than in the spinal cord, even in carnivore mammals.

To clarify the origins of such an unexpected poverty of independent movements led by level *B* in humans, it is necessary to point out what is missing at that level for performance of a larger variety of movements. As we have already seen, the main weakness of the level of synergies is not in monotony or primitive patterns of its movements – quite the opposite, this is the level of "being in control of one's body," able to discover and implement a large variety of movement patterns, varying in their configuration and rhythm, and possessing a unique degree of harmony, inaccessible for other levels. This level, thanks to its particularly intimate and multifaceted links to proprioceptive afferents, is equipped better than other levels for keeping under control reactive forces and creating economical, dynamically stable trajectories. It is quite possible that this is the main reason this level dares to form large multi-muscle synergies. Synergies created by the pallidum are not reflections of its ability to send some kind of ingenious and artful signals, as some neurologists think, but because of its exceptional afferent signals providing for perfect control of the periphery. *The weakness of the level of synergies is not in the quality or quantity of its afferent signals, but in their lopsidedness.* It lacks distant receptors and their synthesis, which are essential for mastering the external space. Its coordinate system is linked to the body, not to the external Eucledian space, which is vast, motionless, and aperiodic. This level is too introverted. It cannot serve as a pilot or a navigator, it is a perfect flight mechanic.

The *background role of the level of synergies* is huge and, in its importance, not inferior to that of level *A*. Because of its disconnect from telereceptors and weak links with the vestibular system and the cerebellum, the level of synergies cannot lead many independent movements; in contrast, it readily accepts responsibility for the intrinsic rough work of any complex movement, if a higher, more extraverted, level ensures its adaptation to the environment and external objects. Hence, level *B* ensures all the intrinsic coordination background of all locomotor movements – walking, running, etc. – by taking care of the kinetics of those enormous synergies but, so to say, in an abstract way, outside the concrete context. Walking should be performed somewhere, over some surface, avoiding obstacles, over rough pavement, steps, turns, etc.; all these factors are beyond the afferent systems of the level of synergies. This is the reason that injuries to the striatum lead to disruption of walking, although its muscle dynamics is realized by the thalamo-pallidar system: Removing the striatum deprives walking from the necessary pilot and confines it to idling. However, when the next level, the level of spatial field *C*, is intact, the synergies and stereotypies of level *B* ensure possibility of various movements controlled by level *C*: locomotions, transfers, transports, ballistic movements, accurate imitations, and many other movements, which will be described in the next chapter (see Figures 4.2 and 4.3).

Without the background provided by the level of synergies, numerous goal-directed motor actions from level *C* become hardly possible. The level of synergies can contribute to these movements in two ways: (1) It can support locomotion at level *C*, as described earlier, while level-*C* locomotion would provide technical background for movements constructed at an even higher level, such as playing tennis or football, working as a railroad car coupler, steel roller, during military attack, etc. (Figure 4.4); or (2) It directly creates background components for one of the higher levels. A very impressive example of this level interaction is quick handwriting, i.e. automated writing, which will be analyzed in more detail later (see Chapter 8). During this action, one can clearly observe two simultaneously working and equally necessary levels (in fact, during writing, the number of levels is larger than two), one of which is leading and the other – technical and subordinate. The subordinate level *B* produces in this action the oscillator synergy of the forearm, hand, and fingers, which typically proceeds with a clear rhythm, and can be seen during preparatory, idling arm motion, which is frequently used by people proud of their handwriting style. The higher levels *D* and *E* (see Chapter 6) impose meaningful modulation on these monotonous rhythmical oscillations, thus turning them into non-cyclical letters and words. To use a metaphor from the area of radio, one can say rather exactly that the level of meanings *E* modulates oscillations

FIGURE 4.2 An instantaneous shot of a posture during figure skating as a typical example of a broad synergy at level *B*.
(Source: from the journal "Physical culture and sport," 1936).

FIGURE 4.3 A typical instantaneous posture – synergy at the end of the backswing before the javelin throw.
(Source: photo by the author, CNIIFK, 1940).

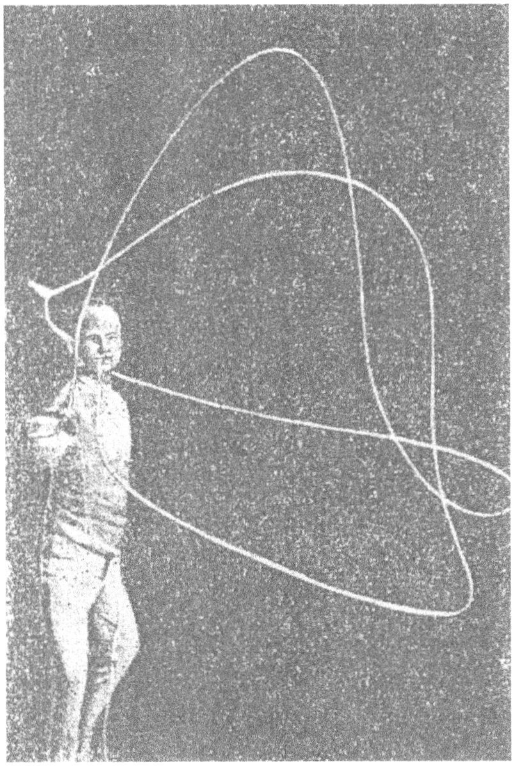

FIGURE 4.4 The primitive cyclogram, taken without the obturator, of a complex fencing gesture without an opponent (gymnastic imitation).
The bulb was affixed to the end of the foil. An example of a synergy at level *B* with a smooth pattern and complex, graceful movements.
(Source: work by V. Fadeev, GCOLIFK, 1930).

produced by the (illiterate) level of synergies, just like the sound of speech, the input into a microphone, modulates monotonous high-frequency oscillations generated by the transmitter, turning them into non-cyclical, meaningful signals.

A similar mode – modulation of monotonous synergies at level *B* by higher levels – is seen in a large number of labor and construction movements, particularly those representing highly automated cyclical processes (Figures 4.5–4.8), which will be appropriately discussed during the description of the corresponding levels.

Synergies of the thalamo-pallidar level play a very important role in the *automation of motor skills*. Automation can proceed at different levels and can be summarized as transition of certain technical coordination components of a motor action to lower levels, which, at that time, are not consciously perceived (as already mentioned, *only the leading level is perceived consciously* for any given movement, independently of its absolute level). Cases of automation related to switching certain components to the level of synergies are particularly typical and impressive. During this process, following motivations of the leading level, either level *B* gradually develops new synergies, which replace the original temporary coordination structures at the leading level, or, in some cases, an ability to use already available automatisms from the rich collection of that level is developed.

In pathological cases, two drastically different clinical pictures can be seen: (1) during hypofunction (or loss) of level *B*, or (2) during hyperfunction of that level, usually due to loss of

FIGURE 4.5 A kimocyclogram of the hand movement during filing as a typical background synergy at level *B*.
Note the extreme rhythmicity and repeatability of the movement pattern, given the significantly higher complexity of the contour compared to sine-like trajectories at level *A*.
(Source: work by the author, 1924).

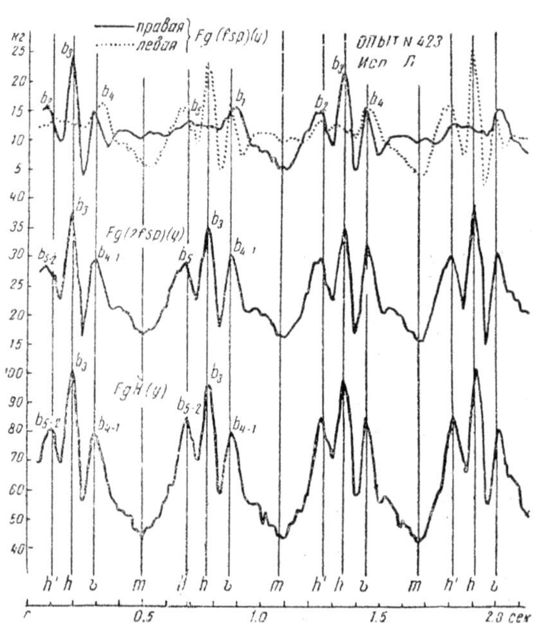

FIGURE 4.6 Complex rhythmical synergic force patterns during normal human walking.
Top: The vertical force components in the center of mass of each of the two legs superimposed on each other. Middle: The vertical force components in the combined center of mass of both legs. Bottom: The vertical force components in the center of mass of the body.
(Source: work by the author, VIEM, 1935).

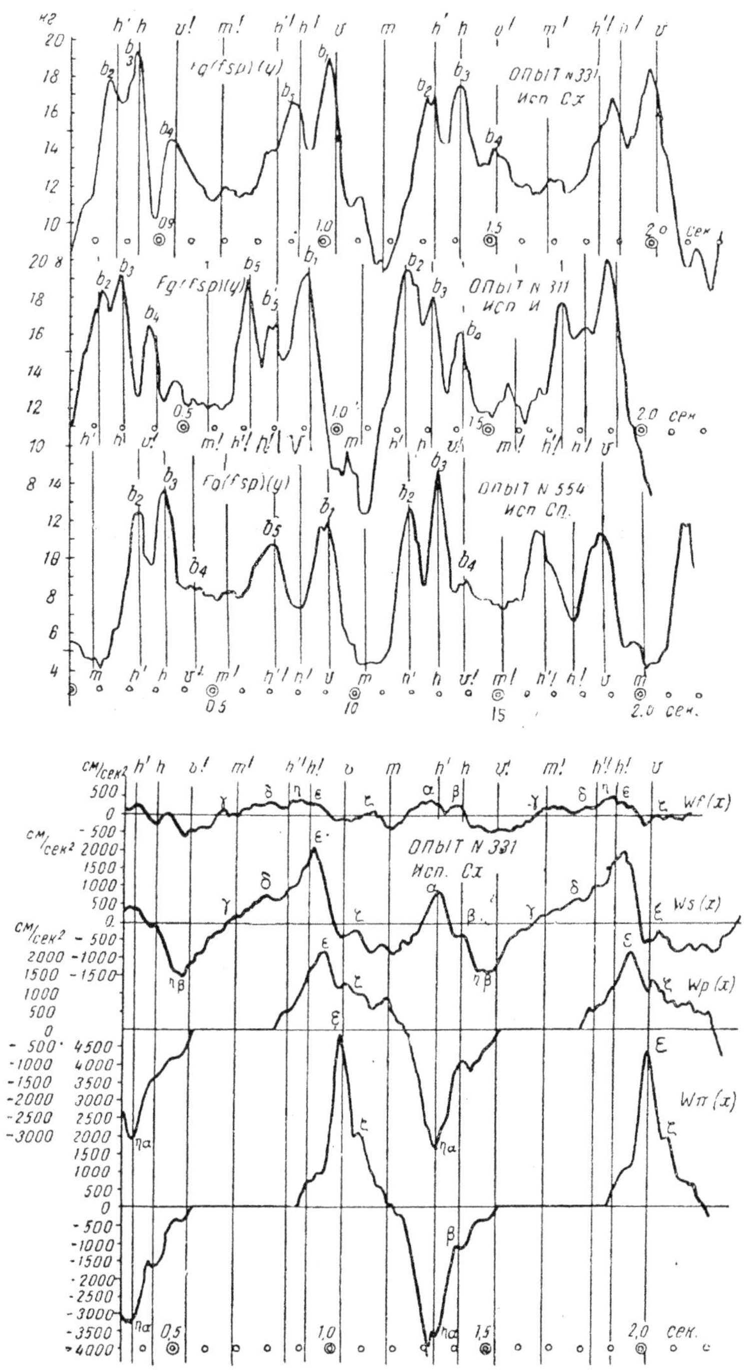

FIGURE 4.7 Top – the curves of the vertical force component in the center of mass of a leg during normal walking by a man (three different subjects). Bottom – the curves of forward-backward accelerations in the main parts of the leg during normal walking. Top-to-bottom: hip joint, knee joint, ankle joint, and the tip of the foot. The curves represent examples of complex rhythmical forces produced by the level of synergies.

(Source: work by the author, VIEM, 1935).

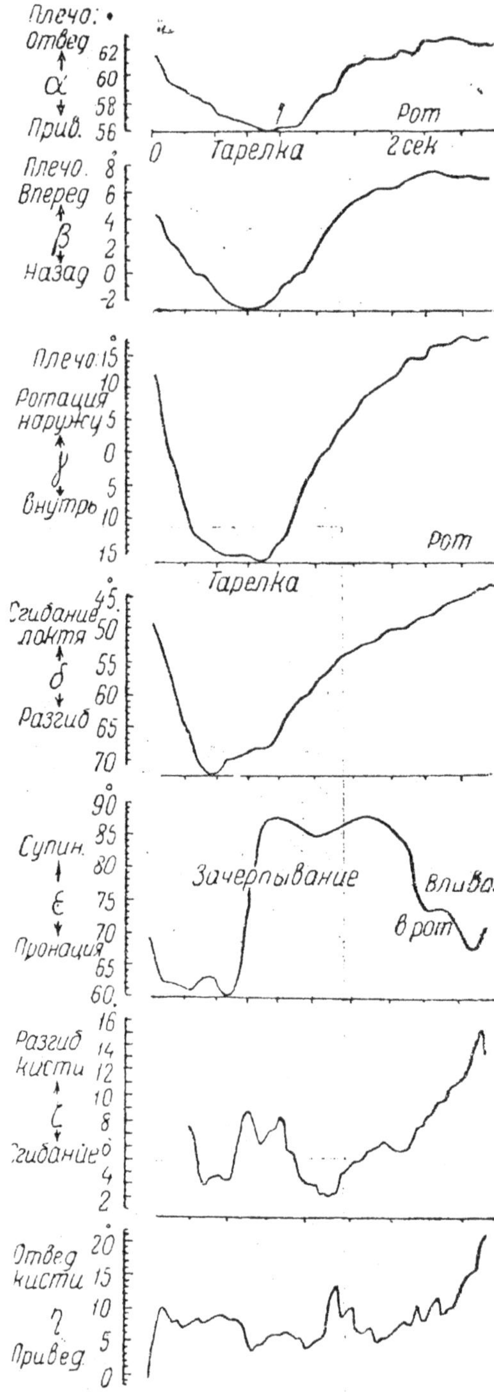

FIGURE 4.8 The curves of the arm angle changes during the automatic action of eating with the spoon. Top to bottom: α – the abduction-adduction shoulder angle, β – the flexion-extension shoulder angle, γ – the external-internal shoulder rotation angle, δ – the elbow angle, ε – the pronation-supination angle, ξ – the wrist flexion-extension angle, η – the wrist abduction-adduction angle. The movement clearly shows a (background) synergy of the first four degrees of freedom (α, β, γ, δ), which shows nearly identical shapes of the curves despite the large differences and mutual independence of the muscular apparatuses. Degrees of freedom ε and ξ show, during the first one-third of the movement, its own direct pattern followed during the rest of the movement by a return synergy.
(Source: work by O. Zaltsgeber and the author, Moscow Research Laboratory of Prosthetics, 1939).

the upper extrapyramidal layer and its regulation of the functions of the thalamo-pallidar level. Loss of level *B* leads to the parkinsonian complex of symptoms, which is composed of switching off functions of the level of synergies itself and loss of its control over level A leading to its over-excitation (hyper-dynamic state). During this syndrome, there is disappearance or significant reduction of movement components representing both independent and background motor functions of level *B*. The loss of the "triade" leads to loss of mimics, postural rigidity, poverty of gestures, and lack of expressive movements (Figure 3.9) in support of the theoretical views of James and Lange; there is also emerging poverty of emotional life of the patient. In addition, there is dramatic loss of automatisms. The loss of background synergies reveals itself in deautomation of both gait and various object-related skills. Disappearance of the aforementioned, typical for that level, mechanisms of rhythmical recoding leads, in turn, to the destruction of many automatized actions and, in addition, contributes to the development of perseverations, which will be discussed in more detail later. Disinhibition of the lower level *A* leads to the development of whole-body rigid hypertonus and resting tremors.

As described in detail in Chapter 9, each of the levels of movement construction has its own, typical perseverations, i.e. impossibility to stop an initiated movement or rhythmical process at one's own will, to modify a posture, etc. At level *A*, such perseverations are displayed as wax-like, cataleptic flexibility, frozen postures, etc. At the level of synergies, most typical perseverations are seen as weakening of triggering and stopping mechanisms, the aforementioned "end-switches." As a result, one can observe the abrupt paralysis of initiative: It is hard to initiate gait and equally hard to stop after gait has been initiated, etc. If such a patient is pushed backward or sideways, he or she would walk backwards or sideways and have major difficulties trying to stop (the, so-called, retropulsion and lateropulsion).

Pathological hyperfunction of the level of synergies is revealed in the emergence of various hyperkinetic syndromes, excessive synergies and synkineses (the last word implies apraxic coordinated movements of a pathological origin). Involuntary rudimentary movements become visible, for example those typical of the so-called Foerster athetosis syndrome: fixations, grasping movements, grasp-and-hold gestures, reactive-expressive movements without a goal or meaning, crawling and climbing movements, etc. Taken together, disorders of this group deserve to be named hyperkinetic dyssynergias (Figures 4.9–4.11).

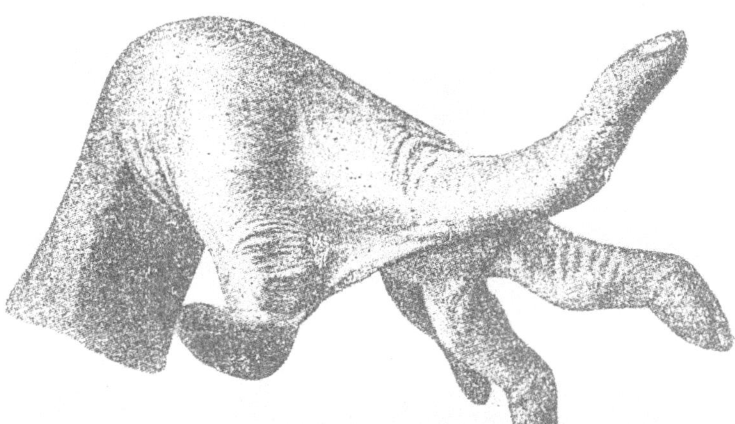

FIGURE 4.9 An instantaneous photograph of the hand during athetosis. (Source: after Monakow, from the manual by L. Mohr and R. Staehellin).

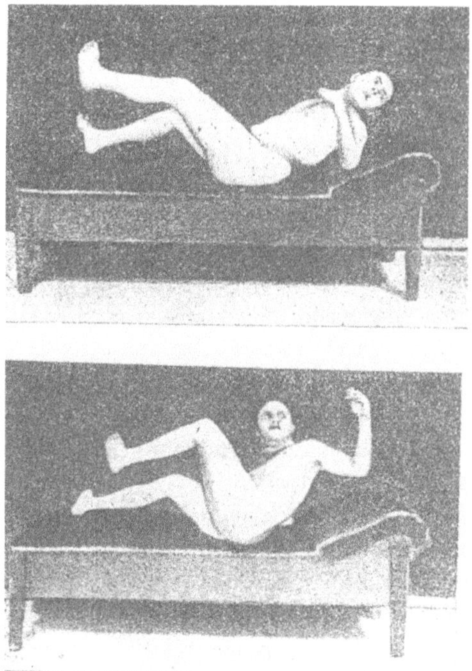

FIGURE 4.10 Two postures during the extrapyramidal seizure induced by overexcitability of the striatal system (after O. Forster).

FIGURE 4.11 An instantaneous photograph of a typical movement in Huntington's chorea. (Source: after Pineles, from the manual by L. Mohr and R. Staehellin).

Hyperfunction of level *B*, emerging due to some disinhibitory pathological processes, in a sense swings the doors wide open for the phylogenetic zoo, hidden in a healthy person. Then, the dungeons of the motor function set free ugly, grotesque backgrounds without forefront images, devoid of meaning and adequacy: Various torsion spasms, fragments of ancient movements, athetoses, choreas, unintentional roars and shrieks – psychomotor chimeras, effectors going mad.

Notes

1 In the English-American literature, stereotypes of this kind are addressed as "patterns" – a very exact word, which, unfortunately, does not have an adequate Russian synonym.
2 In Chapter 8 we present a more detailed description of the phenomenon of dynamical stability and also describe a number of biodynamical examples of such movements.

5

CORTICAL CONSTRUCTION LEVELS

Level C: The Pyramidal-striatal Level of Spatial Field

The level of spatial field, as we will address it based on its most distinguishing feature – also termed the pyramidal-striatal level *C* – is a very complex and, likely, not completely formed object. As we understand it now, it is not a single level, but two different levels, with one of them being to a degree subordinate to the other. This conclusion is supported by the fact that anatomical tools supporting its activity are of different phylogenetic ages. It is beyond doubt, however, that these are not two self-sufficient, independent formations; their functional unity forces us to describe this level as a single object, only with references to sub-levels *C1* and *C2*, when needed. Level *C* is of obvious interest to a psychologist due to the associated striking form of afferent signals, which has been termed (see Chapter 2) sensory syntheses or sensory field; it is also of interest to a physiologist and neurologist due to its merged duality and the specific coordination characteristics such as variability and plasticity.[1]

The complex structure of the level of spatial field reveals itself, first of all, in its two outputs to peripheral effectors: pyramidal and extrapyramidal. The classical neurological separation of all motor phenomena into pyramidal and extrapyramidal does not coincide with the watershed between the levels of construction described in this book, which exists, more or less precisely, between the striatum and pallidum. The border between the pyramidal and striatal (extrapyramidal) systems is seen also in functions of level *C*, and this border reveals itself in the separation of the two mentioned sub-levels, which differ from each other in specific features of the structure of their sensory fields, classes of controlled movements, and phylogenetic biographies.

Level *C* differs dramatically from the previous level in leading afferent systems, the meaning, and the whole row of external characteristics of the associated movements. The leading afferent system of this level is the *synthetic spatial field*.

We have already agreed that, for highly organized construction levels, afferent systems do not mean raw receptor signals. During the description of the synergy level, we had a chance to note that its leading afferent system is formed as a result of synthetic processing of the original sensory material in a system of subcortical nuclei. Afferent systems of posture, reaction dynamics, and angular velocities of joints and body segments represent syntheses of original proprioceptive and tango-receptive sensations, enhanced with local signs and rendered commensurate in a single for the body system of coordinates. The afferent system of the spatial field *C* is much more synthetic, united, and, most importantly, objective. It involves the powerful stream of cortical inflow, although originating from the most peripheral cortical areas, its "input and output gates" according to Monakow. Its *first-order* sensory sources in the cortex of the large hemispheres are as follows: The visual areas (area striata, Brodmann 17 and 18), tactile-proprioceptive areas (the central-posterior gyrus, Brodmann), and probably, to some degree, auditory and vestibular areas. Finally, afferent substrates of this level include the cortex of the hemispheres of the "new brain" (neocerebelli). Hence, tango-receptors contribute to the synthetic afferent systems of this level twice, after they have already contributed to the formation of sensory synthesis at the level of synergies. Here the tango-receptor system acts in a disguise, after the initial processing and filtering in the visual hillocks and including in its route one more additional neuron. This sensory half-baked material coalesces with memory traces, and all the mentioned signals are used over one's individual life to construct an unbreakable synthesis of "spatial field" – a formation well known to psychologists in the processes of orderly perception, but less well known to neurophysiologists as a leading formation coordinating effectors.

The spatial field, where movements by an animal are organized, is of course phylogenetically much more ancient that the multi-cortical level *C*. It is of interest to trace its gradual evolution, which is reflected in the nervous systems of higher vertebrates. Elements of space organization are already seen at the lowest level *A*, where this organization is revealed in the form of primitive tropisms of attraction, which display a nearly perfect polarity of the high-low type. We encounter space as a coordinating factor in such a form even in some completely paleokinetic boneless invertebrates, where fields of attraction are defined and sustained by the body location in space and distribution of muscle tonus (Uexkull). In the first neokinetic, centralizing level *B*, space reveals itself as a system of coordinates of the body, united in a common system of local signs, but not yet projected onto the external distant environment. Transition to level *C* signifies a leap to meanings, which are incomparably more complex and detached from the primary receptor signals.[2]

Spatial field of level *C* is neither perception nor sum of percepts. The process of its formation involves visual perceptions, ocular-motor feelings linked to accommodation and stereoscopic vision, tactile feelings with their local signs, proprioception of the whole body led by vestibular sensations of gravity and acceleration, and, certainly, numerous fragments of other receptor systems. It has room for multiple compensations and vicar interactions; for example, humans who were born blind develop, without the help of vision, an afferent field, which is so similar to the afferent field of those with vision that no contradictions are seen in either geometric representations or in the field of spatial motor events. As soon as this field has been created and elaborated, it abstracts itself from the original receptor signals, which were used to create it, such that it is impossible to notice their signs even with the most intent observation.

The most striking and distinguishing feature from the afferents of the previous level of the spatial field is its *objectivity*. The completely formed spatial field is in full correspondence with the external world and is liberated from the unbreakable bonds to the body, which were so much constraining and devaluating for the spatial synthesis at the level of synergies. Further, the spatial field of level *C* is *vast, immovable, and homogenous*. Its *vast size* is to a large degree due to the participation of telereceptors in its construction; however, the same properties of the field can be seen in animals who were born blind and used *locomotion* in combination with active touch (palpation) in its formation. Helen Keller, who was born without the two main telereceptors, was able to study geometry, understood and liked sculptures, and described space in her literary works similarly to people with vision. The feeling of *immovability* of the spatial field is even more important for its objectivization. When we walk, go upstairs, and turn, we do not only know but feel strongly and convincingly that we are moving, whereas the external space with all the objects in it remains motionless, despite the fact that all the receptors speak to the contrary.[3] Metaphorically speaking, every subject, starting from early childhood, overcomes the egocentric Ptolemean system of world perception and replaces it with the Copernicus one.

Further, the spatial field is *homogenous and aperiodic*, i.e., it is uniform in all its parts and does not contain elements of alternation or cyclical repetition, which are abundant in the coordinate system of the body at the level of synergies.[4]

The next most important properties of the spatial field are its *metrics* and *geometricity*. The field includes accurate and exact estimates of distances, sizes, and shapes, which are also essential characteristics during movements performed at that level: This is the *domain of accuracy and precision*. Geometricity of the spatial field characterizes most strongly the upper, cortical, sub-level *C2*, as revealed in preservation of geometrical shapes and similitudes. With all the enumerated properties, the spatial field of level *C* is the most objective of all the "spaces" associated with different construction levels; further up, at the level of actions (see Chapter 6), it evolves toward schematization, gains in terms of organized meanings, but definitely loses the exact, photographic correspondence to the actual metric relationships.

The space of level *C* is definitely not empty: It is *filled with objects*, which have size, shape, and mass, and *forces* acting among those objects. We will view these forces as external coordinates; they originate from external bodies and are linked to them, not to our body; we project the force field to the external space.

These features of the spatial field, as the leading afferent system, define the main characteristics of movement that the spatial field controls.

Movements performed independently at the level of spatial field form in humans a very large class, which differs dramatically from the meagre menu we observed at the level of synergies. Before we start characterizing these movements, let us note that the class of movements at the spatial field level forms the "ceiling" of motor abilities in: (a) higher reptiles; (b) birds (in migrating and carnivorous birds, the striatal level, in certain aspects, reaches its pinnacle in the phylogenetic development); and (c) in lower mammals, and also in humans during the early ontogenesis – approximately until the second year of life, and dominating during the whole childhood. Higher mammals demonstrate gradual formation of the purely cortical level of actions *D*, but only in humans there is a very intensive increase in the quantity and complexity of object-based and chain actions at this level and the formation of the higher cortical symbolic levels (group *E*), which are alien to all other animals. This development leads to a visible decrease in the relative number of movements at the level of spatial field and, in many aspects, in their quality.

Movements at the level of spatial field possess a clear *goal-directed character*. They move from somewhere, to somewhere, for a certain purpose.[5] These movements are extroverted, directed at the external world to a degree comparable to the introverted character of movements at the level of synergies. Movements at level *C* carry, press, pull, take, tear, and throw. Correspondingly, they have a beginning and an end: approach and reach, swing and throw or strike. Movements in the spatial field, by their nature, are always *transporting*: even if their external form is cyclical due to the design of our extremities (walking and running), according to their meaningful structure, these movements are as aperiodic as the space, where they proceed, itself.

Transport movement, by its very nature, implies adjustment to the external conditions of its execution. This *adaptation to space* and mastering space are the third essential feature of level *C*, which is alien to all the lower construction levels. It reveals itself in two aspects and hues outlining the border between the two aforementioned sub-levels. At the lower (striatal) sub-level, it is seen as adaptation *of the process* of action. For example, the level of synergies is the place where the complex synergic ensemble of walking is formed and completed in childhood, but this is idle or abstract walking. The striatal level adapts this action to the shape and properties of the road, to its slopes, curves, slippery spots, and innumerable little collisions the barefoot leg could tell about to its booted counterpart. All these adjustments and variations are adjustments to the external field, not to one's own body and its dynamics, as it used to be at the level of synergies; they are technically needed to turn walking into a real, practically useful action. This group of adjustments can be defined as projection of the motor process on the external actual space with its forces and objects. At the upper sub-level *C2*, adjustments to the external space become subtler and more specialized, acquiring more obvious goal-directed or *final character* and project the movement onto its final point in the external space with *precision* or *accuracy* requirements. This sub-level is, to a large degree, indifferent to trajectories, methods, and types of intermediate steps in transition, with the emphasis focused on the destination point, which, as the focus, should attract all possible trajectories of a given movement (which is actually observed over repetitive trials due to variability).

Further, since mastering external space is impossible without dealing with *external forces*, movements at the level of spatial field also include forceful actions such as lifting weights and ballistic movements (e.g., striking and throwing), etc.

The next common feature of movements at the described level is tightly linked to the mentioned objective homogeneity of the spatial field; this feature separates these movements from all others in a very salient way. This feature could be called *spatial conditioning* of movements in contrast to *postural* conditioning ruling movements at the level of synergies. Movements at level *C* are clearly linked to their bases in the external space and have very little relation to body position in space as a whole and even to positions of intermediate links within the moving extremity. A violinist who anchors movements of the arm with the bow primarily at the level of synergies[6] would never dare change either posture of the right arm or the standard pattern of its joint angles with respect to the violin or to the upper part of the body. In contrast, the motor coordination of a pianist facilitates various changes in body posture and location of the arms that have no effects on either accuracy and precision of performance or its tempo, particularly during passages including hand transport, which are most tightly linked to the level of spatial field; such postural changes may include, as an experiment or a joke, most unusual, acrobatic body positions. It is obvious that such manipulations lead to very strong effects on the sitting posture and arm position. In contrast, what pianists call "touché," which has most direct relations to the method of producing sound and to its artistic features, is based on the level of synergies and linked directly to sitting posture and arm posture. This is well known to any musician-instructor and, because of this reason, no performer would ever agree to perform under even the smallest errors in quality of the stool by the piano. The element of precision (spatial field) is not violated by postural changes, whereas elements of touché (synergy) is sensitive to them to a similar degree as the bow-holding hand of a violinist.

This contrast between levels *B* and *C* needs further analysis. The lower level of synergies, which represents the phylogenetically oldest example of the cerebral integration of movements and signifies by its emergence the possibility of large synergies encompassing the whole body, at the same time, is pedantically intolerant to slightest changes in the motor composition of its movements and backgrounds and, therefore, is prone to forming stable movement patterns and stereotypies. This is due to the fact that the construction of movements at the level of synergies and all its sensory corrections originates from the *body itself*, from proprioceptive sensation of body posture, joint angles, etc. These postures, which serve as transient points during movements (as well as the reactive forces that always accompany movements), become the movement essence and are fixed as a skill. Obviously, the fixation of such a standard formula is helped by the formation of inherent to this level dynamically stable movement forms, which resist automatically any deviations from the discovered stereotype (Chapter 8). However, even slow movements at the level of synergies, which show no mechanical prerequisites for dynamical stability, also tend to show stereotypes at that level.

Movements at the level of spatial field, which are strikingly different in all the mentioned features from the stereotypy of movements at level *B*, move us close to a large group of phenomena of utmost importance. Here we imply relations between *accuracy and variability*, which also touch upon a few other features of the coordination process.

Depending on the meaning of motor tasks typical of various levels and on the intimately linked quality and content of afferent signals, each level draws its own borders between important and non-important, and equates non-identical movements and their components. Among the innumerable number of movement features that could be corrected by sensory systems, each level singles out those that it views as salient and that can be corrected by its tools, separating them from features that it remains more or less indifferent to. Obviously, it would keep maximal possible accuracy and stability with respect to the former, whereas the latter would be left to chance, or their corrections would be delegated to a background level, which may be more interested in correcting these features and keeping them highly accurate.

For example, the upper sub-level of the spatial field *C2* is focused on *accuracy of outcome or precision*: Precision of a throw or strike, accuracy of pointing or touch, and, in more complex forms, – accuracy in reproducing a visible shape, for example during copying a figure while keeping geometric similitude. During all these actions, sub-level *C2* is rather indifferent toward the means of achieving the desired ultimate precision and to behavior of the effector in intermediate points. The lower sub-level *C1* is equally demanding with respect to accuracy *in the course of the movement*: adjustment of gait to uneven surface, stairs, up and down slopes, adjustments of pencil motion to follow a drawn figure, adjustment of motion of a file to the configuration of the processed surface, etc. The level of actions *D*, which will be discussed later (Chapter 6), is equally demanding with respect to the ultimate goal of a meaningful action chain with visible indifference to the contents of working habits and sequences of the chain links. We could easily present more examples.

There are many movements when it is much harder to establish an objective criterion of accuracy compared to movements related to accuracy of hitting the target. In such cases, we are helped by the fact that, naturally, control levels allow higher *variability* with respect to indifferent aspects and movement components, compared to those aspects that are of primary importance for the level. This defines a characteristic sign of a movement being part of a particular level or of the movement composition, which can be addressed as *signs of specific variability* (Chapter 9). Aside from qualitative features of aspects of variability ensured by different levels, there is a clear *increase in the magnitude of variability in the bottom-to-top direction*, from a level to the next level, which, undoubtedly, is linked to the biological evolution of motor tasks in the direction of increasing variability and adaptability (Chapter 1). Indeed, *variability is the complementary side of accuracy*: If a level did not have adequate reserves of variability and flexibility of movements, it would not be able to adapt it, *accurately and flexibly*, to various types of activity.

Variability and flexibility of movements at the level of synergies is very small,[7] which is partly compensated for by the high accuracy of its meticulous finish and breadth of synergies typical of stereotypical patterns. Variability of sub-level *C1* and, particularly of *C2*, is much higher as will be demonstrated with several examples.[8] It is even higher at the level of actions *D*, although level *D* shows a few characteristic limitations.

The same phenomenon that looks like variability in an involuntary context during meaningful, targeted actions turns into *inter-exchangeability* of motor components and *switching* the action from one organ to another. In stark contrast to movements and sensory corrections of the level of synergies, movements at the level of spatial field display the very pronounced ability to switch. One can point at a target about equally easily and accurately with a right-hand finger and a left-hand finger, and also, if needed, with a toe, nose, etc. One can take an object, lift a pencil from the floor, press the bell button, lift the latch, and pull the rope equally easily with one or the other hand, approaching from different sides, and by different techniques. Stereotypies at the level of synergies, which are anchored at specific positions of the extremities, are certainly tightly linked to the extremity that learned them. At the higher level *D*, one can clearly see the functional differences between the right and left hands due to the dominance of one of the hemispheres; hence, ability to switch reaches its peak at the described level *C*.

The level of spatial field shows easy switching not only among movement trajectories and effectors – there are also *switches of the techniques*. During walking in natural conditions, over unpaved surfaces, during mountain hiking, etc., the goal of the movement – transporting the body to a desired location – can be achieved by a dozen of methods: walking, leaping, crawling, running, and climbing. If a boy has to make 100 steps, he will walk some, run some, hop some, and may even cartwheel the rest. A man who can ride a bicycle is indifferent to what hand in what location and with what technique grabs the handlebar; it is very easy to learn how to ride without grabbing the handlebar and other tricks.

The aforementioned violinist builds movements of the *left* hand with respect to the main motor backgrounds at the level of the spatial field similar to hand movements in a pianist. As it is well known, this results not only in easy changes in hand position or in the selected string (famous virtuosi, after some of the strings had snapped, sometimes finished the very difficult concertos using only the bass string), but also to easy switching from the violin to alto, although the latter has the very different size of the fretboard and, consequently, different fingering, all the required distances between fingertips being nearly twice as big as during playing the violin. A good violinist can very easily master fingering of the cello, although in this case not only the size but also the location and direction of the fretboard are changed.

Another aspect of the ability to switch inherent to the level of spatial field is the notorious *"plasticity" of the nervous system*, which is revealed in numerous experiments with amputations and nerve-muscle anastomoses. It is sufficient to look through the appropriate literature (Bethe, Osborne & Kilvington, Anokhin, Asratyan, and others) to become convinced that all the quick adjustments and switches described by the experimenters belonged to the level of spatial field: They would be bitterly disappointed with movements from the level of synergies. Similar phenomena can be seen clearly in people after the amputation (Figures 5.1, 5.2, and 5.3). For example, persons after bilateral arm amputation easily learn to write with a foot or with the mouth. The author observed in 1941 a girl who had lost both arms and a leg in a railway accident. She studied in the middle school and could write well with: (a) the mouth; (b) the intact leg; and (c) the prosthetic arm. As it is characteristic of coordination at the level of spatial field, the *handwriting*, in all three techniques, retained its characteristic features (Figure 5.5), i.e. one could witness the phenomenon of switching. Figure 5.4 shows handwriting by a man without both arms; one segment was written with the pencil gripped by the teeth, the other one – with the pen gripped by the clippers of the prosthesis. This figure shows very clearly that the spatial contour of a movement finds a way to realize itself via many various recodings. No such phenomena are seen in movements at the level of synergies because of the immutable stereotypies.

We have to note that the handwriting shows similar switching in healthy persons. It does not matter whether we write with small or large letters, with a pen on a horizontal surface or with a piece of chalk on a vertical surface, although, in these conditions, very different muscles are

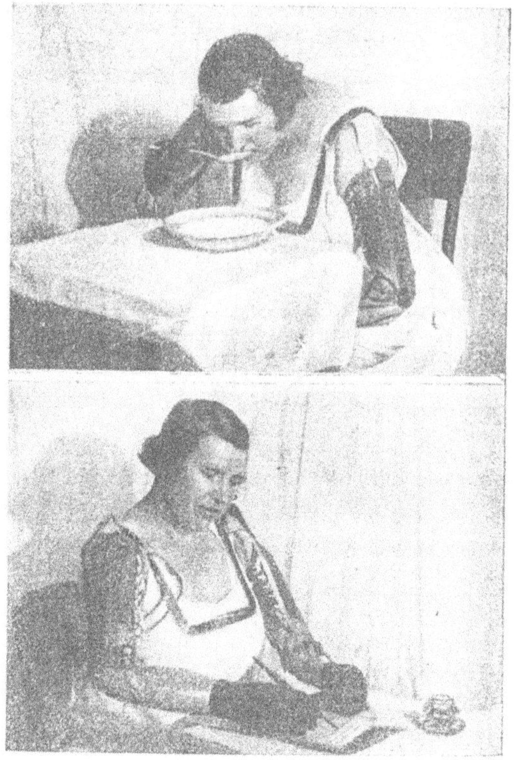

FIGURE 5.1 The subject in the two panels eats and writes after the bilateral shoulder amputation.

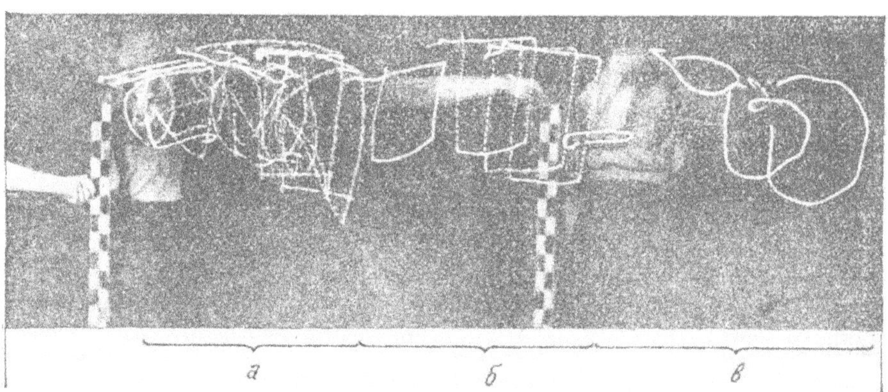

FIGURE 5.2 "Plasticity of the nervous system" in humans.
Left (*a* and *b*): Two consecutive groups of curves during drawing with the prosthesis. In each group, left to right, – the trajectories of the shoulder, elbow, wrist, and finger-like clippers. The subject had amputation at the mid-length of the right shoulder and total replacement of the left shoulder joint. At the second attempt, the square was drawn perfectly (the photograph shows it at an angle). *Right* (*в*) – Drawing of a circle with the prothesis; the subject is the same patient as in Figure 5.1. The trajectories, left to right, – middle of the back, shoulder, elbow joint, and prosthetic fingers. Despite the fact that nearly all the degrees of freedom have been switched (flexion and extension of the elbow was produced by muscles of the contralateral shoulder, finger grip – by muscles of the waist), the complex synergy of drawing the circle was performed not worse than by any healthy subject with the same level of education (elementary school).

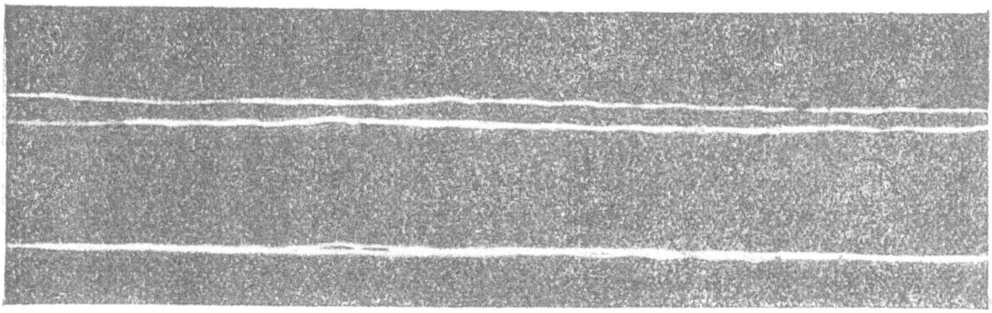

FIGURE 5.3 The kimocyclogram of the writing movement by the hand prosthesis performed by a woman after the bilateral mid-shoulder level amputation.
The lost vibration synergy between the wrist and forearm has been replaced by a similar vibration synergy between the trunk and shoulder muscles visible in both lower interweaving curves (the shoulder and prosthetic elbow joint) and upper curves (the wrist and prosthetic fingers). The patient was fitted with the prosthesis in the Leningrad Institute of Prosthetics (1940).

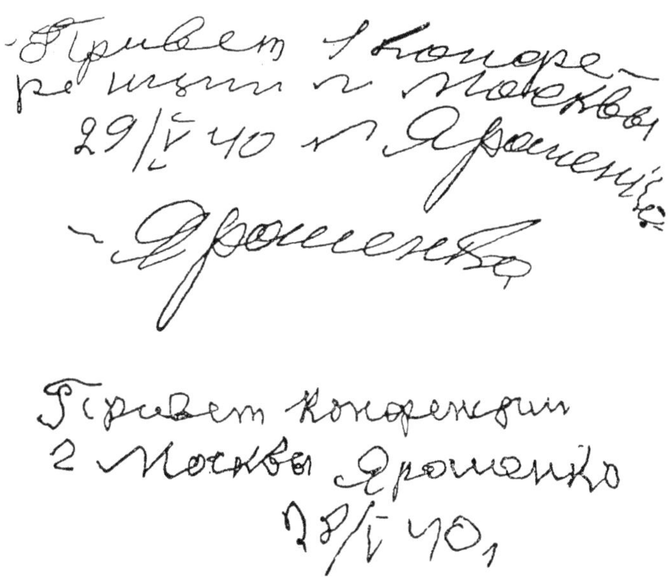

FIGURE 5.4 Writing by a person without both arms Ya.
(see also the left side of Figure 5.2). Top: writing with the pencil gripped by the teeth; bottom – writing with the pen gripped by the clippers of the hand prosthesis. There are clear peculiarities of the handwriting seen in both samples (the end of the word "Привет," groups of letters "конфе," "Ярош," etc.)

involved in very different ways. The author performed an experiment using an educated subject who was asked to write *without any special training* with the pencil attached to: (1) the right wrist; (2) the right elbow; (3) the right upper arm; (4) the tip of the right foot; (5) the tip of the left foot; and also with the pencil (6) gripped by the teeth; and (7) gripped by the left hand (Figure 5.6). In all these switches, handwriting was possible immediately, although with a bit of effort; the results of this experiments are shown in Figure 5.7. It is interesting that the smooth roundness of the handwriting (created by background coordinations at the level of *synergies*) was not preserved in any of

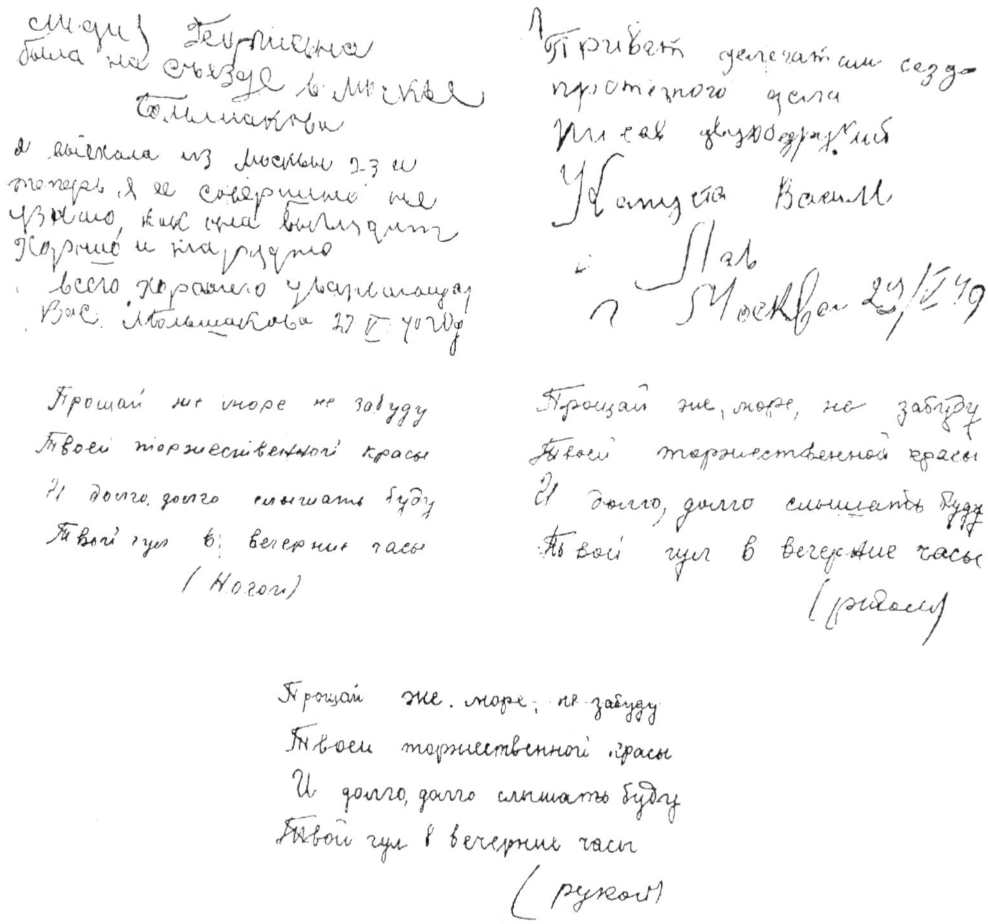

FIGURE 5.5 Samples of writing with the hand prosthesis after the upper arm amputations. Writing by the same patient as in Figures 5.1–5.3; right-top – the writing by patient K. after amputation of the right forearm close to the elbow joint and of the left arm close to the shoulder joint. (Source: Labor Rehabilitation Center in Kharkov, NKSO).

the samples, except the typical writing with the right wrist; at the same time, the shape of separate letters, linked to corrections from level *C*, was clearly retained in all the samples.

All the phenomena of switching discussed in the last group of examples belong to the *upper sub-level* of the spatial field. This sub-level, even more than the lower sub-level, displays features of freedom from the rigid metrics of the lower levels. We do not care not only whether to draw a square, circle, or a letter on a horizontal or vertical surface, with a short or long pencil or even directly with a finger, but also whether the reproduced contours are *large or small*. If the lower sub-level combines postural and joint switching while preserving *spatial identity* (congruency), the upper sub-level *C2* generalizes this ability to switch also to retaining *geometric similitude*. The handwriting example is particularly striking in showing that *the leading geometric image beats the path through any muscle systems, any innervations, at any scales*. When copying a figure or a natural image, the artist reproduces it with precise geometrical similitude, and the degree of difficulty depends the least on the selected scale of the drawing.

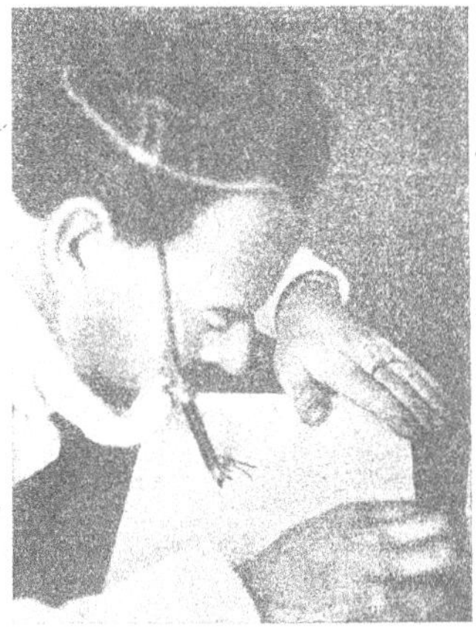

1.

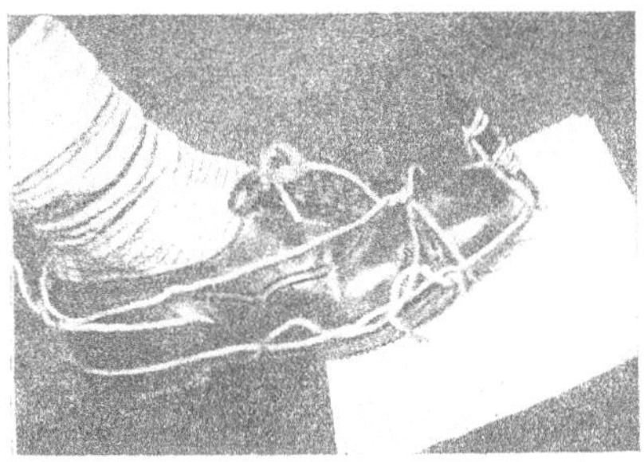

2.

FIGURE 5.6 The experimental setup for Figure 5.7.
1 – writing with the head; 2 – writing with the tip of the right foot.

To summarize, we should note that, in all the described examples, the moving extremity (more exactly, its *working point*) leads to a spatial contour of the movement at sub-level *C1* – as a directly prescribed contour (a typical movement – *tracing* a shape), and at sub-level *C2* – as a prescribed pattern presented as a geometric shape without limitations on its location and scale (a typical movement – *copying* a shape). The lowest variability and highest accuracy are seen at this level exactly with respect to the working point. Particularly remarkable examples are cases when the working point does not correspond to a point on the extremity itself but represents a point on the hand-held tool: the tip of the pen, pencil, or knife, the sweet spot of the tennis racket, the hammer's head, etc. In such cases, during development of the motor skill, corrections of movements of the implement are performed not only by vision, despite the fact that the hand-held tools have no sources that could be used for direct sensory corrections. This is confirmed clearly by the fact that automation of such a skill frequently leads to possibility of acting without looking. This can mean only that the proprioceptive sensory correction system *uses the visual system* to elaborate necessary recodings, which translate spatial feelings in elements of the extremities into the language of corresponding perceptions of the position and motion of the working point.[9]

Finally, the same phenomena of variability and switching, and the described flexible adaptability of level *C* to variations in the environment play the main role in ensuring *extemporaneousness*, which has been mentioned already in Chapter 1 as a feature and direction of development of motor reactions in the phylogenesis. Indeed, levels well-equipped to produce stereotypies are not useful to perform singular, unexpected motor reactions, maybe exactly due to the huge and cumbersome

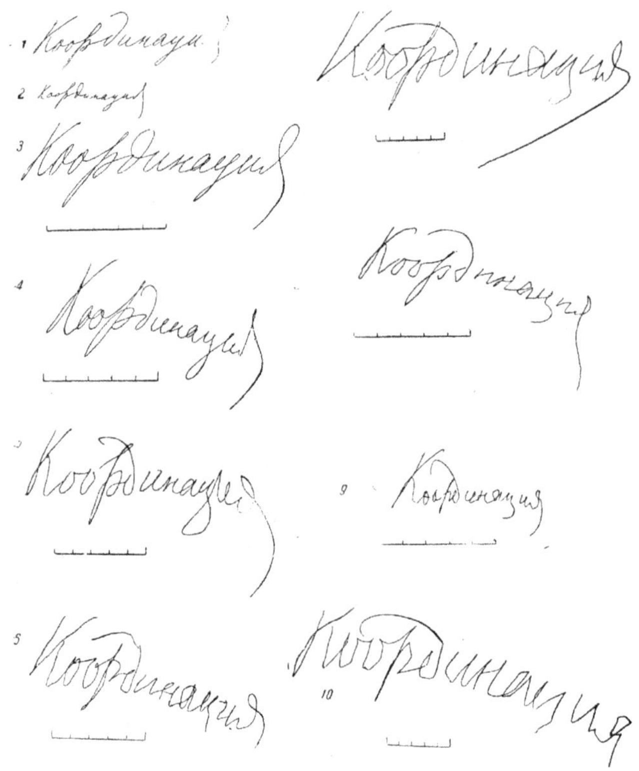

FIGURE 5.7 "Plasticity of the nervous system" during writing with different parts of the effectors. 1 and 2 – normal handwriting with the right hand; 3 – the pencil is gripped by the right hand, the writing is performed by swinging the right hand as a whole; 4 – the pencil is attached to the styloid process of the right ulna; 5 – the pencil is attached close to the medial condyle of the right forearm; the pencil is attached at the right upper arm; 7 – the pencil is attached to the tip of the right shoe; 8 – the pencil is gripped by the teeth and attached to the head with bandages; 9 – the pencil is gripped by the left-hand fingers; 10 – the pencil is attached to the tip of the left shoe.The line below each sample is 5 cm.The subject was an untrained individual with domination of the object level in movements.The figure shows: (a) the substantial reproducibility of the handwriting features; (b) almost complete lack of switching background synergies of the handwriting (smoothness and roundness), which distinguishes samples 1, 2, and 3 from all other samples; (c) extensive switching of the components at the level of spatial field; and (d) the presence of coordination at the object level characterized by the drastic deterioration of the handwriting coordination following switching to the *left side* of the body (samples 9 and 10). (Source: author's work, CNIIFK, 1940).

complexity of the controlled coordinated movements. Patterns produced by the level of synergies may be very accurate, perfectly adjusted, and show excellent teamwork involving sometimes enormous muscle and joint ensembles, but this level has major difficulties in creating new, unexpected motor combinations similar to a large orchestra asked to improvise. In contrast, higher levels, which easily allow *involuntary* variations in different aspects and components of movements, are equally eager to perform *purposeful singular modifications* of its (usually simple and portable) movements, for reasons of comfort or necessity, in response to some unusual changes in the environment. If sensory corrections at a construction level are organized to allow numerous, easily exchangeable, equivalent routes to the same outcome, this ability to switch and create new, varying across trials, paths to a fixed goal also facilitates search for new movement combinations. Extemporaneousness, as a rule, does not reveal itself with respect to movement aspects that the construction level keeps accurate with the help of low variability. Here, the flexibility and adaptability of the motor apparatus reveals itself slowly, as a result of long-term restructuring based on the accumulated experience.

These are the main features of movements and coordinations at the level of spatial field. Let us turn now to its localization.

The brain structures of the afferent part of this level have been discussed earlier in this chapter. There are at least two effector formations at this level, which provide main support in favor of splitting level *C* into two sub-levels: (1) *the corpus striatum*, which is composed of two, anatomically remote, nuclei, the *caudate nucleus* and the *putamen*, which form the upper level of the extrapyramidal system; and (2) the *large pyramidal field 4 of the cortex* of large hemispheres representing the "output cortical gates" of the pyramidal effector system. These two formations have very different phylogenetic ages (Chapter 7); this allows assuming that, with respect to its function, the lower sub-level had been formed prior to the higher one (Figures 5.8, 5.9, 5.10, 5.11, 5.12 and also 4.1a).

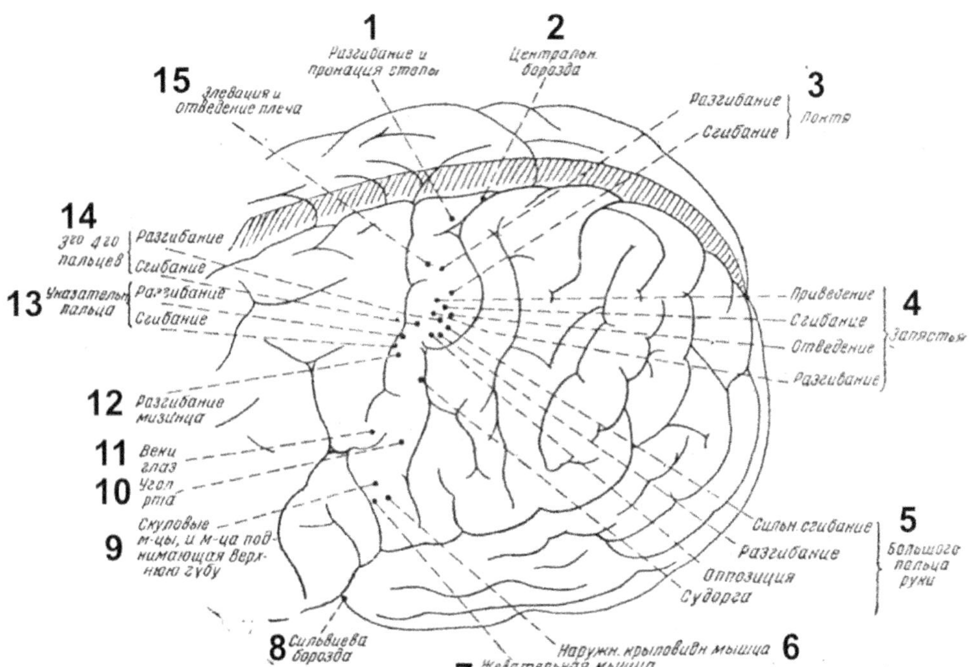

FIGURE 5.8 The left human brain hemisphere with movements observed in response to Faraday stimulation of different cortical locations – after F. Krause, a summary of results from surgeries.

1 – foot extension and pronation; 2 – central fissure; 3 – elbow flexion and extension; 4 – wrist adduction, flexion, abduction, and extension; 5 – thumb forceful flexion, flexion, opposition, spasm; 6 – external pterigoid muscle; 7 – masseter; 8 – Sylvian fissure; 9 – cheek and upper lip muscles; 10 – mouth tip; 11 – eyelids; 12 – little finger flexion; 13 – index finger flexion and extension; 14 – fourth finger flexion and extension; 15 – shoulder elevation and abduction.

(Source: from the manual by L. Mohr and R. Staehelin).

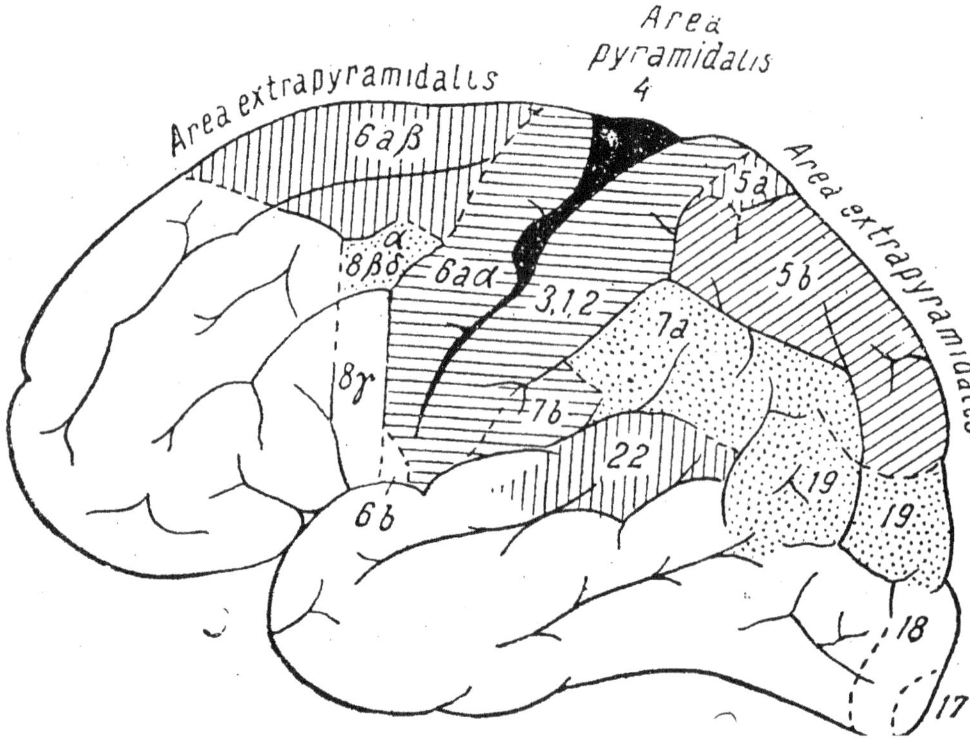

FIGURE 5.9 The map of the left human large hemisphere with the main motor fields. The pyramidal field 4 is shown in black, the primary premotor field 6aα, located directly anterior to field 4, is shown with horizontal stripes (after O. Foerster).

Even given the current insufficient knowledge about afferent inputs into the striatum, it is obvious that they differ in their quality and composition from the afferent inputs into the pyramidal effector system. Taken together with the strikingly different clinical pictures following injuries to the striatum and pyramidal effector system, this factor provides a strong argument in favor of two different sub-levels. A clinician would have had more problems admitting that the two mentioned systems form a single level than accepting its separation into two levels. However, in fact, the difference between the pyramidal effector system and the upper floor of the extrapyramidal effector system is clearly not that large. First, clinicians deal not with activity but with lack of activity, not with functions but with dysfunctions, which are far from being the same thing. Second, most of what is currently understood by neurologists as extrapyramidal movements refer to the thalamo-pallidar level of synergies. These synergies, as shown earlier, are either realized as independent actions at level *B*, or need guidance from the striatum or pyramidal effector system. Difficulties with realization of such background synergies without the striatum led to a (wrong) conclusion that they depended on the striatum, which made the distinction between the pyramidal-cortical and striato-pallidar movements very deep. However, we have to analyze movements more accurately.

If, for the time being, we forget about the truly sharp antithesis between levels *B* and *C*, which have been emphasized sufficiently in the earlier discourse, and look into the clinical literature on *striatal pathology*, we discover that, whereas the thalamo-pallidar level in humans is completely introverted, the striatum with its afferent systems is tightly linked to *realization of movements in external space*. Foerster links it to standing, sitting, and walking. Graham Brown assigns to the

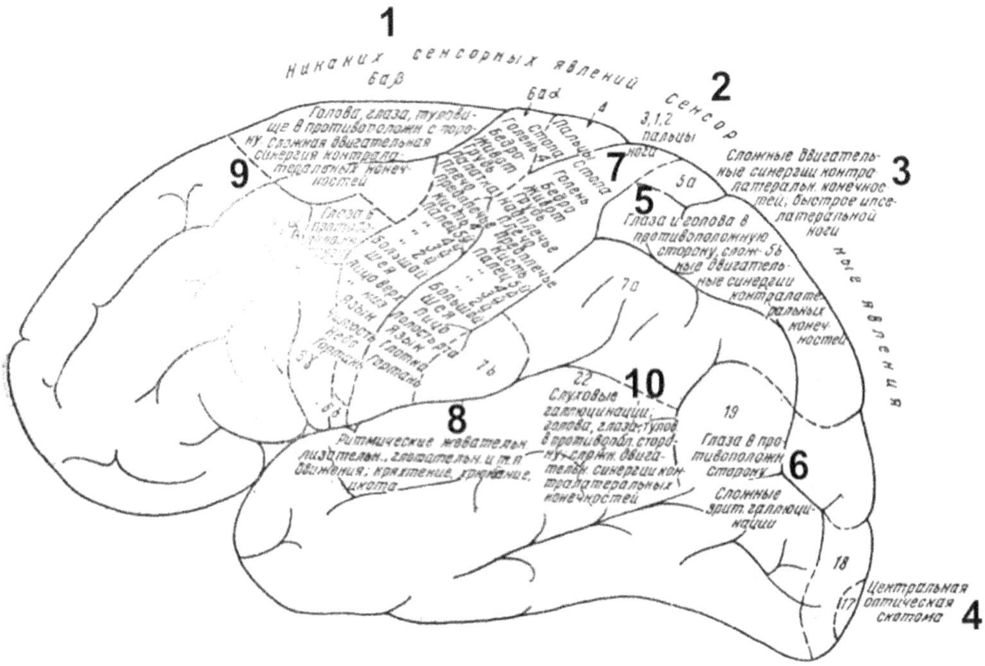

FIGURE 5.10 The exterior surface of the human left large hemisphere showing the movements produced by electrical stimulation of different cortical points (after O. Foerster).
1 – no sensory phenomena; 2 – sensory phenomena; 3 – complex motor synergies of the contralateral extremity, quick motion of the ipsilateral leg; 4 – the central optical system; 5 – head, eyes, trunk moving to the opposite side, complex synergy of the contralateral extremities; 6 – eyes turn in the opposite direction, complex visual hallucinations; 7 – toes, foot, shank, hip, belly, breast, shoulder blade, shoulder, forearm, hand, fingers, ..., thumb, neck, upper face, lower face, tongue, mandible, palate, larynx; 8 – rhythmical chewing, licking, swallowing, etc. movements, grunting, rumbling, hiccups; 9 – eyes and head move to the opposite side, complex motor synergies of the contralateral extremities; 10 – hearing hallucinations, head, eyes, and trunk turn to the opposite side, motor synergies of contralateral extremities.

striatum running, e.g., locomotor movements of a tennis player, i.e., clear examples of movements at the lower sub-level of spatial field, C. & O. Vogt assign to the striatum "highly coordinated motor impulses linked to speech and movements of the trunk and extremities," Jacob views the striatum as the center for expressive and reaction-defensive movements,[10] orienting and positioning movements, motor components and static elements of sitting, standing, and walking – they all belong to movements from the spatial field class. Similar characterization of the striatum is given by Schilder. These descriptions of the striatum by clinicians encompass everything said earlier about projection of the motor process onto the external space as a fluid adjustment to the space and to its force field in the course of the movement.[11]

There are two typical functional differences of the cortical pyramidal effector systems from the striatal system. First, the pyramidal effector systems is much more intimately linked to the visual system, whereas the striatal system – to proprioceptive afferents; therefore, accuracy and precision of motor actions are primarily linked to the pyramidal apparatus and are the first to be lost during injuries to this apparatus. Second, disorders of the pyramidal effector system affect primarily voluntary movements, whereas disorders of the extrapyramidal system affect involuntary

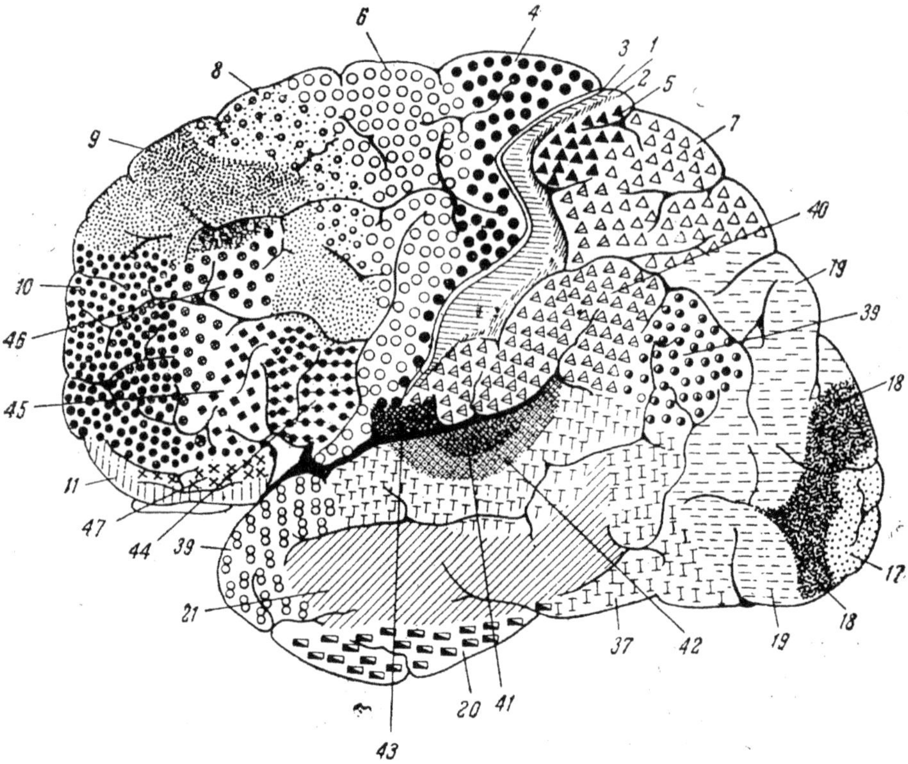

FIGURE 5.11 A citoarchitechtonic map of the external surface of the human left large hemisphere. Different in their microstructure fields are shown with their numbers according to Brodmann and different stripe styles. The pyramidal fields 4 are shown with black circles, premotor fields 6 – with large white circles (after Brodmann).

movements. This distinction, which is very deep and important in itself, belongs to a different class of issues compared to the problem of qualities of motor coordination and has already been discussed (Chapter 2) in light of the gradual increase, bottom-up, of both voluntary movements and conscious perception of movements at each of the subsequent levels.

The list of independent whole movements performed at the level of spatial field is so broad that it is impossible to compile their complete catalogue. All we can feasibly do is to highlight the main types and groups of these movements and to complement this list of the main groups with more or less arbitrarily selected examples.

Example 5.1 Various *locomotor movements*: walking (Figure 5.13), running (Figure 5.14), crawling, climbing, swimming (Figures 5.15, 5.16, and 5.17), tightrope walking, walking on hands, etc. Further, this group includes *locomotion with attachments*: skating (Figure 5.18), skiing (Figure 5.19), walking on stilts, biking, rowing, etc. All these are *cyclical* locomotor processes. They are followed by *noncyclical* locomotion movements: jumps with (Figures 5.20, 5.21, 5.22, and 5.23, and also 9.7a) and without running start; long jump, high jump, downward jump, jumping into a target (e.g., into the window or through the hoop), jumping during horse riding, acrobatic jumps, etc. Finally, the group of locomotion movements includes *walking while pulling or pushing* (a cart,

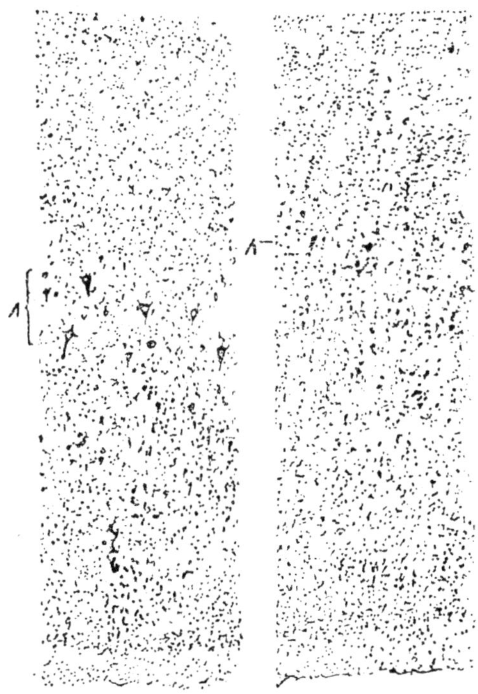

FIGURE 5.12 Microscopic pictures of the effector zones of the human cortex of the large hemispheres at moderate amplification (x 50).

Left – the non-granular frontal field 6ab; right – large pyramidal field 4. The Roman numeral V is both photographs shows the fifth layer, containing giant pyramidal Bets cells in the pyramidal field 4 – the origins of the effector pyramidal neurons (after Brodmann).

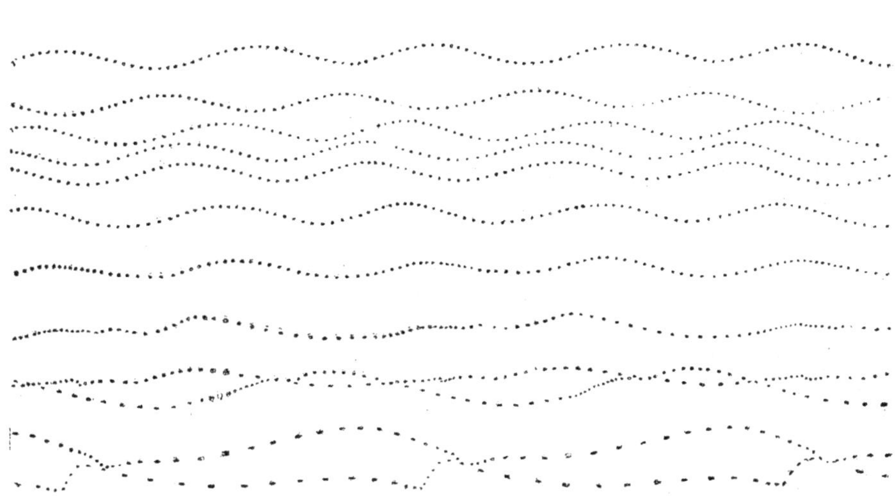

FIGURE 5.13 A typical cyclogram of normal walking. Left side of the body.

The curves of different body parts, similar to the curves in Figure 4.5, show a typical example of the synergic background pattern at level *B*, with its inherent rhythmicity, standard sequence of cycles, and quick convergence of Fourier series representing these curves.

(Source: author's work, VIEM, 1935).

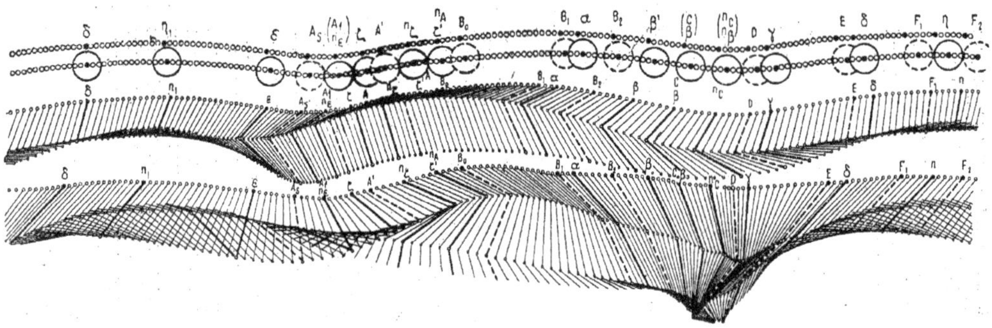

FIGURE 5.14 An example of locomotor movement at the level of spatial field *C1*: Sequential positions of the right side of the body during running with the style mittelstreke (world record holder J. Ladoumegue). The frequency of the positions is 187 per second.
(Source: author's work, VNIIFK, 1936–1939).

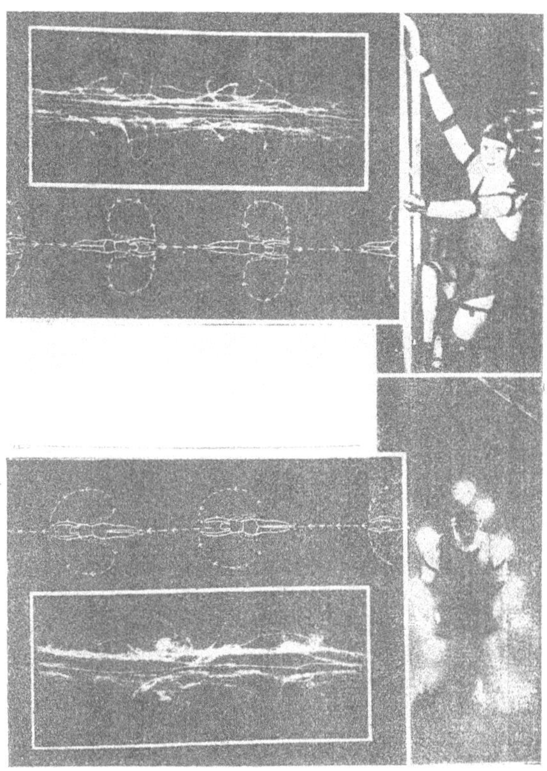

FIGURE 5.15 An experiment with the cyclographic analysis of swimming.
Top and bottom – the cyclograms of butterfly and breaststroke swimming recorded from the diving tower with explanatory schemes. Right – the record-holder of the Soviet Union, L. Meshkov, with the cyclographic bulbs in the pool during the experiment (see Figures 5.16 and 5.17).

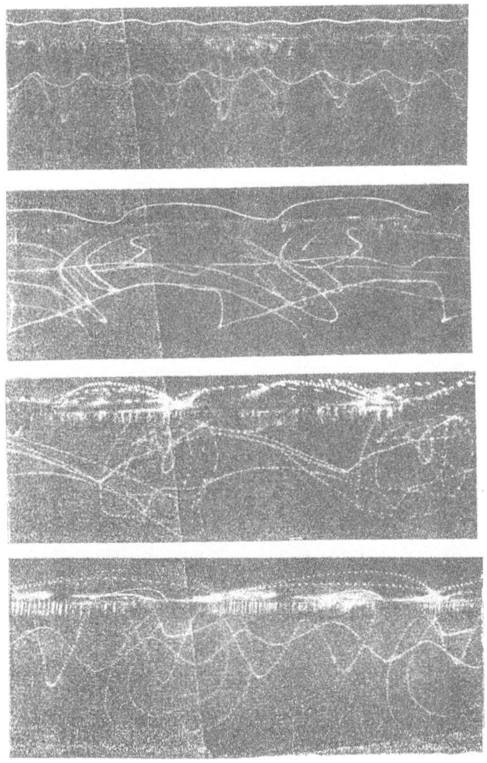

FIGURE 5.16 Types of locomotor movements at the level of spatial field *C1*: Sequences of the cyclograms of swimming movements recorded from the diving tower.
Subjects – the best swimmers of the USSR. Top to bottom: Crawl, breaststroke, butterfly, crawl on the back.
(Source: work by N. Butovich and S. Spitsyn consulted by F. Leontovich, CNIIFK, 1940).

wheelbarrow, dragnet, ship pulling strap, tug of war, etc.). All these actions transport the body in space.

Example 5.2 *Non-locomotor movements of the whole body in space.* These include somersaults, acrobatic movements, exercise on parallel bars, rings, high bar, trapeze, etc. (Figure 5.24).

Example 5.3 Movements of "*manipulation with space*" using different body parts. Within this class, one can encounter movements that could be termed "locomotion of extremities," for example, hand movements of a typist on the typewriter or of a musician on the instrument (the left hand during playing bow instruments, both hands during playing piano, harp, and accordion, leg movements during playing organ, etc.); these movements, even in everyday speech, are commonly described using locomotor terms such as "fingers ran over the strings," etc. (Figure 5.25). The same group includes movements that require finger fluency but without moving the whole hand, such as finger movements during playing wind instruments. Other movements of the same group represent singular goal-directed transfers in precise spatial coordinates: movements of pointing, touch, pinprick, tracing a contour, etc. (Figure 5.26).

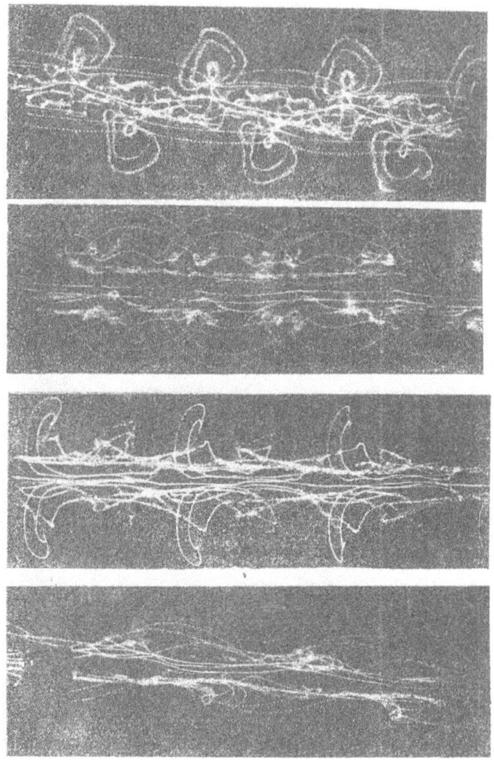

FIGURE 5.17 Types of locomotor movements at the level of spatial field *C1*: Sequences of the cyclograms of swimming movements, sideview recorded from the glass box half-submerged in water (the method of F. Leontovich).
Part of the cyclographic trajectories were seen over the water level, another part – through the water. Top to bottom: Crawl, breaststroke, butterfly, foot work during crawl. The recordings show clear auxiliary background synergies of level *B*, resembling the curves in Figure 4.5.
(Source: work by N. Butovich and S. Spitsyn, CNIIFK, 1940).

Example 5.4 *Movement of objects in space*: Movements of taking, grabbing, catching a flying or moving objects, moving, repositioning, transfer, etc.; inserting, pressing, extracting, winding, and *overcoming external forces*: lifting loads, tightening the bow string or musical string, etc.

The mentioned groups *tend primarily toward the lower sub-level C1*.

Example 5.5 An intermediate group, which is unclear with respect to its relation to the lower or upper sub-level, includes *ballistic movements*. Those that emphasize primarily *force* are likely to be more tightly linked to the lower sub-level *C1*. These include striking and throwing movements: shot put, throwing the grenade, discus, or hammer, weight jerk etc. (Figures 5.28, 5.29, and 5.30). Visual control of movements within this subgroup is secondary, which is confirmed by the fact that these movements can be performed by blind persons. Other ballistic movements that emphasize *accuracy* tend to be related to the upper sub-level: javelin throw, ball pitch, tennis, baseball, skittles, and croquet, movements of a juggler, strike by a blacksmith, bayonet thrust, etc. All these movements require visual control and cannot be performed by the blind (Figures 5.27 and 5.31).

FIGURE 5.18 Schematics of the trajectory of the body center of mass and skate movements during skating along a straight path.
For clarity, orthogonal distances and sizes are exaggerated. The curvature of the center of mass trajectory everywhere (except the double-support, push phases) has the convex part toward the supporting skate; the curvature is larger for larger distances from the projection of the body center of mass to the skate trajectory, which is characteristic of the dynamically balanced movement in the illustrated case.
1 – direction of the run; 2 – trajectory of the projection of the common center of mass; 3 – path of the right skate; 4 – path of the left skate; 5 – right push; 6 – left push.
(Source: author's work, CNIIFK, 1939).

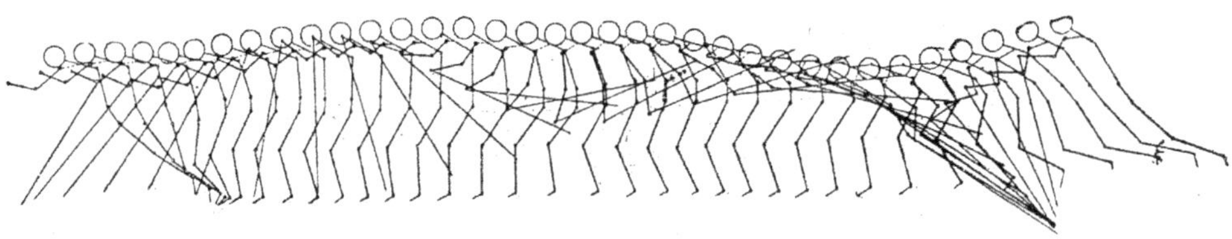

FIGURE 5.19 Sequential stick figures of the body during skiing.
(Source: after the cyclogram by the author and N. Sadchikov, CNIIFK, 1940).

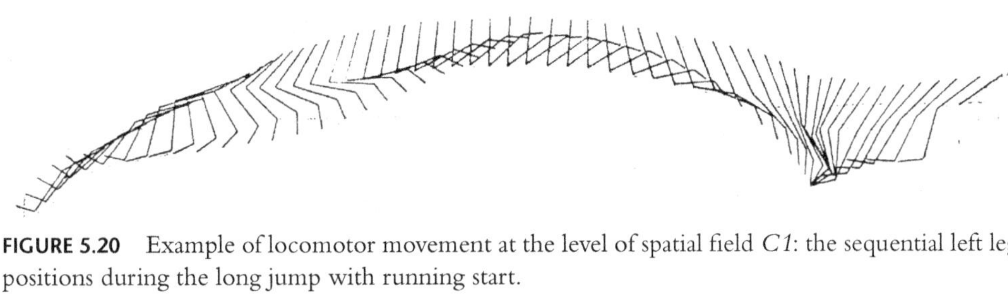

FIGURE 5.20 Example of locomotor movement at the level of spatial field *C1*: the sequential left leg positions during the long jump with running start.
(Source: work by N. Sadchikov, CNIIFK, 1939).

Example 5.6 *Movements of taking aim*: aiming the telescope, diopter, rear sight, etc.; preparatory movements before accurate pinprick (with the needle or compasses) or cut: taking aim with a long-distance weapon, aiming in billiards, in croquet, etc.; postural expectation movements of the goalkeeper during football game; in animals, posture of a carnivore prior to the attacking leap, etc. Given the depth, perspective, and the whole distant geometry of the spatial field, visual control is essential for movements of this group.

Example 5.7 *Imitation and copying movements*: imitation of visually perceived movements and actions by another person; copying (a natural object or a drawing); showing an object or an action with gestures (figurative pantomime in contrast to semi-involuntary, emotional-expressive

FIGURE 5.21 The cyclogram of a pole vault performed by one of the best athletes of the USSR. (Source: author and N. Sadchikov, CNIIFK, 1939).

FIGURE 5.22 Experimental setup of the experiment shown in Figure 5.21.

pantomime at the synergy level); mimicking or parody movements (maybe), etc. The upper sub-level is essentially needed.

We are not going to present here examples of movements where level *C* of the spatial field serves as the *background*. Such movement at higher levels are extremely numerous. The point is that the level of spatial field is unable to perform complex meaningful actions related to *objects and tools* (except their simplest transfers); in such actions, the next highest *level of actions D* takes the stage. The earlier review of movements led at the level of spatial field

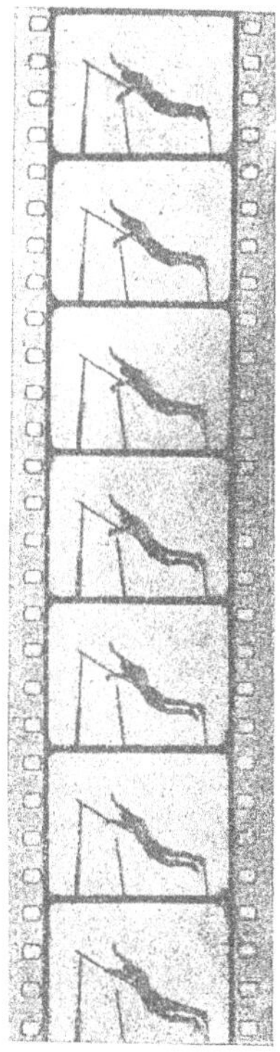

FIGURE 5.23 An excerpt from a high-speed film of the high jump with running start.

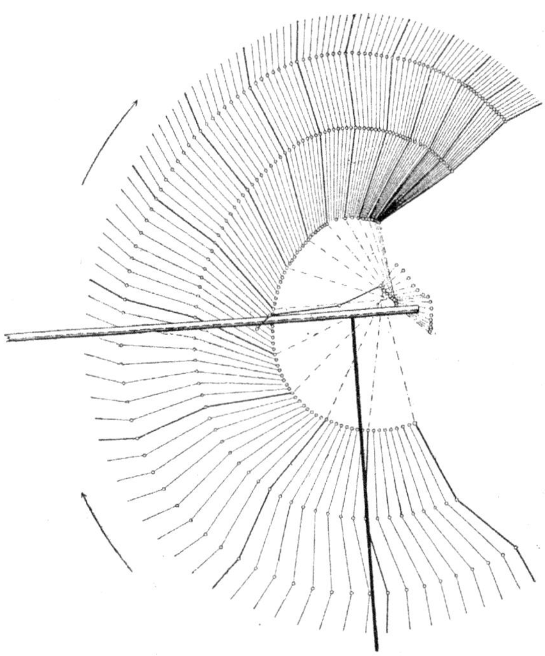

FIGURE 5.24 An example of a non-locomotor movement at the level of spatial field *C1*.
Part of the movement on parallel bars by the USSR champion Rtskhiladze. Sequential positions of the left part of the body recorded with the frequency of 130 postures per second.
(Source: work by the author, CNIIFK, 1937).

FIGURE 5.25 Examples of movements at the level of spatial field.

The kimocyclogram during the performance of the seventh waltz by Chopin. Recording was made from above. Two bulbs were placed on each hand. Sixteen bars are presented (from the 33-d till 48-th). One can see clearly the difference of movement patterns at this level from stereotypical patterns at the level of synergies.

Note the interesting perseverance in the left hand movement close to letter A, which belongs to the full-bar pause. Правая – right; Левая – left.

(Source: work by A. Sheves and the author, The Musical Pedagogy Laboratory of the Moscow Conservatory, 1949).

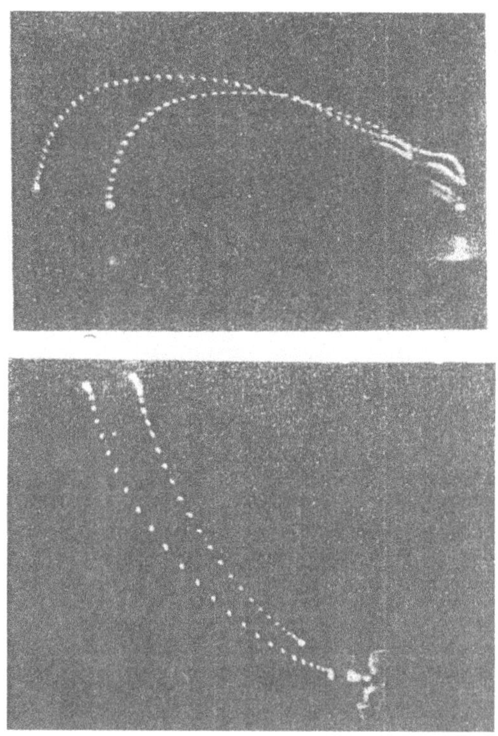

FIGURE 5.26 The cyclogram of pointing at and touching a point (an ataxia test).

а – touching the bulb on the table; *б* – touching the bulb affixed to the chin. The perfectly accurate and fast movements were performed by a patient with a tumor in the parietal part of the brain showing striking disorder of coordination at the higher, object level – see also Figure 6.7.

(Source: the case of Professor A. Luria from a collaborative work with the author, Moscow Institute of Neurosurgery, 1940).

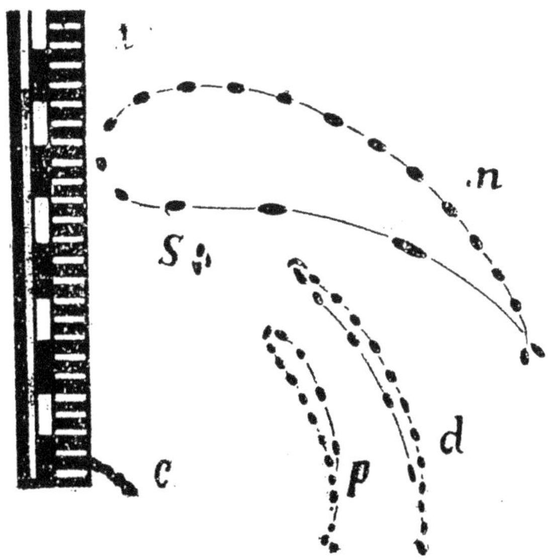

FIGURE 5.27 The cyclogram of the right hand movement during cutting with the chisel – a typical movement of the spatial field, which forms the background for the two-hand labor process at the level of object action.
Trajectories: *m* – the head of the hammer; *d* – the carpo-metacarpal joints; *p* – the wrist joint; *c* – the elbow; *s* – the shoulder joint.
(Source: work by the author, 1924).

FIGURE 5.28 A sequence of vertical strikes with the sledgehammer characterized by high automaticity and low variability of the movements.
(Source: work by the author, CIT, 1924).

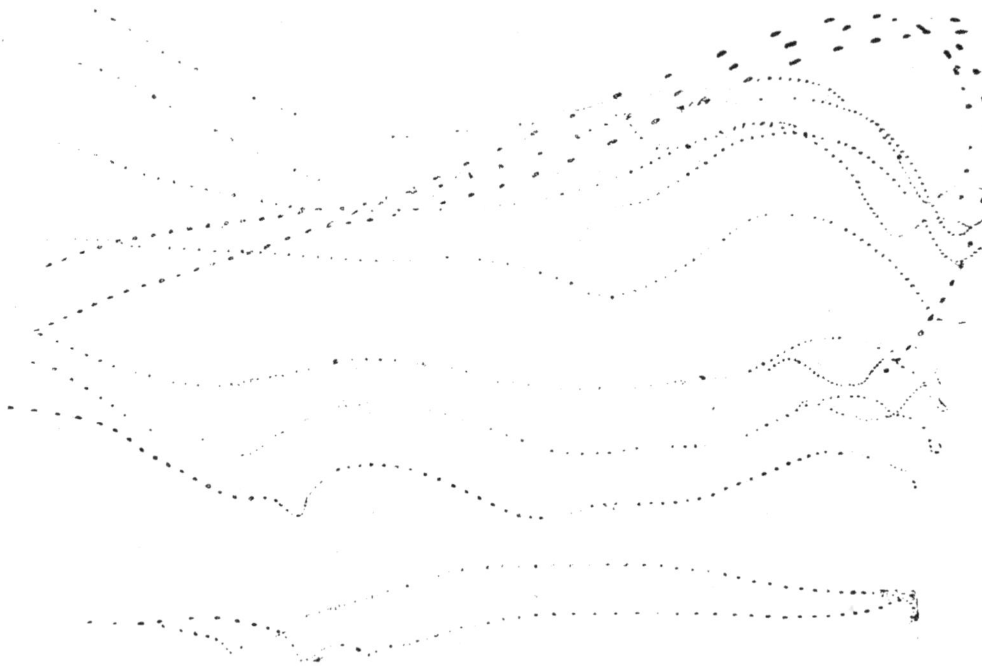

FIGURE 5.29 The cyclogram of throwing the grenade from the back, over the shoulder, with running start – a typical example of a ballistic movement at the level of spatial field.
The figure shows clearly how the background synergy of the running start (sub-level *C1*) is overlaid with the quick, singular throw with the almost straight trajectory.
(Source: work of the author, CNIIFK, 1940).

presented relatively few labor movements and relatively many athletic and acrobatic ones. As a rule, labor and manufacturing deal with an object and, very rarely, with space and its force field. Naturally, the inventory at level *C* is dominated, first, by locomotor movements and, second, – by athletic-gymnastic movements. If would be hard to find a large number of movements at higher levels that would not be linked to level *C* as the background because meaningful manipulation of an object requires, in most situations, mastering space as a prerequisite. As it will be shown later, the level of spatial field provides the base for movements at the level of object manipulations in two ways: (1) as a more or less independent background, which ensures motion of the whole body, accuracy and precision of separate auxiliary movements as technical prerequisites for performing an object-directed action (for example, walking as the background for such actions as mowing, steel rolling, work of a car coupler, etc.); and (2) special coordinations are formed at the level of spatial field motivated and encouraged by a higher level to serve that level – so-called, higher automatisms or skills (Handfertigkeiten), which will be discussed in Chapters 6 and 8.

Let us discuss briefly phenomena associated with dysfunctions of level *C*. If disorders at the rubro-spinal level *A* deserve to be called dystonias, disorders at the thalamo-pallidar level B^{12} – dyssynergias, then disorders at level *C* can be well summarized under the name of *dystaxias* or *ataxias*, or, in other words, what lay-men call "disorders of coordination." All known to clinicians

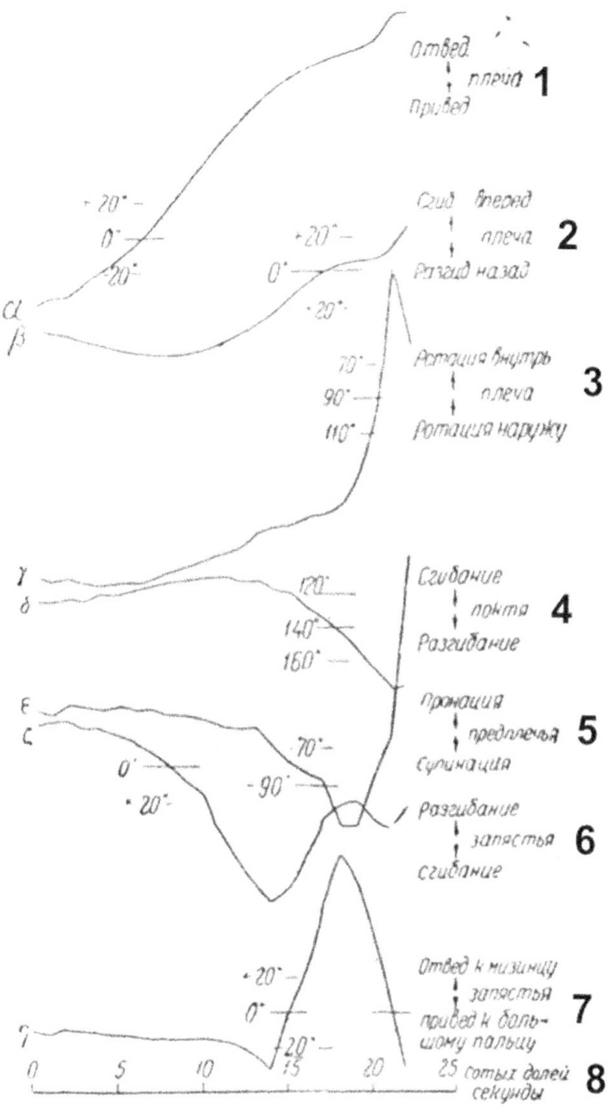

FIGURE 5.30 The joint trajectories of the right arm separated by degrees of freedom for the last ¼ of a second prior to grenade throw after the running start.

The order and marking of the curves are the same as in Figure 4.8. The illustration demonstrates the initial, relatively slow, part of the throwing synergy and the abrupt, typically pyramidal, energy explosion during the last 7–8 one-hundredth of a second prior to the throw.

1 – shoulder abduction-adduction; 2 – shoulder anterior flexion-posterior extension; 3 – shoulder internal-external rotation; 4 – elbow flexion-extension; 5 – forearm pronation-supination; 6 – wrist flexion-extension; 7 – wrist radial adduction-ulnar abduction; 8 – hundredth of a second.

(Source: work by the author, CNIIFK, 1940).

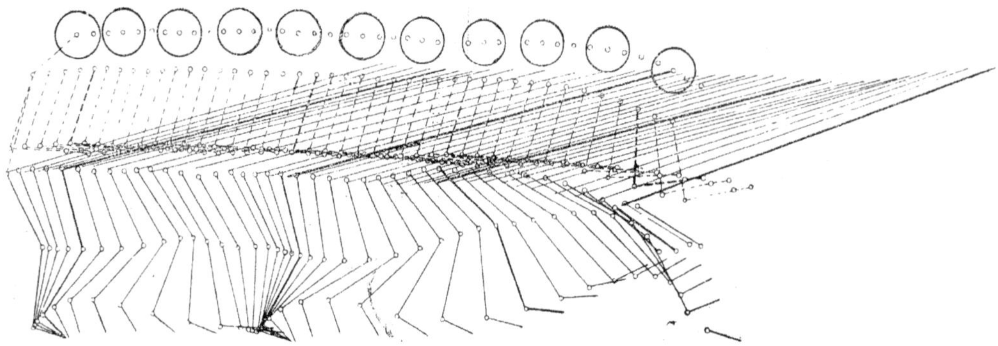

FIGURE 5.31 Images of the sequential positions of the right body part prior to thrusting with the bayonet. An example of an independent locomotor background with the superimposed goal-directed action at level *C2*.
(Source: work by L. Osipov, 1940).

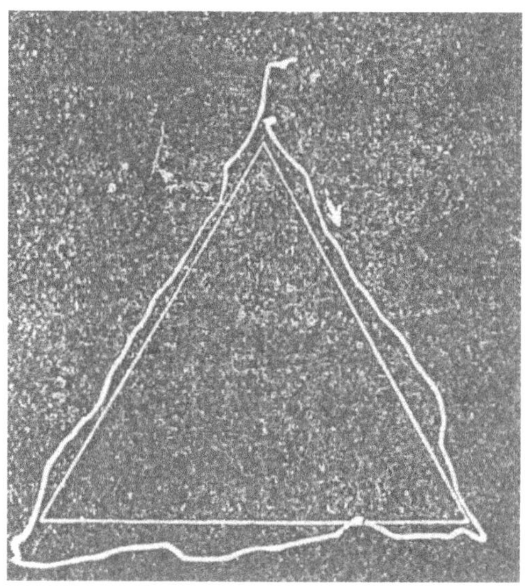

FIGURE 5.32 Cyclograms of walking blindfolded along the prescribed triangular contour. Recording from above. *Left* – normal performance at the level of spatial field; *right* – ataxic disturbance of walking with hypermetria.
(Source: work by N. Ozeretsky and the author, CIT, 1924).

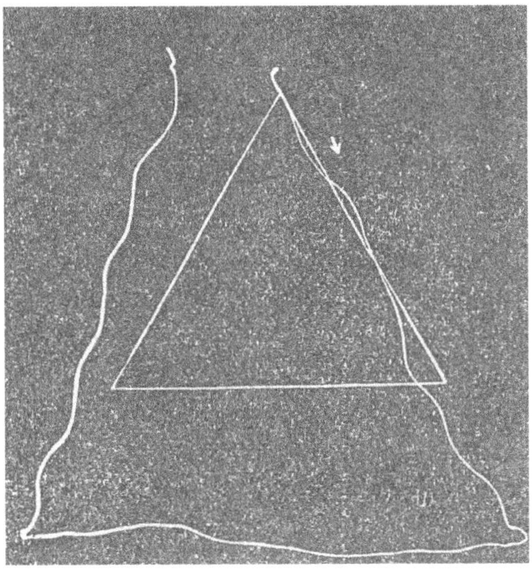

FIGURE 5.33 Cyclograms of walking blindfolded along the prescribed triangular contour. Recording from above. *Left* – normal performance at the level of spatial field; *right* – ataxic disturbance of walking with hypermetria.
(Source: work by N. Ozeretsky and the author, CIT, 1924).

cases of ataxias are in some way associated with injuries to afferent systems of this level. Examples are the *vestibular and cerebellar ataxias*; in essence, the ataxia of tabes dorsalis is of the same nature, although it is complicated, as mentioned earlier, by dystonic disorders at level *A*. All these types of ataxias do not involve the level of synergies and do not affect directly levels higher than level *C*, but they are very selective in their effects on spatial coordination, primarily balance, locomotion, and accuracy (precision). The tabes ataxia associated with the accompanying dystonic syndrome affects indirectly the level of synergies, which is based on a level *A* effector – the red nucleus. Movements at levels higher than *C* suffer from secondary consequences because most of them lose technical backgrounds from the level of spatial field. A clear, albeit transient, ataxic syndrome can be seen in cases of recent, not yet compensated, blindness (Figure 5.32 and 5.33).

In cases of injury to the main effector pathway of this level – the pyramidal tract – the general shock is at first replaced with spastic paralysis, later turning into paresis due to extrapyramidal (partly contralateral) compensation. During this process, selective losses are seen exactly in areas that have been assigned to the spatial field complex: loss or deterioration of quality of accurate actions and difficulty with targeted movements, whereas mimics, expressive, and postural movements are relatively preserved. Injuries to the effector part do not lead to ataxia that would be comparable to the one observed during injuries of afferent systems (see Chapter 9); however, whereas ataxia depends primarily on afferent, not efferent, systems, pyramidal disorders show a clear ataxic component. As far as injuries to the striatum, as an effector, are concerned, they lead to selective losses of movements of the lower sub-level *C1*, typically overshadowed by pathological hyperfunction of the pallidums.

Notes

1 It is very tempting to suggest that the mentioned duality of level *C* substrates in humans is conditioned by the currently proceeding *encephalization transition* from the extrapyramidal system to the pyramidal one. Of course, this process is so slow that we cannot observe it directly; however, likely, after a few millennia, this level, which is purely striatal in birds and lower mammals, will finally switch in humans to the cortex of the large hemispheres.

2 The evolution of the spatial synthesis at the higher level *D* is described in Chapter 6.

3 The conviction on the eternal constancy of the external space is so strong in healthy persons that a feeling of moving or turning external space after a sway or salto is a sign of a pathological change in the vestibulo-cerebellar system ("vertigo").

4 Obviously, there is a link between the aperiodicity of the spatial field and a symptom seen in patients with Parkinson's disease: the inability to walk over a smooth surface without markings described in Chapter 4. This loss of the ability to walk, except following markings on the floor, is typically interpreted as a result of loss of motivation ("Antriebe") for self-initiated movements, and floor markings are seen as an exogenous surrogate of the lost motivation. This explanation is unimpressive because there seems to be no reason to hypothesize that a line on the floor may substitute for motivation to make the next step. Here, one guess is replaced by another. A much more convincing explanation directly follows the described properties of the spatial field. Patients with Parkinson's disease suffer from a selective disorder of the level of synergies, which is particularly well suited, due to its afferents and proprioceptive reflex loops, for rhythmical alternations and repetitions. Such a disorder leads to *deautomation of cyclical processes*, which used to be performed at that level – first of all, of gait. The disorder also affects the mechanism of *recoding*, which normally converts motivation to move the body in the aperiodic spatial field into periodic, cyclical walking movements organized at the level of synergies. Destruction of that level destroys not motivation to move but the ability to find in the aperiodic space prerequisites for periodic actions. Paper shreds or chalk lines on the floor are surrogates of those prerequisites – they *make the spatial field periodic*. For a healthy human, walking over stripes or, for example, over railroad ties, creates additional difficulties by disrupting (deautomating) the habitual mechanism of recoding (see Chapter 8).

5 This does not mean, of course, that we refuse to consider *goal-directedness* of movements at the level of synergies; at any level, a movement properly solving the problem is purposeful, goal-directed. However, movements at level *C*, both singular and repetitive, always lead to a *clear ultimate goal* (moving an object, hammering a nail into the wall, etc.), whereas movements at the level of synergies – do not (twelve gymnastic squats or a smile).

6 More exactly, this level leads the most important of the background movement components, which primarily proceed at the highest cortical levels with inseparable links to movements of the left hand.

7 Our cyclogrammetric measurements of many highly automatized cyclical movements, such as walking, running, filing, etc., have shown that variability of consecutive synergic cycles usually does not exceed a few millimeters along spatial coordinates of the trajectory and a few milliseconds in their duration.

8 The distinguishing feature of interrelations between variability and accuracy is seen from the following example. If a subject repeats a few times a circular arm motion as a gymnastic exercise (at the level of synergies), measurements of cyclograms show that consecutive cycles differ from each other by not more than 10–15 mm – a reflection of very accurate coordination. If the motor task requires pointing at a point target with the needle (the level of spatial field), accuracy of the pointing is much higher than in the previous example, up to fractions of one millimeter, whereas the arm trajectories moving the needle to the target can deviate from each other by many centimeters. Hence, the ultimate accuracy (absent at the level of synergies) is many times higher than the accuracy accessible to level *B*, but the level of spatial field does not care much about trajectories and means used to achieve this high accuracy.

9 Note that the proprioceptive estimation of one's own fingertip position in space gets very little contribution from receptors in the fingertip but is created synthetically based on proprioceptive afferent signals from all parts of the arm; the only difference is that, in this example, the skill of position estimation is automated in childhood.

10 See Chapter 7 for shifts of the striatal and pallidar functions.

11 A very insightful neurophysiologist of the nineteenth century, Hughlings Jackson, at the end of the 1860s, stated that the central anterior cortical fissure, together with the striatum, represents a "middle level" in the control of the motor apparatus and that cortical zones anterior to the pyramidal field are also motor fields: they represent the highest motor centers (H. Jackson 1927).

12 But not movements such as, for example, crossing the legs while sitting, unrelated to spatial coordinates and performed at level *B*.

6

CORTICAL CONSTRUCTION LEVELS

Level D: The Parietal-premotor Level of Actions

All the coordinations and many of the movements described in the previous chapters are available to animals as much as to man. Moreover, at different levels of the evolutionary tree of the vertebrates, we encounter species with better developed aspects of the level of spatial field compared to man: There are animals who can run faster and are more fatigue-resistant, climb better, swim better, have better developed correction function of vision, orient themselves in space better with the help of visual and olfactory system, and have more accurate and precise means of attack. All these facts suggest that the culmination points in the development of level *C* are in the past. It is true that the variety of movements in the spatial field is larger in humans than in any of the animals, but more detailed analysis, partly reflected in this chapter, suggests that many such human movements, seemingly performed at the level of spatial field, are only saturated with numerous backgrounds from that level but in fact belong, according to the leading afferent systems, to a higher construction level. Given the lack of parallelism between the general direction of evolution of the central nervous system and the level of development of the level of spatial field, one has to admit that this level, in its essence, is not yet cortical. It is true that, in humans and higher mammals, this level is linked to the cortex of large hemispheres, but only with its periphery – with the pyramidal fields, which have been very precisely called by someone "anterior horns of the brain," and which work very well in animals who lack both the cortex and pyramidal tract.

In contrast, the level discussed in this chapter is entirely cortical (it has a variety of connections with subcortical systems discussed in more detail later) and almost uniquely human. This striking difference of its evolutionary position from the lower levels is emphasized by the tremendous, still underappreciated, qualitative differences of the cortex from the more ancient, nucleus-based organization of nervous systems. In Chapter 1, it was pointed out that the brain, over a very long interval of its phylogenetic history, performed auxiliary functions for effectors and higher receptors, was led by them in its development, and only recently has assumed a leading position. It seems likely that the principles of cortical functioning opened absolutely new, unexplored possibilities. Based on many facts, currently we witness an intensive growth of the neural-based principles dominating the biological process (unlikely to be observed directly, just like the human eye cannot see motion of the hour arm of the clock), when the brain leads and directs the evolution not only of the whole somatic system but all the life-related processes in both health and pathology. Naturally, one suspects that all these possibilities became available to the brain due to its cortex with its unique structure. Who knows what to expect from these processes in the new few million years.

The theory of coordination described in this book is based on the assumption that qualitatively different afferent signals, linked to various anatomically localized structures within the central nervous system, control and coordinate various groups of movements that may be rather dissimilar according to a number of their features. For the brain stem and subcortical parts of the brain, this assumption created convenient and easy to use classification features because, at those levels, we encountered sharply different afferent systems and equally distinct anatomical, neuronal relationships.

In contrast, both these principles become much more cumbersome and imprecise when we move to cortical structures. Already at the level of spatial field, where afferent systems involve a number of lower, more archaic, cortical fields, we encountered very complex synthesis of afferent signals, which took us far away from the original sensory elements. Afferent systems related to higher levels of movement construction are even more generalized, push even further away from original receptions, and are based to a larger degree on mnestic (stored in memory) reflections of the previous experience. It would be very hard to separate afferent levels and, even more difficult, to relate them to specific cortical fields and conduction systems.

If one considers only the morphological aspect of the cortical structure, its fits well this functional unity and generality. Whereas in lower systems we witness different cellular nuclei linked to each other by bundles of fibers, which are relatively easy to analyze with respect to their neuronal composition and hierarchical interactions, the brain cortex is built as a continuous layer of cells, quilted in all directions by the continuous layer of white matter fibers. With the exception of primary fields and layers ("entrance and exit gates of the cortex"), and tightly linked to them secondary fields (such as area parastriata or the premotor area), all other citoarchitechtonic parts of the cortex show no detectable hierarchical relations and maybe do not have them as permanent features. The problem of cortical localization, undoubtfully, belongs to the most complex, basic problems of the contemporary neurophysiology, which is largely due to the fact that the cortex displays the most complex and functionally plastic links and relations among its anatomical instruments.

These synthetics of functions, uniformity of the morphological structure, and complexity of hierarchical relationships are revealed, of course, in the fact that cortex-related levels of movement construction happen to be most implicit, morphing one into another, and avoiding easy descriptive separation and identification. In some cases, analysis of some movements reveals clearly the existence of a whole bunch of hierarchically organized levels, with each level revealing itself in some specific recoding or peculiar features of selective pathological disorders. In other cases, quite the opposite, even identification of two clearly distinct levels cannot be done with certainty.

Physiological analysis of movements at cortical levels is at its very early stage. Most data for analysis of such movements are still supplied by clinical descriptions of local brain injuries and, to a lesser degree, by experiments with extirpation of cortical areas in higher apes. We have to deal with this material with all the inherent basic drawbacks of descriptions of clinical functional losses.

As it follows from the earlier text, we have no right to assign to cortical coordination levels the same degree of anatomical and functional exactness, which was naturally established for the lower levels (although, already at level *C*, we had to distinguish two accreted sub-levels). Maybe, it would be more prudent, until further, clarifying experiments, to consider levels *D* and *E*, described further in this chapter, as multi-layer complexes of levels with so far elusive, and maybe objectively unstable, sub-levels, while, at the same time, avoiding guesses not supported by available facts.

Level *D* described in this chapter has been nearly monopolized by humans (note that it creates the foundation for the main backgrounds for speech and graphical coordinations) and is far from its culmination. It can barely be seen, as singular phenomena, in highly developed mammals such as the dog, the horse, and the elephant; one can find in books by Brehm many more or less plausible stories. Even in higher monkeys, the representation of this level is very low and (as in humans in early ontogenesis, during the second year of life), it is used only after attempts to solve a motor problem with more familiar, lower levels fail. The human specificity of this level was the main reason for its original identification by clinicians in the negative form of motor losses during specific, localized brain injuries. We have to admit that, so far, experimental human physiology has been unable to add much to clinical findings because stimulation of the exposed cortex, registration of biological potentials of the brain, and similar methods have been unable to establish anything level-*D* specific. We will also start from a negative definition but will try to reach the clearest possible *positive characteristic* of this level.

All the various clinical pictures of motor disorders at the level of actions (these pictures are united under the term *apraxia*, although it would be more adequate to address them as dyspraxias) are characterized by the absence not only of any persistent motor losses – paralyses, pareses, etc. –, but also of any persistent disorders of coordination in the common understanding of this word. In apraxic disorders, problems are *not with the coordination of a motor action but with its realization*. The bridge connecting perfect understanding of the meaning and essence of a motor task (this distinguishes a person with apraxia from one with agnosia, when understanding tasks falls apart) to its motor solution is destroyed. A patient with apraxia is not armless but helpless. In contrast to a patient with ataxia, characterized by disorder of the spatial field leading to an inability to control one's own arms, a patient with apraxia (assuming that it is not complicated by major background disorders) can control all the effectors, but cannot use them to accomplish anything beyond the simplest actions available for highest animals. At the same time, the patient loses hopelessly the ability to learn or recover skills that would move beyond those elementary actions. Understanding the task properly, such a patient has no misconception with respect to his or her failure in solving the task: As a rule, such patients are unhappy with themselves, which distinguishes them from psychiatric patients who, following a similar failure, lose critical evaluation of their actions. So, on the background of intact mobility and elementary kinesthetic control of the peripheral motor apparatus, in the absence of detectable losses in force, velocity, and accuracy of movements, one observes losses of vast groups of motor acts and their learning.

Extensive clinical observations of such patients, started by Nuthnagel and Hughlings Jackson and analyzed deeply by Liepmann in the 1900s, first of all, allow identifying, as a separate coordination level, a *class of actions* (alias, object actions, meaningful chain actions, etc.). This separation is justified by the clear selective losses of motor actions of this class during cortical injuries, caused by a disease or trauma, with different, well-defined localizations. Analysis of lost vs. unaffected motor coordinations allows separating the *level of actions* from the earlier described levels and sketch out its most important positive characteristics.

It would be impossible to claim cortical localization of this level without discussing in detail the general complex problem of cortical localization. All we can do at this time is to separate cortical fields, which are absolutely necessary for the normal functioning of the level of actions; injuries to these fields lead to the described syndromes of apraxic disorders. As a rule, facts of relations between certain localized cortical disorders and syndromes of functional loss prove only that these locations of disorders are obligatory transit points of the underlying neural process – obligatory nodes, which cannot be bypassed by this form of the neural process. Only such nodes of obligatory transit, the apparent "entrance gates" and "exit gates" of the neural process at a particular level, can be identified in the process

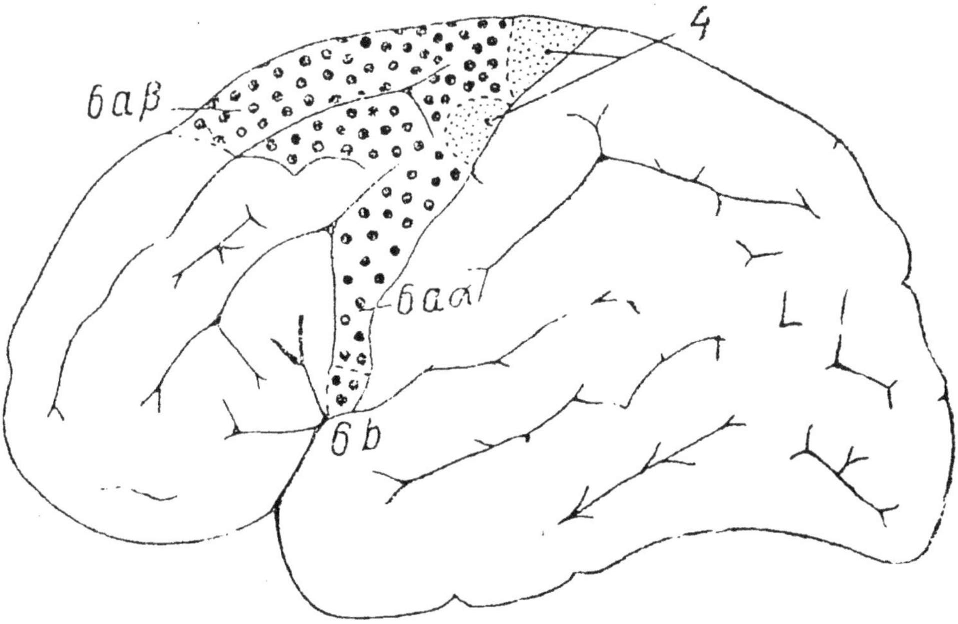

FIGURE 6.1 A scheme of the pyramidal field 4 and premotor fields 6 on the brain hemisphere (after O. Foerster).

of localization of the level of actions. In a distinctive pattern, these fields group in two areas of each of the hemispheres, rather far away from each other. One group is located in the inferior parietal lobe, in-between the fields of sensory visual (occipital lobe) and auditory perception (temporal lobe) and fields of tango-reception (the postcentral gyrus), together with their adjacent, intermediate fields. The second group of fields of this system is located directly anterior of the motor pyramidal field *4*; it is commonly addressed as the premotor zone (Brodmann areas *6a* and *6b*). The bottom edge of the premotor zone borders with the Broca's "motor speech center" (Figures 5.11, 6.1, 6.2, and 6.3).

The first group of fields is tightly linked, both anatomically and functionally, with the primary and secondary *afferent* fields, which surround it. As far as premotor fields are concerned, its functional proximity to *effector* processes is proven by the fact that the electrical stimulation of the naked cortical surface in those areas produces *movements* of separate body parts, which cannot be achieved by stimulation of parietal areas. The effects of electrical stimulation of the premotor fields differ from those observed in response to stimulation of the pyramidal area by higher thresholds, clearly seen phenomena of summation, facilitation, and after-discharge, the significant temporal delay from the stimulation to the induced movement, and finally, by the fact that the evoked motor responses do not represent isolated effects from a single muscle or small muscle group (as it happens during discharges from the pyramidal zone) but rather kind of fragments of complete movements, which involve synergically both mutually antagonistic and protagonist muscle groups. O. Vogt has found that a cut of the cortex along the border between the pyramidal and premotor fields eliminates these effects immediately, which means that motor reactions to stimulation of the premotor areas are possible only via transit from the premotor field to the pyramidal system and the pyramidal effector pathway. Myeloarhitechtonic studies of the brain conduction pathways have shown, however, that premotor fields are connected not only to the pyramidal cortical fields but

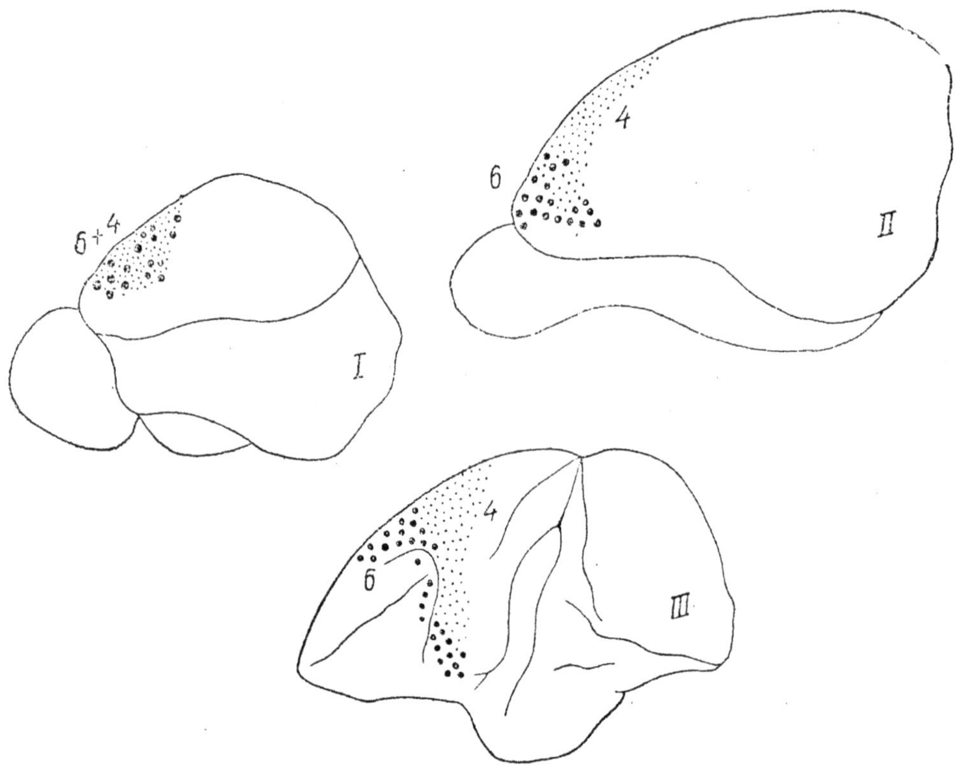

FIGURE 6.2 The location of the pyramidal fields 4 and premotor fields 6 on the large hemispheres in different animals and their gradual phylogenetic differentiation 1 – hedgehog; 2 – rabbit; 3 – marmoset (after O. Foerster).

also to the pallidum and even to lower nuclei of the extrapyramidal effector system (Figure 6.3) (the issue of possible links to the striatum is not clear yet). Therefore, the effector character of premotor fields is supported by their location, by their pathways, and by the effects of experimental stimulation; however, they describe an effector not of such a direct and clear type as, for example, the magnocellular fifth layer of the pyramidal field, but complicated by the hierarchical multi-level structure, similar to that of the striatum. However, as we will show later, the clinical picture of losses during parietal lesions clearly displays afferent functions *sui generis*, whereas functions that suffer most following injury to the premotor fields are similar in their semantics and effector apparatus but reveal this similarity in a new and peculiar way. The functional location of the premotor fields resembles those of fields 18 and 19 in the area of visual perception; these are, so to say, secondary effector zones.

An essay on the cortical apparatus of the level of actions would suffer from an essential omission if we failed to mention one more of its important features. Another zone necessary for the functions of the level of actions is the *left* inferior parietal area linked, according to the law common for the whole cortex, to the *right side of the body*. This is the first feature of the level of actions directly and clearly reflecting the inequality of the two sides of the body, i.e., right-handedness (or left-handedness). At the level of spatial field, and to an even larger degree at the lower levels *A* and *B*, this functional inequality is not noticeable.[1] As mentioned in the previous chapter, at the level of spatial field, one hand can replace the other hand easily. At the level of actions, both overall aptitude

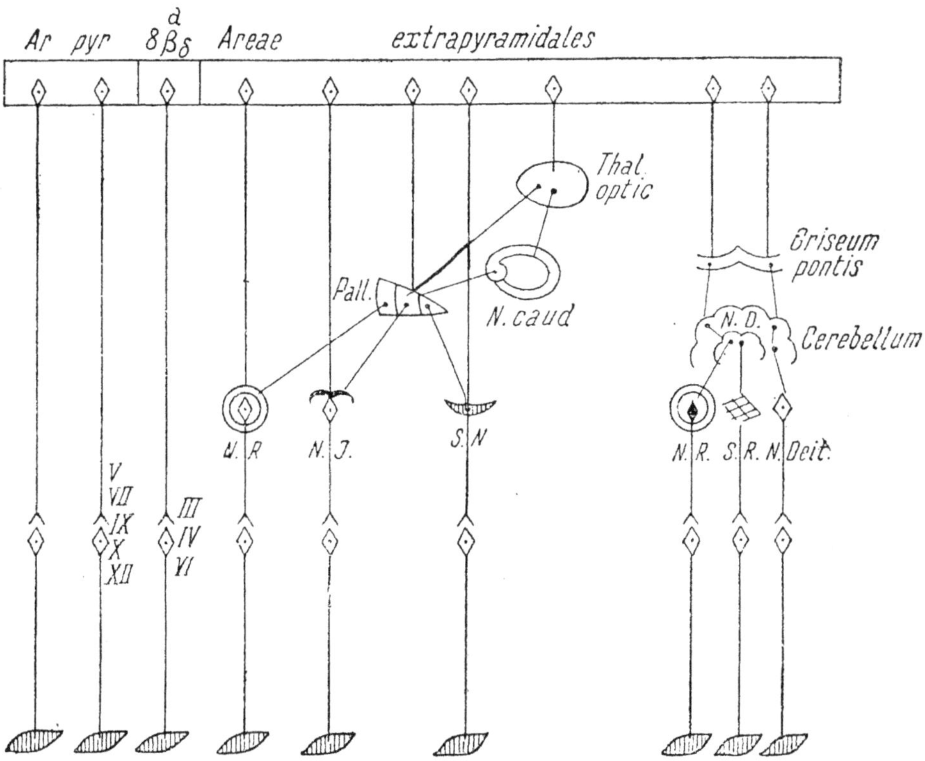

FIGURE 6.3 Scheme of connections between the cortex of the large hemispheres and subcortical nuclei (after O. Foerster).
The *top strip* – the cortex. *Bottom* – peripheral motoneurons connected to muscle units – motor units. Roman numerals from *III* to *XII* stand for the cranial nerve nuclei. *N.R.* – red nucleus; *N.J.* – Lewis nucleus; *S.N.* – substantia nigra; *N.D.* – Dentate cerebellar nucleus; *N. Deit* – Deiters nucleus. *N. caud.* – caudate nucleus (striatum); *Pall* – Pallidum.

of each hand and the acquired over the lifetime various skills can be very different in quality and composition; this may serve as a secondary sign of a motor action being assigned to level *C* or *D*: skills related to the level of action and supported by its main afferent systems, as a rule, do not show easy transfer across the hands.

Let us turn now to functional analysis of the level of actions. As in the previous chapters, we will first offer a description of the relevant afferent systems, and then – a general essay on the main features of motor actions; in the essay on this particular level, it would be worthwhile to review its typical functional losses prior to discussing its normal, intact motor actions.

The *leading afferent system* of the level of actions D is its *object*. Obviously, assigning an object to afferents implies a very broad understanding of the object. Psychological image of an object is a result of much deeper generalizations and much more complex synthetic links between sensory and mnestic components, compared to the synthesis described in the previous chapter under the name of spatial field.

The leading motivation at the level of actions is not the object by itself, as a geometrical shape, as something with a certain mass, consistence, etc. (see about this aspect later), but the meaningful aspect of acting with the object – it does not matter whether the object plays the role of an object of action or of a tool. Exactly this motivation is destroyed in the so-called agnostic (also known as ideation) apraxia, which will be described in detail later. Afferent systems of this level are functional systems, which make sense out of a given object and define what, and in what chain order, can and should be done with this object.

Further analysis would benefit from two notions introduced by neurologists and heuristically valuable: those of the *semantic structure* and *motor composition* of an action. Semantic structure of a motor action is defined by the contents of the motor task and, in turn, defines the sensory or sensory-gnostic synthesis, which is adequate for the task and can lead to its solution; this defines the adequate for the task leading construction level. Motor composition of an action is a result of interaction between the motor task and kinetic means of the organism relevant for the task, which can be compared to substituting these two values into a common equation. Motor composition includes a list of sequential elements if the organism deals with a chain action, definition of motor operations corresponding to those elements, and the background composition of simultaneous components of a complex action. Motor composition is also influenced by the biomechanical design of the levers in the kinematic chains of the body, by its innervation resources, the available sensory corrections, and also by tools that can be used to perform the desired action. Therefore, motor composition is a function of both the task and the performer. Humans solve one and the same problem of fast displacement in space by sprinting (or biking, etc.), horses – by galloping; birds – by flying, etc. The definition and formation of the motor composition of movements and actions will be considered in Chapter 8.

After this brief digression, let us return to semantic structure of actions at the discussed level because it is most directly related to the afferent systems working at this level, similarly to how motor composition of an action is inseparable from its effector apparatus.

The term "object actions" frequently assigned to meaningful chains at level *D* is definitely justified by the dominating percentage of actions at this level dealing, by their very essence, with an object. This fact exemplifies the typical for the psychological hierarchy of levels gradual increase of their objectivization directed at active, modifying interaction with the environment. As a feature of coordinated movement cohorts, object-relation demonstrates relentless "encephalization." The lower levels, parallel to the process of their increasingly auxiliary, background roles, are assigned primarily "propriomotor" actions and components, mobilizing and constraining one's own body, which are getting harder and harder to control given the increase in the mobility and subtle complexity of task-related requirements. Beyond doubt, by themselves, movement cohorts become more and more object-oriented in the process of evolution; the role of the *hand* in this process has been sufficiently emphasized by the founders of the contemporary philosophy of science. Movements at the level of actions deal with manipulations of objects, using them as tools, and, finally, as a symbol, which facilitates and concretizes abstract actions: a drawing, a chess figure, a written letter, a hieroglyph, etc. This level clearly contains also a number of objectless actions (unless one expands the notion of object beyond reason): sport games, war actions, etc.

Theory of coordination would benefit substantially from understanding the underexplored peculiarity of object actions revealed in the leading role occupied by the grasped object during such actions and pointing at very deep, purely unconscious recodings. When an object plays the role of an instrument during highly automated habitual work, it is perceived as the action performer, as part of one's own body up to an illusion of active, leading control of movement emanated by the instrument itself.

Object itself exists in both space and time. Along similar lines, the meaningful aspect of an action with this object contains semantic perception and partition of *space*, where the action is organized, and synthetic feeling of *time* where a sequence within a meaningful chain of the elements is built. Afferent systems at the level of actions include both synthetic space and synthetic time, but in a different representation, with different characteristics compared to those encountered at lower construction levels. A review of changes and complications associated with these psychophysiological categories during ascent to the level of action will help clarify their peculiarities inherent to this level.

In the previous chapter, we have already discussed the *evolution of the "space" synthesis* from the lowest levels to the upper sub-level of the spatial field. The meticulously discussed space of this sub-level – completely projected on the external space, with metrics (e.g., *scale*) and geometry (containing components of geometric *form* and geometric *similitude*) is the most objective of "spaces" inherent to different levels of the construction of movements. It is supported by the phylogenetically youngest and most perfect receptor systems, as compared to lower levels, but is built, however, on the most peripheral parts of the cortex, as compared to higher levels, which are more abstracted and remote from the original sensory material.

Space, where object actions are organized, has a range of peculiarities, which affect not only the structure of the overall leading afferent synthesis of this level but also the coordinated construction of its movements. Already at level *C2*, one can see aspects of separation from an immovable coordinate (possibility to copy rather than trace an object, which would be typical of level *C1*) and from a fixed scale (similitude rather than congruency typical of level *C1*). Abstract transformation of space goes even further at the level of actions. Instead of a geometrical form, we encounter a scheme, i.e., *metric, proportional relationships are replaced with topological, qualitative correspondences*. At the object level, space loses concreteness but gains order, meaning, and extraction of essential features compared to the level of spatial field. At this level, qualitative notions of open and closed shape are formed and organized, as well as such notions as "over," "under," "outside," "inside," "between," etc. Psychologists address these processes as "categorical organization of space," whereas we will emphasize dominance of topological, semantic scheme over geometrical shape.

In any geometrical image, we can identify its *topology* and *metrics*. Topology of a geometrical object is the totality of its *qualitative* features independently of its size, shape, curvature of its outline, etc. Topological properties of a linear shape should include, for example, the following: whether it is open or closed, whether its lines cross themselves (as in figure eight) or not (as in circle), etc. Aside from these features, which have no quantitative measures contained in their definitions, we would also assign to topological features those that include a *number*, but still do not include a *value*. These include, in particular, rectangularity, belonging to the class of five-pronged starts, etc. All images in the top row of Figure 6.4 belong to the same topological class whereas they are quite dissimilar in their metrics. Indeed, they all have five angles or rays, five intersections of their lines, etc. Image 6 in the same figure belongs to a different class because it has four rays and only one intersection; on the other hand, it is similar to the previous images because they all belong to the class of closed figures with intersecting lines. To characterize topological features of this type using a more familiar example, note that *every letter of the printed font* is a separate topological class; note also that a single class of letter "*A*" includes, obviously, capital "*A*" of all font sizes, shapes, and outlines as long as one ignores minor lines with pure calligraphic importance. Every spring, one can see numerous chalk drawings on the pavement for hopscotch – they are also representatives of the same topological class for every variation on the game, which are independent of the scale of the drawing and abilities of the artist. The familiar schematic drawings of a "house" or "man" by a child are also not more than a certain topological class.

Exactly these categories, which belong to topology, not to geometrical metrics, define features of the spatial synthesis at level *D*. Very striking manifestations of the primacy of topology over shape at this level are elements of two of its typical actions: a handwritten letter and an uttered

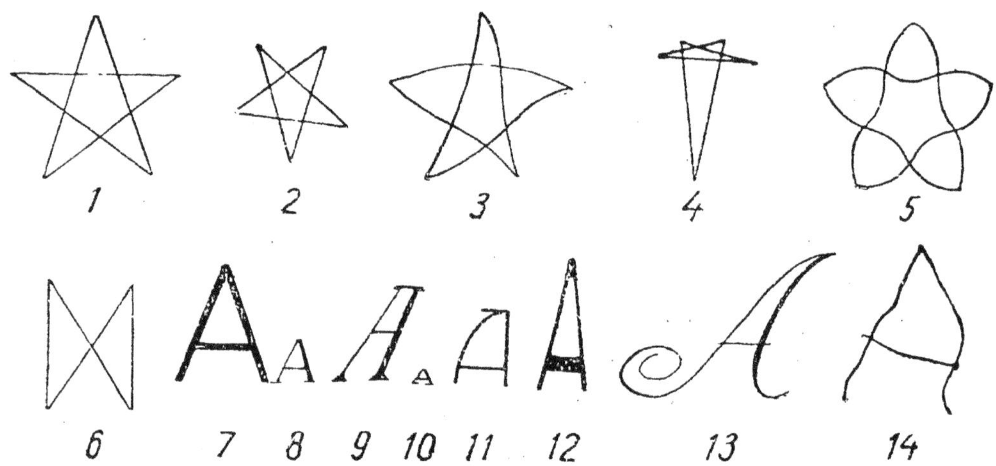

FIGURE 6.4 *Top* – a number of variations within the topological class of a five-pronged star. *Bottom* – a closed figure that does not belong to this class and a number of members of the topological class of letter *A*.

phonetical realization of this letter. Viewing a *letter* as an object, easier to perceive and characterize, note that not only its meaning depends exclusively on topology of the mutual configuration of its strokes, not on their geometrical characteristics, but also movements during writing the letter are equally topological, not constrained by either metrics, shape, or geometrical similitude.

Based on the performed analysis of the evolution of afferent space – note that, at each level, its style defines, via its sensory corrections, style of the performed movement – an interesting phenomenon that deserves our attention. It has already been noted that movement automation consists of switching a number of background coordination components of a movement to lower levels, which is associated with switching them to different *afferent systems*. Hence, in a particular, very common, case of automation of background components to the lower level of synergies, the automation should be accompanied by replacing visual afferent signals with tactile-proprioceptive ones, i.e. *switching off the visual control* of the corresponding component. Logics requires that, if our general understanding of automation is correct, there have to be cases of automation (in this case, linked to transfer from the *level of actions* down into the *spatial field level*) accompanied not by elimination but by *emergence* of visual corrections. Indeed, such cases do exist. A healthy grown-up with the typical pronounced domination of the object level always draws schemes, not forms, – something he or she processes and generalizes, not what he or she sees. Therefore, a person, who is being taught how to draw, has to learn how to see the external world in the way its image emerges on the retina, to learn to see nature and reproduce it in the right light and perspective, not to conceptualize the nature and draw one's own ideograms as it is done by those who cannot draw. Acquisition of the skill to draw nature is a very typical case of the process of automation constructed at the level of actions, which is built essentially on the ability to turn visual control on. Of course, this ability is used by skillful artists unconsciously, as expected from any automated action.[2]

Without multiplying examples, it is necessary to add to the list of characteristics of the spatial synthesis at the level of actions only two more entries.

First, not only with respect to afferent systems but also to the controlled effector apparatus, the space of level *D* is built on the principle of a topological scheme, not of a geometrical shape. Of course, this does not mean that movements elevated to this level necessarily stop being metrical;

this means only that *everything introduced into a movement by this level is topological, not metrical.* If there are metric movements at the level of actions, this means that the level of spatial field in involved in serving one of the background components. The difference from independent movements at level *C* is that these latter movements cannot be non-metrical (otherwise, this would signify disorder of level *C*, ataxia), whereas motor actions at the level of actions and typically non-metrical and show metric characteristics only in the mentioned particular cases. Handwriting is not metrical – this has already been mentioned. There is no metrics in taking one's hat off, lighting-up the cigarette, drawing a house or a man. In its very essence, the following actions are topological, not metrical: tying and untying a knot, putting a rubber band on the box, scooping water with the mug from a reservoir, etc. In all these cases, success of the movement is defined by its qualitative (topological), not metrical result: The knot is tied or not, the ring is on the box or missed it, the mug entered the reservoir or not, spilled the water or not, etc. As a result, in a person with ataxia, following disintegration of the level of spatial field, most movements at the level of actions do not disappear, and only those aspects of movements suffer that rely essentially on the metric backgrounds from level *C*.

Second, the described qualitative, semantic transformation of space typical of the level of actions is intimately linked to the formation of the *notion of object* itself. This transformation is not abstract and proceeds with a goal, it represents one aspect of the evolution of geometrical images (visual or tactile-proprioceptive) – colorful, heavy, movable, etc. – into the generalized semantic image of an object. Shape and metrics are essential for the geometric image; for the object itself, both are secondary. Clearly, metrics is essential for such specially constructed objects as the ruler, the magnifying glass, the knife blade, and the protractor. But what is essential for a cup as an object for meaningful manipulations? Neither its width, nor height, nor the spherical or square shape are essential; it has to have solid sides, solid bottom, and a handle – all these are purely topological features. Based on these features, any child would recognize a cup, even if he has never seen cups with such metrics before, and will be able to use it appropriately. What is essential in the metrics of such objects as the bottle, the fork, the hairpin, the hammer, the button, and the pencil? Of course, none of us would be happy to wear a topological, not metric, shoe – on the other hand, we have already mentioned the significance of metric backgrounds in certain situations.

Time develops comparably from one level to the next one, but its evolution is less explored. At the level of synergies, it is best pronounced as rhythm (i.e., as a temporal pattern); at the level of spatial field – as an instant (e.g., during taking aim), synchronicity, duration, and velocity. At the level of actions, it acts as a semantic or causal sequence, as a link between successive elements of the chain, which produce the action. Time is also organized at the object level not metrically but topologically or categorically; it clarifies such elements as "before," "after," "post hoc" and "propter hoc."

The evolution of the relationships between spatial and time syntheses with afferent and efferent systems at the corresponding levels proceeds in significantly different ways. At all levels, spatial syntheses are tightly linked to afferent signals. At level *C*, they form an objective external field for structured external projection of sensory perceptions. At the level of actions, they create prerequisites for semantic organization of the world and help to extract from it objects for active manipulations. Along these lines, afferent signals contribute to the emergence of a (subjective) space from the external space – an object built on an object – the most general *object-related notions*. In contrast, at all levels, timing syntheses are closer to effector processes. At the level of synergies, the timing syntheses are merged into the movement composition and express its rhythmical dynamics. At the level of spatial field, they define speed, rhythm, and the right instants for an accurate active reaction. At the level of object-related actions, time already converts into a semantic link and chain sequence of actions on the object. As a result, *effector processes* bring about (subjective) time, whereas time brings about meaningful action, and, at the highest levels, the latter brings about behavior; ultimately, the superior synthesis of behavior is the person or the *subject*. It is very interesting that, according to the law of Bell and Magendie, not only in the most ancient brain (the spinal cord and brain stem), the afferent part is represented by its dorsal side; along similar lines, cortical fields linked directly to the raw receptor signals (occipital, temporal, and posterior-central) as well as the intermediate fields of the parietal lobe, linked to the semantic and spatial organization of the categories of space and object, fill the *dorsal parts* of the hemispheres. Hence, the evolutionary chain "effectors-time-subject," at all levels, demonstrates more close systemic affinity with the *ventral*, effector parts of the brain and shows stronger disorders in cases of correspondingly localized injuries. This peculiar generalization of the law of Bell and Magendie is reflected in the localization of the motor apparatuses of the level of actions, as it will be shown soon.

Similar to space and time, *the object* starts playing a role in movements constructed at levels below the level of actions. Indeed, interactions between the moving effector and the object have to happen at all the construction levels; however, they are built differently. This has to be noted to warn against assigning to the level of actions, during systematization of movements and diagnostics of their disorders, phenomena that definitely do not belong to that level.

Grasping and holding an object can be observed in humans at the lowest construction level, the rubro-spinal level *A*, starting with tonic fixations in newborns from the very first days of life (see Chapter 7). Level *A* participates in grasping and holding an object in grown-ups as well; however, not as a leading level but as the provider of background components. This level ensures the perfect fit during grasping with fingers for any object shape, which led Bethe to the practically valuable idea on the design of a grasping hand prosthesis (Figures 6.5 and 6.6), accompanied, however, by a wrong theoretical generalization. He was wrong in associating this mechanism of hand adjustment to the shape of the object, the mechanism of "sliding interface," not to phenomena of the plastic or flowing Uexkull's tonus (as it had to be done), but to mechanisms of compensatory adjustment

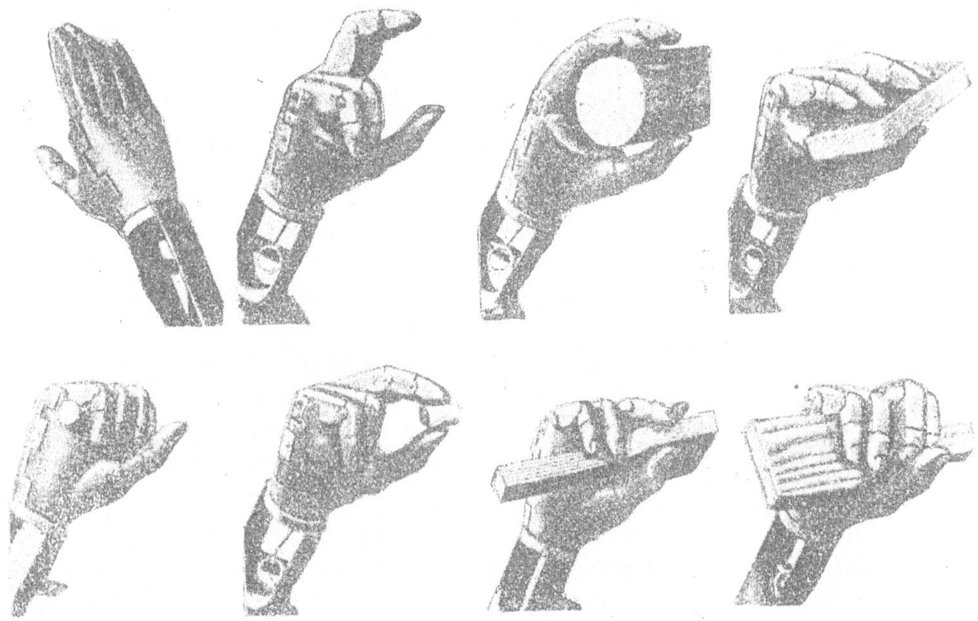

FIGURE 6.5 The active grasping hand prosthesis Bethe built based on the principle of sliding interface.

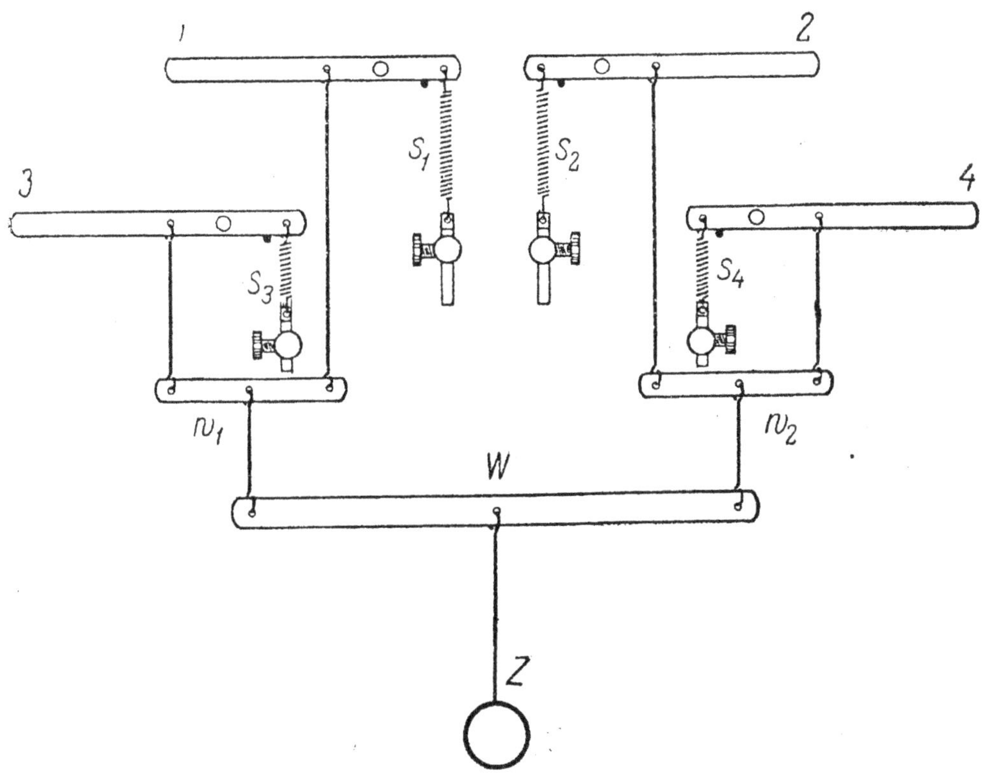

FIGURE 6.6 The scheme by Bethe describing the principle of sliding interface (der gleitenden Kopplebg) used by him for the design of hand prostheses.

during locomotion, which, as we have already seen, are realized at the higher level of spatial field. This led to the attempt at equating two mechanisms, which have nothing anatomical or physiological in common, and moved Bethe away from the correct solution of the problem.

The *level of synergies* plays a background role in a whole range of object actions – either indirectly, via support of locomotion movements playing the background role in an object action, or by ensuring a direct background synergy. There is, however, little interaction between this level and the object. Clear object-related components can be found only at the *level of spatial field*. The object is represented in this level in a variety of ways.

First, it acts as the point of force application, as a physical body, something with weight and ability to move. The ball in a sport game,[3] the shot or discus in the corresponding sport disciplines – are not objects or tools of action at level D, but rather objects at the level of spatial field that have a certain shape and consistency, weight and mobility; in essence, these are material points – as they are commonly analyzed in biomechanics.

Second, an object at the level of spatial field is something that can be taken, placed, carried, thrown, inserted, pulled-out, etc. With respect to this point, it is very interesting to consider observations of animals for whom the level of actions is either impossible or difficult. If a chicken sees food beyond the lattice fence, it is able to react only at the level of spatial field, i.e. trying to reach it unsuccessfully along the shortest straight path. A dog or a monkey can already switch rather easily in such a situation to the level of object (chain) action: It would walk not to the object but away from it where there is a gate in the fence, i.e. includes in its behavior two successive, mutually dependent, actions, with the first of them having not spatial but semantic characteristics. Higher apes, in a similar situation, can even use a tool (for example, to go for a stick). But even an ape, before doing something at the level of actions (e.g., building a tower from a set of empty boxes to reach a fruit on a high branch), will for some time fruitlessly (literally!) jump and try to reach the fruit along a straight line at the level of spatial field.

Third, we have to assign to the level of spatial field certain cases of manipulating an object, i.e., cases, which, at the first glance, should be assigned to the level of actions. When a 1.5-year-old child (in this example, we have to specify the age rather precisely!) or a higher ape gets a wooden detachable egg, they are able to conceptualize it as an object that can be opened. Their afferent systems are, therefore, getting close to the level of actions. But both the child and the ape try to solve the problem in a purely geometrical way: If the egg can be taken apart sideways, its sides have to be pulled exactly sideways. This is exactly what they do with all the might, which frequently leads to the halves of the egg and all its contents flying all over the place. Such movements are led by a geometrical image of the egg, not the object-related experience, which could suggest much more successful methods of opening the egg, which contradict its geometrical image (turning, unscrewing, etc.). Only after such an action has been mastered, one could speak of the action being switched to the object level.

To summarize, a whole range of actions with an object have to be excluded, from the very beginning, from the list of actions led at the level of actions: First of all, *grasping actions*, not only with the simple grasp (of an apple, ring, and the handle of a suitcase) but also with a specialized grasp of an instrument, as well as all the forms where an object is present as a *material point* in space or lead movements as a *geometrical image*. All these movements show disorders following injuries to the corresponding lower levels, whereas movements at the object level *D* are always preserved in such cases if these movements use backgrounds from the spatial field as secondary components, for example during dressing oneself up, washing, using the towel, opening the box, etc.

Before moving to inventories of specific motor actions at the level of actions, we have to summarize briefly their general, common characteristics during unimpaired functioning of the corresponding levels of the central nervous system – general characterization of motor actions that could be called "praxias."

Movements at the level of object-related action represent *semantic actions*, i.e., these are not so much movements but rather elementary behaviors defined by the semantic aspect of the task. To put on and button the coat, to grease the ski, to score a goal in football, to eat an egg in a civilized way, to seal the letter in an envelope, to sharpen the pencil, and to whittle a wooden board – this is a set of examples of the simplest object-related actions; each of them represents a group of movements, which, as a whole, solve a meaningful problem. In most cases, such movements are built as successive chains, more or less complex in their composition and using as links between elements not spatial (kinetic or geometric) variables but semantic motives that cannot be reduced to simple movement of objects in space or generation of forces.

When an action at level *D* represents such a chain process, one can extract, from the movements-links forming its motor composition, *leading movements-links*, which generate essential, meaningful steps of the action and *auxiliary* or background movements-links, which play secondary, but useful roles. Examples of such auxiliary movements-links include various cyclical movements (cutting, whittling, sawing, hatching, etc.) and swinging movements (during hitting labor operations), as well as many independent successive links with secondary roles, e.g., to move an object closer, to take a tool and put it back, to hold an object, to brush away labor waste, etc.

Variability is revealed in actions at this level in a new, specific way. Level *B* was characterized by nearly absent variability of postures and trajectories, i.e. a rigid link between the essence of a movement and its spatial-kinematic pattern supported in many cases by the phenomenon of dynamic stability. At level *C*, we already encountered phenomena of broadly allowed inter-compensation of postures and trajectories and even inter-compensation of effectors ensuring an accurate, invariant, relationship to final, goal-dependent coordinates. At the level of actions, the phenomena of variability and inter-compensation move even further. The cortical localization of the level of actions, which brought into neurophysiology the maximal ability to switch and extemporaneousness, combined with a leading role of the semantic aspect of more or less complex manipulations with an object – taken together, brought about the tremendous growth of adaptive variability. Even during maximally habitual, professional, highly accurate actions, two successive actions do not repeat either the accuracy and nomenclature of successive chain links, or their order, or the number of repetitions of separate action components. In these actions, we see easy mutual replacement not only of trajectories and effectors, but also of whole links of the chain performing the action. If a worker needs to bend in half a narrow tin strip, the chain of his actions is guided by the desired outcome, whereas the individual movements-links of the chain can vary infinitely. He could try to bend with the hands or with plyers, in the wise or without it, using a hammer to strike the bend or not (first, he would have to get the hammer from the drawer with movements from the level of spatial field), press on the strip with one or another object or gesture, etc. We do not mention here that every aspect of an object-related action from the field of topology, not metrics (i.e., the most typical feature of this level), in its essence is variable or indifferent with respect to actual spatial coordinates. The only invariable and precise aspect is the movement essence, its outcome, and, to ensure maximally exact outcome, all the auxiliary, background components maneuver and adjust within a broad range. However, one should mention two very significant and typical limitations in this switching and inter-compensating variability of elements; each of these limitations hides behind it a physiological problem of its own. We will return to this issue a bit later with respect to analysis of the motor composition of actions.

One peculiar feature of movements at the object level, which is intimately linked to the discussed semantic organization of space, endows these movements with a feature, so far only negative, which is very bright and frequently useful to recognize movements at this level and to identify the moment of its ontogenetic ripening on the background of the earlier emerging level of spatial field. The point is that movements at the object level are led not by a spatial image but by a semantic image, and motor components of the chains at this level are dictated by and selected with respect to the semantic essence of the object and what should be done with it. Since this semantic essence is frequently different from the geometrical shape, from the spatial-kinematic properties of the object, movements-links of object-related actions show a large percentage of movements leading in a wrong direction compared to what is called for by the spatial perception and understanding of the object. Earlier, during the discussion of object-related actions at the level of spatial field, we considered a few examples of motor actions where movement is led by a spatial image; for some movements of this kind, the level of spatial field is sufficient and adequate, whereas for other movements it leads to partial or complete failure at the overall task. The action of rinsing a painting brush in the glass with water during painting on *paper*, opening the lid of a box by pressing it *downward*, extracting a screw from a board not by pulling but by turning it, turning a boat counter-clockwise by turning the rudder clockwise – all these are examples of successive components directed the "*wrong way*," diverging from the direct geometrical logic of operating with the object. Designers are absolutely right trying to ensure that, in complex mechanisms, the geometric logics of the controls fit the object-based, technological logics: Excellent examples are given by the design of controls of a contemporary airplane. However, the level of automation of object-based actions saturated with components of the "*wrong way*" class is so large and the aforementioned illusions of fusion of one's organs with parts of the machine or instruments is so powerful, that there were cases, for example, when pilots during World War I, before planes got standardized, had learned to their amazement that the control scheme of the plane they had just flown for the first time in their life was completely opposite to the one they were used to during their lifetime experience.[4] It is much easier, than it looks at the first glance, to learn how to ride a bicycle with crossed hands, with the right hand holding the left handle of the bar and the left hand – the right handle of the bar.

So, the meaning of actions at level *D* with respect to an object is not in its transport but in much more variable and complex forms of changes in the environment. Lighting-up a match, shaving, doing laundry, cooking, and shoe shining – in everyday life; wire isolation, painting, welding, riveting, and polishing – during manufacturing, represent examples of the simplest actions with their outcomes extending beyond moving objects in space. Manufacturing processes are particularly clear examples in this respect and do not need further discussion. Tools, just like objects, can be viewed at levels below the object level (for example, the croquet mallet, the tennis racket, the hockey stick, etc.) and the new aspects introduced by the level of object actions are not in the fact of using the tool but in the mode and goal of this use.

There are comparably unique features of the *motor composition* of movements performed at the level of actions. First of all, these actions are characterized by large and specific *trainability*, their high *ability to automatize* and inherent to this level phenomena, addressed in the literature as high-level automatisms, object-related skills, knacks, Handfertigkeiten, etc. These skills or high-level automatisms are very numerous and inherent to the very broad range of everyday and professional actions. They are so unique in their features that there is a hard-to-resist feeling that they could represent, as a whole, a special coordination level, which plays a background role with respect to the level of meaningful chains-actions, a level located above all the levels described earlier. Clearly, all these skills *are never object actions themselves* because they do not contain a defining semantic component. This is also supported by the fact that, in cases of agnostic and ideation apraxia, skills can be fully preserved but, deprived of semantic control, lead to absurd outcomes. Comparably preserved automatisms can be seen during dementia-related disorders of mind, but with ridiculous results: An elderly demented patient can knit a stocking with twelve heels, a schizophrenic patient writes with slick handwriting meaningless letter perseverations, etc. In object-related actions, automatisms are not semantic elements but technical means and, by this very feature, they are proven to be organized somewhere below the object level.

The hypothesis that high-level automatisms form their own special level subordinate to the object level seems to be supported by their special localization in the cortex; of course, we imply localization in terms of location of injuries leading to their loss – they are destroyed during injuries to the so-called premotor cortical areas (see below). Their loss is what forms the so-called premotor syndrome. Nevertheless, this hypothesis is wrong, and the nature of all the high-level automatisms is completely different.

First of all, let us note that we have already encountered the phenomenon of automation earlier in the book during the description of the level of spatial field, and there we have established that this phenomenon consists of a gradual switch of a set of background components of a complex motor action to lower levels, where they can find most adequate sensory correction syntheses. There are no valid reasons to reject this mechanism for the formation of high-level object-based automatisms.

Further, meticulous analysis of various high-level automatisms (performed by us for a large number of cases with the help of the cyclogrammetric method) – everyday, professional, athletic, military, and calligraphic ones – demonstrates convincingly the crucial fact that none of them contain, neither in its coordination corrections nor in the motor composition, any phenomena that would go beyond those expected from the earlier described levels. None of these automatisms demonstrates a special sensory synthesis, whereas this would be the defining, crucial sign pointing at the existence and composition of a separate independent construction level. None of these high-level automatisms displays novel features of type and style of the coordination process, new peculiarities of variability, switching, transfer, interference, basically anything from the list of properties and features necessary for awarding a motor cohort the name of a "level."

Getting from negative arguments to positive ones, we have gotten convinced that every higher automatism necessarily has clear, unmistakable features of one of the already familiar to us construction levels, starting from level *C* and below. Careful analysis shows that all their coordination properties have features of either the spatial field (for example, the automatisms of hammer strike by the blacksmith, and filing or hacksawing movements by the locksmith), or the level of synergies (for example, movement of knitting needles, violinist's bow, etc.), or even of the rubro-spinal level (grasp of the handle of a tool or machine, rolling pills, etc.). All the criteria we can use at this time to define assignment to a level, the primary being the defining sensory synthesis, lead in all cases to assigning these motor components of object-related actions to one of the earlier, pre-object, levels. In more complex automatisms, one can find a whole hierarchy of involved levels with their own background structures, so to say of a secondary order.

As a rule, the object level does not find such skills or automatisms in a ready to use form in the inventory of lower levels and has to develop them specifically for the learned actions. A characteristic feature of such automatisms is that they are *acquired*. The process of their elaboration has a number of specific features shedding light on their central nervous structure. A motor element of a chain object-related action at the first attempt is built completely of discrete conscious voluntary components, most frequently realized via the pyramidal effector system. At this stage (in a complete novice), the newly developed automatism resembles very much fragments of a similar automatism in a patient with apraxia due to an injury to the premotor cortex: The movement is disjointed, helpless, and forces itself forward on the crutches of the active, conscious attention, frequently accompanied by broad irradiations of effector excitation, useless synkinesias, etc. If the effector tools of the level of actions function properly, gradually cortical afferent systems start to get replaced by afferent systems of lower levels; slowly, extrapyramidal components increase their participation; the movement starts to show components that are unconscious, involuntary, and more adequate. Active attention and the leading level are unloaded and take care only of essential movement features that require its specific afferent systems. The movement *gets automated*.

This process of the active elaboration of automatisms within one's life points at the fact that, although, by itself, the automatism does not contain much beyond components of the spatial field and lower levels and, after being developed, it is realized completely based on those subordinate levels, *neither these levels nor their afferent systems contain motives for the formation and realization* (ecphoria[5]) of the motor and coordination combinations, which we assign to the class of high-level automatisms. These automatisms control all the technical elements of coordination based on some of the lower levels, but the emergence, development, and refinement of these combinations, as well as their recruitment as needed, can only be done along the motives and reasons originating at the level of actions.[6] What could be the reasons to perform certain accurate movements by the needle, plane, chisel, micrometer of the microscope, and by the pen writing these words, if these movements, at some point, would become self-sufficient, completed only in their motor composition, but not in their semantic structure and essence? For example, the technique of hammering a nail into the board is a typical ballistic, targeted movement at the level of spatial field with all its auxiliary backgrounds from the level of synergies (inter-joint coordination, compensation of the reactive forces, etc.), and its main features – force of the strike and accuracy – are corrected by the afferent systems of the spatial field. At the same time, the leading coordinating control of these actions is performed at the level of object-related actions making sense of the whole process as a whole and solving the main meaningful problem: "the nail has to be put into the board." Indeed, this very level defines whether the next strike should be strong or weak, straight or at an angle, or one should strike the side of the bent nail, or to stop striking when the nail got into the board sufficiently deep. This level, and only this level, contains *motives* for the development of the skill of striking with the hammer, which makes sense only with respect to a certain task.

At the beginning of this chapter, we stated with respect to high-level automatisms that the cohort of human movements at the level of spatial field is much richer than in any animals because the animals have no *motives* for the formation of such cohorts, although they possess numerous appropriate coordination abilities as demonstrated by the results of their training. To avoid browsing through the circus programs, recollect, for example, the Elberfeld horses or talking parrots. "A hare, if you beat it up long enough, can learn how to light matches," as mentioned by one of the heroes of Chekhov's stories. The point is that humans light matches for different reasons.

Analysis of high-level automatisms reveals the physiological role of *premotor systems*; their disorders lead to general deautomation of movements at level *D* and to the loss of the ability to use ecphoria and to form new high-level automatisms. Obviously, in normal conditions, premotor centers act as kind of *intermediaries* (one would like to call them "ecphorators" if not for the ugliness of attaching the Latin suffix to the Greek root), establishing and maintaining links between the cortical constellations of the level of actions and lower construction levels. This conclusion is supported by the fact of abundant pathways with well-studied projections from the premotor cortical fields to the pyramidal fields and the pallidum, as well as to the frontal lobes of the hemispheres, which form the connection between the premotor system and the powerful frontal-cerebellar tract. The transmission realized by the premotor systems does not terminate on lower effectors but on construction levels as integral formations with their afferent, central processing, and efferent components. It remains unclear what kind of impulses are sent by the premotor fields, but it is clear that these fields play, in the system of the level of actions, a very special role of a particular kind. Their disorder immediately makes the system of the level of actions unable to connect to peripheral effectors along all its pathways with the exception of a few its own extremely abstract corrections, which have been discussed earlier.

Now it is time to return to the issue of variability in the motor composition of actions and analyze it from a new viewpoint. Earlier, during the analysis of the semantic structure of actions, we had to stress its very profound variability and inter-compensation of elements in its motor composition, in combining separate motor elements into the chain; figuratively speaking, constructing words from elements – letters. Now we have to note variability in elements of the motor composition themselves – in letters themselves. Not only there are action-to-action differences in the number and sequence of stitches by the needle, or scalpel movements by the surgeon, or drill turns in the hands of the machinist, but movements of the hand with the needle, scalpel, scythe, file, drill, etc. never repeat each other perfectly. We discover that individual motor elements (it does not matter, simultaneous or successive) differ from each other very strongly in the degree of their intrinsic variability. One can encounter, side by side, two chain elements, two automatisms, one of them as stable as facial features are, and the other as variable as facial expressions are. Looking at both intently, we can easily establish the cause of this, so to say, "variable variability." Each of these backgrounds of a complex chain action *is variable typically at the level where it is constructed*. Hence, automatisms representing coordinated combinations at the level of synergies are least variable; automatisms organized at the level of spatial field possess the maximal ability to switch and "plasticity," which is always inherent to this level, etc.

In Chapter 5, we have already mentioned the feature of "specific variability" characterizing different construction levels – the feature reflecting both quantitative and, particularly, qualitative characteristics of the observed variations. This feature can be used successfully not only to define the level of independent movements but also to assess the structure and multi-level composition of backgrounds of "higher automatisms" at the level of actions.

The second typical feature of variability of the motor composition of actions is related to the clearly seen differences between the *right* and *left* hand, with the existence of a dominant (leading) and non-dominant side of the body, etc. Neither movements at the level of spatial field nor movements at even lower levels show visible side differences. At the level of synergies, the links of movements and corrections to the practiced extremities is so strong that it is hard to notice functional differences between the right and left hand: They are totally masked by the unmovable stereotypies of this level. At the level of spatial field, however, the left hand in all cases easily replaces the right one, and the differences in agility, accuracy, and speed between the two hands are usually miniscule. It is sufficient to recall the work of both hands during playing piano, harp, etc. This is, of course, directly related to the fact that the inequality between the right and left side of the body starts from substrates of the level of actions (*D*). At this level, the dominant hemisphere is defined, as a rule the one containing *speech centers*, and one can witness noticeable lagging of one body side behind the other with respect to dexterity and learning new skills (which has secondary, reflected effects on the motor apparata of lower levels). Clearly, one should pay attention to the non-random fact that none of the geometric configurations that can be perceived and graphically reproduced at the level of *spatial field* show right-left differences, for example between right-side and left-side mutually symmetrical ornamental figures. It is very easy to remember and recognize later an asymmetrical geometrical figure presented for just a few seconds, but it is very hard to remember its right or left orientation and then recognize it among similar figures if this list contains both right-hand and left-hand versions. At the same time, symbols perceived at the *level of actions* – letters, numbers, and conditional signs are remembered and reproduced with accurate distinction between the right and left versions and, with the exception of children who are inclined to write letters mirror-like exactly because of the under-developed level of actions (Chapter 7), everybody can, without hesitation, to distinguish meaningful signs such as И and N, Я and R, з and ε, б and δ, s and z, etc.

Let us now turn from normal functioning to pathologies, to analysis of clinical syndromes called apraxias and dyspraxias, which have been the main sources supporting the existence of positive features at the level of actions. After all the presented material, definition and systematization of these syndromes should be easy. Of course, we will only describe clinical pictures as far as they help to understand the main messages of this text.

Based on the introduced notions of semantic structure and motor composition of motor actions, one can state very briefly and rather accurately that localized disease or trauma in the *lower-parietal parts of the hemispheres* (more pronounced for injuries to the dominant, typically left, hemisphere) cause disorders of the *semantic structure* of complex motor actions, whereas *disorders of premotor fields cause problems with the motor composition*.

Lower-parietal syndromes (subtle nuances are distinguished among the so-called agnostic, amnestic, and ideation forms) may be united under the common label of *sensory or gnostic apraxias*, thus emphasizing their peculiar afferent-based features; these disorders deserve to be termed

Liepmann's apraxias after the name of the author who studied them in detail first. In all these variations, the semantic structure of object-related actions is affected. In both local and functional proximity, apraxias of this category are frequently accompanied by visual and auditory agnosia, dyslexia, sensory aphasia, etc. Movements selectively affected in such disorders commonly represent complex meaningful chains, successive movement sequences united by a common meaning and goal of the motor problem being solved. That is why, disorders of the semantic structure, typical of Liepmann's apraxias, usually represent semantic distortions of such motor chains. This is accompanied either by total disintegration of the overall movement scheme, or the scheme is kept unchanged in its main contours, but its realization faces various problems. Individual links may fall out of the chain thus rendering the whole chain meaningless or, vice versa, irrelevant links get inserted, or there are violations of the meaningful link order while the links themselves are correct, or the chain stops before its completion, or even slip onto an adjacent chain, frequently not because of their confusing similarity but due to perseveration of elements of the original chain. Naturally, during Liepmann's syndrome, complex semantic chains typically suffer more and show earlier disorders compared to simpler elementary object actions such as dressing up and taking care of oneself, which show milder disorders. Movements of patients suffering from these forms of apraxia are neither discoordinated nor amorphous; they are messed-up and inadequate for the meaning of the action. Violation of the main plan of the action rather than of execution technique of its details is supported best by the fact that *imitating* chain actions performed in front of the patient by another person is easier for the patient than performing such actions spontaneously (self-initiated), because during imitation the patient receives prompts with respect to the nomenclature and sequence of motor links of the chain, and the realization of each of the links by itself is not violated because the motor composition of actions is intact. Disorders of the semantic structure may reveal themselves in more subtle forms reflecting the underlying disorders of the meaningful perception of *space* or problems with conceptualization of the object as a tool or an object for manipulation (Figure 6.7). This already manifests the transition to agnosias, which have no sharp border with the actually observed clinical pictures (not bare schemes) of Liepmann's apraxias.

The tight links of the level of actions to background levels and their easy involvement into subordinate obedience in healthy persons allow one more classification of sensory apraxias introduced by A. Luria. This classification is based on backgrounds supporting specific motor actions that primarily and selectively suffer in particular disorders, and emphasize that, in different cases (maybe due to so far unknown localization nuances), there are selective disorders of the relations between substrates of the level of actions and substrates of various lower levels. Disorders of links to the level of synergies result in what Luria addresses as "postural apraxia," disorders of links to the level of spatial field – "spatial apraxia," whereas some essential disorders in substrates of the level of actions itself – apraxia in the narrow meaning of this word, characterized primarily by disorders of symbolic or linked to gnosis phenomena at this level. This classification is very promising, although it still needs further elaboration.

The second group of apraxic disorders, with injuries localized in *premotor fields*, also includes a few versions, which can be united under the common name of *kinetic or premotor apraxias*, or, in analogy to Liepmann's apraxias, – under the name of Kleist's apraxia.[7] In contrast to the previously described disorders, these syndromes represent problems in the *time course of the motor composition of action* while the basis of its semantic structure is preserved. Similar to cases of sensory apraxia, premotor patients show disorders of mechanisms of realization of actions; the bridge connecting the (intact) understanding of the task with its solution has been blown up but in a different span. According to Kleist, in contrast to sensory apraxias, the sequence of links of a particular action is intact, but the execution of individual links happens to be defective; as a result,

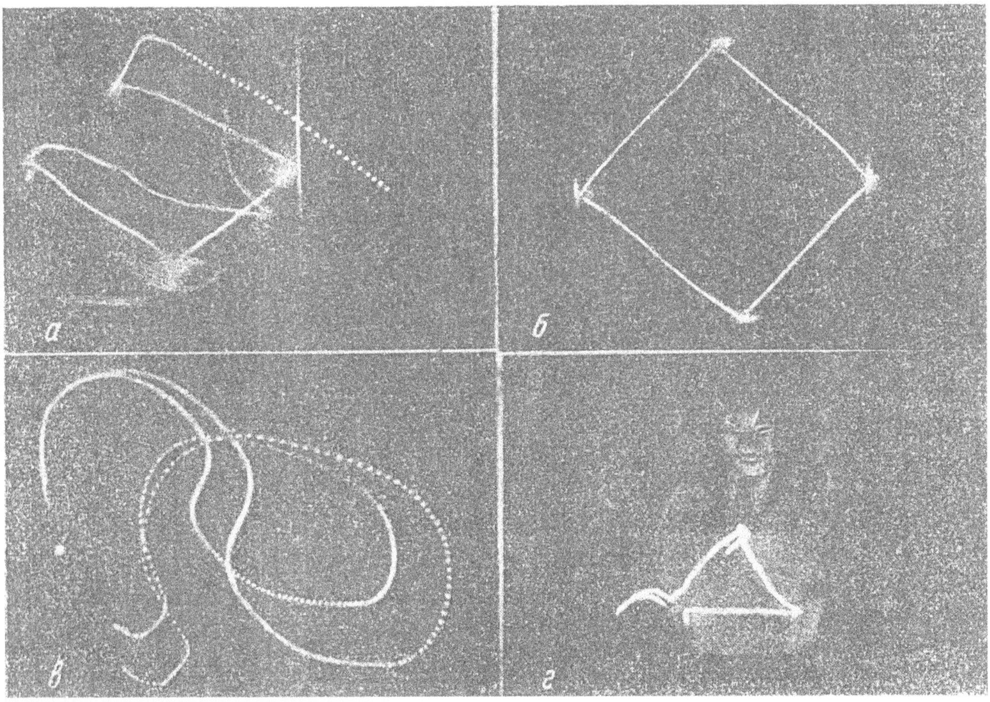

FIGURE 6.7 Drawing a slanted rectangle inside the given straight one.
а – by a patient with disorder of afferent systems of the object level as a result of local disease in the left parietal lobe; б – normal subject; в – drawing by the same patient of a horizontal figure eight with the finger. Both movements display clear coordination disorders pointing at a disorder of the semantic spatial synthesis at the object level. The intact level of spatial field is proven in this patient by the drawing г – tracing a triangle drawn on paper, as well as by tests on ataxia performed during the same experiment (Figure 5.26).

one witnesses "coarsening and crippling of the movement form recognizable by its contours" (Kleist). As mentioned by this author, the disorder is revealed differently for different movement types. In particular, complex meaningful object actions suffer *less* (in ideation apraxia, they suffer most) because the general contour or plan of the movement remains clear for the patient, and he retains the ability to follow a compensatory bypass for disordered specific coordinations led by the meaning and goal of the action.

A patient with a premotor disorder can plan an action properly with respect to its motor composition and even to sketch and write movement scores, which has been mentioned as a metaphor, but he or she becomes helpless with the implementation of its elements. Even superficial observation reveals that such movements are clumsy, slow-witted, and lack natural spontaneity. The patient kind of recites them, reads his movements syllable by syllable. Psychologically passive elements, which alternate in any movement with active elements: lowering an arm after it has been lifted, releasing a button after is has been pressed, relaxing after contraction, etc. – all these elements are equally active and requires focused attention and voluntary effort. According to the characteristic by Luria, the general, system character of movement is affected, which is responsible for uniting individual motor elements automatically into the overall pattern of the action movement composition. The "dynamic background" of the movement – typical, inherent to normal movements, sequences of active and passive elements – is violated and disappear. A single, complete action movement, including both active and passive parts, turns into a sequence of separate, only active, motions; it turns from a smooth movement into a sequence of saccade-like pushes resembling a gymnastic exercise performed following commands. A single "kinetic melody" of a habitual movement, which normally proceeds automatically, turns into a series of separate, deautomated actions. A scheme is replaced with the sum.

Disorders of the time pattern of motor composition of action are reflected particularly vividly in the main sign of premotor apraxia, which may be its best diagnostic sign: The profound disintegration of motor skills, i.e. the *deautomation* of semantic or object-related chains. As noted already by Hughlings Jackson, apraxia of this kind leads to loss not of elemental, subcortical automatisms (i.e., there is no deautomation of movements led by lower construction levels), but of highest, cortical automatisms, i.e., the mechanisms of realization of complex actions according to a unifying dynamical scheme.

Everything performed in everyday life by a healthy person automatically, by itself, subconsciously, and skillfully: buttoning the shirt, lighting up the match, combing one's hair, etc. is dramatically deautomated in premotor disorders, becomes clumsy and inept, as if performed by a child for the first time. More subtle professional and artistic skills are lost hopelessly, and the most dedicated, focused exercise over many months is unable to recover even the simplest of those skills. Deeper analysis shows where exactly the chain is broken in such patients. Patients with premotor apraxia lose the opportunity to put into action the aforementioned scores of movement components assigned to background levels: This is, undoubtfully, the core of the overall deautomation (Figures 6.8 and 6.9).

According to the described concept, automation represents switching of some of the structural components of a motor action to lower, at the moment not consciously perceived, levels, which is

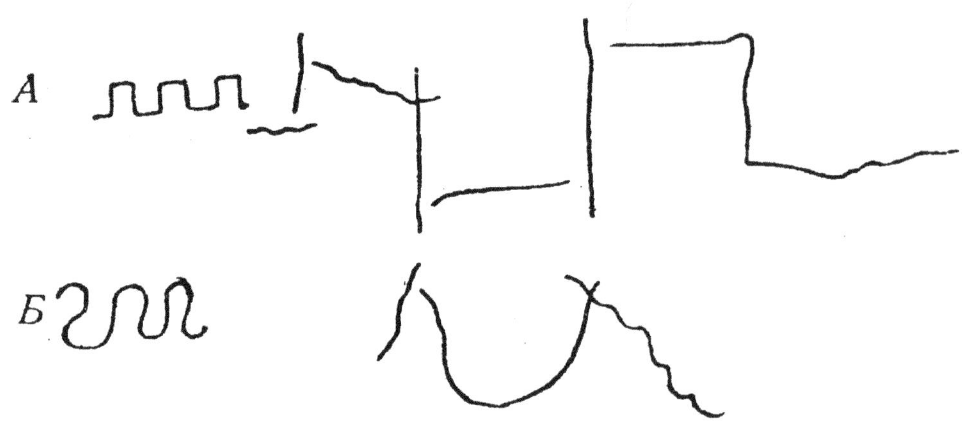

FIGURE 6.8 The problem of reproducing successive structures: Reproduction with the pencil of curvy figures with the samples placed to the left.

A patient suffering from a meningioma in the left premotor zone; the experiment was performed on the fourth day after the surgical removal of the meningioma. "A typical feature of the drawings is that the patient cannot view them as a single dynamical structure. For example, sample *A* turned into the sum of isolated movements used by the patient to build the graphical structure; the same torn apart movement feature and impossibility to master a kinetic melody is particularly well seen in sample *Б*, which turns into a sequence of disjointed strokes. On the twentieth day after the surgery, difficulties of mastering the kinetic melody are still seen very vividly.

(Source: A. Luria. *Research Notes of the Department of Psychology*, Moscow State University, 1945).

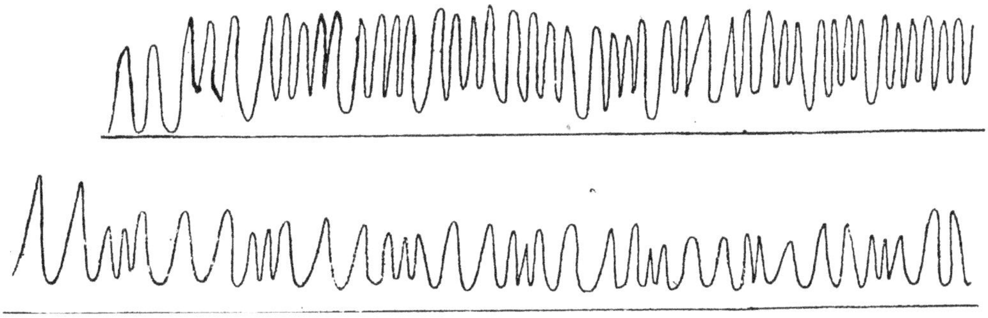

FIGURE 6.9 The performance of a rhythmical test by a subject with a nonpenetrating wound by a shell fragment in the right, and also partly left, premotor area, parasagittally to the posterior portion of the upper frontal gyrus.
Top line – The right hand, 5 months after the injury (soon after the surgery to fix the nonhealing fistula). As reported by the patient himself, typical of the completely preserved self-judgment by such patients, "somehow, it does not work, although it is very simple and understandable; it is harder to perform faster." Low line – The same right hand following the lengthy, fruitless exercise (about three months); accurate accents were only possible with the help of the optimal scheme: II."
(Source: E. Shkolnik-Yarros, Monograph, *Movement Disorders during Injuries to the Premotor Zone*, Institute of Neurology of the Academy of Medical Sciences, 1940).

associated with switching these components to different afferent systems. Therefore, *deautomation* is loss of the possibility of using backgrounds at a lower level, to which some of the movement components were switched. Such a loss may be induced by *disintegration of the lower level itself* or by *violation of the functional links* between anatomical substrates of the leading level and the lower level. Both forms of deautomation may be also observed with respect to levels lower than the level of actions.

A good example of the first form of deautomation is the disintegration of fast handwriting in patients with Parkinson's disease. The origin of the disintegration – the dissolution of substrates at the level of synergies – leads to the loss of the already mentioned main vibration synergy of handwriting; the patients need to compensate for this lost synergy at the level of spatial field, which is preserved. It is very interesting to watch how a patient with Parkinson's disease and strong tremor of both hands, barely capable of writing with the shaky handwriting his or her own name, immediately following this test draws with the steady hand an instructed circle or cross. The handwriting, acquired by the patient long time ago with its structural, background participation of the level of synergies, deautomated and fell apart with the disintegration of that level, whereas drawing a circle, which has never been automated in a similar fashion, remains completely unaffected. It is interesting that, in patients with Parkinson's disease, typical pyramidal innervations suppress their inherent tremor of the extremities much stronger than those originating from the extrapyramidal system.

The second variety of deautomation has already been mentioned with respect to Parkinson's disease: We imply here the typical for this disorder syndrome, which leads to loss of the ability to walk spontaneously over a surface without markings. Earlier (Chapter 5), it has been noted that this syndrome is associated with deautomation of gait caused not by disintegration of the level of synergies but by disorders of mechanisms of *recoding*, which perform in healthy persons translation from the aperiodic language of the level of spatial field to the periodic or cyclical language of the level of synergies. Deautomations typical of the premotor Kleist apraxia should be assigned exactly to this second type – disorders of mechanisms of communication or recoding.

Indeed, during this syndrome, no movements from lower levels are lost including, in particular, the ability to perform voluntary movements; as a result, no effector instrument is destroyed or turned off. Meanwhile, movement disorders typical of this type of apraxia have a pronounced effector, not afferent, character. Obviously, the effector character of a motor disorder, given the intact anatomical effector links to all the lower construction levels, can only be expressed in phenomena of *deautomation* not as a result of disorders of lower levels, but resulting from the *lost ability to control by the level of actions*. This is accompanied by losses of only the background, automated use, whereas these effectors continue to be adequately controlled as reflected in the afferent systems of lower levels. It is possible to refine characteristics of deautomations in the Kleist syndrome by pointing out that the strongest disorders are seen in the automatic links with the level of synergies, milder disorder – with the striatal sub-level (*C1*),[8] and minimal disorders – with the pyramidal sub-level of the spatial field (*C2*). This is a consequence of two factors. First, during transition from the mildest forms of premotor injuries to the most extensive ones, deautomation is typically seen in the mentioned order. Second, this is supported by observations of the character and level structure of the lost automatisms.

As it is easy to imagine at this point, a patient with a premotor cortical injury potentially may keep all the capabilities to plan the motor composition for an action (that is why, in contrast to a patient with ideomotor apraxia, the patient with a premotor disorder would not confuse the order and grouping of links of a motor chain), but the means to realize the planned background pattern are paralyzed. First, this condition deprives the patient of *all the lower recodings*, which ensure in healthy person movement metrics, rhythm, the alteration of innervations and denervations, etc.;[9] this is the reason for the tense and deliberate movements, reading by syllables, as it was expressed metaphorically earlier. Second, the patient loses an essential feature of automatisms – their unconscious control – leading to every minute detail of the movement requiring focused attention and a separate volitional act. Third, this kills all the accumulated over the lifetime skills and knacks; it seems likely that the corresponding coordination forms are still intact but their availability for ecphoria is gone together with the ability to learn new skills. It looks very much depressing when an intellectually healthy person, after two-three months of dedicated daily training, it unable to master the skill of simple rhythmical tapping with a finger, for example of "spondee-anapest." Maybe it would not be necessary to mention this fact if not for the preservation of all the lower levels in such a patient that promise preservation of all skills: The briefcase with all the important papers is safe, but the key has been lost. This metaphor promises hope for treatment; maybe, in not so distant future, we would be able to either find the original key or to make a new one.

Among many motor actions performed by a healthy person at the level of actions (*D*), it is not easy to find convincing principles for classification. The variety of both semantic structures and motor compositions, of essential working points and types of external kinetic appearance is so immense that it is impossible to organize all actions at this level into a table similar to the one presented at the end of the description of the level of spatial field. The most convenient way to introduce some kind of order into the exceptional variety of actions is offered by the typical feature of level *D*: richness of very different and frequently complex *background structures*,[10] which have been described in much detail in the presentation of both healthy movement and disorders at this level. Without claiming completeness or perfect accuracy, I would end this chapter with an essay on types of actions classified according to the mentioned leading feature. In each group, only a handful (selected at random) of examples of corresponding actions will be presented from the following domains: (a) professional; (b) everyday; and (c) athletic.

FIGURE 6.10

Group I. Motor actions with *relatively little participation of technical backgrounds* (higher automatisms). This group includes object-oriented movements, such as feeling, ogling, trying, comparing, selecting, etc. The same group should include various object-related actions by a novice and detour (compensatory) actions in premotor apraxia. Further, the same group will include many chains, which, similar to the corresponding group from level C, could be called topological manipulation of space: drawing on paper a required figure or a sign; schematic drawing, which can be done by all those who cannot draw; orienting an object, turning it upside down, pouring in and out, opening latches and boxes, etc. The same group should probably include elementary actions (e.g., tapping) using complex rhythms. This "all-human" group of object-related actions, not including special skills, naturally presents handy material for diagnostic probes of patients.

Group II. Actions with *domination of backgrounds from the upper sub-level of spatial field C2.* This group includes actions, technical backgrounds, and skills tightly linked to movement accuracy and metrics, with important role played by visual control, such as drafting, engraving, assembly of mechanisms, manipulation of precise instruments (counting ruler, micrometer, rater of precise scales, etc.), actions by a turner, surgeon, optician, cutter, pharmacist, watch maker, etc. From everyday movements, this group includes threading the needle, dropping the medicine, and sharpening the pencil. There seem to be no athletic movements characteristic of this group. Game actions include those with teasers such as balls in a glass-covered box, which have to be moved to their places, cardboard houses, poorly stable objects that have to be put in equilibrium, playing spillikins, etc.

Group III. Actions with *domination of backgrounds from the lower sub-level of spatial field C1.* From professional actions: movements of the car coupler, driver, steam locomotive machinist, filer, sander, blacksmith, furniture upholsterer, laundress, etc. From everyday actions: ironing, combing hair, shaving, tying shoelaces, rolling dough, leafing book pages, turning on electrical appliances, lighting-up a cigarette, etc. From gymnastic and game movements: climbing a rope ladder or tree, balancing objects in unstable equilibrium, playing "diabolo," etc. All these actions contain, to a larger or lesser degree, secondary backgrounds ("second plan") from the level of synergies.

Group IV. Actions with domination of backgrounds from the *level of synergies B.* From professional actions: work of the cropper, smith, harvest binder, digger, and spinner; actions associated with turning the handle (of the hoist, winch, hand-operated typesetter, etc.); and many assembly line operations. From everyday movements: knitting, tying knots, spinning thread, soaping and washing the body, braiding, and dressing-up. Athletic and game actions: French wrestling, judo; to some degree – pole vault; and playing with a rope loop, making different patterns by taking it off the hands of the partner. Further – since science is not subjected to everyday morale – sharper tricks, similar slight-of-hand stunts, and many tricks by pickpockets, where the coordination reaches (unfortunately) very high levels of sophistication.

Group V. Actions with necessary participation of backgrounds from *both the level of spatial field and the level of synergies.* From professional movements, first of all, writing and speaking.[11] Further, worker operations on the rolling mill, sailor's actions on sailboats, and other similar actions inherent to so-called dangerous professions; actions by sawing machine operator, leather and cloth cutter, typesetter, typography folding, and many of the highly developed labor skills used by the masters of Stakhanov's labor movement. From everyday processes: sawing, quilting, wood-cutting, skinning fruits and vegetables, etc. From athletic and game actions: fencing, bayonet fight, archery, throwing complex hunting tools such as the harpoon, boomerang, lasso, etc.; many of the actions during mountaineering, downhill skiing, fancy horse riding, etc.; rowing, assisting in parterre acrobatics and ballet, etc.

Group VI. Actions dominated by backgrounds from the rubro-spinal level *A*: pill rolling, some of the movements during massage, vibrato of the left hand of a violinist, using a hand-held fan, and a number of professional fixations and grasps.

Levels Above the Level of Actions (Group E)

The general characteristics of essential movement features and action at level *D* described in this chapter show clearly that not all the high-level intelligent actions can find a place at that level. For example, the level of actions has no place for symbolic or conditioned meaningful actions. This group includes, first of all, not technical performance-related, but semantically leading coordinations of *speech and writing*: movement chains united not by an object but by a mnestic scheme, abstract task or thought, etc. Examples include artful performance, musical or choreographic; movement imitating object-oriented actions but in the absence of the actual object; and object-related actions when the object is not the direct target of the action but an auxiliary means of illustrating with its help abstract, non-object relations. The existence of such movements and actions demonstrates convincingly the existence, in the inventory of human coordinations, one or more levels, hierarchically higher than level *D*.

It is necessary to note that the presence in humans of motives and psychological conditions for actions, which elevate the actions way above concrete, elementary object manipulations, is beyond doubt. The difficulty is to explore whether, and if so then by how much, such differences in the motive and psychological conditioning of actions affect the *external, coordination pattern* and movement corrections, which is the only topic of the current discourse. When an animal runs because it has to cover a certain distance (sub-level *C1*) vs. it runs to grab a fruit or a pray (background *C1* to the leading level *C2*), the differences in the construction and sensory corrections and in the resulting movements are beyond doubt. But when a man strikes another man with a dagger for self-defense or during robbery (level *D*), we have no strong reasons to expect a different coordination pattern when the subject of this action is Damon, Karl Sand, or Charlotte Corday. First of all, we have to analyze the motor composition of such movements, where we suspect presence of higher coordination levels.

Analysis of some particularly complex and intellectually loaded behavioral acts, such as writing and speech, establishes in them the presence of a large number of hierarchically organized layers or, which is basically the same, the presence of a hierarchy of layers of coordination recodings, more numerous compared to the number of layers we have counted so far, up to the object-related level. For example, during writing, we see the level of synergies providing the main cyclical synergy of handwriting; the level of spatial field *C* ensuring adaptation of the pen movement to the paper surface and preserving geometrical peculiarities of the individual handwriting, whereas allowing plastic variations in size of the letters, location of the paper sheet, posture of the actor, etc.; finally, the level of actions *D* defines topological features of the handwriting, serves as the supreme leader of the high-level handwriting automatism, and ensures proper alphabetical contour of the letters (we addressed this as modulation of the oscillatory handwriting synergy at level *B*). It is easy to see that, over all these levels or recodings, at least two more coordination recodings remain, which have so far failed to find proper place in the mentioned construction levels. First, if we move in the bottom-up direction, there is phonetical and grammatical recoding (one or even two processes, one subordinate to the other), i.e., translation of a phonetic image of speech sound to the language of alphabetical writing, and translation of a phonetic image of a word to the language of grammatically correct letter selection (spelling): writing "Worchester" when you hear "Uuste," etc. Second, there has to be semantic recoding, i.e. converting the gist of a thought or phrase in a familiar, but not mother, language or an uttering, from which you memorized only the general meaning, etc., into a phonetic or graphic image of words, which you plan to write down. An even more distinct example is writing numbers, where over a phonetic recoding ("three" = 3, "two hundred" = 200) there is also am arithmetical-grammatical recoding ("three hundred and seven" = 307, but not 300–7; "einundzwanzig" = 21, but not 1 – 20; "quatre-vingt-dix-huit" = 98, but not 4 – 20 – 10 – 8, etc.) Under each of such hierarchical recodings, one can guess the existence of its own special construction level. Finally, pathological signs, particularly perseveration (Chapter 9) also point at separate levels above level *D*, and each of those levels is revealed in pathological cases by its own, differently constructed perseveration. The following example, based on the already analyzed levels, may be able to clarify the previous statement. A patient, perseverating at levels *B* or *C1*, given the task of drawing a circle, cannot stop after drawing the first contour and continues to draw an endless skein at the same place or a corkscrew-like spiral (see Figure 9.3). If the perseveration is due to an injury at the level of actions, the same tasks brings about drawing a whole family of separate circles, each circle not looking pathological by itself, but together they can fill the whole sheet of paper. Different levels from those discussed earlier produced rather different perseverations in response to the same task.

Along similar lines, a patient with a disorder at the object-related level *D*, able to follow an instruction and draw, for example, figure 8, but prone to persevere, can reproduce the instructed figure as a skein of 8s in the same place (perseveration in higher automatisms of the level of actions) or an endless sequence of 8s: 8 8 8 8 8 8 8 8 … (perseveration in the semantic scheme of level *D* itself). The same patient, given the task of writing 120, writes 122222 …, i.e., falling into perseveration of the later type already at the second link of the correctly initiated action; another patient, given the same task, shows the following perseveration: 120 120 120 … Clearly, in the last pair of observations, the transition from the corkscrew-like perseveration of drawing a circle to an endless series of perfect circles is analogous to the mentioned transition from 122222 … to 120 120 120 …, and, if in the former example, this transition was linked to elevating perseveration to the next highest level, we can expect a similar phenomenon here. We witness a more complex and higher-level type of perseveration, clearly suggesting the involvement of recoding at a level higher than *D*. The same is probably true with respect to a patient who performs the task of writing "120" in the following way: "10020," i.e., in the absence of phenomena of perseveration, this patient shows disintegration in an area where, normally, an arithmetical-grammatical recoding takes place, and this observation corroborates the existence of such an area.

When suggested to draw a house,[12] the patient with perseverance at level *D* either draws a typical schematic picture many times in the same place or a whole street of schematic houses. However, what level of perseveration should be invoked in a patient who performs this task in the following way? First, he draws the roof in the shape of letter "Д," and under the roof an entangled skein of lines clearly showing, however, under the Д-shaped roof, first, circular, O-shaped lines, and then – zigzagged M-shaped lines. This is already not a scheme of a house at level *D*, but some kind of a complex mixture of a schematic drawing, ideographic hieroglyph, and a written word "ДОМ" suggesting that there is at least one more level over level *D*, where object schemes and verbal and written inscriptions meet. Beyond doubt, historically, hieroglyphs of both the Egyptians and the Chinese emerged not as a result of purely intellectually invented conventional symbols, but as merged syncretic mentality of a more primitive type, which, at that time, could reveal itself in corresponding synthetical graphic coordination in healthy persons, and, in our times, it surfaces here and there in pathological cases, like other types of primitive thinking and maybe even of the motor function.

All these facts – the existence of whole motor actions, which do not fit the framework of level *D*, multi-level recodings seen in healthy persons, and multi-level losses and perseverations seen in pathology – speak in favor of at least one more level, dominating over level *D* and, even more likely, a number of such levels. The lack of material is so glaring that the only correct conclusion at present is to unite provisionally all the possible higher levels into a single group *E* because, even under this simplification, we can characterize them only in very general terms. At present, it is impossible to identify for this group both its leading afferent systems and cortical localizations (with the only obviously essential for its effector apparatus frontal cortical lobes, in particular fields 9 and 10 after Brodmann).

First of all, we have to justify the statement that group *E* represents truly a coordination level, not simply psychological superstructures, i.e., that motor actions from this group do not represent summation of movements completely controlled and coordinated by lower levels and only linked to each other by psychological motives of a new type, but are indeed holistic coordinations with special qualities. In spite of the lack of experimental material and related difficulties with finding valid reasons for this statement, it is still possible to formulate a few arguments in its favor.

The first argument stems from understanding the structure of movements at the level of actions and functions of the premotor system, which result from the presented analysis of that level. This analysis has proven that it is possible to control coordinated motor actions "top-down" and allowed establishing that higher automatisms at that level are not summations of actions at levels *B* and *C* but represent qualitatively different coordination combinations controlled via specific directives of the object-related level, via its own effector outlet – the premotor fields. These automated components and backgrounds of the object-related level, these "higher automatisms," proceed due to their automation below the level of consciousness, which always resides in the leading at that moment level. It is natural to conclude that, if we encounter a whole object-related action or a chain of such actions, which proceed in an automatic, subconscious manner leading to a meaningful outcome elevated above the capabilities of the object-related level itself, then we would witness manifestations of an analogous coordination process, localized one step above processes at the level of actions. Such facts exist indeed. First of all, they include actions of speech and writing. As already shown in the previous section, speech process represents a coordination at the level of actions with technical backgrounds from all the lower levels. This is also proven by the close clinical affinity between motor aphasias and apraxias of premotor origin; by the proximity, local and innervational, between the premotor cortical fields and the speech Broca field; by the schematic-topological character of motor speech phenomena and the presence in those phenomena features analogous to those of handwriting – pronunciation or accent, i.e., qualitative features that are not changed during changes in metric speech characteristics (loudness, speed, and voice pitch); and by the pronounced monopoly of the semantic relations with the object at some of the early stages of the ontogenetic development of speech. *Naming* an object, similar to *writing* a letter or *copying* a word, is based at the level of object-related action *D*. When we encounter such coordinations, fully attributable to the object-related level in the subordinate, secondary role during automated or unconscious processes in chain syntheses that cannot be fully motivated by the object-related level, i.e., we encounter *meaningful, coherent speech or similar writing* – we have good reasons to admit that their control mechanisms represent a special coordination level in the most exact meaning of this term. The analogy between the speech motor process and higher automatisms is indeed very strong and, whereas its detailed analysis is beyond the scope of this book, one of its essential features has to be mentioned. Earlier, it has been established that, whereas movements used for the construction of automatisms at the level of actions are always coordinated below that level, these automatisms nevertheless represent movement forms and combinations, which would have never emerged at their lower levels by themselves, without the directive control from above, because of the complete lack, at those lower levels, of motives to form such movement phenomena. Exactly in the same way, although there are sufficient motives at the object-related level for the emergence of *verbal naming* of a concrete experienced object, there are no and cannot be motives at that level for the emergence of higher-organized *semantic* (verbal) forms – verbs, numerals, conjunctions, etc., as well as of higher grammatical forms (declinations, cases, and syntactic speech organization). Therefore, the control of speech, starting from the moment it transfers from level *D* to a higher-level group *E*, cannot be reduced to coupling or threading of already existing (factually or potentially) verbal forms at the object-related level; new forms, both grammatical and semantical, which involve the same motor speech mechanisms as naming of specific perceived objects, are linked with respect to their coordination to level *D* while being alien to it by their origin.[13]

Some of the forms of coordination during musical performance can also be viewed as similar cases of special coordinations at the level of actions emerging and proceeding unconsciously under the control of a higher level. First of all, such actions should include the *coordination of the bow*. Earlier, during the discussion of movements at the level of spatial field (Chapter 5), it has been pointed out that this level practically never participates in the construction of movements of the arm with the bow. In contrast, the level of actions is directly linked to manipulations with this peculiar tool, manipulation that cannot be reduced to the grasp alone or to motion of objects in space. Nevertheless, bow movements were not considered among movements at the level of actions. This is due to the fact that bow movements *are led* not by the level of actions, but by the higher group *E*. Motives to move the hair of the bow over the strings in a particular way cannot emerge at the level of meaningful object-based actions because this movement has no direct meaning related to the object. It is even more essential, and directly linked to our main definition of coordination, that level *D* does not possess *means for adequate sensory corrections* of such a movement; neither the artful sound, nor the expressive dynamics of a sequence of sounds, defined by the holistic artistic concept of the performer, are present in the afferent synthesis of the object-related level. Meanwhile, these very features define the control of all the sensory corrections by the violinist or cellist.

So, the general scheme of the construction of coordination of the bow arm of a violinist is as follows:

- *E* the leading level creating the motive for the motor action and performing its main meaningful correction – matching the sound to the intention.
- *D* manipulation of the object – a "super-high" automated skill.
- *C* does not participate.
- *B* main synergies (secondary backgrounds) supporting the "super-high" automatism at level *D*.
- *A* specific grasp.

It would be wrong to think that, for such construction schemes under the supreme control of the highest group *E*, background participation of level *D* is obligatory. It is essential, for such schemes, that group *E* occupies the leading position, but not a specific background composition. For example, hand movements by a pianist are built according to the following tentative scheme:

E the leading level (see above for its contribution).
D likely, does not participate.
C targeted forces and precise movements in the spatial field.
B background synergies: a) touché, linked to body posture and hand setting; and b) background components for level *C*.

Hence, first, the existence of automatisms, which are controlled by motives and corrections from levels higher than the object-related one and even not always requiring this level as an intermediary, clearly confirm that these higher levels do not only create special, purely psychological, superstructures to motivate movements, but also exert on these movements direct coordinating influence. Second, as mentioned earlier, these higher levels are linked to *recoding and pathological perseveration*, which do not follow the constraints of the lower levels. These observations also speak loudly in favor of actual construction levels with their own coordination mechanisms. Finally, third, the supreme group *E* is characterized by its own, characteristic losses. Disorders of group *E* have to be split into two classes (in close analogy with the level of actions *D*). The first class includes a group of clinical disorders united by Monakow some time ago under the name of asemic disorders: sensory aphasia, alexia, asymbolia, amusia, etc., corresponding to losses of meaningful speech, reading, vocabulary, ability to perceive music, etc. All these types of disorders are united by a common feature: loss, to some degree, of semantic (not object-related) motives and, therefore, approaching in their characteristics losses in the afferent field as in Liepmann's apraxias. The second class of disorders in group *E* produces a specific "frontal lobe syndrome" with its defining disconnect of behavior, loss of links between what has been done and what is to be done, disintegration of correspondence between the situation and action, etc., i.e., a syndrome with an effector picture of losses. Both classes of disorders are induced by brain injuries in areas, which differ in their location from injuries producing apraxic disorders and bring about motor disorders of types different from apraxias.

These are the arguments one can use at this time in favor of an independent system *E* as a separate coordination group.

We are not going to attempt classifying motor actions at the characterized supreme level group. The risk is too high to make an error by assigning to movements coordinated by this group, movements that are only motivated by it but with coordination built, most likely, on lower levels, including level *D*. With certainty, we can assign to coordinations by the supreme group: 1) *All varieties of speech and writing* (oral speech, sign language of the deaf-mute, using Morse code, flag signaling, etc.; hand-writing, typing, stenography, work on printing telegraph and typesetters, etc. 2) *Musical, theatrical and choreographic performance* – non multa, sed multum.

Notes

1 As shown in more detail in Chapter 7, the marked domination of the level of actions and its motor apparatus over other levels of movement construction in grown-up humans reveals itself in the overall advantage of development of the dominant side of the body over the non-dominant side: larger force, speed, etc. of the right hand compared to the left hand, which has secondary effects on movements at lower levels.

2 Artist I. Grabar shares his recollections on a very typical episode in the model drawing class in the Academy of Arts:

"Pavel Petrovich (a well-known artist-pedagogue P.P. Chistjakov) stubbornly refused to give advice to one of the students and avoided looking at his paintings. The student finally dared to ask why the instructor had never talked to him and got the following response:

– There is not much to explain. Do you have a servant in your house?

– Yes.

– Then you do not need anything else. Hire a model and ask your servant to stand behind you and repeat every five minutes: "Master, look at the model." This would be sufficient. You will learn immediately. At present, you draw from yourself, and do not even look at the model."

(Grabar I. My Life, Moscow, 1937, p. 101)

If such comments are possible in the model class of the Academy of Arts, during the domination of realistic trends in art, what can one say about simple mortals?

3 As a whole, a sport game is built at the level of actions; here, we refer to separate, successive episodes of actions with the ball.

4 I am grateful to Professor S.G. Gellerstein for telling me about this case.

5 We use this convenient term, introduced by Semon, which does not force us to share his views.

6 We witnessed similar interactions earlier in the example of interactions between the striatal sub-level and level of synergies during walking.

7 Kleist himself addresses the most characteristic form of premotor apraxia with a very hard to translate term "gliedkinetische Apraxia."

8 The first actions to show deautomation are dexterous movements, skillful manipulations, etc., i.e., the possibility to use background synergies is lost; further, "dynamical adequate schemes," i.e., movements of spatial manipulation (*C1*); finally, – all voluntary movements (pyramidal sub-level *C2*, so-called motor apraxia).

9 Among such low, typically unconscious, recodings, patients with premotor disorders frequently lose the so-called anticipation (Chapter 8), i.e. the ability of preparatory, foreseeing planning of one's own actions. If such a patient is required to walk with a chair in the middle of the path, the patient will not start walking at an angle from the very beginning to avoid the chair but will walk straight to the target and, only after coming very close to the chair, walk around it using sharp turns.

10 Probably, it does not have to be mentioned that all the backgrounds from level *A* are necessary for all movements performed at level *D*.

11 See more on this topic in Chapters 7 and 8.

12 The presented examples of patients have been borrowed from observations of A. Luria. The author is sincerely grateful for these examples.

13 One peculiar detail characterizes the difference between these "super-high" automatisms projecting from *E* to *D* and common ones representing projections from *D* to *C* and lower. The elaboration of an object-oriented automatism, i.e., automated transfer of coordination from the object-related level to the level of spatial field or to the level of synergies, is accompanied – due to the already discussed reasons – by *transformation of a topological movement into a metric one*. Indeed, automatisms at the object-based level are always metric, as it is well-known to anybody watching them. In contrast, automatisms that we have just terms "super-high" (from *E* to *D*), such as speech, writing, and movement of the bow arm of a violinist, can keep the *topological character* at any level of automation.

PART 3

Development and Disintegration

7

EMERGENCE AND DEVELOPMENT OF CONSTRUCTION LEVELS

The previous chapters have shown the great importance for understanding mechanisms of coordination and the structure of movement construction levels of the history of their development, both evolutionary and individual. Issues of the ontogenesis of motor phenomena are reflected in the literature relatively less as compared to multiple publications on the problems of the phylogenesis and morphogenesis of motor apparatuses (R. du Bois-Reymond, Spatz, Veraguth, Rose, Edinger, M.O. Gurevich, and others). Hence, we will discuss the former in more detail.

From the very beginning, it is important to put strict constraints on applicability of the "biogenetic law" of Haeckel to the discussed problems. In the field of movements, there is a range of principled and factual differences between the evolutionary and individual history of development. First, the motor apparatus of an ancient, poorly developed, vertebrate was adequate for the simple motor tasks it encountered in everyday life; the motor system of a newborn, and – even more so – of an embryo, cannot solve any motor problems at all. Second, as a rule, in the ontogenesis, the functional ability of the gradually maturating neural tools is achieved much later than their anatomical maturity and completion of the morphological development. At each successive stage of the ontogenesis, problems at the motor "ceiling" of the individual are solved in the presence and with obvious participation of mature higher-level and phylogenetically younger formations, which are completely absent in lower vertebrates (Spatz and Minkowski). In a child crawling on the floor during the second year of life, the pyramidal system has already been formed anatomically, and it puts its touch affecting the style of the striato-pallidar, i.e., essentially extrapyramidal, action of crawling, whereas in reptiles, who have no pyramidal cortex at all, the same action is performed by the striatum autonomously resulting in unavoidable kinematic differences from child crawling.

Third, the drastic differences between the phylogenesis and ontogenesis of motor phenomena is due to the relentless process of evolutionary encephalization. Not only a newly created center *A* (in the process of "overgrowing" of the nervous system) takes over functions previously located in an ancient nucleus *B*, which becomes a subordinate system; in addition, numerous functions, performed beforehand by even older, lower systems *C*, *D*, etc., now migrate into the vacant nucleus *B*. As a result of this process, older centers, such as, for example, the spinal cord, lose more and more of their independence and the inventory of movements they perform. The spinal cord, which can lead in decerebrate amphibians such a complex motor act as locomotion, or the pallidum, which is able to perform a similarly vicar role in reptiles and birds, in humans are factually useless for any kind of function substitution or for anything else beyond servicing backgrounds of elementary automatisms for higher-level movements. Hence, the spinal cord and the pallidum of a fish or an amphibia are functionally very much unlike their homologues in primates and humans, both grown-ups and babies. Therefore, it is hard to expect any kind of close correspondence between the process of motor development in the phylogenesis and ontogenesis, despite the overall rather accurate reproduction in the ontogenesis of the process of gradual, evolutionary overgrowth of the central nervous system, reflected in the gradual myelinization of the brain systems during the intrauterine and early post-natal period.

Driven by more complex motor problems and, at the same time, acquiring the ability to use ancient movements more efficiently, the central nervous system of vertebrates moves in its evolutionary history over several sequential stages, which are not as strictly separated from each other compared to how this process in commonly presented for the sake of schematization, but nevertheless, representing qualitative leaps forward.

It is necessary to remember that each, more novel, coordinative system, which enriches an animal species with a whole list of new movements and is designated here as the next *level of construction of movements*, introduces into the functioning of the animal's central nervous system, first and foremost, a new class of *sensory corrections*, i.e., if not a qualitatively novel sensitivity to perceived signals, then, necessarily, new methods of perceiving these signals, their estimation, conceptualization, and synthesis with signals from other sensory organs, and related, qualitative changes in the reactive attitude to the environment. The anatomical substrates of coordination levels, formed successively in the phylogenesis, necessarily include both motor and sensory centers, which can be very tightly linked to each other within a given level. For example, the ancient sensory centers of the tegmentum and hypothalamus develop in parallel with the effector centers of the level of paleokinetic regulations, which are represented by the spinal ganglia in lower vertebrates and, as a result of the encephalization process, by the phylogenetically relatively young formations of the red nucleus in mammals. The development of the level of synergies, which is served by the effector mechanisms of the pallidum, proceeds simultaneously with the formation of the main parts of the primary sensory organ of the brain – the thalamus. The pyramido-striatal level of the spatial field, which is partly cortical, corresponds to the emergence of "transit" nuclei of the metathalamic – the external and internal geniculate bodies, the ganglion habenulae, and the vast primary cortical sensory fields. The level of actions is equipped with effector groups of the premotor cortical fields (6, 8, and 42 Brodmann), and sensory (to be more exact, gnostic) groups – by the system of inferior occipital fields surrounded by secondary and tertiary cortical zones of touch, vision, and hearing. The same level, which is also most complex according to the abundance of performed movements, the variety of sensory-gnostic corrections and, finally, the numerous "high-level automatisms," borders with the collaborative, powerful, frontal-pons-cerebellar system, which also branches deep into the extrapyramidal system thus creating a functional connection between the frontal effector centers of this level and the sensory cortex of the cerebellum.

Brain motor centers are formed during the embryogenesis from two embryonic disks, referred to by Spatz as adjacent (Flügelplatte) and main (Grundplatte). The extrapyramidal system is formed primarily from the latter, located ventrally, whereas the pyramidal effector system develops from the former, located dorsally.

It is most convenient to describe the overall sequence of development of parts of both systems and their relation to different embryonic vesicles of the brain using the following Table 7.1.

All the organs, enumerated in Table 7.1, have specific chronological moments of emergence in the complex chronicle of the sequential "overgrowth" of the brain. Besides, many of them split

TABLE 7.1

Plates	*Rhombencephalon*	*Mesencephalon*	*Diencephalon*	*Telencephalon*
Dorsal plate – pyramidal system	Cerebellar cortex	Tectum corpora quadrigemina	Thalamus opticus Corpora geniculate (metathalamus) Ganglion habenulae	Cortex of the large hemispheres
Ventral plate – extrapyramidal system	Dentate nucleus of the cerebellum; n. Deiters	Red nucleus Substantia nigra Nucleus nerve III	Pallidum Hypothalamus Tuber cinereum Corpus Luysi	Striatum a) nucleus caudatus b) putamen A number of cortical fields

into parts, with sharp age contrasts, which differ from each other in the microscopic structure, connections, and functions (Figure 7.1).

The cerebellum is classified into: a) ancient (*paleocerebellum*) and b) new (*neocerebellum*). The former contains the unpaired medial part of the cerebellum – its vermis, and small parts of its hemispheres – flocculus, which are available in all vertebrates, with the exception of the lancelet, and reach high morphological development in birds. The new cerebellum also contains the hemispheres, which emerge only in mammals in the phylogenesis. In ontogenesis, the ancient parts of the cerebellum are mature and myelinated by the time of birth, whereas the cortex of the hemispheres gets myelinated only during the first half-year of life. The dentate nucleus of the cerebellum – its effector part – also contains two parts tightly connected with the ancient and new cerebellum. Both parts give rise to a dense network of pathways ending either in the red nucleus or in the Deiters' nucleus.

The *red nucleus* – the leading representative of the extrapyramidal effector system in the midbrain – in humans and other primates also consists of two parts called the *paleo-* and *neorubrum*. The ancient part of the red nucleus is located caudally and contains large cells; it is the origin of the rubrospinal effector Monakow tract. This tract is found in all mammals, starting from the lowest ones.[1] The younger areas of the red nucleus are located anterior and laterally: The magnocellular part with local pathways, the reticular nucleus, which is the target of fibers from the cerebellum, and finally – the neorubrum itself – a nucleus containing small cells, linked primarily to the frontal cortex and operculum. This latter nucleus is available only in the highest mammals – carnivores and primates – and it grows and develops in the phylogenesis in parallel to the cortex of the large hemispheres.

The substantia nigra, developing close to the red nucleus, is linked to the cortex of the large hemispheres probably tighter than the neorubrum. This largest nucleus of the midbrain is pigmented (and, therefore, deserves the name "nigra") only in humans. It emerges relatively late in the phylogenesis. It has two parts of different age. The older part is not pigmented, contains few cells, and its structure looks similar to that of the pallidum, which has very intimate connections to the substantia nigra. The relatively new part is connected by bi-directional pathways with other centers of the extrapyramidal system as well as with the cortex of the large hemispheres.

The largest of the brain nuclei, representing substrates of the motor coordination function of the central nervous system, develop from the diencephalon: They include the gigantic sensory system of the thalamus with its satellites and the large center of the extrapyramidal effector system – the pallidum.

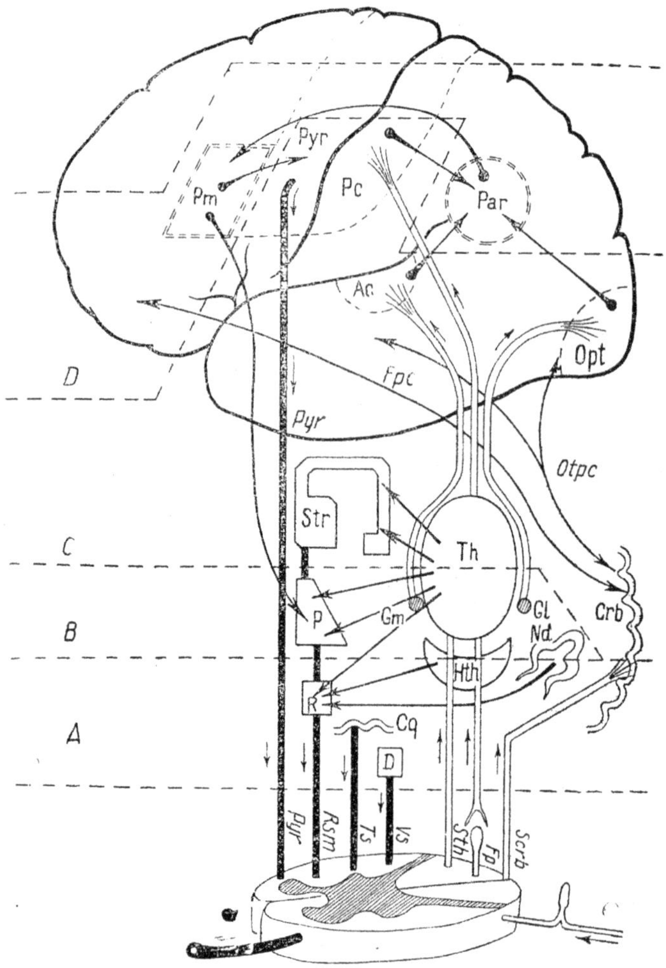

FIGURE 7.1 Scheme of the main brain nuclei and pathways with their assignment to the coordination levels *A–D*.

For better visualization, the actual spatial locations of the nuclei are grossly distorted. Nuclei: *R* – red nucleus; *D* – Deiters' nucleus; *Cq* – corpus quadrigemina; *Hth* – hypothalamus; *Nd* – Dentate nucleus; *P* – pallidum; *Gm* – internal and *Cl* – external geniculate body; *Crb* – cerebellar cortex; *Str* – striatum; *Th* – thalamus; *Pm* – premotor cortex; *Pyr* – pyramidal area; *Pc* – postcentral gyrus; *Ac* – auditory, *Opt* – optical zone; *Par* – occipital area. Pathways: *Pyr* – pyramidal; *Rsm* – rubro-spinal; *Ts* – tectospinal; *Vs* – vestibulospinal; *Sth* – spinothalamic; *Fp* – dorsal column; *Scrb* – spnocerebellar; *Fpc* – frontal-ponto-cerebellar; *Otpc* – occipito-temporal-ponto-cerebellar.

The *thalamic system* includes the thalamus itself as well as its satellites: metathalami, which include the intermediate centers of vision and hearing – geniculate bodies, and epithalami, which include a similar transit nucleus of olfaction – the ganglion habenulae. The thalamus in narrow sense subdivides into: (a) parts, playing the role of targets of second-order neurons of sensory pathways from the body periphery; and (b) parts with bi-directional neural connections to higher centers – the cortex of the large hemispheres. The former, more caudal ones, are the targets of axons of the dorsal column pathway and spino-thalamic pathway of the spinal cord and the axons of the trigeminal nerve, i.e. all the *tangoreceptor* signals, with the exception of the sense of taste: proprioceptive, tactile, nociceptive, and temperature sensation from the whole body. The thalamus possesses exceptional abundance of central neural connections and rich cytoarchitectonic structure (O. & C. Vogt identify in monkeys over 40 zones). Note the abrupt transition of the thalamus in the evolutionary history, which does not have a biogenetic effect on individual human development. Homologues of the thalamus in cold-blooded animals (lobi optici) are the supreme centers of the tangoreceptor system, whereas, at the same time, the same supreme function of telereceptors is realized by the tegmentum nuclei of the midbrain. In birds and mammals, with the development of the sensory cortex of the large hemispheres (much older than the motor cortex), all sensations gradually switch to the two-neuron intra-brain system: In a rather strict sense, the thalamus becomes the transit center for tangoreceptor signals, whereas its younger satellites, meta- and epithalamus, – for telereceptor signals, delivering the signals of both types into the cortex. This transmission is performed after the very profound, initial processing of the signals in the thalamic system. During this process, large components of the signals are filtered out and channeled, with the help of vast synaptic connections, into the extrapyramidal effector system, therefore not reaching the threshold for conscious perception, ensuring adequate coordination reactions and their background elements.

The supreme effector center in lower vertebrates (fish and amphibia), the pallidum, has very simple architectonics. It has no subdivision into fields; it is dominated by one type of cell – large ones with long dendrites, similar to the motor cells of the cortical motor area. There are many myelinated axons inside the pallidum, which make it look pale resulting in its name. The corpus Luysi is a close neighbor of the pallidum according both to its connections and time course of development; it emerges in the phylogenesis, however, much later than the pallidum – only in the mammals. The pallidum itself changes very little in the phylogenesis; it only subdivides by a white layer into two parts starting from primates. Connections of the pallidum are defined by its intermediate position in the extrapyramidal effector system; *it is the target* of large bunches of fibers from its sensory satellite – the thalamus, and from the leading effector center of the pallidum – the striatum. *It is the source* of a massive conduction pathway to the contralateral red nucleus (this is the effector pathway of the pallidum) and to its less conspicuous neighbors. In humans, all these pathways are myelinated and ready to function at birth.

Finally, from the most rostral brain part (telencephalon), the supreme effector instruments of the brain emerge and develop, which are phylogenetically the most recent ones: the corpus striatum from the ventral embryonic plate, and the effector fields of the cortex – pyramidal, premotor, and frontal – from the dorsal plate.

The striatum emerges in the phylogenesis starting from reptiles; its emergence has been linked to the transition to terrestrial forms of locomotion (crawling and climbing rocks and trees) typical of those animals. It reaches the pinnacle of its development in birds, which are still deprived of the pyramidal system and cerebellar hemispheres. In mammals, the striatum is subdivided into two paired nuclei, which are located rather far from each other: the caudate nucleus and the putamen. In humans, the striatal system is not yet developed by the time of birth and is covered with myelin only by the fifth to sixth month of life. The striatum's structure is more complex compared to the pallidum: It contains two types of cells, although its cytoarchitectonics does not show clear structure. In contrast, the putamen is structurally very similar to the caudate nucleus.

Of the two types of cells in the striatum, the numerous small cells with short axons, which do not exit this nucleus, are considered its receptor cells. All the broadly scattered, less numerous large effector cells send their exons to the pallidum; there are no other effector outputs of the striatum. Afferent signals into the striatum almost exclusively come from the thalamus. Connections to the cortex of the large hemispheres have not been documented, in contrast to the pallidum where such connections have been proven beyond doubt.

New movement-related organs in the brain that emerge and develop later than other organs in both phylogenesis and ontogenesis are the *pyramidal effector system* and, in certain sense, its superstructures, the frontal cortical systems. During the phylogenesis, the pyramidal effector system emerges only in mammals. Lower mammals – rodents – have only one giganto-pyramidal field 4, with the effector Betz cells in the fifth layer, but it is already rather large and structured. In more gifted in the movement field rodents, such as the squirrel and the hare, one can see elements of the premotor zone 6 after Vogt (M.O. Gurevich and G.H. Bykhovskaya; see Figure 6.2). The anterior granular area (fields 8 and 12) emerges only in carnivores, who also show significantly developed premotor fields 6. Dog-headed monkeys also display field 9 in the frontal lobes, whereas higher apes – frontal fields 10 and 11. Finally, humans show a whole range of brand-new fields in the frontal lobes (Brodmann areas 44–47), and the premotor zone starts to dominate over the primary pyramidal motor zone. The functional structure of the latter in humans and higher apes is very well developed ensuring precise somatotopic projections for nearly each muscle of the body. This somatotopic clarity is, however, complicated by the abundant dynamical, subtle functional interactions among individual excitable elements of the pyramidal zone and of the adjacent premotor field; numerous phenomena of facilitation, simultaneous and successive induction, etc. mask and smudge the somatotopy in precisely controlled experiments (Leyton and Sherrington, Grünbaum, Lashley, and others).

During ontogenesis, these cortical systems mature much later than the extrapyramidal ones: In humans, the pyramidal effector system and its descending pathways to the spinal cord are myelinated during the second half of the first year after birth, whereas the premotor and frontal systems – only during the second year of life. As it will be shown later, this gradual maturation reveals itself very clearly in the motor development of a baby.

The previous essay on the morphogenesis of effector systems has to be complemented with a review, at least a very brief one, of the phylogenetic history of development of their physiological functions.

The following scheme has been suggested primarily by Foix-Nicolesco and developed by M.O. Gurevich and, later, by the author. It classifies the process of gradual enrichment of movements in vertebrates into four consecutive stages. According to earlier authors, these stages are linked to the orderly recruitment of novel, higher-level *effector nuclei*, but according to the views expressed in this book, they are consequences of novel, interweaving *coordination construction levels*, which include, as substrates, both effector and receptor structural elements. As shown further, the scheme of Foix-Nicolesco if very close to our scheme described in the previous chapters as a hierarchy of levels.

The earliest of these stages corresponds to the domination of the *thalamo-pallidar system* and can be seen in fish and, partly, in amphibia. It is characterized by monotonous, smooth movements involving the whole body, and looking like peristaltic waves. Such movements – of the fins, tail, etc. – do not stop for a single moment, even when the fish resides quietly at the same spot. Nearly 100% of the movement inventory are swimming locomotion movements presenting as generalized, whole-body synergies. Movements are mostly smooth, slow, and rhythmical.

The second stage is characterized by dominant position of the striatum, i.e. the structure corresponding in our scheme to the *lower sub-level C1 of the spatial field*. This stage starts to develop in terrestrial amphibia, can be clearly seen in reptiles, and reaches its pinnacle in birds. The domination of the striatum with accessible features of sensory corrections affords these animals their complex locomotion movements of crawling, climbing, and flying – movements, which are already not body-based but extremity-based. The sense of equilibrium reaches a very high level. Statics of *posture* in equilibrium is not a problem for a fish; in contrast, here, we witness the elaboration of a whole system of stato-kinetic control mechanisms. While keeping all the synergic functions ensured by the thalamo-pallidar system, now pushed to the background, these animals have the means of their inhibition and regulation. They can both freeze and lunge. Movements of the head, neck, and trunk show a peculiar athetoid character, viscous, batter-like, sometimes turning into statue-like motionlessness, which is absolutely alien to the mammals. This style is preserved by movements of this level in humans, when pathological changes lead to its hyperfunction. In parallel to an increase in the complexity and variety of locomotion, a number of different movement categories emerge, with coordination sometimes reaching high levels of perfection and creating a (patently false) illusion of their attribution to the level of actions (e.g., nest building by birds). Birds can perform numerous movements never seen in fish, including grooming, feeding chicks, etc. Here, for the first time, we encounter expressive movements and sounds: the ghastly croaks of frogs are replaced with song and dance. Orientation in space and accuracy of locomotor transitions reach high levels of perfection (in the highly developed carnivorous birds) pointing at the initiated development of the upper sub-level of the spatial field *C2*. There is still not a trace of truly object-related actions – praxies from level *D*. Plasticity and ability to build new motor combinations are both very poor.

The third stage is the domination of the *level of spatial field C in its entirety*, linked to the emergence and development of the pyramidal effector system, which is seen for the first time in mammals. Once again, without losing any qualities of the coordination acquired during the first two stages, the animals acquire, on top of those, an ever-growing percentage of singular, non-stereotypical, aimed movements used for hunting, attack, leaping, etc. There is the continuing increase in plasticity, trainability, ability to improvise ad hoc with new movement combinations. There is a dramatic increase in the variety of self-grooming and toileting movements. As a transition to object-related actions, there is abundance of team play, teaching actions, demonstrations, etc., lacking in more ancient classes of vertebrates. In parallel to expressive *sounds* and intonations, mimics and expressive *movements* start to emerge. In general, movements lose their viscous, gluey style, interleaved with statue-like frozen postures of the whole body typical of the domination of striatal innervations. Movements become elastic, resembling those of a spring: Rest states stop being absolute and become saturated with adverse, wary movements, or habitual rhythmical tail wiggles, or with some other involuntary small actions. Gradually, one can see the emergence of true semantic chains suggesting early stages of the object-related level.

The fourth, *frontal* stage, suggested by M.O. Gurevich, is dominating in primates including humans, but its leap from apes to the human being is truly colossal. If, in comparison to birds, humans made a step backwards in extrapyramidal movements; if in movements from the inventory of the level of spatial field, a large number of mammals are superior in their movements, some – in agility, others – in force, still others – in accuracy, stability, endurance, leap length, perfection of equilibrium, etc., in the realm of semantic chains and object-related actions, the abilities of apes are limited to elementary basics compared to the innumerable psychomotor richness of the human repertoire. In the previous chapter, we discussed the very few examples of actions from level *D* in mammals and the main motor characteristics of this level in humans, so there is not much to add.

Let us now turn to the development of coordination in the *human ontogenesis*. In this chapter, we discuss the *natural process* of movement development during childhood. Everything related to development of new motor skills, learning, and training in adults will be the subject of a special chapter (Chapter 8).

The natural movement ontogenesis consists of two very differently timed phases. The first is the anatomical maturation of the central neural substrates of motor functions, which, as mentioned earlier, is delayed by the moment of birth and completes (at least as far as myelinization is concerned) by the age of 2–2.5 years. This maturation is accompanied by the similarly timed functional recruitment of the brain effector systems; however, this process reaches its full functionality and reliability much later. The second phase, sometimes lasting far beyond adolescence, is the phase of ultimate functional maturation and tuning of the coordination levels. During this phase, the development of motor function does not proceed monotonically and sometimes, with respect to some classes of movements, there may be transient stops and even regresses leading to complex oscillations of proportions and equilibria among coordination levels. It is still unclear what anatomical changes correspond to this functional maturation.

According to Veraguth, intrauterine movements observed starting from the sixth month of gestation point at highly differentiated activity of intermediate (forming reflexes) systems of the spinal cord. The very first movements by the newborn are breathing and screaming. During the first contact with the breast, the newborn is able to turn the head in search of the nipple. Starting from the same time, reflexes of sucking and turning the head toward a finger touching the cheek begin to function.

During the first few days after birth, the newborn shows a range of other coordinated reflexes: for example, using a pinhead to tickle the back of a newborn lying on their tummy leads to moving the spine sideways, away from the stimulus, or to an abrupt contraction of the long paraspinal muscles (m. erectoris trunci) extending the back.

To understand movements of a newborn, it is important to remember that the conduction pathways of the cortical motor systems and of the striatum and, to some degree, of the output link of the extrapyramidal effector system – the red nucleus, are not yet covered with myelin; as a result, only the thalamo-pallidar system of the brain is ready to function. After the "ceiling" level of the newborn, Foerster calls them "Thalamus-Pallidum-Wesen," which by many features of their behavior and postures reveals the phylogenetic descendance from the ape-like ancestors. "Only as a result of maturation of striati, the original pallidar function – crawling-like – is pushed aside, inhibited, and only its purposeful elements are retained due to mechanism of selective inhibition. The tamed pallidum is then used for the motor function of adults." (Foerster).

Since we have started to discuss the issue of taming and inhibiting the pallidum, we have to note that the frequently mentioned similarity of movements of babies and pathological movements in athetosis (Meynert, Freud, Spatz) is definitely related to the fact that, during athetosis, lesions of the striatum lead to "liberation of the normally constrained phylogenetic and ontogenetic lower movement mechanisms" (Economo), performed by the pallidums.

Babies continue to be "thalamo-pallidar creatures" over the first six months of life. During that time, they show "mass, non-differentiated movements of automatic and defensive nature" originating in the pallidum. "During the first months of life, babies show domination of grasping and gripping reflexes, similar to those in monkeys"; "primitive motor reactions, which are inhibited in adults and can be seen only in pathological conditions; these include the mesencephalic reactions of Magnus (Lage- und Bewegungsreflexe)." At the same time, there are "mass motor reactions, sometimes with features resembling elements of climbing and grasping, elementary expressive movements (non-differentiated reactions of fear – Peiper), symptoms of Babinski and Moro, supination of the feet, which only gradually turns into dorsal and plantar flexion, athetoid movements, etc." (Gurevich).

Veraguth notes what he calls Strampelhewegungen (kicking movements) observed in the same time period: internal shoulder rotation and alternating flexion and extension in the hip and knee joints. These movement synergies are frequently linked to movements in distal joints, grasping by the fingers and intensive motion of the toes. Pathways that bring about these motions are the rubro- and vestibulospinal, i.e., lower extrapyramidal ones; stimuli can include proprioceptive signals caused by shifts of the center of mass of body parts.

So, both authors, Gurevuch and Veraguth, agree that there is high percentage of widespread, but completely useless, sometimes reactive, but more commonly spontaneous, synergies.

We have to mention the fact that, during the early postnatal period, human infants lack an abrupt transition typical of mammals born blind. Newborn kittens and puppies, during the whole time prior to opening eyes, completely lack the subordinate regulation of chronaxies and tonus, reflected in their slow, shaky, "reptile-looking" movements on the creeping sideways legs. Chronaxies of the muscles in the extremities are kept, during this period, at a very high level. The moment of gaining sight is accompanied, synchronously or nearly synchronously, by the abrupt recruitment of subordination mechanisms, the equally abrupt drop of muscle chronaxies to their normal values (Arshavsky), and the involvement into reflex activity of the proprioceptive sensory system, dysfunctional until that moment.[2]

The age of 5–6 months is a very important critical moment in movements of an infant. By that time, more or less simultaneously, the anatomical maturation of two highly important systems is complete: myelinization and recruitment of (1) the *red nucleus group* with its pathways, ensuring functioning of the lowest level of paleokinetic regulations *A*; and (2) the *striatum* with its effector pathways to the pallidum, which is the substrate of the effector part of the spatial field sub-level *C1*.

To summarize, the main functional gains provided by this morphological enrichment are as follows.

As far as *statics* is concerned, by the second half-year of life, babies acquire *posture*. Until that time, the baby was supine, heavy and motionless, whereas the attached short and weak extremities could only produce various dysfunctional kicking movements without any useful load or any other goal that would unite all these movements together. The *striatum* (and maturing in parallel system of the red nucleus) brings about *posture*: level *A* – its tonic and reflex foundations and capabilities; level *C1* – its synergic form. The baby acquires the ability to sit, lie down, turn on the tummy, and a bit later – to stand up and keep vertical posture.

In dynamics, by that time, the modifications can be summarized as transition from *synkinesias* to *synergies*. Under the former term, we imply simultaneous movements lacking meaningful links, which are always a sign of pathology in adults; under the latter term – simultaneous movements or movement components, which together are directed at solving a particular motor problem. During the first six months, the baby's posture itself did not allow anything beyond idle movements of the extremities. Now, the trunk turns from dead weight into an organ of movable support for posture and movement; extremities begin to work in loaded conditions, frequently used as supports.[3] This change is supported by the recruitment of the neck-trunk reflex tonus (i.e., based on the properly functioning level *A*), and by the progress in the purely anatomical development of the skeleton and muscles of the extremities.[4]

Although the definition of a newborn as a "thalamo-pallidar being" is generally correct, and, by the time of birth, its thalamo-pallidar level is ready to work, during the first period of life, despite the pallidum having monopoly on synergies, useful synergies are few: until the complete maturation of the rubro-spinal system, the level of synergies has only meagre, indirect pathways to ventral root cells.

At the time of its recruitment, level *A* brings about proper functioning of the vestibular system, both otoliths and semicircular canals, and this provides the baby with the opportunity to maintain dynamic equilibrium during sitting, standing-up and turning, regulates muscle tonus, and leads to germs of active synthetic exploration by the baby, first of the spatial contours of its own body (Körperschema), and then – of the external spatial field.

At the same crucial time, progress in the area of *sound production* begins. The tongue and articulators are an instrument used for exercise by all the coordination levels over the process of ontogenesis. The thalamo-pallidar level of synergies, available to the infant at the time of birth, is only able to extract from the articulators unimpressive sounds: mutter, babble, saying "agu" without any meaning, etc. Over the first months of life, the infant cannot cry, only scream. Along identical lines, mimics is completely absent: There are only grimaces – synkinesias, representing free-form muscle play and expressing nothing.

The striatal sub-level brings with it two very important sound and mimics synergies – laugh and cry; expressive mimics emerge reflecting elementary emotions of happiness, suffering, scare, interest, and anger. The following recruitment of the pyramidal, higher sub-level of the spatial field *C2* produces (we are a bit ahead of ourselves here) still inarticulate but already purposeful sounds such as those expressing demand. Finally, the maturation of the premotor fields (including field *D*) and the system of the level of actions affords the child the first speech sounds during the second year of life.

The whole second half-a-year of life is dominated by the gradual maturation of the level of spatial field with the slowly completing recruitment of the pyramidal mechanisms of the upper sub-level replacing older, pallidar surrogates. A bright example is the development in early ontogenesis of one of the simplest psychomotor functions – *object grasping*. During the very first weeks of life, the infant is able to flex the digits and grasp the object placed in the hand and stimulating the palm's sensory endings. This is a more or less clean spinal reflex, maybe slightly colored by the participation of the rubro-spinal level *A*. Starting from the fourth to fifth month of life, attempts at grasping a visually perceived object begin (e.g., a bright toy suspended in the visual field). These attempts look like rather diffuse, disordered, and irradiating synkinesias, resembling intense bursts of floundering leading to alternating, rhythmical rocking movements of all four extremities and spreading to muscles of the face, neck, and trunk. Such a burst of irradiating motor excitation can lead, by chance, to collision of the palm with the desired object ending up in successful grasping; then, the action stops. If, however, such a lucky outcome does not happen, the burst fades away by itself and, after 10–20 seconds, it is followed by another similar fit of activity. During the second half-a-year of life, such hyperkinetic, typically thalamo-pallidar bursts start to be interrupted by alternating singular simple targeted movements of one hand toward the object at the level of spatial field – first, inaccurate, ataxic, with large percentage of failures, and later – more and more adequate. By the end of the first year, a normally developing child switches the grasping process to the pyramidal system – to the most adequate level of spatial field. This sequence in the development of the grasping action shows as, in the course of the ontogenetic development and maturation of coordination levels and kinetic systems, a motor act crawls up the system of levels until it reaches the level where it would be realized in adulthood.

It is interesting that grasping a visible object by a 4–5-month-old child continues sometimes with the described irradiations and hyperkineses even after the time when dragging the hand-held object into the mouth is performed with a well-coordinated, simple and singular flexor movement. Likely, this is explained by the fact that hand movement toward the mouth, with or without an object, corresponds by its direction to the natural tendency of the child, whereas grasping and possessing an object suspended in the visual field requires making an extensor movement away from the body, which is against the primitive tendency of self-centered actions and, as a result, emerges later and is more difficult.

The whole second half-a-year of life represents a *pre-locomotor period* of motor development: Preparation to walking and running, and broad use of locomotion surrogates such as crawling.

TABLE 7.2

Level	*Type of movement*	*Afferent system*
A	Latching (Sperrung)	Tactile
B	Synkinesias, hyperkinesis	Visual diffuse stimulation
C	Targeted singular action; ataxia, later – accurate action	Visual image of the object in spatial field coordinates

To clarify the nature of this preparation to locomotion movements, let us recall that the coordination structure of walking and running involves collaborative work of all the construction levels, from the bottom to the top. The rubro-spinal level produces mechanisms for: (1) the dynamic control of tonus of both the trunk and extremities; (2) direct and crossed reciprocal innervation; and (3) vestibulo-cerebellar control of equilibrium. The thalamo-pallidar level ensures the main huge synergy of walking including rhythmical alternation of nearly 100% of skeletal muscles. The striatal sub-level ensures adaptation of the general pallidar synergy, which is not related to the external space by level *B*, to the actual conditions of walking: properties and uneven parts of the ground, steps, slopes, grooves, etc. And finally, the upper, pyramidal level of spatial field overlays this already meaningful process of motion with what endows it with goal-directedness: to go somewhere, for something, on the way to return and take something, to throw a ball or a grenade after the runup, etc.

Based on this schematic, it is easier to orient oneself in the intensive preparatory work performed during the pre-locomotor half-a-year in the movement domain of a child.

Even before the ultimate expulsion of the seemingly aimless "kicking" movements, the child forms very important for locomotion basic movements participating in such actions as sitting and standing and linked to work of the trunk and hip extensor muscles – the main stabilizers of the whole-body equilibrium. During this time, the body learns to maintain and support all its mass over the minimal support area. To integrate all these balancing movements, processes performed by the extrapyramidal effector system are needed – particularly, impulses it transmits from the vestibulo-cerebellar system (Veraguth).

During this very early period of preparation to walking, the child faces a number of additional, purely anthropometric, problems, which disappear at more advanced age. The lower extremities, particularly the hip muscles, are very weak. The legs are short and, in addition, semi-flexed because of the unfinished formation of the normal lumbar lordosis. Compared to adults, the overall body center of mass is located higher over the hip axis, because of the relatively larger trunk and head mass and the aforementioned short legs. This generates a rather large moment of force of the upper body with respect to the hip axis and, given the weak muscles in that area, leads to continuous flexions of the hip joints. It is not surprising that a one-year-old child flops on the buttocks frequently; it is fortunate that the height of the flop is not too large. Compared to adults, the relatively small foot area of a child creates additional difficulties. All the observations identify two main difficulties of this period: maintaining equilibrium and fighting the moment of the force of gravity with respect to the hip axis.

Further, locomotion movements require, as a prerequisite, reliability of intermediate systems: The subtle, balancing action of foot muscles during standing, the contralateral walking synergy ("stepping"), etc. There are also highly differentiated cerebellar regulations, mediated by the red nucleus: Fighting the force of gravity, being able to move the center of mass in a goal-related way, positioning and movements of the legs, etc. All these acquisitions point at the high

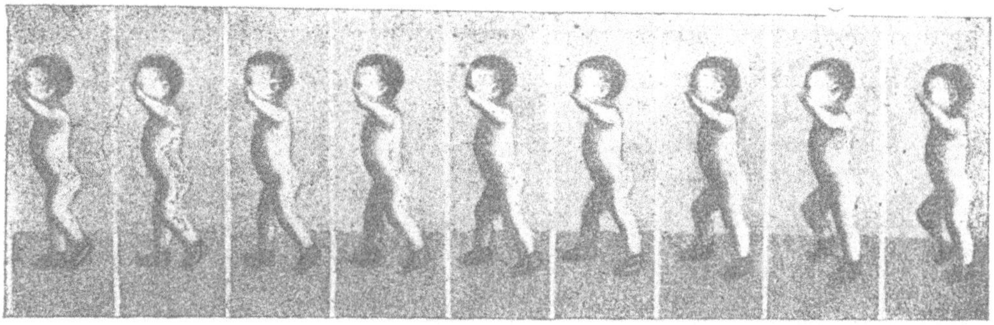

FIGURE 7.2 The fast-speed recording (50 frames per second) of a single step by a child during the first days of independent stepping.
(Source: T. Popova, VIEM, 1935).

level of regulations from the cerebellum, red nucleus, and the thalamo-pallidar system achieved by that time.

The amplitudes and tempi of locomotor movements are defined primarily by structures of the extrapyramidal effector system and its signals delivered via the red nucleus and the rubro-spinal tract.

Genuine bipedal locomotion develops at the beginning of the second year of life (Figure 7.2). "Until that time, on top of the underdeveloped neural system, the muscles of the lower extremities and even their weight, compared to the body weight, are insufficient to support body statics. At the age of 1–2 years, movements are clumsy and unstable due to the insufficient differentiation and lack of the necessary control of tonus. Children of that age show adjustments of expressive and defensive actions and the emergence of everyday movements (i.e., movements of everyday life and self-care, NB). Therefore, the striatal functions, in their static and kinetic manifestations, display significant development, whereas the pyramidal functions are still developed poorly, movements are very inaccurate, and there are numerous cases of synkinesias. Body posture is characterized by some degree of lordosis." (M.O. Gurevih).

During the second year of life, toddler's locomotion is neither walking, nor running, but something not yet differentiated (T.S. Popova). The divergence of walking from running begins not before the third year of life (Figure 7.3). The complex biodynamical structure of walking typical of adults is completely absent in a child who is just starting to walk. Instead of a vast, harmonious system of impulses, which show fixed order and configuration in the force curves of the leg segments during a single stride in adults, a 12–18-month-old child shows only one pair of reciprocal impulses, one directed forward and the other – backward, corresponding to the pattern observed, for example, during the stepping reflex in decerebrate cats. This stage of innervational primitives lasts for about a year, i.e. approximately until the beginning of the third year of life.

The full inventory of dynamic impulses of walking takes time to develop and is complete by about the age of five. Very gradually, individual elements of the force curves migrate from the group of accidental, seen not at every step and showing a tendency to disappear with an increase in stepping tempo, into the category of emerging reliably during slow walking, and finally – into the group of those seen consistently across conditions. Hence, the gradual emergence and fixation of new structural elements is unrelated to the development of elementary coordination and equilibrium during walking: at the age of 3–4, the child is not only walking perfectly well, but running, hopping on one leg, riding a scooter or a tricycle, etc. This means that mechanisms of coordination

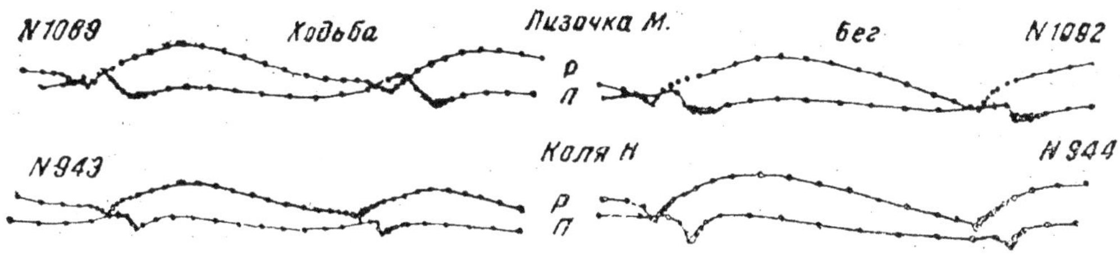

FIGURE 7.3 Cyclograms of movements of the ankle joint and the tip of the foot during "walking" and "running" of two children, 1 year and 4 months old. The difference is nearly exclusively seen in the small step length deviations.
(Source: T. Popova).

of different types of locomotion and maintaining equilibrium have been developing for a long time and become reliable by that age: The same structural elements, discussed here, obviously have different meanings and are related to more subtle aspects of the motor action. As far as the neural structure of walking is concerned, dynamic elements reflect complex synergic work of the thalamo-pallidar level and, as it should be clear from the previous text, their elaboration lags behind by years compared to the time of anatomical maturation not only of the pallidum, but also of the striatum, and the pyramidal system.

The following cyclogrammetric observations by T.S. Popova show that the development of the dynamic structure of walking proceeds in the ontogenesis not along the shortest route. Approximately between 5 and 6 years of age, there is sometimes an immense overproduction of dynamic impulses in the leg force curves accompanied by their complete lack of a consistent pattern. After 8 years of age, these "childish" elements show, one after another, reverse development, and the curves slowly acquire their characteristic patterns typical of an adult. The involution of these excessive curves, associated with transformation of the shapeless, toothlike patterns into typical configurations, is a result of the selective, shaping inhibition by the striatum, reflected in the aforementioned quotation by Foerster. It is possible that some of the "extra" impulses reflects "fending" of the perturbing reactive forces by phasic impulses originating from the pyramidal system (see Chapter 8). The simplification of the shape of dynamic curves and the elimination of "childish" excessive elements are dictated, in terms of biodynamics, by the transition to more perfect methods of counteracting the reactive forces emerging during movements of multi-link kinematic chains of the extremities. This transition to a more economical and subtle method of coordination is likely related, on the one hand, to the recruitment of the highly organized anatomical substrate and more differentiated functional level, and on the other hand, – to the process of automation moving in the opposite direction, i.e., switching of the mentioned cortical reactive components to the lower, but more adequate, construction level.

The biodynamic divergence of running from walking starts not before the third year of life from the organization of the *flight phase* of running, originally absent. More or less subtle changes in innervation are accompanied by the resultant biodynamic refinement of child's running (Figure 7.5). Step length growth continuously: by the age of five it doubles, by the age of 8 – triples, and by the age of 10 it is about 4.5 times larger during sprint compared to a toddler who has just started running. Obviously, leg length does not increase in the same proportion, and the primary reason for the increased step length is an increase in the joint excursion amplitude and prolongation of the flight phase. The mean running speed also increases continuously: it doubles by the age of 3–4, triples by the age of 5, and is five times larger by the age of 10 reaching 5.75 m/s (about 20.7 km/h) and becoming four times higher than the walking speed by the same child.

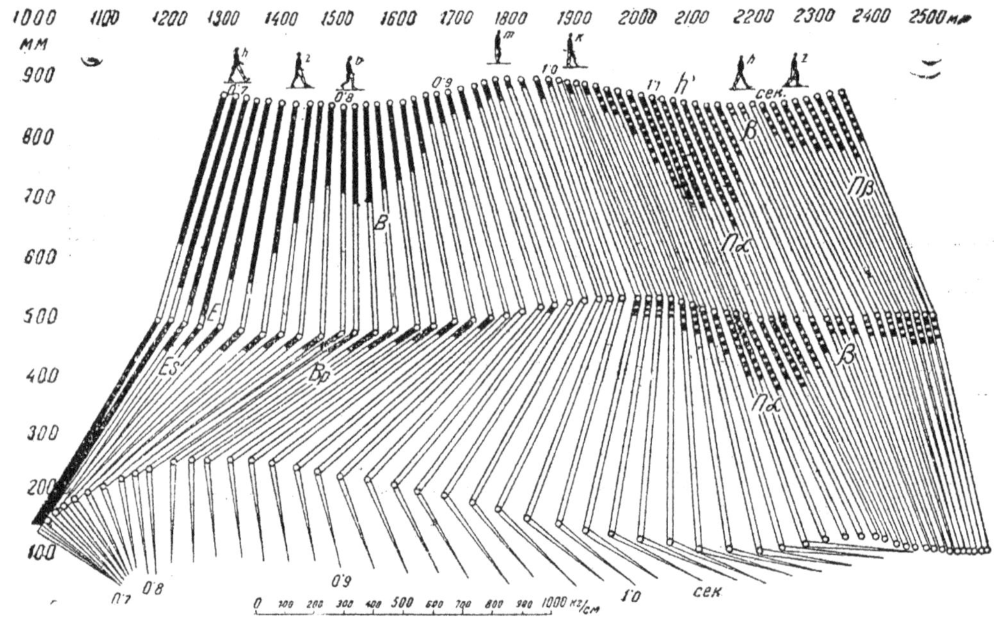

FIGURE 7.4 Consecutive positions of the right leg during the swing phase of walking with designations of phases of the main force waves and with marks showing the resultant of the moments of muscle forces in the leg joints.
Moments of hip muscles are shown with strips of different length along the upper leg, moments at the knee joint – by strips along the shin; moments directed forward (extension) are shown with black, moments directed backwards (flexion) – with hatched strips.

The whole second year of child's life is the year of recruitment of the level of spatial field and mastering locomotion. The next developmental stage, spanning the end of the second year and the whole third year of life, represents the period of completion of the anatomical maturation of higher psychomotor systems. During this period, the child shows the emergence and abrupt growth in both quantity and degree of success of actions from the *object-related level*. As described in Chapter 6, this group of motor actions primarily involves movements of two categories: proper object-oriented, such as manipulation of objects and tools, and symbolic, including elementary coordination of speech and writing. With respect to the first category of proper object-related actions, children master during this time a whole range of self-service actions, play with toys successfully and break them, build constructions from small bricks and sand, and start to draw something with the pencil. Mastering speech begins typically by the end of the second year of life. Problems of speech and its development, however, form a special independent, comprehensive problem, which is beyond the scope of this book.

As far as the general character of movements of the growing child is concerned, there are very impressive changes in the inventory and type of movements proceeding in parallel with the rapid change in the overall body structure. In contrast to clumsy two-year-old laggards, children from 3 till 7–8 are usually agile, adroit, and graceful. This is the time of the functional prime of the lower (striatal) sub-level of the spatial field, which is in firm control of the synergic thalamo-pallidar backgrounds. The functional weakness of the pyramidal sub-level *C2* and minimal participation of the level of actions *D* lead to two phenomena typical of that age period.

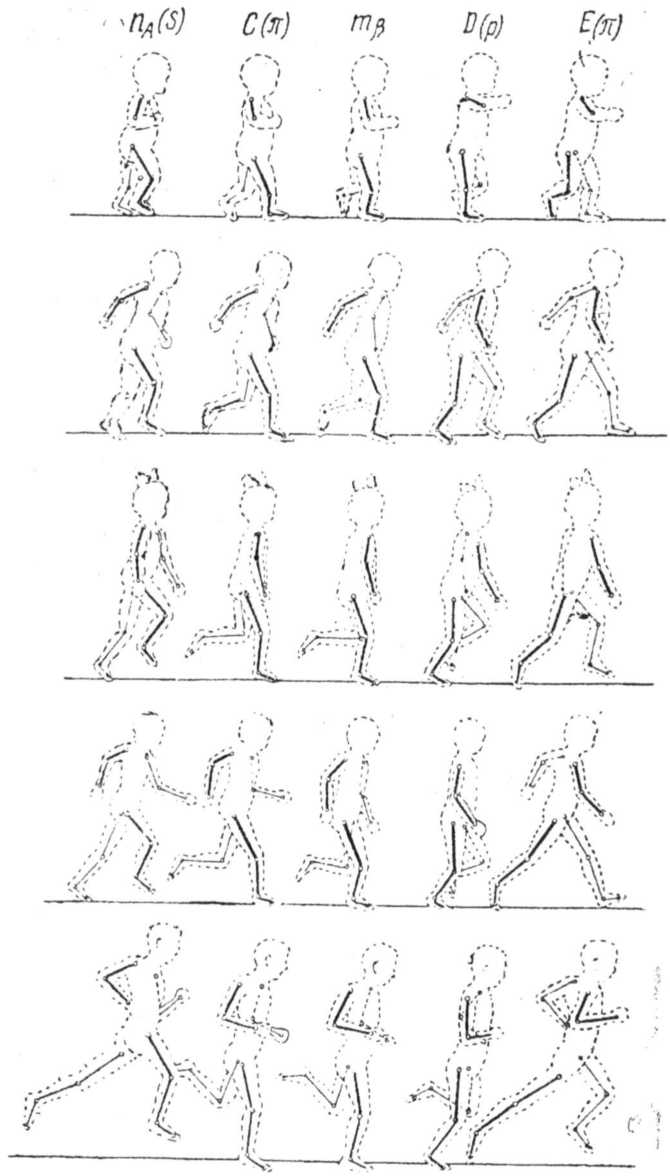

FIGURE 7.5 The body postures of several children of different age in typical phases of running. Left to right: 1 – the main kick by the swing hip downward; 2 – the forward kick; 3 – maximal shin flexion of the trailing leg; 4 – backward kick; 5 – the last phase of push-off of the supporting foot. The age top to bottom: 1 year 4.5 months, 3 years 3 months, 4 years 9 months, 6 years 1 month, and 10 years. (Source: T. Popova).

First, children who may be relentless in various locomotor actions get tired quickly and try to escape if they are given a task requiring accurate movements and overcoming resistance. As correctly noted by M.O. Gurevich, the relentlessness, revealed in games, is to a large degree illusory, because games do not require productive labor movements. All the motor activity is relaxed and natural; it starts and end depending on their psychophysical needs. Likely, this ease is to a large degree responsible for the gracefulness of their movements. Restlessness of children, during attempts to force them to concentrate on some kind of work associated with actions at levels *C2* and *D*, is also due to the initially small number of skills associated with those levels to make such an occupation interesting and engaging.

Second, typically for the discussed age, gracefulness and agility of locomotor movements coexist in children with very low levels of the overall body dexterity (not to mention hand dexterity). This very typical "graceful clumsiness" is explained by the fact that dexterity, representing a combination of motor maneuverability, resourcefulness, and ability to combine quickly ad hoc adequate movement formulas, requires high functional development of the cortical levels *C2* and *D*, and their saturation with certain locomotor experience. The lack of maturity of these coordination systems leads to the 3–7-year-old children performing rather poorly in navigating obstacles, throwing at the target, climbing trees or rope ladders, etc. All this comes later, at adolescence.

There are very few trustworthy observations on the development of motor behavior during the next time period that would fit the structure of this book; therefore, we are going to limit this topic to a few additional remarks. Analysis of the development of writing action will be presented in Chapter 8.

First, it is necessary to recall the fact, mentioned in Chapter 6, that the functional differences between the right and left hands are inherent not to all the construction levels but reveal themselves in an essential way starting from the purely cortical level *D*. This qualitative difference of the level of actions from the lower coordination levels is masked by secondary complicating factors emerging during the ontogenetic development and, therefore, remains mainly unnoticed. Therefore, we have to dwell on this phenomenon.

A little child does not yet show differences between the right and left sides of the body, either in muscle strength or in coordination dexterity. These differences become noticeable only later, following the increase in the number of movements at the level of actions in the child's repertoire. This does not mean, of course, that the child does not have constitutional prerequisites for righthandedness or lefthandedness. Quite the opposite, all data (for example, multiple genealogical observations) speak in favor of an important role of inborn features in identification of the brain dominant hemisphere, but this dominance reveals itself only when there is the corresponding developer for this, so far latent, inequality. Disagreements among different authors (Baldwin, Dix, Stier, Brann, and others) on the issue of the age of identification of this asymmetry reach the situation when some authors estimate this age to be 2 years and others, 14 years; this broad range is obviously defined by specific movements of the child observed by different authors. The more attention authors paid to object-related actions, the earlier they had to document the emergence of right-left inequality.

The tight, inseparable functional entanglement of all types of motor actions at each age inevitably leads to modifying effects of the construction levels leading at the given stage of development on levels of secondary importance. As a result, two interesting factors emerge.

In early childhood, when extrapyramidal movements and cortical movements, not exceeding the so far symmetrical spatial field, are dominating, they are not only deprived of any sign of asymmetry, but even a few, secondary for the child, actions from higher levels also exhibit the same nuance of side interchangeability. As a result, there is continuous mixing of the right and left sides. The child is reminded non-stop by the grown-ups to take the fork, spoon, needle, scissors, etc., into the correct hand; the abundance of mirror-like letters and even whole words and reading phrases right-to-left is stunning. As a rule, at that time, there are no anthropometric or dynamometric differences between the hands.[5]

During the second half of adolescence and youth, the ever-increasing dominance of object-related actions in the total number of movements leads to an opposite effect. As a secondary consequence, there is predominant development of the right hand (of the left hand – in the inborn lefthanders), not only in its functional-coordination aspect but even directly anatomically. At this time, the availability in this hand of greater dexterity, force, and endurance is inevitably reflected in success of movements at any level: distance of a throw, force of a strike, strength of compressing the dynamometer, etc. However, endurance during static loading (primarily, level *A*) in a very large percentage of cases remains for the lifetime in the left extremities of the righthanders, compared to the right extremities, and vice versa in the lefthanders (Farfel).

General development of the motor function in the second half of adolescence, between 7 and 10 years, proceeds in the direction of gradual mastering of the coordination capabilities that have been formed in the child as a result of complete anatomical maturation of movement-related structures (as mentioned earlier, approximately by the age of three). By the beginning of the second decade of life, the child shows gradual recruitment of the higher cortical levels with all the features related to their capabilities, as described in earlier chapters: On the one hand, there is progressive improvement of accurate and forceful movements; on the other hand – there is an increase in the number of mastered skills and object manipulations with the right hand. Due to the mentioned reasons, there is more and more dexterity in throwing, climbing, sports, and track-and-field movements. M.O. Gurevich assigns the limited ability for productive work to the still insufficient development of frontal cortical mechanisms.

This clear psychomotor progress is doomed to undergo, during the development of the adolescent, a complex period of restructuring related to the general reorganization of the organism during puberty and leading, sometimes, to a transient regress in coordination.

Most frequently, disproportions shown by adolescents during puberty are: (a) the striking domination of the pyramido-cortical apparatus at the expense of extrapyramidal, background levels; this is the cause of the observed awkwardness, clumsiness of global movements, and instability of the tonus; (b) disorders of the normal, just starting to be established, relations between the level of actions *D* and background levels, which supply the adolescent with technical skills and "higher automatisms" necessary for most actions. This leads to deautomation, lack of movement precision, and high fatigability. With time, these disproportions are corrected, and the youth demonstrates an individual psychomotor profile of an adult. These shifts in the motor function during puberty, accompanied by various changes in mentality, as well as in the endocrine apparatus and nervous system, can be very different; they move us close to a very important question, which we can only touch upon briefly. Since these variations and transient regresses are not caused by an organic disorder and are not accompanied by actual losses of already acquired skills, obviously, they depend on purely functional changes in the already achieved proportions among the coordination levels. Differences in these phenomena could be due to individual differences, which are impossible to pin down, and, to a large degree, they are defined by the *constitutional* proportions among the levels reached before puberty and what these proportions would look like by the end of the restructuring. The issue of changes in the proportions represent only one side of a wider and deeper problem, which can be termed the *problem of individual motor profiles*, i.e., individual, constitutionally predetermined relations between levels of perfection and ability to develop separate construction levels. In turn, this problem is closely related to the frequently asked question on motor talent, which is much more complex and, at this time, likely not solvable with proper methodological tools because it involves the very difficult issue of objective quantitative measure. The issue of motor profiles allows more reliable qualitative interpretation and, at the same time, is not less practically important.

Different, perfectly normal, persons can show very different relative levels of development of individual coordination levels. There are people endowed with gracefulness and harmony of body movements (level *B*) with the rather helpless hands, unable to operate even the most primitive tools. Others show exceptional accuracy of small movements needed, for example, for watchmakers and engravers, but they are clumsy, stumble while walking over the even floor, and knock down chairs; in these persons, levels *C2* and *D* dominate over level B.

The existence of such individual qualitative differences in psychomotorics has been also known for as long a time as the direct proportional correlation between development of different aspects of motor function and the ability to develop various new motor skills. One person learns quickly and very well precise, small-amplitude manipulations but has real problems, for example, trying to learn how to play piano. Another person is very good at learning gymnastic and acrobatic dexterous exercises but has troubles with track-and-field and sport games. Still another one, even within a single movement area, is very good at scaling dynamics and quality of touché in playing piano, and has real problems with finger agility and fluency; opposite problems may be seen in another person. So far, all the empirical attempts, made in the absence of a theoretical background, failed due to the enormous variety of the motley material. It seems that the described theory of coordination levels can bring some clarity into this problem. First, it is obvious that a degree of development and trainability is inherent not to individual motor actions in all their variety but to whole contingencies of movements forming the inventory of one of the construction levels. The number of levels is not so large, and it is much easier to characterize each one of them for a given person than to study and characterize each motor action. Second, it is more than likely that, even within the short list of construction levels, not all of their hues and proportions are possible, but only a handful of typical proportions. Such a description would be fully analogous to constitutional profiles of body composition and character, which have been accepted in the literature and are definitely, heuristically valuable. Third, one could upfront expect to see strong correlations between, on the one hand, the explored motor profiles and, on the other hand, constitutional profiles. All these factors set the stage for the problem of elaboration of methods of objective estimation of both current levels of development of different levels and their plasticity during development and training.

Notes

1 There are no established homologues in reptiles and birds for the red nucleus and corpus Luysi in mammals. The red nucleus itself changes rather dramatically during the phylogenesis of the mammals. For substantia nigra, the likely homologues in birds are nuclei ectopedinculares, for corpus Luysi – nucleus rotundus (Ariens, Kappers, Spatz).

2 Until the time when the baby is able to see, typically one cannot induce the phenomenon of decerebrate rigidity.

3 To stand up, infants use a rather complex action termed by Schaltenbrand "quadrupedales Syndrom" Turning from the back to the tummy, standing on all four, extending the knees while standing on all four, and finally – standing up, instead of the typical for an adult standing-up from a supine position by lifting the trunk by flexion of the hip joints. In an infant, the moment of the force of gravity of the trunk and head is too high in comparison with moments generated by the legs to allow using the grown-up method.

4 Kleist points at possible localizational separation of static and dynamic functions in the extrapyramidal system: statics is performed by the internal nucleus of the pallidum and putamen, dynamics – by the external nucleus of the pallidum and caudate nucleus.

5 According to some authors, among school-age children, there are more left-handers than among grown-ups (where the number is 4–5% of apparent and up to 20% of concealed left-handers). For younger schoolchildren, this percentage is higher. This is a wrong conclusion. Among younger schoolchildren not the number of lefthanders is higher but the number of left-sided movements due to the mentioned in the text indifference. It would be more accurate to speak about childhood ambidexterity.

8

DEVELOPMENT OF MOTOR SKILLS

The motor skill theory passed through a number of successive stages, which have exerted crucial influence on the practice of training various movements. According to the prevailing early views, motor exercisability is localized in the peripheral, skeletal-muscular apparatus (which is true only with respect to muscle strength training). Based on this opinion, instructors of labor, physical culture, and art paid attention primarily to training of the peripheral muscular-ligament apparatus. This was the origin of a whole system of actions directed at "development" of that apparatus: stretching ligaments and tendons, muscle massage, a whole range of gymnastic exercises of massage type and purpose, etc.

The conviction that motor skills were located in the periphery brought about the impetus to begin skill elaboration as early as possible, from childhood, when the process of skeletal formation is far from completion and all tissues are elastic and pliable. This led to numerous excesses and harmful effects.

By the beginning of the twentieth century, it had started to become clear that the central nervous system plays the leading role in skill elaboration and fixation. However, the physiology at the beginning of the century was still very far from understanding the crucial role of afferent signals in the coordination process. With the formation of the view that motor skill was something developed, fixated, and localized in the central nervous system, attention became focused on *effector brain systems* and, first of all – on the previously explored cortical effector (pyramidal) apparatus. The view that, during any motor training, whether it is a gymnastic exercise or learning a piano piece, the exercised organ is not the hands but the brain at first looked paradoxical and had major difficulties being incorporated into the thinking of instructors. The new view received significant support from the theory of conditioned reflexes, which had its main concepts developed and published at about the same time (the first decade of the twentieth century). The hypotheses leading to the formation of its theoretical concepts were not novel by themselves; they had been introduced by materialist psycho-neurologists of the end of the nineteenth century led by Meynert, but the solid experimental foundation generated by I.P. Pavlov provided them with distinctive credibility and clarity.

The fact of fixation of traces within the central nervous system during individual's lifetime was given, although a hypothetical, but a concrete and definite interpretation as *forcing connection pathways in the brain* by numerous persistent repetitions of a conditioned combination. The impressive power of the new facts and offered theoretical interpretations was so strong that the theory of conditioned reflexes started to be relied upon in hopes to erect on its foundation the whole structure of materialistic psychology. It was most natural to use this theory for understanding a much more narrow and specific issue of the motor skill and its genesis. Naturally, a comparison arises between fixation of conditioned connections in animals and fixation of a motor skill in humans – both as a result of multiple repetitions and lengthy exercise.

The first major difference is the fact that the design of experiments with the elaboration of conditioned connections in animals required them to be completely passive with respect to the presented stimuli. This is well confirmed by the frequent occurrence of sleepy inhibition in experimental animals, which represents a true curse of experiments with conditioned reflexes.[1] In contrast, every stage in the elaboration of a motor skill represents not a passive "submission" to external stimuli, even if they come from one's own proprioceptors, but *active psychomotor phenomena* that form both the external appearance and essence of motor exercise. This intense activity constructs a motor skill similarly to constructing a building, which justify the title of this book: "Construction of Movements."

Second, the hypothesis on forcing conduction pathways describes this process as something monotonous, producing in its development a phenomenon of purely quantitative, homogenous growth. Meanwhile, as we are going to show later, the formation of a motor skill represents a chain of sequential, replacing each other, phases with different meaning and qualitatively different mechanisms. The process of developing a motor skill is a *meaningful chain action* where it is impossible to omit individual semantic links or to modify their order, similar to any of the object-related chain from level *D*. By itself, motor skill is a very complex structure: It always has leading and background levels, leading and auxiliary chains, backgrounds in the direct meaning of this word, automatisms and recodings of different complexity, etc. The process of its formation is also saturated with qualitative structural difficulties. The interpretation of the formation of a motor skill as forcing conditioned connections was visibly harmful, primarily because it justified monotonous, passive memorization, "jagging," which emphasized quantity of repetitions. The very low practical effectiveness of this method forces its critical reconsideration and major reassessment.

The forcing theory of motor skill elaboration has one more unclear element. Champions of the theory of conditioned reflexes accepted, as something obvious, that *fixation of conditioned connections was very slow*, which was then naturally compared to the long duration of physical exercise. All this slowness was ascribed to very low time-related coefficients of responsiveness of the central nervous system – at least, this is how they looked during conditioned reflex experiments on dogs. Meanwhile, these multi-week time intervals, if we assume that this is the time required for a single monotonous process of the gradual forcing of one or a few pathways through the central nervous system, are in stark contrast to the facts of fixation of non-indifferent percepts after a single presentation (in dogs and, even more so, in humans), and also to all the physiological velocity indices of the nervous system, which consistently characterize it as a lightning-like, quick bioelectrical instrument. The situation changes dramatically if, instead of such a monotonous fixation in a neural center, we consider the actually occurring phenomena: An active, multi-phase construction in the sensory-motor brain levels. This view leads to a completely different formulation of the problem.

As it follows from the contents of this book, motor skill cannot represent a *fixed effector formula* of a sequence of neuro-muscular impulses. We have already established that such a standard formula cannot emerge within the central nervous system since the essence of motor coordination is in the continuous adjustments of effector impulses to external conditions, which change all the time and have to be continuously monitored by receptor systems. The level of synergies *B*, which could possibly mislead us by its propensity toward standard patterns, can keep those patterns exactly due to its exceptional sensory apparatus, i.e., due to its inherent ability to adjust, trial to trial, its active impulses to the variable external and reactive forces. Movements at the level of synergies are standard exactly because its effector signals are not.

It is equally impossible to expect, as the foundation for a motor skill, any kind of standard sensory corrections, which are not less variable than effector impulses are, exactly due to their adaptability. Neither effector, nor sensory, nor any other brain systems and centers can serve as sites for localization of stable forced, or fixed by any other means, traces of a motor skill.

Note that, if exercise or training a skill, could be reduced to forcing or beating through of anything based on meaningless repetitions, this would not lead to anything useful because, at early stages of skill development, when movements are wrong and clumsy, there is nothing to be "jagged." This factor one more time emphasizes the fallacy of "forcing" concepts. The dialectics of skill development is exactly that, where there is *development*, every next execution is better than the previous one, i.e., *the next movement does not repeat the previous one*; therefore, exercise, in essence, is *repetition without repetition*. The solution of this seeming paradox is that exercise represents *not repetition or forcing of movement but its construction*. Correctly designed exercise repeats not the means used to solve a given motor problem but the *process of solving this problem*, which improves the means across repetitions.

According to the currently available, most exact definition, *motor skill is a coordinative structure, representing a mastered ability to solve a particular type of motor problem*. Since motor problems can have very different semantic contents, clearly, a skill can be elaborated with respect to any voluntary motor process, independently of its level and composition.

The suggested definition reveals two actual reasons for the long duration of motor skill acquisition. One of them defines primarily the first, initial stage of working on a skill, the other one – primarily the second stage.

First, the process taking place within the central nervous system during exercise and skill elaboration represents active search for more and more adequate, in all aspects, solutions for the exercised motor problem. The central nervous system is actively moving through a large number of trials, errors, adjustments, adaptive modulations, etc., which ultimately ensure the best, fastest, most rational, and at the same time dexterous, execution of solutions. This search for optimal solutions is, undoubtfully, one of the reasons for the long duration of work on the motor skill. This is particularly true with respect to its early stages, before completing the automation, which represents some kind of final proofs of the movement before it is sent to the printer.

Second, the motor problem itself and, even to a larger extent, the overall background where the problem is formulated are also as non-standard as the movements, coordinations, and corrections, which form adequate solutions for the problem. While mastering a new type of problem, the central nervous system receives and absorbs vast streams of receptor signals typical of problems of this type and defining adequate sensory corrections. During this process, the central nervous system encounters a wide variety of versions of the problem. To use an expression common among psychologists, one could say that, during the formation of a skill, there takes place rehearsal of all these versions. Clearly, for more fully and reliably built motor skills, there is a broader range of versions and complications of the motor problem, not leading to disorientation and deautomation, for which the subject can find adequate coordination resources.

So, instead of monotonous forcing a path, we witness a complex multi-phase process, as repeatedly mentioned by theoreticians in pedagogy, albeit without much detail. Based on available observations, it is already possible to identify, with high confidence, successive phases of motor skill development. Their actual existence is well confirmed by clinical observations of selective losses of each of the phases separately. It goes without saying that such an identification of qualitatively peculiar phases of development leads to certain schematization. In reality, the borders and transitions between phases are fuzzy: it is impossible to establish either abrupt time borders between phases or their chronological sequences. Sometimes, one observes overlaps between the edges of two phases or even coexistence of two phases within the same time interval. Here we witness not so much chronological but causal sequence, and the latter one is followed very precisely.

It makes sense to separate the whole lengthy process of skill development into two periods with the border corresponding to completion of the assignment of background components of the constructed movement to background levels and the beginning of mastering by these levels of the components and collaborative work of the levels needed to realize the given movement.

The first period involves: 1) Establishing the leading level; 2) Defining motor composition of the movement; and 3) Identifying adequate corrections for all the details and components of the movement, type and degree of precision required from the corrections, and the nomenclature of responsible background levels. The fourth phase of this period includes actual switching of the background corrections to corresponding lower levels, i.e., the process of automation as we defined it earlier.

A typical characteristic of all the processes during the first period is the establishment of the properties and features primarily responsible for every detail of the motor action, which have to show most specific and strict corrections. These processes define qualities of corrections that would be able to ensure precision and stability of each particular detail required for the movement as a whole; this process is the beginning of the aforementioned assignment of backgrounds. In parallel to the identification of these highly important features, other features are identified, which may be regarded by a coordination level with a degree of indifference thus allowing their higher or lower variability. Thus, during this period of construction, two groups of features are identified: The first group includes features with high demands toward their qualitative characteristics of *precision and standardization* defined by the meaning of the motor task; the second group unites features that can be allowed to vary either by chance or as a result of adaptive processes.

The *second period* of skill construction may be called period of *stabilization*. During this period: 1) The background levels master components of the motor composition switched as a result of automation and – likely a more difficult task – *mutual functional adjustment* of individual background levels takes place with the leading level and with each other; 2) A component of the automation process completes, which can be addressed as *standardization* of motor composition and its components; and 3) Actual *stabilization* of the motor action takes place – an improvement of the resistance of its features and details against perturbing influences.

The most important change in a motor skill during the second period of its construction is broadening of the range of external and internal conditions that define the borders of a range where realization of the skill is not in danger of being destroyed. If, at the beginning of this period, the practiced movement can sometimes create an impression of perfect execution, this only happens in sterile conditions. But even in cases of advanced coordination exercisability: 1) Different types and qualities of perturbing influences keep being considerably different in their efficacy with respect to different background components; and 2) Different elements and features of the motor action keep demonstrating dramatically different tolerance to the introduction of changes of different quality and magnitude, such that, even for a well settled and stable skill, there remain influences and shifts dangerous for its particular aspects and able to cause its deautomation. These selective features of various perturbing influences will be discussed in Chapter 9.

To begin the structured phase-by-phase analysis of the construction of a motor skill, let us point out that this analysis will address skill acquisition by an adult. Everything that could be said about the genesis of motor coordination in the childhood has been covered in Chapter 7; in this chapter, we will make relevant remarks as needed.

The physiological specificity inherent to each of the construction levels is also revealed in their different characteristics seen during skill acquisition. Relevant observations will be described in parallel with characterization of the corresponding phases.

We can claim with confidence that each motor skill represents a *multi-level structure*. Only the most elementary movements can be confined with all their corrections to a single leading level, but such simplest actions do not need skill elaboration with exercise. Obviously, such a multi-level structure, as a holistic coordinated motor action with its leading and background levels, which are all performing their roles properly and ensuring the perfectly correct movement in their specific domain, cannot emerge instantaneously. The first attempt at performing an unfamiliar movement, the embryos of a future skill, are built as *single-level* actions, unless they find, from the very first steps, old, previously elaborated and stored in memory backgrounds. However, even if a person has somewhat useful ready backgrounds, at first, he or she is usually unable to recruit them during a new action and use in new combinations. It is still necessary, at first, to perform the whole movement, nearly in its entirety, under the control and corrections of the leading level.

Of course, none of the construction levels is equipped with an absolutely comprehensive all-around set of sensory qualities and corrections, which could be used for adequate performance of any movement. As a result, at first, the movement is performed with the help of *surrogate corrections*, similar to temporary wooden scaffolding used to build a future stone building. Within the central nervous system, only the cortical sensory-motor systems, *C2* and higher, possess differentiation sufficient for building new motor combinations ad hoc. The cortex of the large hemispheres has the most developed links to sensory fields of all modalities. The cortex is the only structure able to generate an adequate motor reaction to new, unfamiliar stimuli. Of course, at the beginning of skill acquisition, the cortex is forced to perform atypical motor tasks. For example, if a practiced movement is an unusual, difficult labor movement, e.g. mowing, sensory systems needed for its *leading* corrections, with the domination of visual and vestibular contributions, have their projections in the cortex. In contrast, the musculo-articular, proprioceptive sensitivity is poorly reflected in the cortex, and the leading level during mowing is more helpless in correcting positions of body parts and their reactive interactions compared to lower, subcortical construction levels.

In the process of gradual mastering a skill, the central nervous system finds the means to delegate background movement corrections to corresponding background construction levels – those that are best equipped, according to qualities of their afferent systems, for executing those particular corrections. For example, during learning a throwing movement, the general synergy including turning of the trunk on the legs, screw-like motion of the trunk and head, and a broad, sweeping movement of the shoulder girdle and arm is at first corrected by cortical systems, which, as already mentioned, have only weak connections to proprioceptive signals. When, as a result of exercise, this synergy is taken over by the thalamo-pallidar level, which has a very intimate and well-developed system of connections to organs of tactile and proprioceptive sensitivity, the whole coordination of that particular movement aspect becomes more precise, stronger, easier, and more economical. As a result: 1) The leading level is relieved of additional, secondary work; 2) Corrections of the mentioned synergies are moved beyond consciousness; and 3) These corrections are not only improved quantitatively, but also undergo qualitative changes because they switch to most adequate features of sensory corrections.

The very first semantic phase of the elaboration of a motor skill – *definition of its leading level* – takes no time, because this issue is predetermined. If not all, but a by far predominant number of skills that can be led in humans independently by the level of spatial field *C*, at least in their general features, have been learned during childhood and adolescence. Problems an adult encounters for the first time are overwhelmingly dominated by those associated with the level of actions *D*. Besides, problems from this level dominate everyday movement repertoire. As a result, level *D* holds a firm monopoly on the leading position with respect to all new motor skills. Observations show that this happens even when a practiced motor action obviously is related to the competency of level *C*, for example if this is some kind of unfamiliar locomotion, such as biking or swimming. We have to mention that, in contrast to easy and quick switching of background levels, switching the leading level of an already learned skill is a hard and painful process. Since, after learning a locomotor action with the help of level *D*, the central nervous system, at some point, will have to switch to leading control by the adequate sub-level *C1*, the mentioned problem dooms it to experience lots of unnecessary difficulties. Clearly, this explains why children or adolescents master such locomotion tasks much faster and easier: from the very beginning, they put it under the control of sub-level *C1*.

An opposite mismatch happens in children and in higher animals, who do not have a functional level *D* and try to master skills from the level of actions. Examples include failure by monkeys in their imitation movements if they try to mimic problems, which attract them with their appearance but semantically are beyond their abilities.

The second phase of skill acquisition, which is the beginning of actual work on the skill, has been designated as *defining motor composition* for the skill. Recall that under motor composition of movements at the level of spatial field *C* we imply all features related to the form and extrinsic pattern of movement or, to use another term, its *construction*. This is the totality of what has been united under the symbol "Mot" in the reflex loop scheme in Chapter 2. In sport and gymnastic skills, motor composition more or less accurately coincides with what is addressed as movement *style* (for example, different styles of swimming – crawl, breaststroke, and butterfly; and styles of high jump – scissors, Eastern cut-off, etc.). In complex chain actions at level *D*, motor composition involves both structures of individual movement-links and the lists of those links.

The phase of defining motor composition of a skill, in most cases, is not the most effortful one. Many actions (for example, riding a bicycle or driving a car) have been observed by everyone hundreds of times since childhood. For other skills, students frequently have the benefit of demonstration by an instructor. Nevertheless, in the course of this phase, there may be difficulties, which require active intervention. Most frequently, this happens when a complex (e.g., craft) skill is mastered in the process of self-education – recollect the case of Robinson. Further, in many sport-gymnastic and professional skills, there are very important elements within the motor composition, which may be so fleeting that they are hard to identify even in the course of multiple demonstrations. Many movement details, even those well watched, are very hard to reproduce. However, aside from this purely imitating aspect, the phase of defining motor composition always has active and creative contents, no matter whether this phase is conscious and voluntary. Here we mean the individual adjustments of peculiarities of the motor composition to the personal features of the person practicing the skill: body composition, physiological characteristics of the motor periphery, coordination profile, etc. It goes without saying that this phase offers much freedom for conscious rationalization and inventive creativity demonstrated so brilliantly by the Soviet Stakhanov workers.

Everybody knows very well the illusion of being able to reproduce, from the first attempt, movements of a skilled, experienced master performing his or her professional actions. Performed in front of our own eyes and envisioned with perfect clarity in full detail of the motor composition, the movement looks to us so simple and trivial that it is perceived as something easy to reproduce. Children who are not yet under the control of "inhibition centers," which are developed in adults, get caught particularly easily by this illusion and try overconfidently to imitate risky manipulations such as shaving, sewing, and turning on the stove. Everyone who has ever tried to test the impression created by this illusion remembers the peculiar and very bright feeling of confused bewilderment produced by the unexpected disobedience and clumsiness of the hand.[2]

The cause of this phenomenon is clear. Everyone has accumulated during the lifetime a vast amount of skills for the right hand; movements by that hand are performed all the time without getting outside the scope of those habitual skills. Each of these movements has a range of generalization and transfer to similar motor actions, but this does not imply overall coordination development. Being obedient within the scope of habitual everyday actions, the hand seems deceptively obedient in general, which is far from reality.

The point is that, no matter how clear an image of the motor composition of a movement is, the image gives no information on *corrections and recodings* necessary for its execution. The difference between the second phase (defining the motor composition) and the third phase, always following the second one, exactly is in the fact that, during the second phase, the student learns how movements contributing to the skill would *look* (from outside); in the third phase, the student gets (internal) *feelings* of these movements and their sensory corrections. Hence, we should designate the third phase as the *phase of identification of requisite sensory corrections*. During this phase, the central nervous system intensely accumulates streams of receptor (proprioceptor, in the wide, functional meaning of this term, see Chapter 2) information characteristic of the whole variety of external and internal variations of the movement – streams of sensations that cannot be observed in another person. It is obvious that conscious introspection accompanied by critical analysis from the instructor can yield important advantages as compared, for example, with a huge number of mechanical, thoughtless repetitions.

The phase of identification of sensory corrections is, without doubt, the most effortful of the initial, so to say, planning phases of skill construction. From the very first steps, parallel with accumulation of all those proprioceptive sensations, their internal sorting begins. The central nervous system starts to figure out what kind of corrections and of what quality are required by various components of the motor action. After defining what has to be corrected, the central nervous system discovers what sensory systems and what background levels possess the most adequate means for these corrections. Therefore, the phase of identification of sensory corrections imperceptibly merges into the *phase of assignment of sensory corrections* across adequate construction levels.

At the beginning of skill learning, two essentially different scenarios can happen. As already mentioned, at first the movement is built on a single level and has at its disposal corrective resources from the leading level only. By the time the basic, most essential, corrections have been defined, the leading level either is able (well or poorly) to ensure approximately those vital corrections, or it is completely deprived of features needed for some of them.

In the former case, these temporary "crutches" of cortical corrections give an opportunity to move somehow, with difficulty, based on surrogate corrections. Meanwhile, background levels have time to select and form truly adequate background corrections discussed later. This happens, for example, in the already discussed example of the throwing synergy. In the other case, the leading level may not have any, even auxiliary, highly needed corrections, and their absence is equivalent to destruction of the whole movement. Likely, this happens when required, highly important corrections represent large muscle synergies from the thalamo-pallidar level *B*. This is exactly what happens with locomotor skills of swimming, biking, walking the tightrope, etc. In these cases, the leading level of all locomotor movements *C* is well-equipped with all meaningful corrections necessary for piloting the action, but it lacks the ability to control necessary synergies, without which the movement cannot be performed, although they are unrelated to piloting and to the ultimate meaningful goal of the motor action. In such cases, at first, the movement simply does not work: The student falls down with the bicycle or stubbornly gets submerged in water.

It is well known that there is a general, without exceptions, law related to movements of the mentioned type (definitely, not only locomotions). Its origins remained puzzling, and only the theory of sensory corrections offers a simple, exhaustive solution. This law claims, first, that all skills of this type at some moment of learning are mastered at once, in a single leap, as if following an insight. Second, it claims that, once such a skill has been learned, its is never lost throughout lifetime, no matter how long a break happens in its application and how poorly trained the person becomes quantitatively. Learning how to stay on water, on the bicycle, or on the tightrope is impossible to forget, similar to a bright sensation or impression experienced once in the lifetime.

This sudden leap in skill acquisition means that, at that particular moment, a *crucial background synergic correction* has been recruited after being elaborated at level *B*. The movement did not work out until that crucial moment because, for movements of this type, the leading level (neither *C* nor *D*) has no adequate corrections of the required quality, even surrogate ones. The main fact is that the "secret" of swimming, biking, etc. skills is *not in some special body movements but in special sensations and corrections*. This fact explains why these "secrets" cannot be explained with any demonstration (any *movement* can be demonstrated) and why they are impossible to forget over the lifetime.

The issue of recruiting *background levels as experts* in specific types and qualities of corrections emerges straight after the phase of identification of correction requirements.[3] This moves us to the crucial of all the phases of motor skill development – *the automation phase*.

According to the already presented definition, *automation* of a motor action is switching some of its components to lower construction levels, i.e., *switching a set of coordination corrections of a motor action to afferent systems of lower levels, which are most adequate for those specific corrections*. This definition of automation, particularly its italicized part, seems to us most concrete from all the definitions offered until now and also compatible with all the actually observed cases of automation. Movement automation may consist of using already available backgrounds (backgrounds themselves and appropriate automatisms developed earlier for a different skill) and in developing special new background automatisms. Both terms will be explained later.

The well-known fact that automated movement components leave consciousness leads to an important generalization: *In each motor action, only the composition of the leading level is consciously perceived*. In particular, if the leading level is the highest level of group *E*, for example if the action is typing during dictation, unconscious action components include not only all the simultaneous background components from *D*, *C*, and *B* (which accompany typing the meaningful letters of the text), but also successive elements of the semantic chain of actions, which represent, at the time of their execution, independent movements-automatisms at level *D*, for example the movement changing the line on the typewriter. If the leading level is among the highest ones, unconscious successive background chains can sometimes be very complex in their composition and proceed over long time intervals. In such situations, they are sometimes addressed as *reflexive (machine-like) actions* and, in pathological cases, can sometimes last for several days (so-called, hysterical fugues).

The border between *voluntary and involuntary* movements does not coincide with the border between conscious and unconscious ones. If we consider only the leading levels of movements (because technical backgrounds are, as a rule, unconscious), these leading levels are always consciously perceived but can bring about both voluntary and involuntary movements. It seems likely that movements associated with cortical systems (level *C2* and higher) are normally all voluntary; at level *C1*, there is a percentage of involuntary movements, they are very numerous at level *B*, whereas at level *A*, nearly the whole movement inventory is involuntary. Of course, the complete loss of voluntary movements during pyramidal paralysis does not give us the right to equate normal functioning of the pyramidal apparatus with voluntary movements and the extrapyramidal apparatus – with involuntary movements, because this pathological case is associated with very complex changes compared to the normal functioning linked to changes in "triggering mechanisms" of movements, interactions of the patient with the environment, etc., which, taken together, make the whole issue extremely complicated.

In this book, we have encountered two kinds of technical backgrounds of movements, which can be addressed as *proper backgrounds and automatisms*. Proper backgrounds are movement components, which, in other situations, could be seen as *independent movements built at the corresponding lower levels* playing the role of a leading level, i.e., movements available in the inventory of the lower level and able to find in afferent systems of that level motives to emerge and show ecphoria. Movement components created by the leading construction level overlay or, using a metaphorical expression by Graham Brown, "clap" onto these proper backgrounds, of course controlling them in a different way but without creating in the background level any essentially new motor combinations. Examples of such backgrounds include: running as a run-off to high jump or long jump, using walking during mowing, sweeping, etc., or using various types of standing and locomotion at the striatal level *C1* by a tennis player as backgrounds for the main leading coordination at the pyramidal level *C2* – moving the racket.[4]

The nature of *automatisms* is more complex. Automatisms are coordinations realized at lower levels, which by themselves do not exist in the contingents of these levels, do not have motives to emerge and show ecphoria, and get such motives from higher levels. In all situations when the genesis of an automatism can be traced down, it happens to be acquired during the lifetime, i.e., as a result of moving at some point through the stage of automation. As a result of this process, contingents of movements at different construction levels show the emergence of special background components emerging under the coordinating action by higher levels and receiving motives for ecphoria from these levels only. Briefly, *automatism is a background coordination at level N created and showing ecphoria at this level by a higher level P*. In this algebraic formula, different specific levels can play the role of *N* and *P*. Examples[5] can be seen in Table 8.1.

Sometimes, automatisms are classified into lower, serving the leading level of spatial field *C*, and higher, representing background components for the level of actions *D*, but this classification is not important. Of course, automatisms performing special skills at the level of actions *D* are much more complex and variable compared to the "lower" ones: They canthemselves represent rather complex structures with their own backgrounds at even lower levels.

It is necessary to note that it would be completely wrong to view automatisms as stable stereotypes. *Automatisms possess adaptive variability and plasticity* within the ranges defined by the properties

TABLE 8.1

Leading level P	*Level of background automatism N*	*Movement at the leading level*	*Background automatism*
C1	*B*	Walking	Walking synergy
C2	*C1* and *B*	Throwing the grenade	Run-off and swing
D	*C* and *B*	Mowing	Walking and mowing synergy
D	*C*	Steering (in a car, boat, etc.)	Spatial skill of steering
E	*D*, *C*, and *B*	Meaningful speech	Articular coordination

and means of their level; their essential feature is only that they do not need consciousness for their action.

It follows from the definition of automatization that it can and actually does proceed in *several separate steps* corresponding to the number of components that have to undergo automation switching during the skill construction. These separate automation switchings can be spread over time and cover a substantial portion of the whole exercise time. At the same time, each individual switching is always realized not continuously but as a definite *leap* – a novel ability the student suddenly discovers after resuming exercise. Finally, as it also follows from the previous material, automation always represents a *qualitative leap* – both as reflected in a sudden improvement in the performance of one of the aspects of movement and in the sense of a *change in the structure* and character of the controlling corrections. The latter is explained, of course, by the fact that switching of a movement component to another level means switching its control to a different afferent system with all the inherent qualitative subtleties of accuracy, variability, differences between corrected and indifferent components, etc. In particular, a very common version of automation switching is from cortical levels equipped with telereceptors to the proprioceptive level of synergies *B* accompanied by the well-known, characteristic phenomenon of *withdrawing visual control*. The person suddenly discovers that he or she can perform a certain movement element *without looking* and that returning to visual control starts to hurt and disrupt the action.

Two successive phases can be seen in the process of automation of components of a motor action. The first – the switching phase, accompanied by the main characteristic of automation, namely moving the action component outside the field of consciousness – has just been described. The second phase of the automation process ends much later, already during the second stage of skill elaboration; it is addressed here as *standardization* of the component. Standardization represents the fixation and strengthening of both motor appearance of a component and the borders and types of its allowed variability. During this phase, the background level uses one of the typical mechanisms for this level physiological to ensure the required stability and invariance of this movement feature on the background of variable and perturbing external stimuli. We will return to a detailed description of the standardization phase when we discuss the second stage of elaborating a skill.

Different scenarios can be encountered with respect to both technical backgrounds with independent genesis and automatisms emerging during the elaboration of a new skill.

Independent backgrounds required for a practiced movement can frequently be found by the central nervous system in a ready to use form, kept by memory in the storage space of corresponding levels. If one is allowed to stretch the etymology a bit, these storage spaces could be addressed as "libraries of backgrounds" at the brain levels. With full appreciation of the major sin committed against the language, we would like to keep this expression because of the lack of a different, equally brief and expressive one.

Such ready, or almost ready, backgrounds for motor skills at the level of actions can be found, for example, at the level of spatial field, if the learned movement is a locomotion or another independent motor action at that level; such backgrounds can be found among a huge number of those learned since childhood. For example, if the task is to turn *running into a run-up* (which is obviously not the same action), the leading level *D* or *C2* only has to impose on this well learnt, habitual action a few adaptive correctional touches and convert its independent nature into a background. Of course, there are cases when technical backgrounds extracted from a lower level "library of backgrounds" require substantial adjustment and processing, and, finally, there are cases requiring, for a given skill, elaboration of special backgrounds of this category.

Similar to independent backgrounds, *automatisms* can also be found, in some quantity and degree of adequacy, in memory storages of the brain. Since automatisms, by definition, cannot have an independent origin, one can only speak about coordinations developed earlier in the process of mastering a different skill. Of course, all possible gradations are possible, from perfectly matching automatisms to their complete absence.

The phenomenon of using automatisms developed earlier for skill A to build a new skill B is nothing else but the well-known, although poorly studied, phenomenon of *skill transfer*. It was discovered long time ago that exercising one particular type of motor activity produces, in addition to an improvement of the practiced movements, a more or less wide *effect of generalization*. Until now, there has not been much success in either the analysis of basic causes of generalization, or causes of different effects of irradiation of different types of exercise, or studies of regularities defining what motor acts would lead to most pronounced effects of transfer. Thorndike's theory of identical elements tried to explain transfer phenomena, but unfortunately, it failed completely in attempts to predict facts, i.e., turned out to be essentially false. There is an additional difficulty that, sometimes, movements very similar in their motor composition do not show any visible transfer, whereas rather dissimilar movements (e.g., riding a bicycle and skating or even marksmanship) show strong transfer effects.

The theory of sensory corrections turns out to be much more powerful that the Thorndike theory in its ability to reveal errors of the latter theory and better predict phenomena of transfer and generalization. As it follows from the previous text, exercise transfer is defined by the ability to use automatisms previously developed for another skill; however, *automatisms are not movements, but corrections*, and this is the root of the main difference. Obviously, the movements of filing, ironing, and moving the bow during playing double-bass, with all their external similarity, have virtually no shared corrections; as a result, they do not lead to transfer. And vice versa, when empirical observations identify facts of substantial transfer, it is easy to reveal close similarity of sensory corrections.

For example, the skills of riding the bicycle and skating, very different in their motor composition, are closely related by *maintaining dynamical stability over the support surface of zero width*. As a result, the principle of corrections used in both skills, namely *driving up* the movable support under the deviated projections of the center of mass, is the same. In such dramatically different in their motor composition skills as figure skating and marksmanship one can discover common, most important, background corrections: visual precision, confident stability of the gesture, and accurate identification of the crucial time moment. It is quite possible that the gesture of "aiming" with the stone, used by many boys, by holding it in front of the leading eye, while keeping the other eye squinted, is based on a peculiar modification of skill transfer at sub-level *C2*. Although, by themselves, the coordination of the typical boy's sweeping movement of the throw has nothing to do with such sighted aiming, it would have made sense if the throw were performed directly in front of the person, similar to pushing a billiard ball. So, the "sighted aiming" may be interpreted as a meaningful successive element of the throw *P* (not actually used because of its synergetic disadvantage), switched, as a result of transfer, into the chain of the actual throwing synergy *Q*.

Under the term exercise transfer, another phenomenon is understood related to the one we have just described; to distinguish it, we will address it as *transfer by the executive effector*. Under this term, we imply an improvement in a skill practiced by one effector *A* (for example, the left hand) in another, non-exercised, effector *B* (for example, in the right hand). This type of transfer will be characterized later during the analysis of the phase of skill stabilization.

The different roles of automatisms at different levels during skill transfer are clearly characterized by the fact of strong perturbing effects on a highly automated skill of a change, for example, in the familiar handle of the tool, e.g., the shovel handle, familiar ruling pen, scissors, pen, in other words of a familiar grip. These components from the lowest levels are least suited for transfer, whereas similar components from the level of spatial field, for example movement of fingers during typing on typewriters of different brands and models are very easily transferred.

The opposite side of skill transfer is the well-known phenomenon of interference, which we will address as "*successive interference*" to distinguish it from another phenomenon to be described later. In accordance with our interpretation of transfer, successive interference is ecphoria of backgrounds and automatisms not appropriate for the learned skill and either recruited by false associations or (most frequently) involved in the ecphoria together with a whole range of adequate automatisms from the same former skill. Typical successive interference emerges, for example, during steering a row-boat when the movement is exactly opposite to the familiar action of steering a bicycle or a car. According to observations by Gellerstein, during training pilots, professional drivers are, at first, too abrupt in their movements, former glider pilot – too sweeping, former blacksmith – "clamp the control," according to their instructors, etc. All these are interference traces of their former skills.

In all cases, over the whole duration of the automation phase – it does not matter whether it is related to the elaboration of new backgrounds and automatisms or uses to some degree previous ones – there is continuous process of *decreasing thresholds* for the receptor mechanisms that ensure necessary sensory corrections. It follows from the principle of coordinating reflex loop that, if a movement starts to deviate from the desired by the person path, this deviation is not perceived immediately. Excitation produced by the accumulating deviation has to reach a suprathreshold intensity to become perceived and to turn on the correction "trigger," which will introduce corrections into the efferent impulses. When a novice, who has already mastered the main secret of maintaining equilibrium on the bicycle, tries to ride along a straight line, this does not work out for a long time, and the path is sharply winding. The shape of the path is explained by the person's ability to perceive a deviation of the body and bicycle from the vertical only when the deviation reaches a large magnitude. Only at those moments, the crude thresholds of the vestibular and proprioceptive sensitivity are reached and corrective signals to the arms are generated leading to turning the handles. In contrast, an experienced biker feels the deviations so quickly and sensitively that correction signals emerge even during the smallest deviations from the equilibrium and the path of the bicycle shows very small deviations from the straight line. This example is a good illustration of threshold decrease during skill development. Of course, this decrease does not follow an uninterrupted straight path; its graph may show episodes of constant threshold magnitudes and even transient relapses of threshold increase. Such transient deteriorations may happen as results of fatigue, alcohol intoxication (drunken gait), etc. In small children, during mastering the skill of walking (see Chapter 7), medio-lateral body deviations, similar to those in a drunk person, are overcome very early in the process, but they can still be seen along the anterior-posterior axis in cyclographic recordings as the large variations of progression velocity, and even more – of acceleration, which can last until the age of 4–5 years.

No doubt, and this is supported by our extensive experimental material, the described decrease of the threshold of correction "triggers" results not only from their quantitative changes but also from large qualitative changes, namely replacing one group of receptors with another, more sensitive and quickly responding. Exactly this type of trigger designates a switch to another afferent system (for example, from visual to proprioceptive-tactile one), but there are definitely cases of switches within a single afferent system from more crude instruments in its inventory to more precise ones.

It is convenient to end the aforementioned period of motor skill acquisition with the automation phase. Along with the recruitment and formation of backgrounds and automatisms, a new, highly active period begins involving complex synthetic work with those backgrounds and automatisms. The second period of skill construction is characterized by a number of qualitatively different phases, which do not display identifiable time borders. These various processes typically run in parallel, at all times exerting mutual effects on each other.

We are going to identify *three main* processes within the second period, but there may well be other construction processes, which have not yet been identified. The three main processes or phases are: 1) The phase of *mutual adjustment* of coordination elements of the skill; 2) The phase of *standardization*; and 3) The phase of *stabilization*.

If the former phases of skill development, which formed its first period, could be compared to distributing the play parts among actors, writing the parts down, and learning by each actor of the part, which had been read by the director – the leading level, the mutual adjustment phase corresponds to the period of rehearsals of the piece. This is the phase of intrinsic linkages of movements, ensuring the joint, negotiated work of corrections both among different background levels and between them, taken as a whole, and the leading level of the skill. It is necessary to take into consideration that all the levels have to share both the single final common path of each motor unit and the very limited number of effector conduction pathways in the spinal cord. Impulses from all three levels of the extrapyramidal system ultimately are transmitted by the centrifugal tracts of the red nucleus group; the same line gets a substantial proportion of impulses of the premotor system, sent from level *D*. The pyramidal projection system also serves levels *C2*, *D*, and the higher cortical group.

However, there is no doubt that, with respect to each more or less complex action, major difficulties for mutual adjustment of the levels and the main "bottleneck" is formed not by neural pathways or centers. A much more serious factor is that all these volleys of impulses and correction loop processes have to act ultimately on the same executive system: the same muscles, the same inertial masses of body segments, and the same joint degrees of freedom. Exactly this necessity for a few riders to use a shared horse creates difficulties that are not easy to overcome. It would be best to designate them as *simultaneous interferences*, in contrast to the aforementioned ones. A typical example of such interference is the contradiction between the strong grip of the handles of the bicycle produced at the beginning of skill acquisition by the rubro-spinal level *A*, and the sensitive and subtle corrective actions on the steering bar in response to vestibular signals controlled by the striatal sub-level *C1*. Of course, simultaneous interferences can take place across different effectors, for example, the interferences emerging at the beginning of skill acquisition between the automatisms of runup and push-off and the automatism of precise placement of the pole during learning pole vault. Sometimes, simultaneous interferences are overcome so easily that they are not noticed. If they do create significant difficulties, there is an accurate and typical symptom of *plateau* known to all coaches and instructors – a transient stop or even regression and discordance of a movement that has just started to work out in an automated form. Such difficulties can always be resolved, but they require careful and thoughtful tactics. No doubt, switching temporarily to different types of exercise or even introducing a complete stop in the skill training process are more rational than trying to fix those interferences by "brute force." As shown by observations, if the nervous system is forced to solve a motor problem while being caught in-between incompatible requirements of two different corrections, it has no choice but to *compromise quality*. This happens, for example, when there is a contradiction between corrections of precision and requirement of fluency while playing a musical instrument. If the coordination systems are not given an opportunity to use the plateau to search for an adequate solution for the problem, either by mutual adjustment of the contradicting automatisms or by bypassing them and elaborating more mutually tolerant new automatisms, there is a risk of nudging the system along the path of least resistance, which could hurt both precision and smooth, evenly sounding fluency. An essential damage caused by such compromises of quality is that re-learning a new movement form from a closely related form is always difficult, and the difficulty increases when there is more similarity between the two. This resembles the increase in difficulty of conditioned differentiations with a drop in the contrast between the two stimuli that have to be distinguished. It is very hard to get rid of such compromises.

The processes that we unite under the name of *standardization processes* develop in parallel with the processes of mutual adjustment of skill components. The notion of standardization requires an exact definition.

Any motor action proceeds in a complex, time-varying force field generated by both external and reactive forces. Given the large excess of degrees of freedom, these forces generate a variety of disruptive influences on the organism, which are handled with the help of sensory corrections. This struggle proceeds in two directions. Every movement aspect has a degree of *acceptable variability*,[6] which the construction levels cope with either because of their indifference or in order to use this variability for purposes of adaptation. Fighting the disruptive effects is directed, first, at preventing the movement from getting outside the range of acceptable variability and, second, at ensuring that various disruptive factors do not exert destructive effects on the movement, do not cause its disruption and deautomation. The first line of defense tries to keep stable, standard movement forms – this is the line of standardization. The second line tries to endow the motor process with resistance against deautomating factors; this is fight for *stability*, a feature to be described later.

We already know that standard movements do not mean standard innervation formulas or patterns. Everywhere, where one encounters standard features of movements, the standardization does not emerge by itself: The central nervous system fights to achieve it and elaborates, for this purpose, special corrections and automatisms. Therefore, clearly, standardization cannot be a goal in itself, it is fought for where it is essential to ensure movement precision, economy, or stability. Moreover, standardization of movements is more of an exception, it creates an impression of an old form, which, in some sense, is close to extinction. In contrast, as we have already seen, functioning of the higher, most differentiated, coordination systems of the brain is characterized by adaptive movement variability. Nevertheless, standardization, as the means of keeping movements within the borders of reasonable, acceptable variability is inherent to all brain systems and levels, albeit with different major characteristics.

Processes of standardization proceed in the most peculiar form in movements with complex synergic automated backgrounds from the thalamo-pallidar level *B*: locomotion movements, ballistic movements, etc. In skills of this group, standardization is inseparably merged with processes of stabilization in the joint aspiration to reach maximal coherence, economy, and stability of movements. We have accumulated a large amount of relevant experimental material, which provides a more or less detailed account of the process of such skill acquisition. From the very first steps of practicing such a skill, we observed *three* clearly identifiable and qualitatively different stages.

According to the earlier definition, *movement coordination is overcoming excessive degrees of freedom of moving organs* or *its transformation into a controllable system*. This process can take different routes, which is also revealed in the case we consider. Comparing the materials on the development of the biodynamic structure of walking and running with the cyclogrammetric data on running of world record holders (work by T. Popova and the author) has shown how, in the course of skill development, there is a qualitative change in the attitude of the organism to the numerous degrees of freedom of its motor apparatus. Quite naturally, the excess of the degrees of freedom and the associated extreme difficulty in the control of motor periphery lead, in a beginner, who tries to master movement of a complex multi-link chain (for example, the whole body) for the first time, to instinctive attempts to *decrease the number* of degrees of freedom he has to deal with. To achieve this goal, the novice *fixes joints* by co-activating all their muscles. This is the simplest way to solve the problem: the student eliminates excessive, interfering degrees of freedom by absolute fixation of all of them except the one that is directly involved in the current segment of the movement. Such co-activation of all the antagonistic pairs of the chain is frequently accompanied by broad irradiation (grimacing, popping the tongue, etc. in exercising children), and effectively protects the student from the development of reactive forces. But it is easy to see that this method of coordination is extremely non-economical, leads to spending a lot of expensive static tetanic contractions and, because of that, fatiguing. The fixation leads to the appearance of stiffness or constraints, which are very typical of novice's movements during synergic motor actions, in particular in many sport and gymnastic exercises.

As the skill is being gradually mastered and correction triggers become more and more sensitive, this stage is replaced by the next one. As shown in Chapter 2, each link in a kinematic chain acts during its movement on the adjacent links, connected to the first one directly or indirectly, and causes in those links a variety of reaction forces. Each active muscle force applied to a point within a kinematic chain induces a whole array of complex reactive consequences in all other links of the chain. A novice tries to extinguish those numerous reactive forces, directly related to the redundancy of degrees of freedom, by fixation of all the joints. As the skill develops, the novice, gradually and very carefully, *releases* degrees of freedom, one after another, and learns to fight reactive forces not by continuous tetanic fixation, in a sense anticipatory and non-selective, but by an ability to parry these forces in a timely manner by *brief physical impulses* sent at proper times to proper muscles. Therefore, at this stage, movement becomes much more economical compared to the previous stage. During the development of walking and running, children pass this stage at about 5–7 years of age. This leads to a dramatic change in the patterns of curves characterizing dynamic efforts in the centers of mass of the leg and its segments. In early age, corresponding to the first, fixation, stage, during a full step cycle of a leg, these curves show only two force waves: a direct (forward) one and a reverse (backward) one, as one should expect from an oscillating system without reactive dynamics. At the second stage, the gradual release of mobility causes a huge overproduction of brief, intensive, phasic waves in the leg force curves as compared to the curves of both younger children and grown-ups; in grown-ups, these force waves are represented in a modest number, they have very different amplitudes and emerge with high reproducibility. Obviously, during the second stage, literally every reactive force emerging in the moving organ causes the central nervous system to generate parrying or equilibrating impulses, which produce the ragged and shapeless appearance of the force curves seen at that stage.

In the course of development of a motor skill, sooner or later, a *third* stage emerges, which is even more precise in dealing with reactive forces: The organism learns literally to *use* them. In such universal skills as, for example, walking, the last stage is achieved by all people with no exception, whereas in special skills, it can be frequently observed in masterful experts of sport or physical labor. It corresponds to the full recruitment of the level of synergies *B* into background work.

Movement can be constructed in such a way that a dominant proportion of the forces emerging in the periphery not only fail to disrupt it but contribute to it. If the central nervous system finds how to use reactive forces in such a way, this leads to large economy of muscle activity. Besides, a movement allowed to follow a trajectory required by the biomechanical nature of the moving organ happens to be particularly smooth, easy, and good looking.

For many movements, particularly for movements with large amplitude and displacement of large inertial masses, such as locomotion (running) or ballistic movements (blacksmith movements, throwing movements, etc.), the organism finds forms when even a slight deviation from the correct trajectory immediately brings about reactive forces that tend to move the effector back to the original trajectory. Two opposite movement structures – those with disruptive reactive forces and those with reactive forces supporting the movement and contributing to its stability – are very close to the phenomena of stable and unstable equilibrium from basic mechanics. Reactive force caused by a particular movement form will either disrupt the movement and it will only be possible with the help of active damping of those reactive forces (running child, novice skater, etc.), or reactive forces will contribute to the movement, guard it against deviations, and restore equilibrium at the very early phases of perturbations. In the latter case, movement trajectory with the mentioned properties can be compared to a gutter with the moving organ rolling down the gutter as a small ball, and the walls immediately push the ball into the gutter when it starts to deviate toward one of the walls. The role of the walls in the gutter during movements with stable structure is played by reactive (and, sometimes, also external) forces. Hence, such a movement possesses *dynamical stability*. If such a dynamically stable form is discovered and learned, clearly, during large portions of the movements, it would not only be unnecessary to dampen the reactive forces with tonic or phasic means, but, vice versa, effector impulses and active muscle forces would be saved as a result of substitution with the reactive forces.

Most commonly, a person who has reached this level of development of a skill perceives subjectively only one aspect of the task without an ability to report on it consciously. The subject does not perceive the mastered reactive forces that help him or her to move and perceives only one's own active muscle efforts, which can now be spent in much smaller amounts. Besides, the ability and desire to use most of the reactive forces allow a true master to *release all the degrees of freedom* of the kinematic chains, i.e. the proprioception leads to the feeling of eliminated fixation in all the joints. The skill to use fully the reactive forces and the associated maximal economy in spending active muscle work is perceived subjectively by masters as some kind of *relaxation*. Of course, this "relaxation," representing the highest form of coordination at the level of synergies, refers to inactive rather than active muscles and, consequently, has nothing to do with asthenia.

We observed dynamically stable forms of the described type in many movements and their background components. Clear dynamically stable structures can already be seen among motor actions at the lowest level *A* – these are quick vibrations following the pattern of forced oscillations (Figure 8.1). Further, such forms are clearly seen during movements of the arm with the hammer during cutting with the chisel (see earlier Figures 2.4 and 5.27). During the analysis of this process, we could use the large body of material to witness with extreme clarity how variability of one of the trajectories (in one person or between two persons) drops with an increase in speed or mass of the effector moving along that trajectory. The most reproducible, varying very little in all subjects, were the hitting phases of the hammer head and the center of mass of the arm with the hammer; in contrast, the most unstable and variable were the trajectories of slowly moving and not carrying additional mass centers of the elbow and shoulder joints. In such quick ballistic movements, measures of freedom of mobility of the moving organ happen to differ for kinematic and dynamic analyses. *Kinematically*, the extremities of experts in the described movements are maximally free, all the degrees of freedom are independent; in contrast, *dynamically*, the movements are of forced nature, which pegs them to well-defined trajectories, i.e. those having only one degree of freedom. Therefore, *in dynamically stable movements all the degrees of freedom, except one, are constrained by reactive forces* inherent to the particular movement structure. Since reactive forces are not under the control of the central nervous system and are not produced by effector impulses, it is clear that movement corrections with reactive forces proceeds *without active corrective impulses*, i.e., that movements with such structure are economical not only from the point of view of muscles and neural *effector processes*, but also for central neural *coordination mechanisms based on signals from receptors*. When a movement enters a dynamically stable track, it rolls down the track as a train over the railway, and its leading level – the driver – does not have to control the steering wheel.

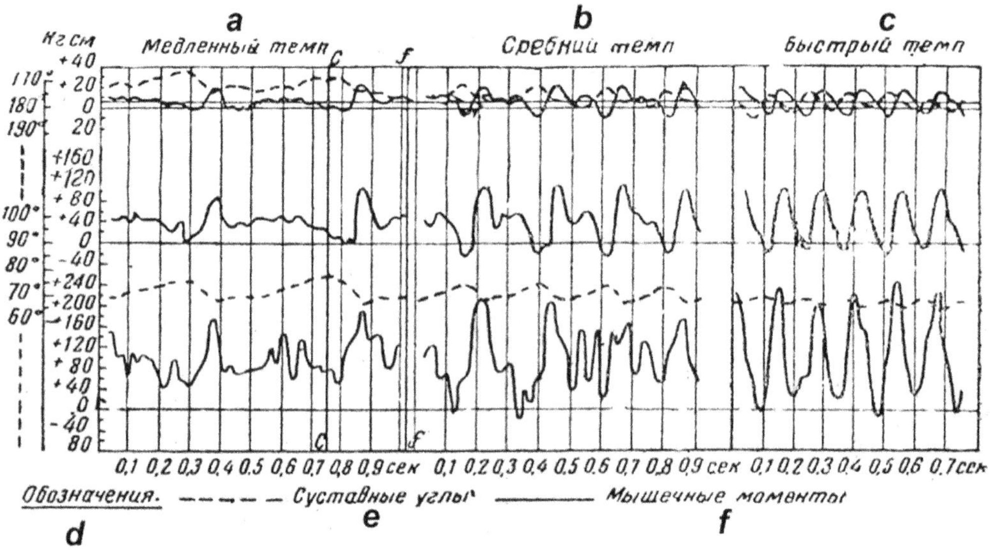

FIGURE 8.1 Curves of the joint angle changes (top to bottom: wrist and elbow) and of the moments of muscle forces.

Top to bottom: the wrist, elbow and shoulder muscles in three episodes of continuously accelerating octave vibrato by a world-known pianist. The figure clearly shows that, during the slow and medium tempi, regular rhythmical movements (joint changes) are accompanied by the very irregular course of the elbow, and particularly shoulder, curves of the moment of muscle force due to the continuous fight against the arhythmical, reactive forces of the multi-link pendulum of the arm. Only during the highest tempo (about 7 strikes per second), these irregularities of the muscle curves became smooth: partly – because of the insufficient for this high rhythm recording frequency, but mostly – due to the transition of the movement structure into a dynamically stable form of forces elastic oscillations.

a – slow tempo; *b* – medium tempo, *c* – fast tempo, *d* – caption, *e* – joint angles, *f* – muscle moments. (Source: work by the author and T. Popova, Laboratory of Musical Pedagogy, 1927).

A striking example of using reactive forces is presented by the biodynamics of running by the world record-holder J. Ladoumegue; we recorded and studied his movements in 1934. In fact, even less famous runners can demonstrate very impressive examples of meaningful exploitation of the reactive dynamics. Here is a typical example.

When the swing leg is already in front of the body and its knee joint angle is close to 90°, the supporting leg loses contact with the ground (Figure 8.4). The loss of contact is not produced by muscles of the supporting leg but in a very different way. During the whole support time interval, the supporting leg generated efforts directed upward and produced the common center of mass acceleration directed upward, and at the end of the support phase – upward and forward. At the last moment, the *swing leg* generates an active muscle impulse directed downward, i.e., a fast, vigorous contraction of the hip flexor muscles. This impulse generates a large force (Figure 8.5) in the center of mass of the swing leg, reaching 40–60 kg during running at a moderate speed and sometimes exceeding 100 kg during sprint. This downward force causes a simultaneous *reactive force* directed upward and applied to the common center of mass of the body minus the active leg, and this latter force jerks the body and the other leg upward, when the other leg finishes its support function. However, this is not the end of the reactive dynamics produced by the flexion impulse in the swing leg.

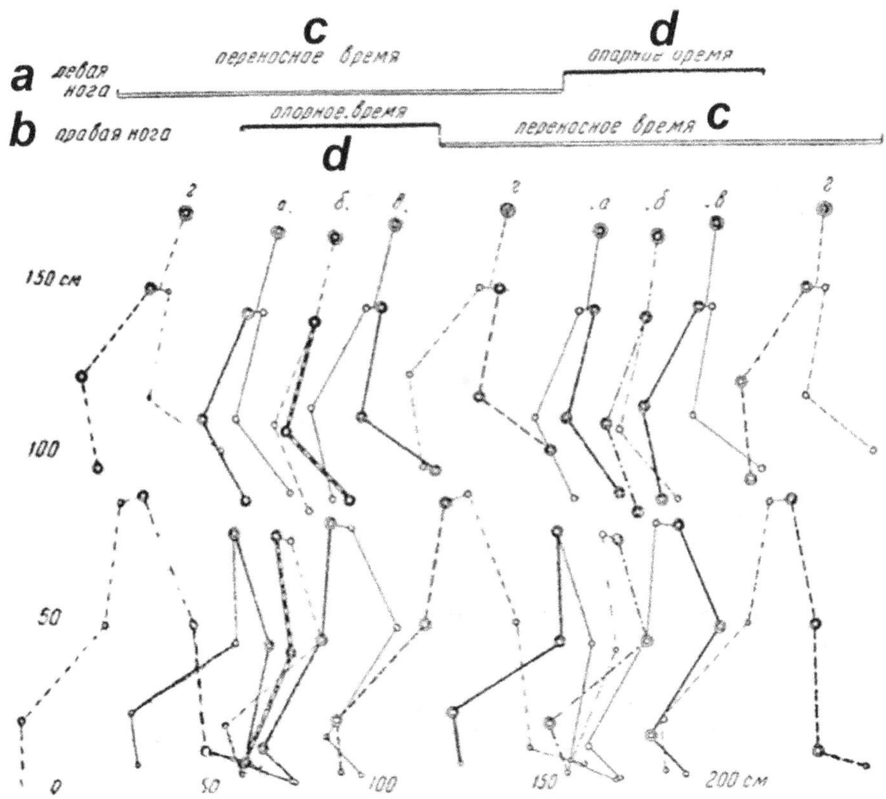

FIGURE 8.2 The main dynamic phases of running of a novice (a middle-age person with a sedentary profession).
А – б – в – the period of support push-off, г – the flight phase. Compare to the running of the world record holder shown in Figure 5.14.
a – left leg, *b* – right leg, *c* – swing time, *d* – support time.
(Source: work by the author, NTK NKPS, 1926).

As shown by our measurements, the speed of foot movement upwards after the end of the support phase is so large that a free material point object thrown with this speed would have reached the height of 4–5 m. Actually, the runner's foot lifts by only 0.5–0.6 m. The excess of its kinetic energy is spent on the energetic knee flexion and movement of the foot and shin forward along the forced trajectory. As a result, one can find in the initial impulse of the swing leg, similarly to features of a sprout in a seed, causes of the whole complex chain of reactive coordination that pre-defines movement of the leg, which has just left the support, over a few tens of percent of the step duration.

A large number of more subtle examples of using reactive dynamics are found in the analysis of movements performed by high-level experts. Many of those have been reported in our experimental studies of walking, running, and military marching. Without dwelling on these examples, let me note a characteristic observable sign of the described stage of skill development, which is the abundance of very subtle force transmissions across segments of the moving kinematic chain clearly seen in the cyclogrammetric curves. The point is that, in most cases, reactive forces are accompanied by the emergence in adjacent segments of wave couples of opposite signs

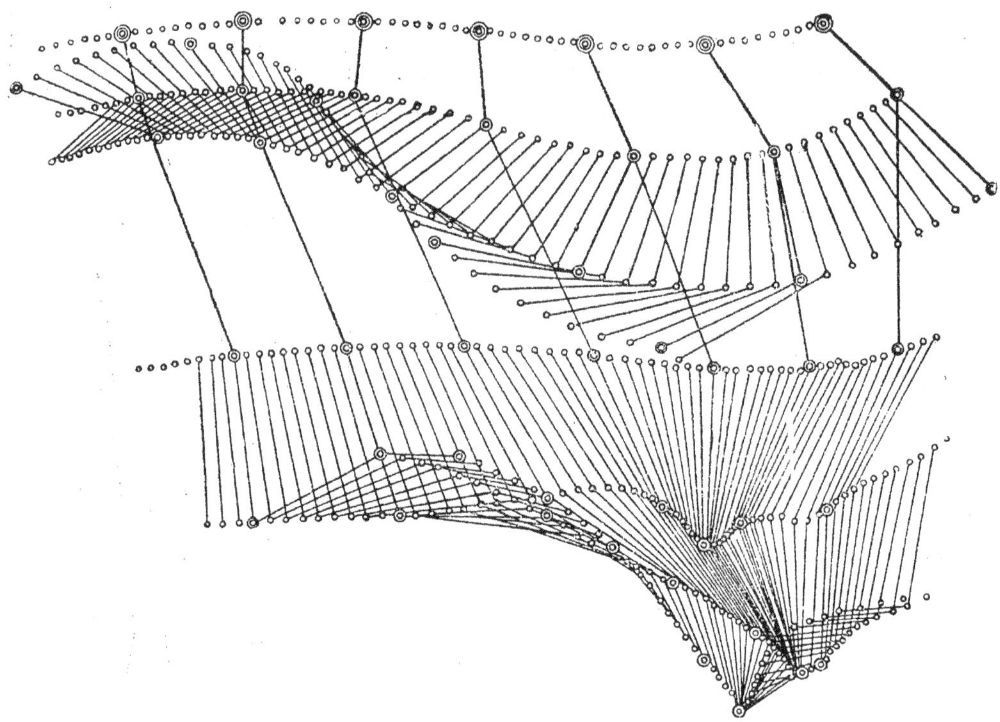

FIGURE 8.3 Consecutive positions of the left side of the body during long jump with the runup by a person who cannot jump.
Compare to the movements of an expert – Figure 5.20. Note the landing prior to the push-off on the half-flexed in the knee leg, which very strongly attenuated the push-off, and the nearly horizontal path of the hip and head following the push-off, in stark contrast to the expert's movement.
(Source: L. Osipov, GCOLIFK, 1940).

in the acceleration curves. This is natural: When one segment, under the action of an external or muscle force, experiences acceleration in a certain direction, the adjacent segment, under the action of reactive force, shows symmetrical acceleration in the opposite direction. This is the opposite-sign couple. Whereas, during early stages of skill development (for example, during running by small children), the force waves are high-amplitude, poorly differentiated and, in most cases, show the same sign over all the limb segments, experts, such as Ladoumegue, display similar waves commonly split into families of subtly differentiated waves with a large number of opposite-sign couples, directly suggesting reactive force transmission. Since the kinematic leg scheme and the number and distribution of degrees of freedom are the same in an expert and in a child, clearly, reactive forces emerging in the legs of these two persons during approximately the same movement have to be qualitatively similar. The absence in the child's curves any detailed differentiation, which is common in Ladoumegue's curves (Figures 8.6 and 8.7), can only mean that the child uses means to dampen the reactive forces whereas Ladoumegue gives them all the freedom.

A very impressive addition to the previous text is the observation by Popova of a late stage of childhood running development. When a 7–8-year-old child masters the aforementioned second stage characterized by over-development of physical waves and when his or her dynamical curves gradually start to show adult-like patterns, the curves start to show the mentioned differentiation,

FIGURE 8.4 The running phase corresponding to the reactive loss of contact by the foot of the supporting leg. One can very well see the contraction of the hip flexors of the right leg.
(Source: J. Ladoumegue, work by the author, CNIIFK, 1934–1939).

force wave transmissions, and opposite-sign couples. If one of two children of the same age, during running at the same speed, shows in the cyclograms more differentiation compared to the other child, then force *amplitude* in the former child are always *smaller* than in the latter one. In other words, given the same ultimate performance, active force expenditures are smaller for larger differentiation of dynamical curves. It is hard to find more convincing support for the economy of dynamically stable patterns.

In Chapter 4, in the context of general characterization of movements and backgrounds at the thalamo-pallidar level, we have already explained and justified the characteristics of *discreteness* and *standardization* of dynamically stable movement forms typical of this level. Without repeating these arguments here, let us only recall that the paucity of actually possible solutions for the problem of finding dynamically stable forms of meaningful movements by the skeletal-motor apparatus necessarily leads to the divergence and standardization of these forms. Besides, the similarity of all humans with respect to the structure of the motor organs explains the inter-individual similarity of dynamically stable standards not only for such ancient and common actions as walking and running, but also of various types of hopping, throwing, swimming, striking, etc.

There are only a few comments left to be added to the presented story on dynamically stable movement forms.

First, it is clear from the previous text that dynamically stable movements are not always cyclical. For example, a range of singular motor actions such as various jumps, large throwing chain

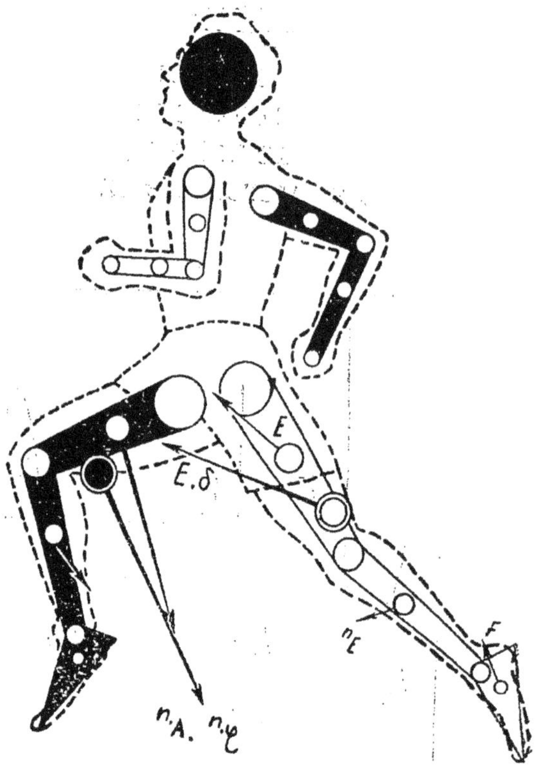

FIGURE 8.5 A scheme of efforts in the center of mass of the leg segments and whole legs during the phase illustrated in Figure 8.4. The monotonic flexion effort ($n_A - n_\varepsilon$) is the source of the reactive force pulling the left leg upward.
(Source: J. Ladoumegue, work by the author, CNIIFK, 1934–1939).

synergies (javelin, discus, hammer, shot-putting, etc.), and sweeping-striking movements by the axe or sledgehammer are well suited for constructing dynamically stable forms.

Second, let us emphasize that we have never encountered movements with dynamical stability that would not be based on synergies of the thalamo-pallidar level – in other words, movements that would not be completely deautomated and deeply disrupted during disorders of that level (e.g., in parkinsonism). This implies that the construction and stable maintenance of these forms, obviously, needs the qualitatively special synthetic tactile-proprioceptive afferent signals that characterize this construction level. As already mentioned in Chapter 4, the achieved standardization contributes to both strengths and drawbacks of this level.

Indeed, standardization of complete motor actions is very much atypical of any of the higher levels. At the level of spatial field *C*, high standardization is maintained only for some of the crucial movement aspects or phases, in particular phases associated with exceptional precision. In such cases, level *C* corrections very easily allow broad indifferent variability in not-so-crucial parts of the trajectory and focus them, carefully and precisely, at crucial points of the movement. As far as the problem of movement stability is concerned, i.e., protecting it against disruption, the tactics of higher levels is exactly opposite to that of level *B*. At level *B*, the main weapon against disruption is dynamical stability with all its consequences. At levels *C* and *D*, the same problem is solved primarily with the help of adaptive variability.

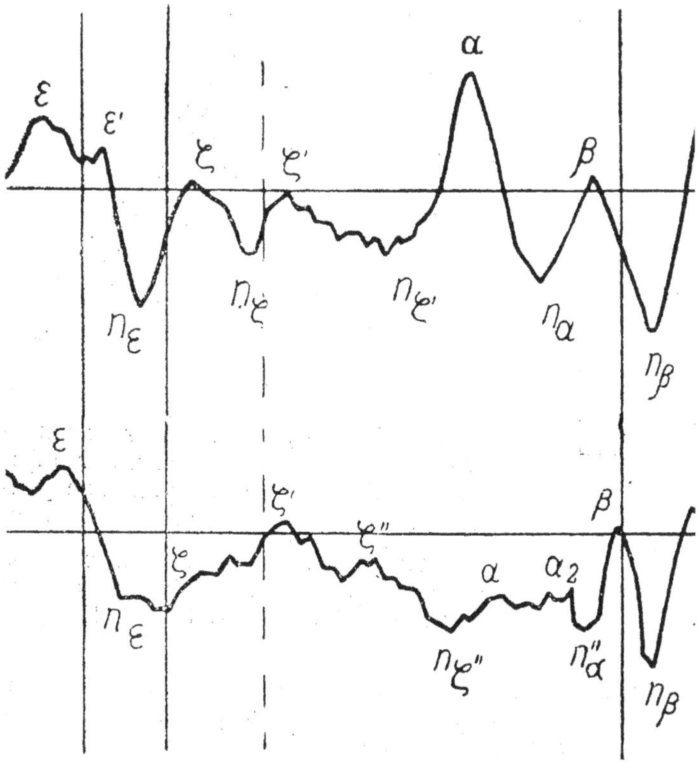

FIGURE 8.6 Curves of the longitudinal force component in the center of mass of the upper leg (top) and shin (bottom) during the second half of the swing interval during running in the interval from ε to n_β.
The subject was J. Ladoumegue. The figure shows very well the strikingly disjointed pattern of the curves suggesting the richness of the liberated reaction forces. Compare to Figure 8.7.
(Source: work by the author,VNIIFK, 1936–1939).

The last phase of the construction of motor skill to consider is the *stabilization phase*, which we will start to characterize now. Let us begin with an example.

Two different performers, *A* and *B*, can perform the same motor action with equal degree of perfection and show the same characteristics of precision and automation when the action proceeds in an optimal, stable environment. However, if a complicating, disrupting factor emerges, of an external origin or due to variations introduced into the ongoing movement by the performer, a dramatic difference between the motor skills of the two performers can be revealed. These external factors or intentional modifications of the movement could lead to a complete or partial disruption of one of the movement components, i.e. its *deautomation*, in performer *A*; meanwhile, performer *B* could show no movement disruption under the influence of the same factors. In the modified conditions, *B* would show much higher resistance against disruptions, higher robustness of the movement automation. We will address this *resistance against disruptive factors* with the work "*stability*" of the motor skill or its components and features.

It is convenient to classify factors causing disruption into the group of interferences and complicating factors of *collateral nature* and those emerging *within* the performed motor action itself. Both groups can involve both exogenous and endogenous factors.

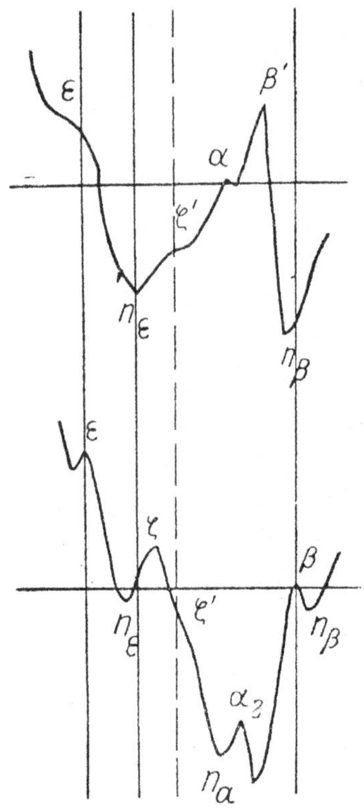

FIGURE 8.7 Curves of the longitudinal accelerations of the knee (top) and ankle (bottom) during running of a 3.5-year-old child.
The second half of the swing interval, from ε to n_β. Note the very primitive pattern and absence of wave differentiation. Compare to Figure 8.6.
(Source: work of T.S. Popova, VIEM-CNIIFK, 1937).

The first group includes such exogenous factors as distracting noise, dazzling light, chill, vibrations, pushes, etc. Examples of endogenous factors from the same group are fatigue, headache, intoxication, malaise, problems with functioning of certain receptors, etc. It is well known that these factors cause a wide range of disruptive, deautomating effects in persons with different degrees of skill elaboration and stabilization.

The same group of causes of deautomation should also include corresponding *organic disorders* of the central nervous system discussed in Chapters 6 and 9.

In healthy persons, it is much more interesting to consider complicating factors *directly related to the motor problem* for the phenomenon of disruption. These factors can also be subdivided into two groups, exogenous and endogenous.

The former group involves complications in the motor task itself: initial conditions; switching from a familiar tool or material; slipperiness, viscosity, uneven surface – during locomotion; malfunction or atypical situations – during labor actions, etc. These are the complications that highlight the degree of mastering a skill. Active exploration of complications occurring during the practiced task takes place over the whole duration of the second period of skill elaboration to ensure that none of the complication can catch the performer off guard. It goes without saying that conscious, systematic selection of task variants and complications during this phase by the student or their instructor can speed-up substantially the process of skill acquisition and improve its stability.[7]

The endogenous group of task-related disruptive factors should include *changes in the execution of motor solutions* that render the task unsolvable. First of all, these will include *excessive variation* in the motor composition, beyond the limits of acceptable variability. Much more important, however, are *changes introduced into the level structure* of the performed action.

Deautomation happens when: (1) For whatever reason, a subordinate level of automated backgrounds becomes the leading level; and (2) When a movement led by level *N*, with an elaborated automated background, switches to another, unusual level *P*. The first case has a well-known, outstanding illustration in the classical literature – in the description of mowing by Levin (*Anna Karenina*). The two mowers, the experienced elderly peasant Tit and the gentleman-novice Levin, very closely reproduce the two metaphorical persons, *B* and *A*, in the introductory example. Whereas the mowing skill in the older peasant is characterized by high stability, Levin's work is deautomated by a whole range of disruptive factors. In particular, there is a vivid description of deautomating effects produced by switching background levels to the leading position. The skill of mowing is built on the level of actions *D*. As long as Levin kept level *D* as the leading level, i.e., the operation with the grass had a leading semantic motive – everything went smoothly and the synergic background worked perfectly.[8] As soon as Levin started to pay attention to movements of his body, i.e. turned the background level of synergies into a consciously perceived leading position, deautomation emerged immediately. This type of deautomation is also known from the funny story about a centipede who was once asked in what order it moved the legs and what the leg number 17 was doing when the leg number 28 was lifted; the centipede could not make a step. Of course, this is true the least for actual centipedes.

The second aforementioned type of deautomation can happen, for example, when a student who has learned a piano passage as a purely transporting motor action at the level of spatial field *C* is asked to perform it under the control of artistic auditory perception, i.e. at the leading level from group *E* – as musicians and instructors say briefly: "to play and to hear." As soon as the student shifts to this new afferent system, the rather smooth passage becomes immediately and completely disrupted. Another example of such a switch to a non-practiced level is the deautomation of gait during walking over railway sleepers, which has already been mentioned for a different reason. The danger of emergence of such deautomations during learning a skill, i.e. fruitless losses of time and effort, has to encourage teachers and trainers in sports, labor, musical performance, etc. to design training tasks and exercises with much care. If such an exercise is itself built or practiced by mistake at a level different from the one, which has to lead the learned meaningful skill, automation developed during this exercise will be disrupted anyway as soon as the student tries to switch to the leading level of the fully developed skill. Among such errors are, for example, practicing playing piano on the mute keyboard or (even worse) on the drawn keyboard, practicing skating or swimming by exercising on the floor or on a bench, etc. The same danger of deautomation is present for all imitations of actual movements such as practicing the movement of cutting with the chisel while using a small, wooden hammer or practicing preparatory movements to shot put without the shot in the hand. At best, such pseudo-training will be useless; in the worst-case scenario, it may disrupt automatisms at the proper level, which could have accumulated in the students beforehand. Let us remind that switching or complete replacement of the leading level is always a very difficult and strongly deautomating process.

So, *disruption* can be caused, first, by *complicating influences or changes* able to violate or deautomate the movement and, second, by emergence of excessive movement variations that cannot be handled successfully. At the beginning of the second period of skill construction, stability of background components that have just been switched to corresponding lower levels is close to zero, and the whole construction can be destroyed as a house of cards by the first blow of the wind. During the second, stabilization period, different details and aspects of the motor action gradually increase their stability with respect to the disruption phenomena. These various features and details of the movement reach their highest level of stability at very different rates depending on properties of each given correction and on individual features of the psychomotor profile of the student. Further, different aspects and details of a motor action can reach very different ultimate degrees of durability and *stability*, different across movement characteristics and also, within each characteristic, with respect to different disrupting factors. Each background component of a movement and each sensory correction have their own particularly vulnerable aspects that persist through the process, their Achilles' heels, where stability is stuck hopelessly at a relatively low level. This endows the phenomenon of disruption with *very specific selectivity* depending, on the one hand, on the nature and character of background components and corrections and, on the other hand, on features of the disrupting factors.

Coordination systems of the central nervous system react to changes in external influences either with *compensations*, corrections directed at keeping the corresponding movement details that took the hit *standard* (for example, this is a typical behavior of the level of synergies), or with introduction of *adaptive variations*. In the latter case, degree of stability of a movement detail is directly defined by the breadth of variations, which are still harmless for the movement from the point of view of its disruption.

In Chapter 5, we have shown that movement variability, both accidental and purposeful (when it displays such features as switching, plasticity, extemporaneousness, vicar interchangeability, etc.), has specific qualitative features typical of each of the levels, which reveal themselves clearly on the general background of quantitatively increasing possible variability of movement details in the bottom-up direction. During the process of skill acquisition, the degree and variety of variability, which the student has the courage to allow with respect to each background movement component, is gradually increasing. The border between harmless for the movement variations and those that already exert disruptive effects is slowly being shifted.

The process of *standardization* of movement details, which is the essence of the second stage of automation, and the just mentioned process of *broadening the range of admissible variability* move toward each other in the direction of a single, common for both, goal – *stabilization* of the motor action. At the beginning of skill acquisition, movement stability against disruptive influences introduced by variability is low, while variability caused by the inability of background levels to introduce compensatory corrections to the varying external conditions is substantial. Naturally, this leads to a large number of disruptions – consequences of destructive deautomation. With progress in the process of skill acquisition, background levels become gradually better in ensuring standard and stable movement details assigned to these levels and, at the same time, the range of variations, which the central nervous system can afford as harmless, increases. As a result, the number of situations when involuntary, undercompensated variations go beyond the allowed range gradually approaches zero and, usually, reaches it sooner or later.

Whereas the main means to stabilize skills and their components at level *B* are movement standardization, maintaining dynamically stable movement forms, and sticking to patterns and stereotypes, at higher levels we encounter an opposite picture. A single phenomenon dominates the hierarchy of construction levels, namely an increase in the bottom-up direction of the phenomenon of *switching* with an increase in breadth of the range of skill application. This phenomenon is directly related to the similar increase in the adaptive and compensatory variability of all movements described earlier. However, this feature of generalizability does not emerge by itself from the very first steps during elaboration of a new skill at one of the upper construction levels: Its development is the essence of the *stabilization phase*.

As it has already been demonstrated with earlier examples, adaptive skill switching at the upper construction levels and the range of admissible variability both increase in the stabilization phase. Skills at the level of spatial field show an increase in variability of the shape of trajectories, interchangeability of effectors, and switching of techniques to solve one and the same ultimate problem. This switching forms the foundation of both the so-called plasticity of the nervous system and the broad range of *compensations and substitutions* typical of movements and skills at the level of spatial field. Adaptive variability and plasticity reach even higher levels for skills at the level of actions *D*. The composition of movement chains and techniques can switch within a wide range for a highly developed skill. No object-related skill is realized identically twice in a row; this can be seen in any type of action, from complex professional skills to lighting-up the cigarette or tying the shoelaces. As far as individual movements-links of a chain action are concerned, variability and switching of details within each of them is directly defined by the background level performing this automatism. One can produce an index of switching as the product of variability of consecutive movement automatisms by variability of the whole chain, and this result would be much larger than any value accessible for level *C*.

This fact of the high and increasing across levels skill plasticity is obviously directly related to the general process of continuous development of the motor function in the phylogenesis. In becomes particularly clear if the development of the construction levels is viewed as a biological process of adaptation of the organism to more and more complex motor problems. As it has already been mentioned, each level emerges and develops as a response of the organism to the emergence of new in their content and quality problems. More complex problems, containing more elements of adjustment to the environment and to the variable conditions, require from motor coordination more flexibility, adaptability and, therefore, switching. Naturally, every higher construction level displays an ever increasing maneuverability and higher flexibility in selection of motor reactions.

The interpretation of skill automation presented in this chapter allows offering an interpretation of mechanisms performing this generalization and switching of skills. The higher we move along the construction levels, the more general and broad *sensory syntheses* can be seen playing the role of afferent controllers. In the description of the corresponding levels, we could already see in their sensory syntheses, during the bottom-up transition, the orderly disconnects first from the coordinate system of one's own body, then from precise space coordinates, then from the executive organ, scale, metrics, geometrical similitude, etc.

Here we encounter the same phenomenon, which played such an important role in the initial phases of skill formation – the phenomenon of *transfer*. In the phase of automation, transfer of effects of training was seen in the form of using previously developed adequate automatisms – what we have addressed as *skill transfer*; in contrast, during the phase of stabilization, highly similar mechanisms function in much more variable forms.

More generalized and synthetic sensory fields are more independent of the working organ due to the domination of telereceptors located outside the organ and show broader ranges of identical sensory corrections shared among different variants of the movement.[9] Tangoreceptors and proprioceptors have to be located on the working organ itself and, as a result, their sensory signals are always forever linked to the organ itself. If corrections are based on telereceptors and, in addition, if they are synthesized into a holistic spatial field not linked to a specific extremity, then, clearly, such corrections, as long as they have been elaborated, can control movements of very different effectors with equal ease and precision. This fact explains mechanisms of the so-called transfer by organ and also reveal the cause of selectively different pliability of different levels to transfers of this type.

Due to these reasons, the level of synergies, which is most tightly linked to tangoreceptors and proprioceptors, shows very little pliability to organ transfer, although its sensory field is already united to a degree into a single coordinate system of the body. We have already emphasized the vast generality of the synthetic field of the spatial level *C* (particularly, of its higher sub-level). The generality of the sensory-gnostic synthesis at the level of actions *D* goes even further: its space is ordered by "qualitative geometry" (topology), whereas its kinematics – by some kind of "qualitative phoronomia." As a result, after elaborating a complex motor skill in a specific organ, there are highly and strictly selective effects of transfer of its components to another, non-exercised, organ: Backgrounds and automatisms from level *B* – to a minimal degree, the same elements from the level of spatial field and from the level of actions itself – to a large degree, with a single limitation relevant to the level of actions related to inequalities in that level of the right and left arms.

The increasing generalization of skill components based on synthetic sensory fields forms an essential part of the stabilization phase at levels *C* and *D*. In parallel to transfer by organ, many other forms of training transfer can be developed. In essence, everything that we perceive in a snapshot as *switching* of skills and components, from the point of view of dynamics, is always based on a corresponding type of *transfer*. Both phenomena are directly based on generalization of the corresponding sensory field and on the increasing ability to use it. As a result, a highly developed motor skill reveals switching and, therefore, transfer by the form and structure of elements of the motor composition, by motor technique, and finally, by the structure of the chain of actions itself.

We will finish this essay on the development of movement skills by a few additional comments on issues closely related to the theme of skill development. The first if related to a particular type of sensory corrections, which is distinctly seen at later phases of skill development and deserves to be mentioned due to its frequent occurrence.

On quite a few occasions, we have persistently emphasized the irregularity and unpredictable variability of the reactive and external forces, which are not under the control of the organism. Now, it is appropriate to state that, with respect to standardized aspects and elements of a movement – such as large locomotor synergies, dynamically stable movements, precision episodes, etc. – the accumulated experience allows the gradual isolation of some of external influences, which can be more or less precisely *known in advance*. This creates the possibility of using advance or *preliminary corrections* included into the very early stages of that specific movement episode. In such cases, preliminary corrections replace the earlier used secondary corrections or post factum corrections, which introduced movement corrections with the actual accumulation of deviations.

Such preliminary corrections are elaborated, for example, during the development of walking in a toddler (T.S. Popova) with respect to the leg swing time. During the second-third year of life, after the main extending impulse at the very beginning of the leg swing, there is always a second impulse in the same direction during the second half of the swing time ("childhood" ξ). Magnitude of the second impulse is highly variable and corresponds to variations in step length, which are also very large at that age. This "childhood" additional impulse, clearly playing the role of a secondary correction to step length, completely disappears at a later age. At the same time, step length reproducibility in older children and adults is very high; it is primarily ensured by a preliminary force wave ε. At the time of the "childhood" secondary correction wave and force curves, older children and adults show only small, highly variable cogs dealing with final, residual corrections.

Similar preliminary corrections take place at the initiation of leg swing during running. As mentioned earlier, in a skilled runner, leg swing proceeds as a dynamically stable movement virtually without active contribution by corrective dynamics. The initial impulse triggers that movement phase and introduces preliminary corrections available by that time; further, the movement evolves by itself as a precisely thrown ball.

Naturally, preliminary corrections play a visible role in *ballistic movements*: strike, push, and throw. Experienced instructors of locksmithing point out that precision of a striking movement during chisel cutting is defined during the initial backswing phase of the movement. Preliminary corrections are even more important in throwing movements, when the projectile cannot be controlled after the moment of release from the hand. Finally, they plan a similar role during jumping, which can be rightfully viewed as throwing one's own body, and when any control of motion of the common center of mass of the body is impossible after the beginning of the flight phase.

Preliminary corrections can also be formed at higher levels of development of chain object skills at the level of actions *D*. In such cases, corrections of this type always originate at the leading level *D* itself and, as shown by their disruptions in pathologies, their realization is directly related to normal functioning of premotor cortical systems. Psychoneurologists address this or a closely related phenomenon with the term "anticipation." Observations suggest that preliminary corrections are formed in skilled performers; for example, during bimanual labor actions. Such corrections make it possible for the performer to start movement of a hand during one of the action links and then to concentrate attention on anticipatory preparation to initiating the next action link by the other hand, etc. Common clinical pictures suggest that anticipations are close to cortical "trigger mechanisms" of movements and to phenomena of neuro-psychic "setting," i.e., to a range of phenomena that are still far from comprehensive exploration and clarity.

We have to supplement the essay on skill acquisition with analysis of one of the most complex skills realized at the level of actions – *handwriting*. The action of *cursive handwriting* in its developed form has a much more complex coordination construction compared to actions of grasping and locomotion discussed in Chapter 7; it is of no surprise that disruptions of handwriting during localized and systemic brain disorders are highly variable. The *level of paleokinetic regulations A* ensures, first, the overall background tonus in the writing extremity and its overall posture and, second, the main vibration innervation of forearm muscles (pronators, supinators, and also flexors and extensors of the wrist and digits). As any action organized at level *A*, this vibration is monotonous, highly rhythmical, and proceeds as a forced elastic oscillation following a nearly perfect sine wave – the most elementary of all curves of cyclical processes. *The level of synergies B* ensures the smooth roundness of the movement and its temporal (rhythmical) pattern. The former is created by a very subtle but robust synergy of all the muscles of the hand and forearm leading to a delicate, gradual transition of forces among the muscles. The group of background coordinations realized by level *B* can be easily observed because the backgrounds of this level never transfer to other extremities and even to other points on the same extremity. In contrast, movement components realized by the striatal sub-level *C1* and cortical levels can easily transfer as an example of "plasticity of the nervous system." If you ask a subject without any special training to write with different points on the right arm, with the left arm, tip of the foot, etc. (see Figure 5.7), and watch the process of writing and changes in the handwriting, it is possible to identify clearly background components from level *B* and distinguish them from all other components. As a rule, during such transfers, the general pattern of the "handwriting" is fully preserved, but the smooth roundness of movements and letter shapes, which form the characteristic feature of cursive handwriting and directly contribute to its speed, disappear altogether. They are replaced with inhibited clumsiness and slowness of movements resembling those of persons who have just started to learn how to write.

The *striatal sub-level C1* introduces into the handwriting action the same elements of adjustment to space in the course of the process as those seen during walking: The realization of moving the tip of the pen over the paper surface, along actual or imagined lines, adequate grasp of the implement, etc. It is harder to formulate the participation of the upper, *pyramidal sub-level of spatial field C2*. As we can conclude at this time, based on the clinical material with disruptions of elements of writing, this level is responsible for the geometrical, drawing aspects of handwriting: Execution of letter contours and preservation of features that form the essential components of individual handwriting, i.e., ensuring geometrical similitude of the written letters to some standard features typical of this particular individual, which can be reproduced similarly for any scale of handwriting, different postures (writing with the pen while sitting, writing with a piece of chalk while standing, etc.), and different implements.[10] These characteristic standards are preserved during switching to any extremity. Another highly important technical detail of the handwriting actions, which commonly escapes the attention of observers, is likely realized by the level of spatial field as a whole.

A complex movement in a plane, which is the *trajectory of the tip* of the pen or pencil, produces the trace left on the paper as an outcome of the handwriting action. To make sure that a point, which is a few centimeters away from the fingertips (this distance is about the size of two dozen of individual letters), could produce all the required details with the accuracy on the order of a few tenth of millimeter (both during writing and drawing), other points on the complex kinematic chain of the hand with the writing implement should produce in space very different but also equally precise trajectories. As shown by precise cyclogrammetric observations of the action of handwriting, even the fingertips, which are the closest points to the tip of the implement, show non-planar movements, which are so different from the movement of the tip of the implement that these trajectories cannot be read (Figure 8.8). The trajectories of the carpo-metacarpal joints differ even more from the written letters, and one cannot even guess the text by looking at those trajectories. Therefore, none of the points on the extremity draws in space a single letter but only their dramatically, although lawfully, distorted shapes (anamorphoses), and, as a rule, the subjects themselves have no conscious understanding of the appearance of those anamorphoses. This recoding is performed by level *C*, and (getting a bit ahead of ourselves) its automation represents one of the main difficulties when a child learns how to write. Since the construction of this recoding requires the control under visual corrections, this skill is never perfectly mastered by those born blind, which leaves its mark on the peculiar characteristics of their handwriting.

Finally, the level of actions *D*, with its cortical anatomical substrates including the centers of writing known to clinicians, together with the highest-level group *E*, generate the semantic contents of handwriting. At this level, letters are not simply geometrical configurations but semantic schemes associated with their sounds and with the written contours of words. These are the levels that "modulate" monotonous background vibrations generated by lower levels as described in Chapter 4.

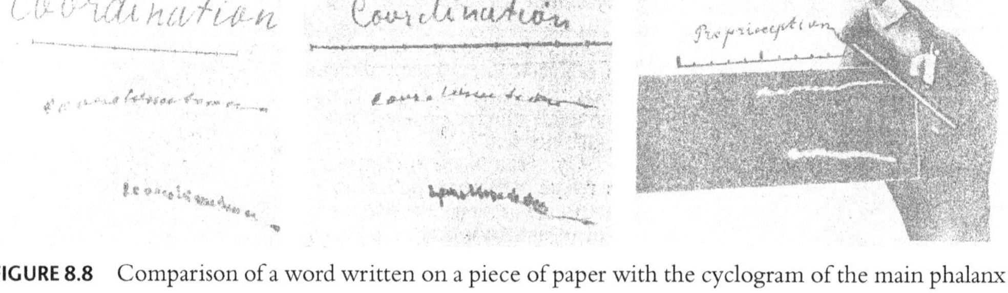

FIGURE 8.8 Comparison of a word written on a piece of paper with the cyclogram of the main phalanx of the index finger and carpo-metacarpal joints during writing this word.
Scales are in centimeters. 91 – large-scale handwriting. 92 – medium-scale handwriting. 93 – small-scale cursive writing.

The *role of the premotor system* in the organization of handwriting is well defined by its properties characterized in Chapter 4, confirmed fully by observations of handwriting disorders during dysfunctions of the premotor system. Disruptions of premotor functions lead to deautomation of handwriting, i.e., regress of its technical characteristics to the level of elementary school students: Instead of a smooth, quick process based on the automatic oscillatory synergy, there emerge disjointed movements and changes in pen pressure on the paper. According to the terminology of A. Luria, this leads to destruction of the kinetic melody of alternating active and passive elements. These conditions destroy not the geometrical, topological letter image (as it happens in cases of local sensory and gnostic agraphia, Chapter 6) but the automatic skill of cursive handwriting: Letters in patients with premotor disorders are not distorted or meaningless, but only rough and clumsy in their contour and slow in their execution.

Hence, during handwriting, similar to other skills, the premotor system does not introduce new correction elements but ensures the possibility of full, automatized use of background levels, primarily those from the level of synergies.

There is no doubt that *cortical levels higher than D* exist and influence the process of handwriting. Their existence is revealed and proven by the variety of forms of pathological disorders of the semantic aspect of handwriting (prolapse, perseverance, paragraphia, etc.) during brain disorders of different localization already described in Chapter 6. Here we witness a barely perceptible transition to the realm of pure psychology; therefore, it would be prudent to stop at what has already been said about the motor coordination structure of handwriting in order to trace briefly the development of this skill in a student.

There is significant difference between the process of mastering handwriting and the process of development of grasping objects described in Chapter 7 – desire to grasp emerges from the very first weeks of baby's life when only the substrates of the lowest levels are fully anatomically developed. At first, the grasping action is performed in a surrogate way with the available coordination resources; further, it continuously moves *upward*, in the direction of encephalization, as the corresponding levels mature, up to the most adequate level for this action where it resides for the rest of life. The process of handwriting acquisition is opposite. Students begin to learn this skill after the formation of all the levels is mostly complete; as a result, it proceeds similarly to typical skill acquisition processes in adults, i.e., with gradual transition of components *downward*, from higher to lower levels. A child has an advantage compared to an adult: the child's central nervous system is much more plastic and flexible; in contrast, during new skill learning, adults have the advantage of larger libraries of corrections accumulated during the preceding life, i.e. auxiliary background automatisms, which the adult can use in a new required combination as a transfer phenomenon.

According to the typical skill acquisition scheme, realization of writing in a novice begins with the *highest*, leading level of this action; at least (ignoring for now possible role of higher gnostic levels) from the highest level dealing with the purely motor, coordination aspect of handwriting, which is level *D*.

A child who knows letters and starts to study grammar draws images of (block) letters as their basic schemes with only very rough spatial geometrical coordination. Two issues are worth mentioning that shed light on some aspects of the coordination construction of handwriting.

Even an illiterate child who can draw with a pencil and is interested in books is willingly copying letters drawing them with all the noticeable details of the typographical font – dashes, protrusions, etc., because the child cannot identify essential from non-essential in letters. At a later phase of writing meaningful texts with block letters, all these dashes and protrusions disappear, and the graphics looks sloppier. But this is a false impression. The child simply moves away from the geometrical image of a letter, which he copied at the level of spatial field as a pattern, and switches to a semantic scheme of the letter image at the level of actions, where embellishments have no meaning anymore.

Another interesting detail is the abundance in the described phase of mirror-written letters and, sometimes, even words (frequently accompanied by analogous mistakes in reading the words). This phenomenon depends on the *dominance of the level of spatial field*, which is indifferent to the right and left sides, during this phase of the ontogenesis. In future, when the level of actions takes over most motor actions and defines their meaning, these errors disappear altogether.

Let us mention a few phases, which every adolescent moves through, independently of the method of education. During the first stage, the child writes with large letters and this is defined not only by the relative quantitative inaccuracy of his or her coordination. Likely, the main reason is that, for larger letters, the relative difference between movements of the tip of the pen and of points on the arm itself is smaller, and the described recoding is more accessible and simpler. This is directly confirmed by cyclographical observations. Gradually, in the process of mastering this recording, the child learns to transfer to the tip of the pen first – visual, and later – proprioceptive corrections, thus mastering the skill of ensuring any trajectory of the tip of the pen automatically. This progressive mastering of handwriting allows reducing gradually the letter size. The described recoding and its acquisition deserve special attention because it represents a general phenomenon common across all actions performed with a tool: knife, needle, soldering iron, plyers, welding apparatus, etc. A grown-up learning how to use such a tool has at his or her disposal many analogous, acquired earlier skills, which are frequently based on the prototype of handwriting; a child without such an experience has much more serious difficulties.

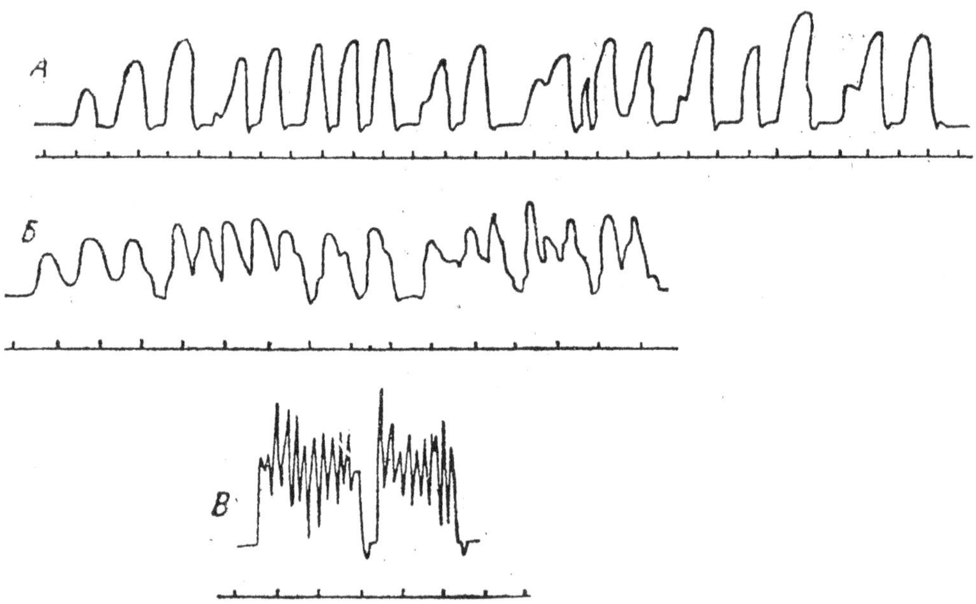

FIGURE 8.9 A kimographic recording of the pen pressure during writing words "Маша несла" ("Masha carried") by students.

А – first grade, *Б* – second grade, and *В* – college student. Note the merging of graphical elements and transition into cursive writing.

(Source: E. Gurjanov).

The automation of the mentioned recoding consists of switching this coordination component from the leading level of actions *D* down to the level of spatial field *C*.

Automation of a different kind – learning how to write on lines – takes place simultaneously with the described process. Forearm movement leading the implement along the line adjusts to the requirements of the split plane located in front of the eyes of the performer and gradually shifts from the competence of visual control to proprioception, which allows keeping evenly spread and directed lines on plain paper. This background component switches from the leading level to the striatal sub-level *C1*.

Finally, there is another gradual, but slower and more difficult, process of learning cursive writing linked to the elaboration of background automatisms in both lower sub-levels – thalamo-pallidar *B* and rubro-spinal *A*. This process leads to mastering proper distribution of the pen pressure, i.e., the control of efforts along the third coordinate, perpendicular to the paper surface. The implement stops clinging to the paper, making blots, and using exaggerated pen pressure changes. The skill of merged letter writing is mastered and, even later, the cursive writing oscillatory synergy. The philogenetically oldest and first to develop in the ontogenesis levels are the hardest and latest to harness; as a result, true cursive writing develops only after very long practice when the child exits adolescence.

Let me finish this chapter with a few notes related to deautomation, which has practical significance.

When deautomation is not caused by an organic disorder and, consequently, is not irreversible, movement disruptions may be relatively easily eliminated. The process of recovering a pre-existent but, for some reason, temporarily lost or weakened movement automation deserves to be addressed as *re-automation*. This process includes neither elaboration of new background automatisms nor creation of some vicar patches on the emerging coordination holes, but, as a rule, is limited to turning functional the same automatisms the central nervous system had at its disposal prior to the deautomation. In some cases, when deautomation happened as a response to an excessive disruptive effect from external factors or from inadequately large allowed movement variability (Chapter 9), it is sufficient to remove the disruptive influences to achieve immediate re-automation. In other cases, the situation may be more complicated, and then the character of re-automation depends on conditions that had caused deautomation and on its characteristics. One of such cases is interesting by itself and deserves an independent, brief exposition.

Every elaborated motor skill in its fully automated and ready to be used form may be viewed as a *functional constellation* in the central nervous system without yet deciding what to imply under this term in mechanistic sense.

Everybody admits that such working constellations are characterized by instability and lack of robustness, which causes them to fall apart, at least partly, immediately after the work process is over. Clearly, all working constellations corresponding to an active, functional ability to perform an elaborated skill quickly show disruptions following periods of inactivity due to the action of time itself and time-related perturbing factors, like a sensitive instrument transported over a bumpy road.

It seems very likely that elements used to construct constellations are very few and in such a high demand that, after the need for constellation *A* is over, separate elements are getting extracted from it as they are needed to construct next working constellations *B*, *C*, etc. Constellations are disrupted, even stronger than by time, by recruitment of similar constellations, which share identical elements.

Large time intervals (on the order of weeks or months) of disuse of one of the skill constellations, i.e. lack of their exercise, deautomate the skill very strongly so that persistent exercise – *retraining* – is needed for their re-automation. It is important to note that deautomation begins not simultaneously and proceeds at different rates in different skill components. It is known that some components show pronounced instability and quick disruption following lack of exercise, whereas other components, in particular those forming the "secrets" of swimming, riding a bicycle, etc., described earlier, seem to possess lifetime indestructability.

An action based on a *short time scale* skill constellation, for example a day-to-day one, is qualitatively different. Partial disruption of the constellations happens even over those short time intervals and, during all labor processes, the first minutes after the work has resumed are spent on their adjustment and recovery. This is the widely known process of "warm-up" following work initiation. The physiological structure and mechanisms of warm-up are far from being explored: Very likely, in different situations, the involved physiological processes may be very different. In some cases, for example during precise manipulations, the emphasis of the warm-up is on some subtle adjustments and regulations of coordination background mechanisms. In other cases – a crucial role is played by getting into the rhythm, and then introductory gymnastics is very much helpful when it targets the rhythm of the following work movements. In still other cases, the warm-up resembles mostly a run-up, obtaining some work inertia, kind of turning a motor on or pushing a heavy pendulum. It is clear that success in the described case of re-automation – the warm-up – should depend directly on adequacy of analysis of this process in each particular case and on adequacy of the used facilitation methods. One universal feature of the warm-up deserves particularly focused attention during its analysis: The degree and character of particular dis-regulations and deautomations, which require warm-up, do not depend on the time since skill acquisition and skill stability. Due to this factor, duration and techniques of warm-up are only minimally defined by the time since skill acquisition, although this dependence varies across different skills and different level compositions. This makes deautomation, which is to be overcome by warm-up, substantially different from de-training, which is due to a more or less lengthy gap in exercising a given skill: After long interruptions, old, well-learned skills, as a rule, recover faster.

The role of time in some cases of deautomation is revealed by the disruptive effects of small micro-pauses on some types of labor actions, which require warm-up after each occurrence. Composing detailed time-sheets and schedules should be done taking into consideration degree of deautomation of different types of labor actions caused by even the shortest intervals between work periods.[11]

Notes

1 See, for example, "Twenty years of experience with studies of the higher nervous activity" by I.P. Pavlov, 1928, pp. 107, 139, 225, etc.

2 It is very easy to reproduce this feeling of hand "bewilderment" with the following easy experiment: Put a mirror in front of your eyes, block vision of the right hand by a piece of cardboard or paper so that you can only see it in the mirror, and then draw a square and a cross inside it, or something of that kind.

3 We classify corrections as *leading* when their semantic contents are defined and demanded by the leading level, corrections that define the meaningful success or failure at the motor task as a whole. Not all corrections performed by the leading level are leading corrections. If the leading level happens to have all the sensory corrections necessary for background trim and finish, it can, together with lower siblings, become occupied with background work.

4 We have already noted (Chapter 3) that such independent backgrounds are independent only by their origin, whereas their actual appearance can be changed essentially under the influence of the leading level.

5 Backgrounds from level *A* are not shown in the table because their participation in each of the enumerated movements is not associated with clear automatisms in the aforementioned meaning of this word.

6 Further, we are going to use the term "acceptable variability" for a range of variations that do not yet produce disrupting actions.
7 In contrast, during the first stage of skill elaboration, during the selection of motor composition and sensory corrections, introducing task variations and disturbances too early can only hurt the process.
8 The leading role played at those moments by the object-related level is very well characterized by Tolstoy's description of the focus of Levin's attention: "He did not think about anything else and did not want anything except to keep up with the peasants and *do the job at his best.*" And further: "Somewhere in the middle of their work, there were minutes when he forgot what he was doing, (i.e., he forgot to concentrate on his movements – N.B.), and everything felt easy, and during those very minutes, his row was as straight and good as that of Tit (this means that he focused on the quality of the row – N.B.) (*Anna Karenina*, part 3, chapter 4). Further, in chapter 5:

> The longer Levin mowed, the more and more frequently he felt minutes of oblivion, when not his arms were moving the scythe, but rather the *scythe itself moved the whole body*, full of self-consciousness and life, and, as if by magic, without thinking, mowing, proper and accurate, was done *by itself.*

Here we have a very vivid description of: The object-related character of the leading motive, the automation of the whole process, and the already mentioned (chapter 6) illusion of the movement being led by the tool. Further, it becomes clear how automation in poorly trained Levin is *disrupted* during the same semantic changes in the situation, which have well-automated motor corrections in experienced Tit:

> It was difficult only when it was necessary to stop the unconscious movements and to think: When it was necessary to mow around a hummock or a patch of sorrel. *The old man did it very easily.* When a hummock emerged, he changed the movement, and, sometimes with the front and sometimes with the back of his scythe, mowed the hummock from all sides with brief strikes. While doing it, he looked ahead at further field ... Such changes in the movement were difficult for both Levin and the young lad behind him. Both of them, after arranging a single tense movement, were in the excitement of the work, and were unable to modify the movement and, at the same time, watch what was lying ahead.

This remarkable text has many subtle observations of deautomation and the phenomenon of disruption. The whole example is very valuable as an autobiographic sketch of the nature.
9 For animals, who use their muzzle equipped with telereceptors as the main working organ for various motor actions, this factor definitely acts against the mentioned sharing of sensory fields and emphasizes the importance of transferring many operations from the muzzle to the hand.
10 We have mentioned in Chapter 4 that individual handwriting, similarly to gait, touché, habits, etc., are defined by level *B*; this is correct but has to be qualified. Level *B* imparts on handwriting its general *expressive guise* reflected in its smoothness, angularity, resoluteness, etc., whereas level *C2* defines standard geometrical contours of individual letters and written signs.
11 L. Osipov and E. Babaeva performed a series of special studies in the Central Scientific Research Institute of Physical Culture (1934–1940) of the process of warm-up and of the role of workplace gymnastics and exercise pauses. According to their data:
    1. "Lengthy warm-up (over one hour and more) typically takes place in types of manufacturing actions where movements of the worker are not dictated by the speed of machines; however, warm-up may be needed when working on an assembly line. Lengthy warm-up is primarily seen during manual operations associated with modest physical efforts and when the type of work requires high degree of movement coordination."
    2. "stable state of organism efficiency, which emerges after more or less lengthy period of a warm-up, is conditioned not only by adaptation of the circulatory and respiratory systems, but primarily by the corresponding adjustment of the central nervous system to perform the given work."
    3. Using warm-up gymnastics developed based on specific features of labor operations, in particular with their tempo, helps workers to become alert and ready to work quicker and shortens the whole warm-up duration (for example, in a shoe factory, with the warm-up gymnastics, workers-stakhanovists

spent on the first set of workpieces, on average, 19.4 s less and produced during the first 1.5 hours, 1–2 workpieces more, as compared to the control days without the warm-up gymnastics.)

4. Three- of four-minute exercise pauses (active rest) during their correct organization produced positive recovery effect in tens of factories in the USSR. On the Kaluga Electromechanical Factory, the work intensity of workers performing assembly of telephone apparatuses increased by 9.3%; on the L.M. Kaganovich Bearing Factory, in the sander shop, – by 6.3%; on the shoe factory "Paris Commune" – by 3.6%, etc. (L. Osipov and E. Babaeva. *Industrial Gymnastics*. Moscow, pp. 4–7, 1940).

9

FEATURES OF THE LEVEL STRUCTURE IN HEALTH AND PATHOLOGY

In previous chapters, we tried our best to identify applications of the described theory to everyday situations using illustrations of the main concepts with specific examples from daily life. In particular, the whole of Chapter 8 represented such an application to the practice of pedagogy – to issues related to development of skills in sports, labor, and performance.

Now we have to summarize, as accurately as possible, applications of the theory of coordination to the field of *movement pathology*, in particular, to pathology of the nervous system, which has been touched upon only fleetingly during the characterization of individual construction levels. We do not plan to infringe on the field of neuropathological semiotics, but will try to present a brief essay on changed interpretations of local and systemic movement disorders and associated compensations according to the theory of sensory corrections. Our main emphasis in on the vital question, important for any reader, on *criteria* that would allow drawing conclusions on the levels of construction of motor actions, on *objective specific features* able to establish different traits and characteristics of the construction, which could be useful in future for differential diagnostics.

Our search for such features is based on the same basic concept of *selective contingencies* of movements controlled by each of the levels of the central nervous system. We focus not on the question *what is disordered in movements* (all the movements) during specific localizations of injury, but *what specific movements are disordered* in each of such cases. As far as the former question is concerned, qualitative disorders of particular traits and details of movement finish observed during localized injuries in most movements are interpreted within our approach as disorders in the processes of *specific technical backgrounds* controlled by the injured level, i.e., once again, taking into consideration specific contingencies and level of these backgrounds. Indeed, as it frequently happens, if a movement disorder caused by a local injury is not limited to a narrow list of movements that belong to a single construction level but affects the whole motor repertoire of the patient, even in this case it is possible to establish that either the injury affected anatomical substrates or conduction pathways of all coordination levels or different movements suffer differently and to different degrees, with obvious selective specificity, which can be interpreted clearly and vividly even at our time.

To enable confident practical application of the theory of coordination to both health and pathology, we need specific signs that would be unambiguously selective with respect to different construction levels and also provide answers to the following three questions:

1. *What is the leading level* of a given motor action, which is defined by the contents of the motor problem, solved by the action and, in turn, defines its semantic structure?
2. *What is the level composition* of a given motor action, i.e., what levels act significantly in its motor composition, what are their relative proportions and relative significance in the action as a whole and in its successive chain components?

3a. For a given detail, feature, or component of a movement: *What background level* does ensure its existence?

3b. For a given background level: *What details* or features of an observed movement are defined and led by this level?

These questions, crucial for classification of a normal movement, can be easily reformulated for pathology. Their goal with respect to both normal and pathologically changed movements, as well as the practical utility of analysis of movement construction, are in the possibility to develop rational approaches and tools for labor and athletic pedagogy, for labor-and kinesiotherapy, for therapeutic exercise, etc., and also in facilitating decision-making in the search for resources to create therapeutic vicariates and substitutions. Their goals for the neurological clinic is facilitation and refinement of topical diagnostics in cases of localized injuries; in using more sensitive and accurate methods of patient study for early diagnostics, which improves dramatically chances of recovery; and finally, for the possibility of differential exploration of the dynamics of vicariates and compensations following an acute local injury, which is of high scientific value.

We are not going to review here aspects (maybe most promising in future) related to the physiological specificity of the neural process at each of the construction levels, specific features of effector "bursts" of excitation characterized by bioelectric potentials in the brain and muscles, changes in the tissue and synaptic excitability, interactions between the effector process and functioning of the autonomic nervous system, the humoral contributions, etc. Currently, all these issues are at very early phases of investigation.

Presentation of our research materials related to these issues will form the subject of a separate publication, currently in preparation. Here we will limit the discourse to description of features that can be observed without special equipment in the gym, in a conservatory cell, and at bedside. Some of these features are already obvious during simple observation of the subject's spontaneous motor behavior; others can be seen during special tests.

As already mentioned in Chapter 6, we distinguish in each motor action its semantic structure and motor composition. The *semantic structure*, directly related to the motor task contents, fully defines the adequate synthetic sensory field and may be viewed as the most reliable source of information on the *leading construction level* for the action. Typically, the semantic structure does not contain convincing information on the background level composition, i.e. on the construction of movement in the complete meaning of this concept.

In contrast, the *motor composition* of the action, at least during its crude macroscopic observation with the naked eye, does not allow deciding on the leading level because (as we have observed earlier, from Chapter 2 onwards) frequently, movements that look very similar in their patterns can proceed at various construction levels. A more sensitive (e.g., cyclogrammetric) analysis of motor composition can reveal certain level peculiarities and differences between apparently similar movements because different sensory corrections of the leading level, naturally, impart somewhat different appearances upon movements at each of the construction levels.

Answers to the second and third of the aforementioned questions are given by the motor composition, and its analysis is facilitated by features described in this chapter. What technical backgrounds, at what tempo, in what order, and how adequately could the subject recruit to solve the current problem? To what degree could these backgrounds work together, what form and degree of automation could they ensure, and what kind of stability against deautomation under the influence of various perturbing factors could the central nervous system create? To answer all these questions, we have to develop reliable and objective criteria.

Some of the features vital for drawing analytical conclusions of the motor composition and its disruptions are relevant primarily to motor pathologies and can be used for analysis of healthy movements only in cases of transient functional disturbances: e.g., during fatigue, intoxication, etc. Other features are equally applicable to both movement disorders and perfectly normal movements, thus showing maximal extensive universality. We will begin with the former group of features.

Every spatially well defined, acute injury to the central nervous system with no complications (we will begin with this most clear and unambiguous situation) leads to certain changes in all aspects of life. It would be better to say that such an injury brings about not phenomena but modified, as compared to normal, riverbeds or rail tracks for the time evolution of phenomena. Typically, the goal of a clinician is to perform diagnostic induction; he or she should base conclusions on the available observable phenomena and their causes, injury location, and its features. Among these numerous phenomena, the clinician should strive to identify *pathognomonic features*, i.e. those most specific and meaningful for the involved levels.

Complexity of syndromes depends, of course, on the huge variety of factors that define the modified time course of phenomena.

These phenomena are defined primarily by the location of the injury, or, to be more precise, the involved organ or functional system of the brain; important factors include also extent of the damage and its somatotopic properties if the injury is located in one of the projection parts of the central nervous system. Further, for a given location of the injury, direction of the riverbed for changes in the phenomena is linked to *sign and type* of the injury. In some cases, the injury can lead to *hypodynamic* phenomena: inhibition, hypofunction and even complete loss of function; in other cases, the injury can show *hyperdynamic* consequences expressed in quantitative irritation of the organ, its hyperfunction, and accompanying qualitative perversions. Further complications of the stream of consequences are defined by *interaction of the locus of injury with other brain organs and systems*. This can lead to various (excitatory or inhibitory) effects on tonically adjacent brain areas, similar to a fire engulfing adjacent buildings, or to systemically adjacent areas linked to the injury site by synaptic connections via conduction pathways, which can be spatially far away from the injury locus. Finally, if we consider the *time* coordinate, we have to take into account effects of the locus dynamics (single aseptic trauma without consequences, slowly developing gumma, actively developing abscess, etc.), its recent or old time of occurrence, i.e. the degree of development of processes of compensation and vicar substitution of the lost functions. All this chaos of causal dependences, complicated by the fact that each level of the central nervous system reacts in its own way to these causes and influences, makes it very hard to establish any type of a systematic order.

To obtain, as signs, phenomena with maximal level specificity, we have to consider one more variable – the *location of an injury along the reflex loop*. Leaving aside injuries of mixed peripheral nerves, which lead to loss of function of the *ultimate effector, not of the motor function*, schematically, there are five possible injury locations on the reflex loop at a given level (Figure 9.1): (1) The central afferent pathway (in the spinal cord); (2) The sensory (afferent) nucleus of that level; (3) The central switching systems; (4) The effector nucleus of the level; and (5) The effector cerebrospinal pathway.

The general principle of the dependence of a phenomenon on location along the reflex loop can be expressed as follows. For lower injuries (those closer to the periphery), there is stronger selectivity of the disorder with respect to the peripheral effector or body part and weaker

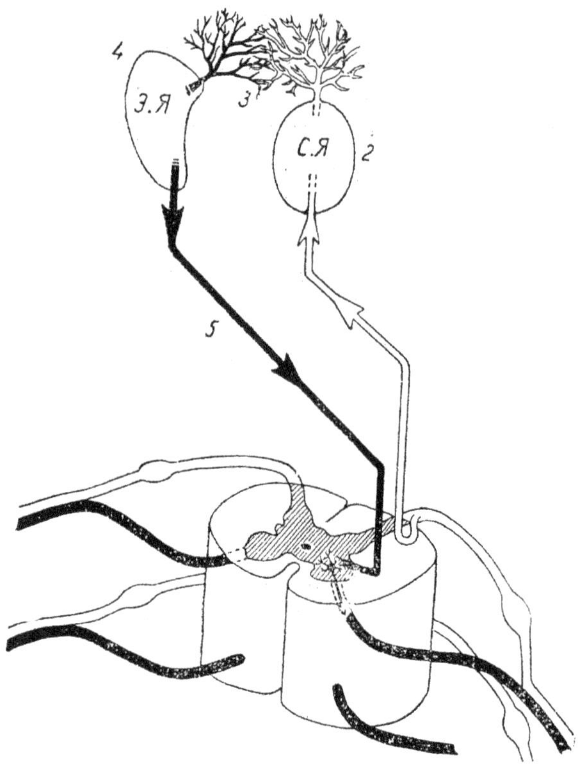

FIGURE 9.1 A scheme showing five possible injury loci in the reflex loop.

expression with respect to the contingent of disordered movements. The opposite is also true: For higher (more central) locations of an injury, disorders are more selective with respect to movement contingents and more general with resect to peripheral organs.

For example, an injury to point *1* (Figure 9.1) affects all the original, unprocessed signals from the given peripheral receptor before it has been processed and synthesized in sensory fields together with other receptor modalities. These original signals participate in processes in almost all the fields of coordination levels; therefore, movement disorders can be seen at nearly all levels and there is no selectivity of this injury sign. In particular, this is how the syndrome of tabes ataxia emerges as a result of the degeneration of the spinal cord posterior columns, which carry in healthy persons all the proprioceptive and nearly all the tactile-nociceptive signals. As shown in Chapters 3–5, both these receptor modalities are intimately linked to levels *A*, *B*, and *C* resulting in the well-known complex of the posterior column ataxia. At level *C2* (going in the top-to-bottom direction), there are disorders of spatial movement accuracy, precision of getting into the target, grading effort, etc. At level *C1*, there are disorders of locomotion and all the movements of touching, feeling, and tracing, which were assigned to this level in Chapter 5. Disorders of cooperation of level *C1* with level *A* lead to problems with equilibrium and astasia (the Rhomberg symptom). The thalamo-pallidar level of synergies, which is particularly tightly linked to proprioception, responds to loss of its signals by mastering the reactive dynamics, which is its forte during normal movements; the normal ability to create dynamically stable movements is lost completely. Finally, disorders at level *A* by themselves lead to signs of its hypofunction mentioned in Chapter 3: intentional tremor as a result of the loss of reciprocal innervation and denervation, hypotonia as a result of the lost reflex contributions to tonus, and hypermetria following disruption of the mechanisms of alternating recruitment of antagonists during cyclical actions. There is also deautomation of movements at all levels, particularly at the object level *D*, due to destruction of all the backgrounds relying on proprioception. Taken together, we get the general clinical picture, which is characteristic of the movement disorder but is non-specific with respect to virtually all types of motor actions.

Injuries of *spinal effector neurons*, for example pyramidal neurons (we will discuss this condition later from a different viewpoint), lead to similarly comprehensive movement disruptions, which result from an inability to reach specific, well-defined muscle groups by effector impulses of very different origin and composition, and originating from all the cortical levels.

Therefore, level-specific selectivity of symptoms can be seen, as a rule, only in cases of injuries localized in points 3 and 4 of the reflex loop (Figure 9.1).

Most clear forms of selective disorders emerge during injuries to central afferent nuclei (2 in Figure 9.1). In such cases, disorders represent *disruptions of the semantic structure* or meaning of motor actions at the affected level, i.e., of their most essential and difficult to replace features. Here one should note the characteristic dependence between the hierarchical position of the injured construction level and ability to recover disordered movements.

For *hierarchically lower levels, it is harder* to compensate for the deficits associated with *disruptions of the motor composition*, and vice versa, *it is easier* to compensate for deficits brought about by *disruptions of the movement semantic structure*. For hierarchically higher levels, an opposite statement is true.

To explain this statement, we have to consider that semantic structure is always completely defined by the leading level of a given movement. To recover a movement following disruption of its leading level *P* means an ability to reproduce at *higher construction levels* the semantic structure of the disordered movement. For higher levels, there are few superior resources for such a surrogate reconstruction and, vice versa, more intact lower levels are available to a certain degree for substitution of the movement background composition. In contrast, substitution of the simple semantic structure of movements that can be handled by lower levels is overshadowed by the much more difficult to recover losses in the motor composition, which accompany disruptions of lower levels, which show low plasticity and are highly specialized by the quality and composition of their corrections.

It is necessary to explain the differences between disruptions of the semantic structure and disorders in the motor composition. Disruptions of the semantic structure reveal themselves in inadequacy of realized motor solutions to the meaning of a motor task independently of whether the patient understands failure of the solution or not. Solutions that do not achieve a required outcome or achieve it poorly because of motor coordination deficits but which do not distort the meaning of the problem represent cases of motor composition disorders with an intact semantic structure. In fact, unfortunately, it is frequently not easy to recognize which of the two deficits takes place. The two forms show most clear differences during the performance of semantic chain actions at the level of actions *D*.

With all the qualifications and reservations, we can now approach directly pathognomonic signs of different levels, which naturally form two classes. The first class includes phenomena of *hypodynamia*, i.e. a decrease of a function, from its complete loss to phenomena of disruption, i.e. ability of the neural apparatus to ensure reasonably normal performance but only under conditions of extreme protection, elimination of any perturbations or complications, which are disruptive for such vulnerable functions. At the other end of the spectrum, there are *hyperdynamic* phenomena brought about by overexcitation of an injured center or a center that is affected by secondary changes produced by a clinical syndrome. First of all, the hyperdynamic class involves quantitative phenomena of irritation revealed as increased excitability or exaggerated amplitude, speed, or forces during movements. A hyperdynamic state of a level may further lead to overproduction of spontaneous movements, excessive in their essence, i.e. not solving any problem and lacking meaningful completeness in their motor composition; these would be various *hyperkinesias* including one of their most common type, tremors. If excessive movements emerge not successively but simultaneously, they turn into *synkinesias*, which we distinguish from synergies and syndynamias or mutually assistive movements (and efforts) exactly because of their inadequacy. There is a static variety of synkinesias representing various involuntary irradiating over-exertions, both tonic and tetanic, such as the known risus sardonicus during tetany. Finally, a very common hyperdynamic phenomena, which produce, particularly for higher levels, a number of useful level-specific signs, are *hyperdynamic movement perseverations*, i.e. loss of the ability to stop at will a motor process. The clearest signs are typically produced by phenomena of loss and perseveration. Let us begin with the former.

Every localized injury causing loss of substrates of a specific level produces a rather complex picture of *selective motor losses*. The solid array of movements available to a healthy person starts to show gaps, classes of impossible movements, on the background of other, unaffected movements or movements with partial loss with a specific pattern. The next problem is to assemble a *list of typical combinations* of motor losses pathognomonic for injuries to successive levels, which would allow their unmistakable identification.

Complete loss of a level P results in a selective complete loss of movements that used to be led by that level because of the loss of supporting afferent systems. There are partial disorders of movements organized at higher levels; they show deautomation of various degrees depending on whether the lost level was part of their background composition. Clearly, the qualitative picture of deautomation of motor actions organized at higher levels is highly characteristic of the lost level. Movements performed at levels lower than level P may display disorders of two types after level P has been lost. On the one hand, if the subject had been able to develop lower-level automatisms to serve motor actions at level P, loss of this level leads to an irreversible loss of ecphoria of those automatisms because higher levels, $(P + 1)$, $(P + 2)$, etc., which start forming vicar substitutions for the lost motor actions, as a rule, have no keys to recruit those automatisms, which used to respond to requests only from their customer level P. Lower levels, which do not lose any independent movements, show a loss of many of their background motor functions. Besides, in many cases, the hierarchical relations between adjacent levels in the central nervous system make substrates of a higher level exert specific inhibitory actions on effectors of inferior lower levels. This inhibitory action (Foerster addresses it, with respect to effects of striatum on pallidum, as "*bridling*") is eliminated with loss of the corresponding level. As a consequence, a peculiar form of hyperdynamia emerges, which gets contribution from the released lower level and which could be called, following Foerster's metaphor, "unbridling" or *effrenation*. Such effrenation takes place, for example, at the rubro-spinal level *A* during the parkinsonian inhibition of the pallidum or, at the same level, in the anterior horn neurons of the spinal cord following loss of the pyramidal system following capsular hemiplegia, etc.

We have to emphasize that, at least in some cases or for some types of movements (at this time, it is still very difficult to identify necessary and sufficient features of these cases and types), it is nevertheless possible to switch ecphoria of at least the main of the backgrounds to an intact higher level, which accepts the responsibility for functional recovery of lost movements and is able, to a certain degree, to adopt them with their backgrounds. Sometimes, such a recovery switch can be achieved easily by a simple change of the formulation of the motor task. The patient cannot follow the command to lift the hand to the forehead level but, can easily perform this movement when asked to take off (even an imagined) hat or to perform military salute. A. Luria asked a patient with a premotor lesion to draw on the paper the following polygonal line (Figure 9.2, top). After the

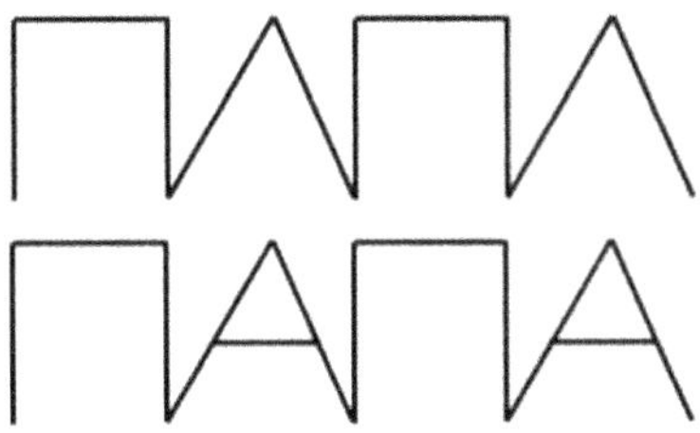

FIGURE 9.2 Two tasks described in the text

patient failed completely at the task, the experimenter modified it by adding two extra dashes to the polygonal line (Figure 9.2, bottom), and the movement immediately became possible. There are many similar examples.

Getting back to phenomena that accompany localized injuries according to points 2, 3, and 4 (Figure 9.1) with respect to each construction level, one at a time, we will first exclude from detailed analysis conduction disorders related to point 3. Similar to all conduction pathway disorders, they are very similar to phenomena accompanying injuries of the corresponding ganglionic nuclei but in a more *direct, crude, and downright* way. Clearly, conduction pathway injuries can produce only: (a) alternative, selective *losses*; and (b) *disinhibition* syndromes in efferent processes of lower centers, while being unable to produce hyperdynamic syndromes of the injured level itself. There is nothing new in selectively specific symptoms produced by injuries of conduction pathways compared to what can be seen during localized injuries of central neural nuclei.

Loss of *afferent systems of the rubro-spinal level A* leads to a clinical picture similar to the well-known Luciani-Orbeli "cerebellar syndrome." First of all, this picture is characterized by pronounced *hypotonia* accompanied by disorders of the *reflex modulation of tonus*. Disruption of what we have previously addressed as the primitive spatial synthesis of level *A* (Chapter 5), i.e., the *top-down* tropism controlled by the vestibular (otolith) system, leads to loss of equilibrium during standing (*astasia*), and, in cases of irritation of the injured centers, falling over, arena movements, etc. Loss of main components of proprioceptive signals leads to phenomena of general, widespread disorders of coordination, *ataxia*, because loss of level *A* backgrounds leads to crude deautomation of movements from all the higher levels. Finally, as emphasized in Chapter 3, the intimate link between level *A* substrates and the autonomic nervous system, which communicated with the central nervous system almost exclusively via substrates of level *A*, leads to a range of autonomic disorders. With respect to the neuro-muscular motor apparatus, such disorders bring about loss of the so-called adaptive-trophic function of the sympathetic system leading to both general disorder of the neuro-muscular periphery and its quick fatiguability, apathy, etc. – symptoms that are united under the umbrella of "muscular asthenia."

Loss of the *effector system* at level A, i.e., of what has been defined in Chapter 3 as the red nucleus group, interrupts impulse conduction along both major extrapyramidal effector pathways of the spinal cord – the rubro-spinal and vestibulo-spinal pathways. Neurons in the anterior horns of the spinal cord experience only impulses from pyramidal neurons and respond to these phenomena with *general tonic inhibition* of the corresponding motoneurons. Later we will see that the roles of the pyramidal and extrapyramidal spinal cord tracts are antagonistic with respect to tonus, and pyramidal neurons exert inhibitory effects on motoneurons. This leads to a syndrome of hypotonia with particularly pronounced loss of the tonus elastic component (see Figures 3.8 and 3.9). Further phenomena characteristic of lost effector impulses from the red nucleus group have been already enumerated in the final section of Chapter 3: disorder of subordinate regulation, intentional tremor as its direct consequence, hypermetria, frequently observed joint hyperextension, etc.

We are not going to dwell on the syndrome accompanying *hyperdynamic states* of the red nucleus group. Briefly, we can state that the emerging phenomena form half of the typical symptoms of parkinsonism: typical disturbances of body posture, catalepsy-like phenomena, i.e., freezing in a spontaneously achieved or artificially imposed posture, general *viscous* (rigid) *hypertonia*, rest tremor (so-called, "pill-rolling"), and movement *hypometria*, i.e., undershooting the target. The second half of parkinsonian symptoms is linked to hypodynamia of the thalamo-pallidar level *B*, which will be addressed later.

Hyperdynamic states of the red nucleus system emerge most frequently as consequences of effrenation, i.e. disinhibition caused by the lost descending regulation; in particular, this happens in parkinsonism.

Disorders of the *central afferent nucleus of the thalamo-pallidar level B* – the thalamus, which is the unavoidable transmitter of all the proprioception and tango-reception to level *B*, as well as to all the higher levels – leads to such dramatic and widespread coordination disorders that it may be defined as the ultimate failure of the motor function. *Moderate hypodynamic syndromes* originating in the thalamus lead to disorders of what has been defined in Chapter 4 as the "triad" of independent movements at the level of synergies: *expressive movements* of mimics, pantomime, and plastic movements. These syndromes lead to more or less pronounced asynergia or dyssynergia, i.e. disruption of global extrapyramidal movements and, what is more important for the overall balance of human motor function, of extrapyramidal backgrounds; their character and importance for movements at levels *C* and *D* have been covered in detail in Chapters 4–6. There are relatively minor disruptions of movements at the higher levels that proceed without backgrounds: Small finger movements at the level of spatial field *C* and some of the elementary chain actions at level, *D*, such as orienting movements, simplest self-care movements, reproduction of rhythms with tapping, etc. As far as the motor function of speech is concerned, speech by itself is not strongly affected, but voice, prosody, and expressive components are. However, the disruption of backgrounds can frequently lead to indistinctness, lisp, twang, etc. The intimate relations between the thalamic system and nociception lead to the emergence of a whole range of painful and sometimes symptoms very torturous for the patient, but we are not going to discuss them because they are not directly related to the motor function. However, the possibility of strong painful sensations (e.g., hyperpathia) sometimes caused by thalamic injuries makes it worth mentioning that *perception of pain*, whether of peripheral or central origin, if it is associated with specific postures or movement elements, brings about a *special class of corrections*, which may be addressed as *antalgic* or anti-pain corrections. Such corrections used by the patient involuntarily, nearly in a reflex fashion, to protect the painful organ from painful sensations can lead to gross distortions of proper coordination leading to the false impression in an observer that there are organic coordination disorders. It is possible to identify antalgic disruptions and to distinguish them from true coordination disorders (without even interviewing the patient who may produce insufficiently objective statements) using the feature described earlier of generality of peripheral disorders over movement contingents and their selectivity with respect to the effector organ. Antalgic corrections reveal their peripheral localization, whereas they can be easily differentiated from organic peripheral disorders of mobility. Clearly, *diffuse* pain of a central origin cannot lead to any consistent pictures of antalgic corrections.

Localized injuries in the afferent nucleus of the level of synergies – the pallidum – result, in most cases, in *hypodynamic* syndromes. The loss of the pallidar function brings about the second half of parkinsonian symptoms: General asynergia, loss of mimics as a consequence of the disruption of the "triad," constrained poverty and stinginess of movements, and the replacement of rhythmic smoothness with discreteness and angularity of movements. Skills of dynamically stable movements and the corresponding backgrounds are lost, the cyclogrammetric curves show the lack of effort transitions of different signs, which are replaced with the typical for hypertonic rigidity single-sign efforts characteristic of the initial stage of skill development (Chapter 8). As mentioned earlier, there are disruptions of rhythmical background recoding for movements at the level of spatial field. Finally, there is a dramatic drop in the percentage of involuntary movements and components; voluntary movements are performed stingily because of the decrease in initiative. Frequently, one can also see perseverative complications – propulsions, retropulsions, the aforementioned postural perseverations – cataleptic symptoms, etc. For clarity, we will discuss phenomena of perseveration at all levels in a special section closer to the end of this chapter.

Hyperdynamic forms of the pallidum output typically belong to the effrenation type and are brought about by the loss of inhibitory, constraining action of the striatum. These forms are many and varied but they are relatively poorly studied. This list contains everything from the rich catalog of hyperdynamic signs: hyperkineses, synkinesias, perseverances, as well as innumerable obsessive movements. With the exception of dynamically stable forms and patterns, which cannot be formed outside normal conditions of exercise and repetition, the rest of the arsenal of motor abilities of the thalamo-pallidar level is revealed in these hyperdynamic effects as unstoppable and ugly synergic molds or fragments of movements. We are not going to delve into deeper analysis of differential-diagnostic differences among the syndromes of athetosis, chorea, torsion spasm, etc., which all have, as their basis, phenomena of pallidar hyperdynamia (see Figures 4.9–4.11). Let us point out only that the differences in the clinical pictures within this group of syndromes depend primarily on the relations between the main fact – hyperexcitability of the pallidum – and dysfunctions of other, adjacent levels, not on differences in the pathological origin on the hyperexcitability. For example, the syndrome of athetosis is related to a hyperdynamic state of the rubro-spinal level, while the syndrome of chorea – to a normal or even hypodynamic state of that level; the clinical picture of torsion spasm involves, in addition to increased excitability of the pallidum, effects of general, widespread hyperdynamic state of the whole extrapyramidal effector system, etc.

Deafferentation of the level of spatial field C is unlikely to happen in pure form. We still do not know if there are, somewhere in the cortex, concentrated fields that could be seen as "centers" of the sensory synthesis of spatial field; turning off the primary or secondary cortical fields of separate modalities of sensitivity always lead to clinical pictures of isolated losses of those specific sensory properties, frequently within the somatotopic borders of the loss. More or less general disorders of the main sensory components of the "spatial field" due to an injury below the cortex – in the thalamic system of spinal conduction pathways – lead, as mentioned earlier, to disorders of a range of levels, starting from *B* or even *A*. It is quite probable that the so far unsuccessful search for a cortical synthetic center of the "spatial field" (it does not matter how we address it) is due to the very high degree of inter-compensation among spatial field components, which leads to an impossibility of turning this whole center off in a pure, isolated way. Disorders of perception and comprehension of space, addressed as astereognoses, which have specific known localizations in the cortex represent violations of semantic, topological perception of space at the level of actions *D* (note the word "gnosis") and, therefore, do not belong to this group.

If we start not from injury localizations (which cannot be identified for the discussed case) but from coordination disorders, it is very easy to characterize features that have to be classified as those occurring at level *C* in complex forms of coordination disorders, which are produced by disorders of space-forming afferent signals. All such components are parts of syndromes of various *ataxias*. The level of spatial field contributes to these syndromes losses of accuracy and precision of movements, effort grading, the control of timely initiation, velocity, and tempo – according to the upper sub-level *C2*, and disorders of the general "command of space" resulting is disruptions of locomotion – according to the lower sub-level. Since the level of spatial field is responsible for most small, accurate hand manipulations (either as independent movements or as backgrounds for level *D*), its afferent, ataxic disorders can be best described by the patient that "my hands do not obey me." This hand disobedience may, but does not have to, be accompanied by one of the tremors, which always represents an extraneous, background phenomenon at the lower level *A*. "Hand disobedience," as both a peculiar feeling and an objective fact, leads to a dramatic drop in the quality and tempo of precise actions and always accompanies tactile-proprioceptive hand deafferentation of any origin and is well-known in a moderate, transient version to everybody who had to perform precise actions following hand overcooling, "falling asleep," etc.

Losses in the *effector segment of level C* require a whole new classification. Hypodynamic states of the striatum, which are accompanied by unbridling of the thalamo-pallidar system, have already been mentioned. It is well-known that *losses of the pyramidal segment* produce essentially different consequences depending on whether the injury involved only gigantopyramidal Betz cells (e.g., during hemiplegias due to hemorrhagic accident in the white matter of the brain internal capsule) or the whole pyramidal cortex with all its layers (e.g., during gunshot wounds, abscesses, etc.).

It has already been mentioned that the pyramidal neurons and extrapyramidal neurons of the rubro-spinal (and vestibulo-spinal) tract exert on cells in the anterior horns of the spinal cord mutually opposite actions *with respect to tonus*. More exactly, this antagonism can be formulated as follows: *discharges of the red nucleus group increase all types of tonus, both elastic and viscous; pyramidal discharge exerts inhibitory, suppressive action primarily on the elastic tonus, while remaining neutral with respect to the viscous one.*

In normal conditions, the mutually antagonistic effects of the rubro-spinal and pyramidal systems complement and balance each other. Loss of pyramidal neurons, which used to inhibit the elastic tonus component, leads immediately to an effrenation increase in that component – to an elastic or, as it is more commonly addressed, spastic hypertonus always accompanying the described capsular hemiplegia. It is very typical that the elastic tonus increase produced by the *unaffected* red nucleus does not spread indiscriminately to all the muscles but preserves the subordination appearance typical of the rubro-spinal level. In humans, hemiplegic hypertonia affects selectively flexors in the arm and extensors in the leg (the so-called Wernicke-Mann posture) – exactly those muscle systems that are normally characterized by shorter subordination chronaxie, i.e. higher excitability.

The described elastic hypertonia is frequently accompanied by a specific, elastic type of *tremor*, the so-called *clonus*, which is a reaction representing perseverance of the tendon tap reflex in hypertonic muscle groups.

There is another type of hypertonia, with rather different properties, which is produced by a different version of domination of the red nucleus group – when this group works normally but the rubro-spinal level is in a state of hyperexcitation, e.g., as in the parkinsonian syndrome. The hyperexcited red nucleus group tries to increase all types of tonus, but the normally working pyramidal system resists this action with respect to the elastic tonus, and the rubro-spinal level can only increase the rigid, viscous tonus component. In such cases, the general, widespread hyperdynamia bring about a comparably general and widespread increase in the viscous tonus, with no difference between agonists and antagonists, as it is well known from the clinical picture of parkinsonian rigidity.

Let us also mention that now we have reached a rational explanation of *hypotonia accompanying loss of function of level A*. This hypotonia, characterized primarily by a drop in the elastic component, is produced by the inhibitory action of the pyramidal neurons on this tonus component, which does not meet resistance from the malfunctioning red nucleus.

Loss of level *C2* induced by an interruption of the pyramidal pathway in the capsule is to a large degree masked and muffled by the aforementioned crude tonus disorders and the loss of all voluntary motor functions (at initial stages), which is typical of hemiplegia. During the process of following adaptive compensations, leading to an ultimate or residual stage when hemiparalysis turns into hemiparesis, the selective losses become more transparent. All the semi-involuntary motor repertoire of the level of synergies *B* recovers gradually as long as this process is not limited by the muscle paresis; movements at the level of action, which normally recruit the extrapyramidal system together with the pyramidal one, slowly recover by switching to the extrapyramidal system some of the background corrections that used to be performed at the pyramidal sub-level. Clearly, the situation is much worse with respect to recovery of small, precise movements. It is worth mentioning that involuntary, purely extrapyramidal, movements are frequently possible even at the initial stage of deep paralysis of an extremity. During the process of development of compensations during the recovery of the affected side of the body the percentage of voluntary movements in the overall motor repertoire of the patient increases, and the recovery of motor deficits in humans can go very far. In mammals, even in monkeys, because of the relatively smaller role of the pyramidal effector system compared to the extrapyramidal effector system, experimental transections of the pyramidal pathway usually produce a rather quick and full movement recovery, exclusively due to compensations, because the pyramidal pathway never regenerates.

The main differences of pyramidal disorders produced by cortical local injuries from those produced by the described capsular injuries require making a few general comments on specificity of phenomena during local injuries of cortical effector centers. First of all, we should note that all the cortical fields belong to the class of mixed formations containing both centripetal and centrifugal neurons, while purely afferent nuclei can undoubtedly be encountered in both the cortical and subcortical formations. Maybe, this is one of the reasons of less clear and expressive phenomena of coordination losses during local injuries to those areas compared to injuries of afferent nuclei. Let us note the following facts.

An injury of an effector center always leads to disorders in the *motor composition* of a motor action, and very rarely affects its semantic structure. Disorders of the motor composition affect either movements and their components, which are directly linked to the injured level, or motor components from background levels, which normally undergo ecphoria by the effector nucleus of the injured level. For example, local injuries of premotor cortical fields, which are the effector centers of the level of actions *D*, do not produce paralysis or disorders of elementary active mobility, but they lead to a dramatic and widespread disruption of the background motor composition of actions, i.e., their deautomation with relatively preserved semantic structure, which is typical for Kleist's kinetic apraxias. The so-called motor aphasia (motor disorder of speech) linked to the anatomical locus in the left hemisphere, in the Broca area, is perfectly homologous to the described syndrome. Motor aphasia is accompanied by neither paralysis of tongue and larynx muscles, nor by a decrease in the volume, force, speed, and even accuracy of tongue movements; all voluntary tongue movements are intact as well as actions of chewing, swallowing, spitting, etc. However, the articulation of speech movements is lost, i.e., there is loss of *all speech background automatisms*. Local injuries in the *cortical pyramidal fields* lead to motor disorders including both pyramidal and extrapyramidal symptoms, which clearly speaks in favor of the normal pyramidal sub-level having extrapyramidal backgrounds, which cannot undergo ecphoria in response to demands from the pyramidal cortex due to local injuries of the latter. This mixture of pyramidal and extrapyramidal symptoms represents one of the expressive and significant differences between cortical hemiplegia from the conduction, capsular, hemiplegia. Another confirmation of the mixture of backgrounds is the fact that cortical hemiplegias are frequently not associated with spastic hypertonia. During capsular hemiplegia, only the pyramidal component is lost, such that the rubral component, released from the inhibitory action of its antagonist, can bring about the full-fledged picture of hypertonia. In contrast, localized injuries of the cortex lead to a hypodynamic state of both antagonists, pyramidal and extrapyramidal, which does not give crucial advantage to either one.

With respect to deautomations induced by disruptions of ecphoria due to a local injury to the effector center, we have to mention that there is a very convenient method to distinguish deautomations induced by disruption of one of the background levels from deautomations produced by disorders of ecphoria by the leading level effector. Loss of a background level *P* (e.g., the level of synergies during paralysis with tremor), leads to deautomation at *all* higher levels, (*P*+1), (*P*+2), etc., but only those that had backgrounds from the lost level *P*. If deautomation emerged due to destruction of the center – "ecphorator" –, it affects movements of *only* the leading level P but with respect to background automatisms from all lower levels, (*P*–1), (*P*–2), etc.

With respect to *localized injuries of effector centers*, we have to note that such injuries typically show *higher degree of somatotopy*, i.e., their reflection in a more or less well-defined peripheral area of the body (arm, tongue, etc.), in contrast to injuries in afferent centers. If we recall the rule of dependence of phenomena on the site of an injury within the reflex loop, we can note that the effector nucleus of each of the coordination reflex loops (starting from level *C1* and higher) is located, metaphorically speaking, lower, closer to the body periphery, compared to the corresponding afferent nucleus.

We have already considered, starting from level-specific syndrome characteristics, the characteristic feature of localized injuries of effector centers, compared to injuries of conduction pathways, namely their ability to produce hypodynamias.

Consequences of local disorders in the motor domain caused by injuries of substrates of the *level of action D* have already been described in Chapter 6 where this level has been characterized. To summarize, local injuries in cortical fields related to this level lead to clinical pictures of *apraxic disorders*, which form the following groups: (1) afferent or gnostic apraxias (Liepmann); and (2) effector, a.k.a. kinetic, a.k.a. premotor apraxias (Kleist). Both groups belong to the group of hypodynamic disorders with typical movement losses. Over-excitation and hyperdynamia in the premotor system typically present not as apraxias but as obsessive hyperkineses of perseverative type, with the exception of brief periods of irradiation, which can be seen during the first days following an injury or surgery, mildly expressed due to their irradiation. Specific combinations of hypo- and hyperdynamic phenomena, such as *perseverations*, are very common and typical of injuries at this level; we will address those during the following analysis of perseverations across all levels.

Perseveration is loss of the ability to stop voluntarily a process in the motor domain once it has started when the semantic aspect of the problem has been exhausted and does not require continuation of the action. In normal conditions, motivations to initiate and terminate a motor process always reside in the afferentation of its leading level. Perseveration emerges when signals from leading afferent systems stop reaching its effector components and are unable to stop the process when its semantic aspect has been exhausted. This can happen when: (1) the afferent mechanism of switching-off ceases to function; of (2) the switching-off signals are available but they cannot overcome the inertia of a hyperdynamic effector process. In both cases, the effectors and ecphorators of the leading level continue to work stubbornly, the background motor composition of the process continues to be recruited, and the patient has no ability to stop it.

The aforementioned classification allows addressing the two types of perseverations as *sensory* or hypodynamic and *effector* or hyperdynamic, respectively. Mixed forms are also possible.

One should distinguish perseveration from "jamming" (Sperrung). "Jamming" is dysfunction of the mechanism, which normally makes the common motor pathway available for the realization of future motor actions after completion of an ongoing action, similar to a telephone operator disconnecting the person with whom we have just stopped talking and making the telephone line available for future calls. During "jamming," the motor action itself shows no tendency to be prolonged, as it does during perseverations, but all future stimuli bring it about again and again, even though it is not an adequate reaction to the stimuli. The difference between perseveration and "jamming" can be understood best from the following example.

If a patient, who has been asked to draw a cross, fills the whole sheet of paper with crosses and cannot stop, this is perseveration. If, however, a patient completes the task of drawing a cross properly and then continues to draw crosses in response to following tasks of drawing a circle, a square, etc., this defect is "jamming." One may say that perseveration is a disorder of effector function of the *endpoint switches* at a given level, while "jamming" is malfunctioning of its *switching mechanisms*. The phenomenon of "jamming" at higher levels comes closely to the symptom of catatonic stereotypy and, in such cases, suggests a deeper dysfunction of psychic activity compared to perseverations.

Perseverations produce many bright features selective for specific levels, likely because of the following reason. As in the aforementioned types of localized injuries, pathological perseverations selectively involve specific contingents of independent or background motor actions without affecting other movements; at the same time, the specific mechanisms of endpoint switching, which stop functioning in cases of perseverations, have level-specific features. It is possible to say that, depending on the localization and character of an injury, the patient perseverates *in different actions and in a different way* compared to other injury locations. We have to admit that this area of neurosemiotics has many white pages.

Moving bottom-up, as in earlier sections, we encounter definite cases of pathological perseveration in the exact meaning of this term in motoneurons themselves, i.e. below all the cerebral construction levels. These are *clonuses* of muscles that normally show tendon-tap reflexes. In spastic, hypertonic paresis, the reflex response of the triceps surae or quadriceps muscle to a single stimulation of their spindle and tendon receptors leads not to a single contraction but produces a whole rhythmical sequence, when each contraction, on the background of the general overexcitability of neurons in the anterior horns, becomes a stimulus sufficient for inducing the next contraction. Hence, this is an example of effector perseveration as it has been defined earlier.

At the low *rubro-spinal level A*, perseverations of both types can be seen in different clinical forms. In cases of *hypodynamia* at the red nucleus level, one can nearly always see the symptom of *hypermetria*, excessive character of movements, and target overshoots, which are definitely phenomena of perseverative character emerging because of the lacking recruitment of antagonist muscles stopping the movement, i.e., because of the weakening or loss of the myotatic reflex to stretch (myotatic reflex or stretch-reflex). Therefore, the hypermetric symptom directly caused by disorders of proprioceptive signals can be attributed to the group of sensory perseverations. We have to note that this symptom, which is frequently seen within the group of phenomena accompanying hypodynamia of level *A* and characterizing ataxic disorders at the level of spatial field, is commonly attributed by mistake to the syndrome of ataxia itself; in fact, this symptom, as well as intentional tremor, is always only a frequent, but secondary background satellite of ataxia.

The opposite, *hyperdynamic*, form at the rubro-spinal level *A* also contains perseverative phenomena among its symptoms. In Chapter 3, during the description of level *A* pathologies, we have mentioned that the extrapyramidal, rubral, rigidity frequently morphs into a state of cataleptic, so-called wax flexibility (flexibilitas cerea). Since posture is a particular case of movement (proceeding with zero velocities), then, undoubtedly, the patient's inability to change body posture voluntarily represents "posture perseveration"; independently of clinical syndromes that contain posture perseveration, it is pathognomonic for hyperdynamia of level *A*. Mechanisms of this type of perseveration are nearly opposite to those during hypermetria: In hypermetria, proprioceptive reflex responses to muscle stretch are delayed and happen to be too weak; in the discussed case, they emerge too early, induce an exaggerated effect and "fix," as hardening concrete, any posture imposed on the patient.

Interestingly, the difference mentioned earlier between the mechanisms and forms of hypertonia emerging in motor units: (a) due to loss of the pyramidal component; and (b) due to hyperfunction of the red nucleus, emerges here as well presenting an even brighter qualitative contrast. Essentially the same mechanism of the tendon-tap proprioceptive reflex, produced in response to stimulation of tendon sensory endings by their stretch, acting on muscles with elevated elasticity and low viscosity leads to perseveration in the form of clonic elastic-oscillatory process; when the same reflex acts on muscles with a predominant increase in viscosity (elasticity cannot increase due to the preserved action of pyramidal neurons), it turns the muscle into a batter-like substance leading to a tonic version of perseveration; using the language of mechanics, it leads to an overdamped oscillation.

A typical "jamming"-type phenomenon at level *A* is the so-called grasping reflex (forced grasping). Even the smallest stimulation of the palm produces in the patient clenching the fingers into a fist, and no voluntary effort by the patient can release the grip. The most interesting phenomenon, most directly demonstrating that the grasping reflex belongs to the class of "jamming" is that other tactile stimuli of any kind and applied to any area of the body lead to strengthening the grip, which turns into a true dominant and absorbs the excitatory effects of all these stimuli.

It is very hard to find convincing examples of sensory perseverations at the leading level of synergies *B*, possibly because the meagre list of independent actions at that level in normal conditions includes many involuntary and repetitive actions. However, when playing a background role, the level of synergies produces clear perseverations. Since we are talking not about a complete loss of the level of synergies (otherwise, there would be no perseverations to mention) but about a hypodynamic decrease of the tonus of its afferent signals, the overall picture of the disorder involves a drop in both quantity and scale of the backgrounds for all the higher levels – a widespread partial deautomation of movements in all their synergic aspects. If the switching mechanisms of the afferent component of level *B* are weakened more that the effectors, this may bring about a picture of *perseveration of the residuals of synergic backgrounds*. This phenomenon does not have a commonly accepted name and is addressed with respect to walking as "*propulsion*," "*retropulsion*," or "*lateropulsion*," but it can also be generalized to other motor actions. A patient with tremulous paralysis (paralysis agitans) shows hyperdynamic phenomena at the lower extrapyramidal level, rubrospinal level *A*, brought about by effrenation caused by a hypodynamic state of the middle level of the extrapyramidal effector system, the thalamo-pallidar level *B*. The general hypertonic state of stiffness of a patient, on the background of completely intact otolith apparatus of equilibrium (related to level *A*), facilitates standing and walking, but the extreme muscle rigidity and poverty of synergies makes walking a difficult and fatiguing action for the nervous system. Nevertheless, if the patient is pushed forward and faces an inevitable loss of balance – unless there is protective stepping –, the patient *involuntarily* engages into the process of walking using the residuals of stepping synergies and have *major troubles stopping voluntarily*. Similar perseverations of stepping emerge also as consequences of pushing the patient sideways or backward. We should also mention the similarity of the defensive role of hypertonicity at level *A* during the syndrome of *deautomation* and the protective voluntary widespread fixation of all the joints typical of novices prior to *automation* of a new skill at the same level of synergies (Chapter 8).

Perseverations of the effector type at the level of synergies produced by hyperdynamic state of its effectors – the pallidums – sometimes form a substantial part of various hyperkineses, which are abundant in syndromes of chorea, athetosis, tic, etc. Very frequently, these hyperkineses show intentional characteristics: As long as the patient is relaxed, he or she can to a degree control or delay these involuntary hyperkineses, but initiating any active movement leads to their emergence and overflow. In such cases, one can easily see the emergence of effector perseverations: The awakened hyperkineses repeat the initiated movement or its initial fragment in a distorted, caricature appearance. This is how grotesque backgrounds without the main action emerge, mentioned in Chapter 4. A typical feature of perseverations and hyperkinetic syndromes at the level of synergies in general is their distortion, which can be called confusion of sensory corrections. At the level of spatial field, which is our next topic, the essence of corrections is in coordination of the ongoing motor process with the external objective space. Their disorders lead to problems with accuracy, directedness, and timeliness of movements – their purposeful relations with the external world; as a result, pathologically changed corrections know exactly, so to speak, by how much and in what direction they erred in each particular case. During movements at the level of synergies, the main goal of corrections is to arrange an intrinsic coordination of a large movement by co-adjusting its components to each other. If only torn apart pieces are left from a global movement, the correction apparatus is totally deprived of motives and directives on how and for what purpose to correct these fragments. This is probably the origin of the motor anarchy, which is most typical of hyperkinetic syndromes at this level.

Sensory perseverations at the level of spatial field are not rare as consequences of cortical traumas in the postcentral and parietal areas. In the simplest test, using paper and pencil, such a patient, after being asked to draw a circle, responds with drawing numerous circles filling the whole sheet (a patient with Parkinson's disease and phenomena of propulsion, reacts to such a task with non-stop movement of the tip of the pencil over the same spot producing a whole ball). Similar phenomena of perseveration at the level of spatial field emerge also during the tapping test: The task to tap "dot-dot" produces a never-ending sequence of dots or colons.

Phenomena of sensory, hypodynamic perseverations at level *C* frequently alternate with "jamming" phenomena, which can be easily discovered with the help of drawing tests similar to the one used as an example to define the notion of "jamming"[1] (Figures 9.3, 9.4).

We have never seen hyperdynamic perseverations at level *C*. As it has already been mentioned, states of effector irritation at this level with any types of hyperdynamic phenomena are hard to observe.

In contrast, the clinics of level *D* is very rich with perseverations of all hues, which smoothly move close to the highest, symbolic levels of group *E*. In the field of sensory or gnostic apraxias, perseverations happen easily in any of the elements of a chain action making the whole chain meaningless. It is also possible to observe switches from the original chain to a previously realized chain, closely resembling "jamming." In the field of writing, one can see all types of deficiencies, from simple sensory agraphia to indisputable "jamming."

Writing tests are very convenient for recognizing all these varieties.[2]

A general feature of perseverations of the sensory group at level *D* is the obligatory distortion of the semantic structure, making the action meaningless. Normally, the trigger for terminating an element in the chain is semantic realization of the specific problem the element was designed to solve. During gnostic apraxia, with disruption of the overall comprehension of the problem the whole chain action is expected to solve, and hence of the semantic structure, the motivation

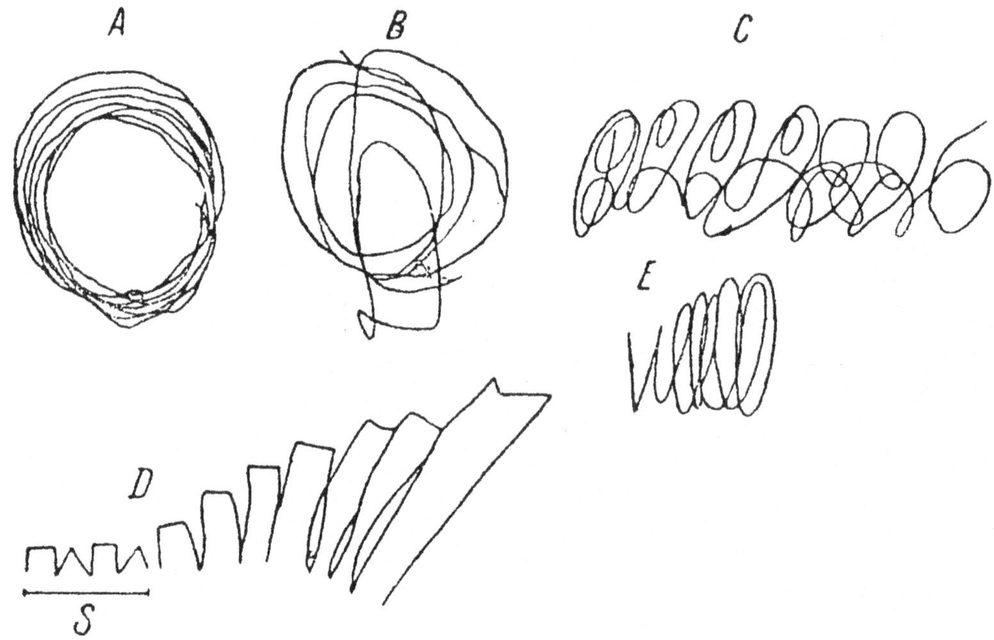

FIGURE 9.3 Perseverative writing and drawing by a patient with the syndrome of premotor coordination disorder (later resolved) due to hemorrhage into the brain white matter close to the premotor zone. Task *A* – draw a circle; *B* – draw a square, *C* – write 122, *D* – copy the angular line shown with the letter *S*, *E* – write 101. Patient's report after task *D*: "After starting, I continued ... Noticed that something was wrong but could not change it."
(Source: after A. Luria).

to terminate an active element disappears. Its interminable continuation reflects the tendency to follow the path of least resistance: It is easier to continue doing the same element than to switch to another one, which has nothing to attract the actor.

It is very important to emphasize the pronounced selectivity with respect to movement contingents, which is typical of perseverations at the level of actions. To be more exact, we have to speak not about selectivity of movements (in a sense of their external, kinetic appearance) but about selectivity with respect to the meaning of problems of different type.

For example, a patient can, without any difficulties, draw a required circle or triangle but cannot overcome perseverations when he starts performing the task of drawing the "sun" or a "house," which he draws indistinguishable from the circle and triangle. In other cases, the situation may be exactly opposite, which makes such selective perseverations highly valuable signs for topical diagnostics (Figures 9.5, 9.6).[3]

Effector perseverations at the level of actions, in contrast to the sensory ones, display a hyperdynamic, or, more narrowly, hyperkinetic origin. These reflections of irritation of an organ, which ensures realization of the action motor composition, are very frequently seen as obsessive, involuntary, *machine-like actions*, or better to say, fragments of actions, because they typically represent breaking out, repeated endlessly, background automatisms from the motor composition. These obsessive, machine-like actions differ from sensory perseverations at level *D* by their emergence on the background of a general speech and motor state of excitation of the patient and by the fact that *sensory* perseverations emerge as errors in the helpless attempts

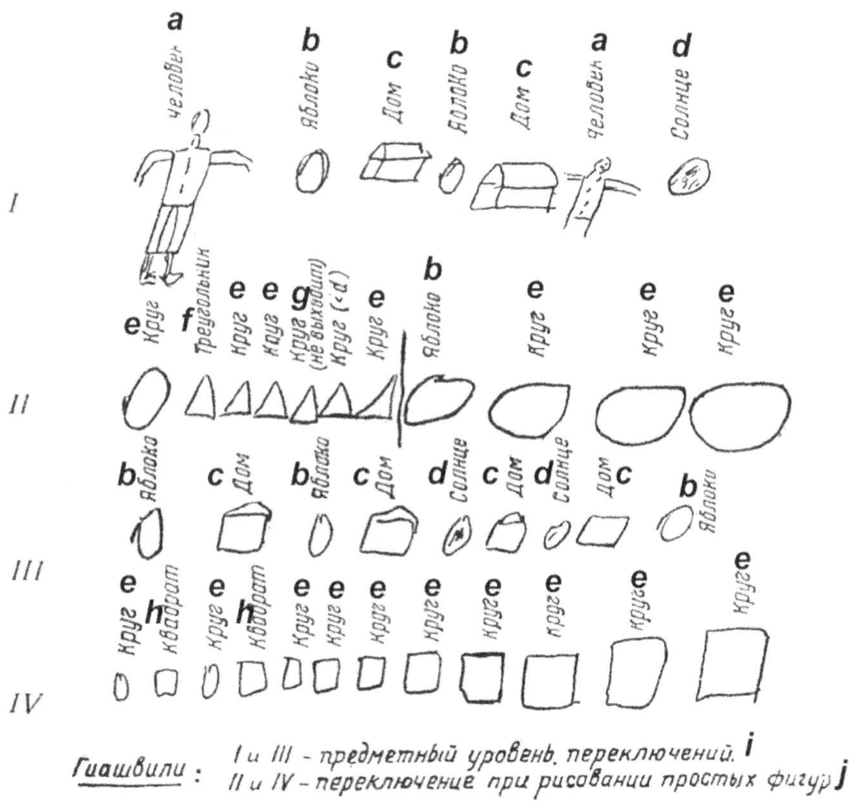

FIGURE 9.4 A case of selective perseveration at the level of spatial field without any perseveration at the object level.
The notes over each drawing were made by the experimenter after the termination of the experiment and reproduce the verbal tasks given to the patient. Patient G with a deep tumor in the frontal lobe.
a – man; *b* – apple; *c* – house; *d* – sun; *e* – circle; *f* – triangle; *g* – cannot draw the circle; *h* – square; *i* – object level of switches; *j* – switches during drawing simple shapes.
(Source: after the materials of Prof. A. Luria).

by an apraxic patient to solve a meaningful motor problem. In contrast, *effector* perseverations, machine-like actions, emerge without a link to a specific problem, as responses to an accidental stimulus or even without a stimulus, as a reflection of general hyperkinesis, which reveals itself in repetition of the same automatisms. These obsessive, machine-like actions differ from subcortical, extrapyramidal perseverations and general hyperkineses of the thalamo-pallidar level by the lack of chaotic disfigurement: They are quite harmonious and intrinsically coordinated, as expected from the backgrounds originating from unaffected construction levels; if the patient were able to master and recruit them in meaningful action chains, they would have performed their roles perfectly well. Such machine-like actions are not discoordinated but only inappropriate.

To finish the description of the phenomenon of perseveration, we have to indicate that it can be seen not only in organic, localized brain injuries, but there are cases when perseveration reveals itself in normal persons without any signs of pathology. In such cases, both types of perseveration are possible – sensory or hypodynamic and effector or hyperdynamic.

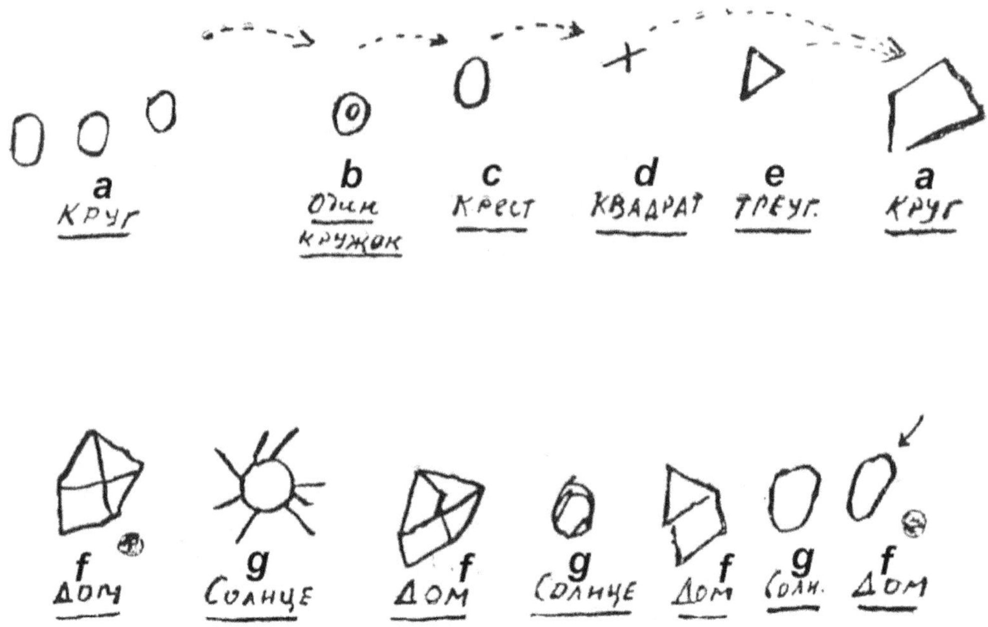

FIGURE 9.5 Phenomena of perseveration at level *C2* without perseveration at level *D* in a patient with a tumor of the left frontal lobe.
a – circle; *b* – one circle; *c* – cross; *d* – square; *e* – triangle; *f* – house; *g* – sun.
(Source: A. Luria).

Hypodynamic perseverative phenomena are possible primarily on the background of an overall weakening of the nervous system, during profound fatigue, sleepiness, intoxication, etc. During normal states of the organism, sensory perseverations (or "jammings") can emerge: (1) when it is necessary to switch quickly to a new form or tempo of the movement; (2) when the motor process is too difficult or too novel for the actor; and (3) when an intense factor distracts or perturbs the action. Let us consider a few examples. Speech perseverations are typical in children learning to speak ("papapa" instead of "papa" – daddy; "lampapa" instead of "lampa" – lamp; "bakaka" instead of "sobaka" – dog, etc.) or mastered speech only recently (mistakes such as "gigopotam," "papjatnik," "kakaratitsa," "mama, momogi," "papa, popogi," etc.) Grown-ups show similar mistakes when it is hard to pronounce the text, which requires subtle switches to maneuver among similar sounds; specifically, perseverative errors emerge most frequently in cases (e.g., tongue-twisters) when the required text itself represents nearly a perseveration.[4] Here come a few examples of perseverative errors originating from words borrowed from a different language: "tartarary" from the Latin "tartarus," the French "trésor" with two "r"s from the Greek "thesauros" with one "r," etc.; doubling of syllabi of perseverative nature in words from archaic languages: themo – tithemi, tango – tetigi, etc.

There are frequent perseverative errors during learning hard passages on a musical instrument. Sometimes, it is easier to use perseveration and hit an extra note on the piano than to make the prescribed by the text out-of-rhythm skip, a syncope. Cyclographic recordings of piano playing clearly show perseverative hand movements in the air during such syncope skips because this is easier for the pianist than to stop the hand (see Figure 5.25). Under fatigue, perseverative errors can sometimes be seen during writing.

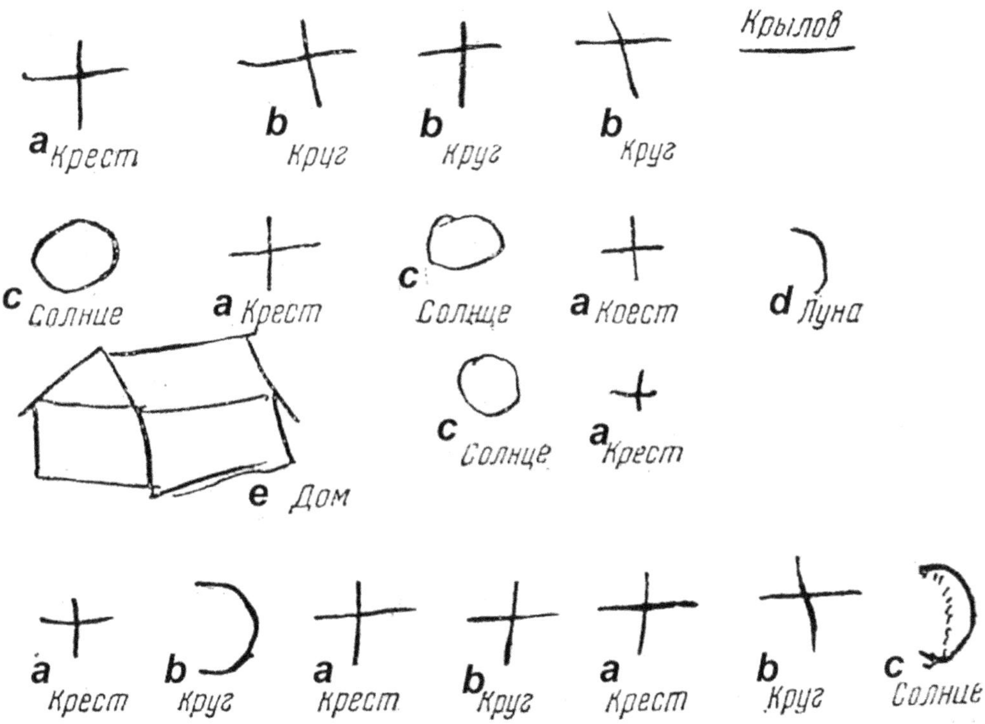

FIGURE 9.6 Perseveration in a patient with a frontal lobe complex of symptoms.
Drawings following tasks (the verbal tasks by the experimenter are shown with legends under each sample.) Top row: After the correct execution of the first task – drawing a cross – the patient perseverates it three times despite the repetitive requests to draw a circle. In the bottom row, the task to draw a cross results in a cross, whereas the task to draw a circle produces a half-circle, and all the following tasks once again produce perseveration of crosses. It is very interesting and typical of the selectivity of perseverations that it is sufficient to convert verbal tasks from the abstract field into an object field, i.e., to switch the patient to another construction level, to eliminate the perseveration completely – see the second and third lines and also the end of line four.
a – cross; *b* – circle; *c* – sun; *d* – moon; *e* – house.
(Source: the figure has been provided by Prof. A. Luria).

During *locomotor movements*, we encounter perseverations when the cyclical process undergoes a rapid switch to a discrete action requiring much effort and attention. During long jump with runup, many masters, after the push-off, in the aerial phase of the jump, continue to move legs in the air in the rhythm of the preceding run (so-called scissors), which can be clearly seen in chronophotographic recordings (Figure 9.7a). Frequently, an opinion is expressed that "scissors" contribute mechanically to the result of the jump. This is wrong because no movement in the absence of an external point of force application can affect the aerial phase. But maybe, for the jumper, it is easier during the abrupt effort at the push-off, to yield to perseveration than to distract and try to fight it, particularly because it cannot hurt the outcome either.

Perseverations of the effector, hyperdynamic type in normal conditions suggest an excess of efforts, which look for a satisfactory outlet and frequently display a direct or indirect playful hue. They can be seen from early childhood in all those endless repetitions, sometimes with an absolutely meaningless text, which K. Chukovsky addresses aptly as "shouts"; emerge in half-verses

(Gazelles) at primitive levels of culture and consisting of endless repetitions of nearly unchangeable one or two lines; can be seen in various alliterations and assonances, numerous humorous word doublings, which can be encountered in all languages: Wiztli-Putzli, pique-nique, Humpty-Dumpty, Handy-Spandy, Heetum-Peetum, shakher-makher, tary-bary, gogol'-mogol', figli-migli, ten'-ten'-poteten', etc. There is only one step from here to explaining the attractiveness, and maybe even the genesis, of rhymes. In the general motor repertoire, voluntary hyperdynamic perseveration of locomotion movements based on "too much energy" leads to *dancing*. Interestingly, such "euphoric" perseverations-hyperkineses can be seen at levels not lower than the striatal sublevel *C1* of the spatial field; with respect to phylogenesis, the same observation is confirmed by the fact that songs and dances emerge not earlier than in birds.

Now we turn to *specific features of healthy normal movements*, i.e., to those selected phenomena that are able to ensure the highest accuracy and specificity with respect to the four main questions on the leading level and on the level composition of a normal holistic motor action.

Of course, all the characteristics of the general appearance of movements at different levels and their specific features, described in the corresponding Chapters 3–6, should play the main role in recognizing the leading levels. Now we discuss selection and grouping of auxiliary features, which are easy to systematize, can be turned into functional tests, and can help to identify the levels of construction and motor composition of actions in difficult and ambiguous cases. They are related to the descriptions given in Chapters 3–6 similarly to how features are related to portraits.

To select and group such features, it is convenient and useful to take, as the basis, the *construction dynamics of a motor skill* within the central nervous system, which has been characterized in detail in Chapter 8.

The process of motor skill elaboration has been divided into *two periods* separated by the phase of automatizing switches.

The most specific feature of processes typical of the first period is the identification of properties and components of the motor action, which require particularly elaborate mechanisms of corrections, and their separation from less crucial components that allow variability, random or adaptive.

Naturally, this first period is characterized most eloquently by features reflecting specific to each level characteristics of *accuracy* and of its underside – *variability*, i.e., features of *specific precision and variability*.

The *second period* of motor skill construction has been termed period of *stabilization* (Chapter 8). It includes overlapping in time phases of the involvement of different levels, standardization and stabilization of the skill components. The most essential feature of this period is broadening the range of conditions (external and internal) where stability of the skill is guaranteed. However, no matter how far efforts by the central nervous system have progressed toward strengthening skill stability against perturbing factors, there still remains a difference between effects of different perturbing factors with respect to different construction levels. Along the same lines, the deautomating actions of excessive variability remain highly selective with respect to different levels within the central nervous system.

Some types of movement variations may be achieved by exercising within very broad ranges of allowed variability, whereas other types always remain bottlenecks, weak, inflexible, and endowed with low adaptability. So, with respect to both allowed variability and effects of various external factors, there is *substantial selectivity of resistance to perturbing factors*: It is possible to identify variations and perturbations, which are particularly efficient in their destructive effects for each of the background levels.

This means that, for analytical estimation of the *second period of skill construction*, most informative is a group of *features of specific stability* and its underside – *destruction and deautomation*, which reflect both the level structure of the analyzed complex movement and the achieved degree of mastering the skill related to this movement.

So, to analyze the level composition and structure of a healthy normal movement and to answer the four questions opening this chapter, we can use two paired groups of features, which have the highest selectivity. These pairs of level-specific features, the first related to the definition and recruitment of the motor composition, and the second – to its fixation and stabilization, can provide more reliable and exact answers to the questions if we can explore with much accuracy and in much detail properties of each of the levels with respect to the mentioned features. When there is reliable experimental material on these issues, these features should be sufficient to compile a table of answers to the following questions, with respect to each of the coordination levels separately.

1. For what elements or features of a movement or movement component does the given level maintain *maximal precision* and how precise is it?[5]
2. For what elements or features of a movement or movement component does the given level allow *maximal range of variability* and how large is this allowed variability?
3. With respect to what types of stimuli or movement variations do movements and their components, assigned to the given level, show *maximal stability*?
4. What type of stimuli or movement variations do produce *maximal destructive effects* on movements and components controlled by the given level?

The first two questions are related to the group of features of precision-variability, i.e., to the first period of skill construction; the other two questions are related to the group of features of stability, i.e., to the second period. On the other hand, the first two questions are most closely related to goals and quality of *solutions of the motor task*; the other two – to goals and qualities of the *time course of movement itself* and the degree of its automation.

Unfortunately, the table of answers to these four questions still contains many white spots and ambiguities, which does not allow moving beyond a rather general descriptive essay. Assuming that the formulation of the questions is correct and legitimate, precise experimental and clinical answers have to emerge in a near future and fill all the corresponding white spots.

The feature of precision, reflected in the first question, requires an explanation regarding its relations to phenomena and features of variability and destructability.

There are not so many movements where precision has easily identifiable, objective quantitative criteria. In all these cases, whether we deal with a holistic movement or its component, the issue of *precision at the target* directly related to success in solving the motor problem. Such precision, in some cases, can be *ultimate*, i.e., related to the final moment of time or spatial location of the movement. In such cases, precision becomes nearly synonymous with marksmanship: precision of pointing at a point, of touching, of getting into the target with a ballistic – striking or throwing – action, etc. In the other group, the same precision at the target has a *process character*, which is seen during the course of a movement: tracing a visual or tactile contour, writing accurately along the lines, following an exactly straight line while riding a bicycle, etc.[6] All these cases of target-related precision are always characterized by the availability of *objective, extrinsic criteria*, which allow *estimating objectively* errors and deviations. It is easy to imagine such criteria for movements such as threading the needle, shooting from a rifle or a bow, returning the tennis ball so that it flies over the net, hitting the nail with the hammer, drawing a circle by eye, throwing the grenade into the target or through an opening, etc.

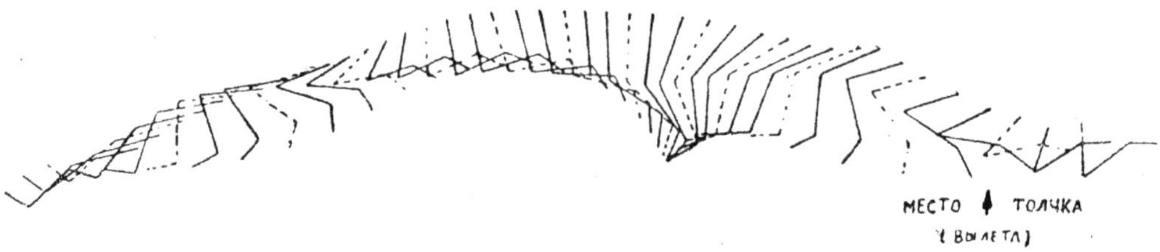

FIGURE 9.7A Consecutive positions of the left (push-off) leg during the long-jump with the runup – a good illustration of a biomechanically useless perseveration: The left leg makes in the air, during the aerial phase, a full step cycle, with the fictitious support stop in the air followed by the swing. (Source: word by N. Sadchikova, CNIIFK, 1938).

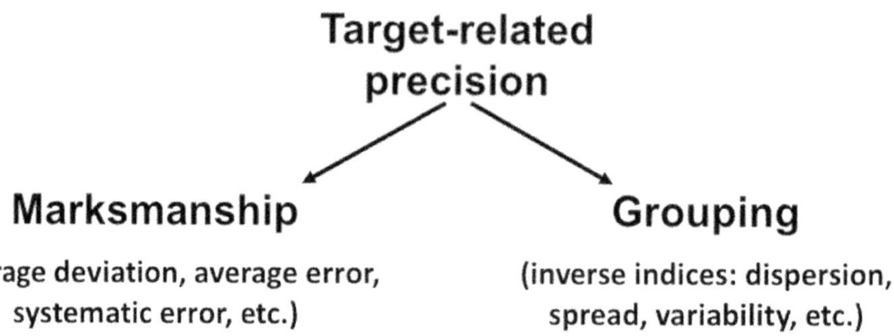

FIGURE 9.7B Estimation of target-related precision.

However, an objectively correct estimation of target-related precision cannot be obtained from observation of a single movement performance because such an estimate has to be free of the *element of chance*: by chance, a very precise or, vice versa, a very deficiently performed movement compared to what is typical of the subject. An objective and reliable estimate of target-related precision can only be obtained from a *series* of movements, by constructing the distribution curve of errors or deviations from the accepted for the given movement criterion: for example, deviations of the thread tip from the center of the needle's eye, etc. Analysis of such a distribution curve yields *two indices* for the target-related precision: (1) *Average deviation*, i.e., the mean magnitude of inaccuracy over all the measured trials; in future, we will expand the meaning of this term and address it as the index of *marksmanship*; (2) An index of *dispersion* of the distribution curve or, an inverse index of *accuracy or grouping* of the curve. The term "grouping" (as well as "marksmanship") we borrowed from a dictionary related to shooting as related to density of shooting at as narrow as possible target area. We feel that it expresses very well a notion opposite to deviation or dispersion of the distribution curve of biological movement.

It is obvious that dispersion of the distribution curve is nothing more than variability of movements according to the selected parameter; hence, variability is one of the two indices of target-related precision. Schematically, estimation of target-related precision looks as shown in Figure 9.7b.

Both indices of *marksmanship*, or the systematic error, and of *grouping* may have very different magnitudes across different conditions. As far as grouping is concerned, in most movements, variable deviations are distributed according to the *law of random numbers*, i.e., the distribution curve has the shape of the Gauss curve.

As an example, let us mention studies of L. Bril under the direction of the author (1944–1945). These studies explored precision of repetitive transversal rhythmical hand movements over 50 cm under proprioceptive control following brief practice under visual control. There was a stop on the left side of the table, and the hand was puncturing a piece of paper with the needle. The abscissa coordinates of all the pinpricks and their time dynamics was measured and then subjected to statistical dispersions analysis.

Figure 9.8 clearly shows both shifts of the systematic error and changes in dispersion of the distribution curves under the influence of fatigue and rest.

There are no movements with semantic content that requires target-related precision, which would have much higher *demands* toward grouping than toward marksmanship, although there is a whole row of movements demonstrating much higher indices of grouping compared to those of marksmanship. Requirements for high marksmanship in the absence of requirements for grouping are very rare but possible (for example, movements of tracing with the planimeter or, to a degree, steering movements).[7]

However, in most movements, there are no high requirements to target-related precision and no criterion for estimating the systematic error of performance. What can be considered an index of coordination precision during walking, choreographic movements, writing, forceful athletic movements, shot put, disk throw, etc.? Clearly, the notion of precision cannot be viewed as in principle inapplicable to movements similar to the mentioned ones. They possess coordination elements related to precision, but they are very hard to quantify.

One can move along the path of making the mentioned movements a bit more complicated to move them closer to actions of the target-related group, for example to ask a person to walk

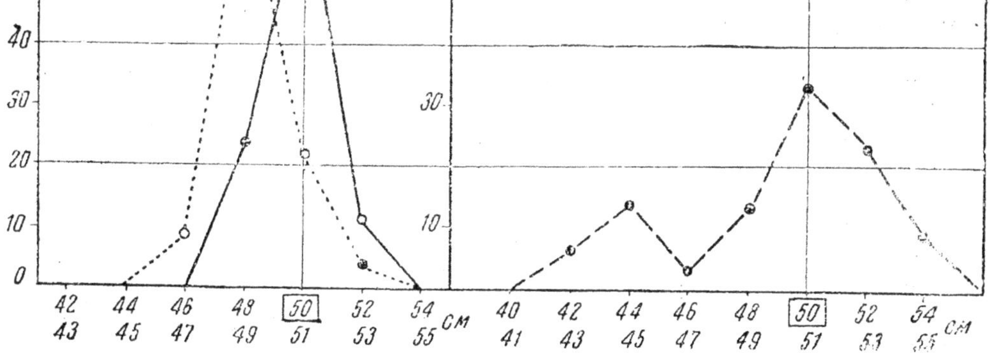

FIGURE 9.8 The distribution curves of pointing under the proprioceptive control of movement distance. Left – in the non-fatigued state; right – under fatigue. 1 – the first 5 mins of work; 2 – the second 5 mins of work; 3 – the first 5 mins of work after the 1-min. rest. Typical data in a single experiment.

a – number of observations.

(Source: L. Bril, Laboratory of K. H. Kekcheev, Institute of Psychology, 1945).

over a striped surface and insist that he steps accurately on the consecutive stripes, or to ask him to walk along a narrow beam or trampoline, or to give the task of "running over the hummocks" (an exercise used in orienteering), or to write not on plain paper but following horizontal lines, etc. It is easy to see that all these complications *change essentially the formulation of the motor task*, introduce changes into the physiological movement structure, encourage the subject to recruit unusual for the movement afferent systems, i.e., *delegate the movement to other construction levels*. This is not what we want.

It is clear, however, that even though a measure of precision could be inapplicable to movements deprived of the element of target-related precision, the second index, which characterizes their *variability*, can be universally applied to all possible kinds of movements and does not require introducing their changes or complications. Such an index of variability can be used conveniently for various cyclical actions: walking, running, swimming, rowing, writing, filing, sawing, rhythmical gymnastic exercises, etc., but it can also be used for singular actions as long as they can be repeated a few times. The idea of a measure or *criterion of variability*, which we have already defined during the description of the distribution curves of deviations consists in *comparing movements among themselves* rather than to an external control event, which can sometimes be an essential part of the movement and, sometimes, absent altogether.

Precision is never encountered in holistic movements or movement components as the *goal in itself*, and there should be a significant reason for the development and refinement of corrections at corresponding coordination levels that show in some directions elevated precision demands. Such reasons are always either a requirement of *target-related precision*, which can stimulate the development and increase in indices of both marksmanship and grouping, or the requirement of stability – the necessity to protect the movement against excessive destructive variations. In this case, the absence of an external criterion (inherent to movements with target-related precision) makes the notion of marksmanship clearly inapplicable, and development of precision follows exclusively the path of *increasing grouping*, i.e. in the direction of a *decrease in the range of allowed variability*. Therefore, in this aspect, the variable of precision connects closely with phenomena of movement perturbations and, in particular, to the fourth of the aforementioned questions.

So, the presented analysis splits the issue of precision into three issues: characteristics of *goal-related precision* according to two metrics: I) marksmanship and II) grouping, and III) characteristics of the degree of grouping, which are dictated not by goal-related precision but by *stability* of the movement itself.

Now let us turn to the necessarily brief, summarizing review of level characteristics and relations with respect to the maximally possible at this time clarification of the problem of specific traits.

The trait of *goal-related precision* can be seen starting from the lowest levels with respect to both mean values and indices of dispersion or grouping. In particular, the rubro-spinal level[1] *A* according to all the available data, represents a level of *high and strict precision*. However, given that this level works in deeply concealed and masked physiological and biomechanical backgrounds, revealing indices of this precision and making them measurable by currently available methods seems very complicated. Indirect data suggest that level *A* keeps high precision and stability of absolute magnitudes of *tonus*, which it sets and regulates, in both neck and trunk muscles and antagonistic muscle pairs of the extremities. In normal conditions, it keeps similar precision of the *timing* of turning on antagonist muscles at crucial points in the movement, in particular when this antagonist recruitment proceeds as a myotatic stretch reflex. Our observations of the *dynamics of running*, for example the timing of turning on the force myotatic wave n_A of the hip flexors in the swing leg is kept within grouping parameters corresponding to the dispersion of under one millisecond. The precision of this level's reactions to deviations of the head and neck from the vertical is better known: It is seen in all types and cases of motor, tonic and tetanic, reactions to vestibular (otolith) stimuli. This level also shows high precision in producing *dosed excitability* of extremity muscles. Although conditions for heterochronism required for the reciprocal innervation and denervation of antagonists are known to be very imprecise, it is important that the result of this subordinate heterochronism is unambiguous in its precision. Unfortunately, the deeply concealed nature of most of the regulation processes performed by level *A* and the related lack of information have prevented us so far from distinguishing the two metrics of goal-related precision and even from estimating, with any degree of certainty, the order of their absolute magnitudes.

In stark contrast to the preceding level *A*, the *level of synergies B* shows absolutely no cases of goal-related precision in movements and their components controlled by this level. These movements and background automatisms are always so self-contained and detached from external metrics that an attempt to measure precision, for example, of a smile deserves nothing more than a smile.

Moving to the sublevels of the *system of spatial field C* transfers us again into the field of goal-related precision. At the *lower sublevel C1*, indices of the absolute magnitude of precision ("marksmanship" according to our expanded understanding of this term) show particularly high values using motor metrics of: a) *extent*, i.e., measures of length; b) movement *direction*; and c) dosage of *force* magnitudes of pressure and friction. Both extents, in other words scales, and directions of movements are kept during actions controlled by this sublevel within the range of a few millimeters. About the same order of goal-directed precision is seen in movements controlled by this sublevel that can be denoted as "adjoining" movements: Movements of tracing a figure, winding, surrounding, motor tracing a point moving independently within the visual or tactile field, etc. In all movements of this kind, sublevel *C* ensures high goal-related precision of both overlap and congruency. As far as the index of *grouping* is concerned, it shows the best values (typically, on the order of a few millimeters for hand actions) for: a) scales; and b) directions of the performed movement.

The *C2 sublevel* based on the *pyramidal* effector system is already known to us as the ultimate level for goal-related precision. At this level, the intensive use of visual control allows reaching maximal values for both indices of precision. The "strong suits" of sublevel *C2* include: First, the *final point* of a movement such as pointing, touching, pricking, etc. yields precision of a fraction of a millimeter; Second, marksmanship and grouping of *ballistic movements* of throwing and striking (showing very large individual variations and strong dependence on the reliability and perfection of background automatisms at the level of synergies); Third, precision of preserving *geometric shape and geometric similitude*; and Fourth, precision of defining the time of the initiation of singular, target-oriented movements. This precision, known as "the time of simple motor reaction," is on the order of one hundred milliseconds with very high grouping.

Making a step up to the *level of actions D*, we once again enter the realm of corrections, which have little to do with measurable goal-related precision. As it has already been emphasized in Chapter 6, all the requirements for precision of actions at level *D* are fully satisfied by special background automatisms from lower levels; corrections at level *D* are switched entirely to the system of concepts and indices of "qualitative geometry" – *topology*, which is deeply alien to metrics. Hence, goal-related precision, in its both components, cannot offer significant information about the level of actions *D*.

To review features and properties of movements, which are kept by the control levels at minimal variability for the sake of movement stability, we would like to note the following.

At the rubro-spinal level of paleokinetic regulations *A*, most pronounced perturbing actions originate, first or all, from various *deviations of the head and trunk*, which can produce additional stimuli for the vestibular (otolith) apparatuses. As a result, even slight variations of posture, in particular the neck-trunk posture – lateral head deviations, leaning the trunk with the head, and of course transitions from the vertical posture to a horizontal one – exert strong perturbing actions on components controlled by this level. These components are also very sensitive to even minimal changes in *grasping posture* of the hand and fingers, no matter whether these are induced by a change in the tool handle or by a volitional variation.

Motor performance at the level of synergies *B* is literally thick with various examples of elevated vulnerability to introduced variations. The overall complex precision, typical of movements and backgrounds at the level of synergies, is based on the surrounding threatening danger of moving outside the range of permissible variations.

Causes of this high vulnerability are, first, that movement corrections at the level of synergies are built exclusively on proprioceptive and tactile sensations, which are inseparable from elements of the movement itself: sequential postures, time patterns of joint angles, segment velocities, their experienced accelerations and efforts, etc. Even slightest variations of all these variables cause changes in proprioceptive and tango-receptive percepts, and even a small increase in these variations can easily push the moving organ into the realm of uncharted, undeveloped receptions, i.e., cause coordination confusion and deautomation. For example, the level of spatial field, with its objectivized and externally projected space built primarily on signal from telereceptors, which are far from the working organs and do not move with them, the area of spatial field in front of the eyes – for the most habitual hand manipulations – has long ago been studied in each of its cubic centimeters and allows shifting trajectories easily in any possible way within its volume. In contrast, in a situation when receptors move together with the working organ and perceive only the organ itself, every new variation is a new percept and, clearly, only a narrow strip of such percepts is explored and mastered for each skill, possibly with an increase in the area of this strip with the exercise time for a given particular skill. Essentially, here we face the very group of reasons making the level of synergies so inflexible with respect to organ transfer (Chapter 8). Second, the narrow range of allowed variations in movements and backgrounds at the level of synergies is conditioned by the phenomenon of *dynamic stability*, which has been discussed in detail in Chapter 8. In that chapter, we pointed out that the riverbeds of dynamically stable movements are narrow, discrete, and separated from each other by areas of unstable movement forms, which self-destruct under the reaction dynamics. These are the reasons for the level of synergies to display its characteristic tendency to show patterns – luckily discovered movement time profiles that are immune to the reactive forces; the stable grouping of such patterns, i.e., the purposefully and masterfully maintained low variability of movements at the level of synergies is nothing else but the method of fighting the perturbing and destructive action of the reactive dynamics inherent to this level.

As a result, the level of synergies *B* maintains minimal variability (i.e., high degree of grouping), first of all, of posture, patterns of joint angles, shapes of trajectories and movement speeds along those trajectories. There is similar vulnerability of movement tempo and intrinsic rhythm with respect to motor variations. It is easy to see how a change in the tempo perturbs the movement using, as an example, the sharp differences in the movement structure during short-distance and middle-distance running, where the tempo difference is on the order of 2:3, or the forced transition of walking into running at a critical tempo of about 100 steps per minute. For comparison, let us note the miniscule changes in the movement structure and the corresponding miniscule deautomation effects, for example, during an increase of the tempo by a factor of 1.5 or even 2 of a quick piano passage (level *C*).

Let us emphasize that the degree of standardization of trajectories, amplitudes, and movement velocities at the level of synergies increases when one considers systems with larger moments of inertia, as it has been shown in our studies of chisel cutting and walking movements. The degree of grouping increases parallel to the rate of convergence of the Fourier series imitating a rhythmical biological movement, i.e., the degree of simplicity and harmony of the movement. For example, during cutting with the chisel, the minimal grouping and minimal convergence of the trigonometric series were shown by the trajectories of the elbow and shoulder joints, whereas the highest values of the same indices were seen in the trajectories of: a). The center of mass of the system from the elbow to the tip of the hammer; and particularly b). The center of mass of the whole arm with the hammer.

The quantitative index of precision (grouping) ensured by the level of synergies can be roughly estimated as being on the order of centimeters in space and centiseconds in time, as it can be seen from our cyclogrammetric materials. Hence, the grouping is overall lower compared to the two adjacent levels, but it extends to a very large number of indices of performance.

We have less to say about the sublevels of the *system of spatial field C*. With respect to the lower, striatal sublevel *C1*, most destructive effects are produced by variations in: a). Movement *scale* or amplitude metrics; and b). Movement *direction*. If a habitual, automated movement of tracing a contour or sliding along the rail is rotated, this introduces very much visible deautomation.

As far as the upper, pyramidal sublevel *C2* is concerned, its inherently high, non-destructive variability and switching have allowed to discover only one weak spot as far as variability is concerned: If one has to reproduce a geometrical figure, the weakness is revealed by turning the figure by 90° or 180°, if we consider a less primitive movement built at that sublevel – a similar effect can be seen using a mirror image of the movement, i.e., mutual exchange of the right and left sides.[8]

The even higher variability typical of movements at the *level of actions D* has already been emphasized in the general description of that level (Chapter 6). There seems to be only one weakness related to destructive variability at level *D*; it is related to the *disappearance of interchangeability between*

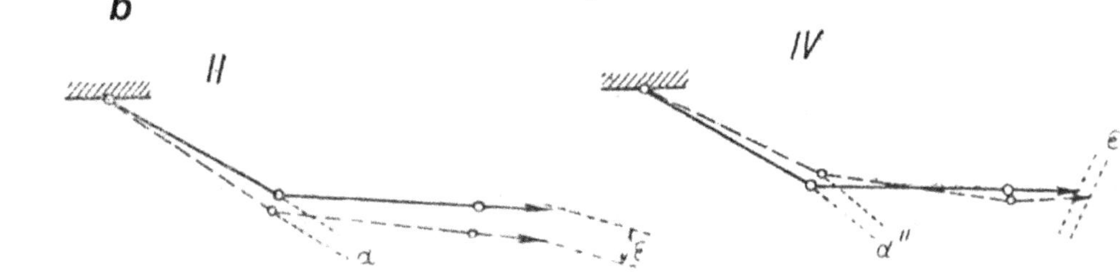

FIGURE 9.9 A scheme of the dependence of changes in precision on qualities of sensory corrections. α – allowed shoulder deviation; ε – resultant deviation of the fingertip. During proprioceptive corrections based on the muscle-joint sensitivity in the shoulder (postures I and II), the deviation changes proportionally to the radius of displacement (the distance from the shoulder to the fingertip) because it is defined by constancy of α. During visual correction (postures III and IV), precision at the fingertip depends neither on posture nor on the elbow joint value, nor on the shoulder allowed deviation: compare α, α', α," etc. *a* – shoulder; *b* – elbow; *c* – wrist; *d* – finger.

the right and left hands, which is present at the level of spatial field. Therefore, if changing the right hand to the left hand or vice versa exerts a disruptive, deautomating effect, this is a nearly certain sign that the action is performed under the control of level *D*.[9]

The issue of features and properties of movements that are the *safest with respect to tolerable variability* can now be addressed only in most general terms. For the two lower levels based on the tango- and proprioception, both *adaptive plastic variability* and "*variability due to indifference*" are expressed very weakly, similar to effects of transfer. At the level of synergies, among the aforementioned variables, the highest variability is seen in movement amplitude, which allows preserving the habitual, automatized motor structure in a rather wide range.

Tolerable variability, both adaptive and indifferent, starts to increase dramatically for a wide range of indices at the sublevel *C1*. This level allows detaching from a certain spatial coordinate, i.e., large variability in the location of the trajectories of points on the moving organ with the emergence of a possibility of various vicar interchangeabilities. At the sublevel *C2*, this list is complemented by easy variability of amplitude and scale, variability of locations and forms, and the possibility to rotate trajectories in various planes (with the exception of mirror-like movement transformation, see earlier). Variability of semantic chains at the level of actions extends even further and allows not only switching the working point and the effector organ, but a whole range of variations in the list and order of links within the chain, which do not change the meaning of the action; the only regression of variability is seen with respect to the right and left hand.

Comparably little can be said at this time on the issue of the *least perturbing external stimuli,* i.e. on features of external conditions that definitely complicate the general conditions of movement execution but are handled by corrections at the given level with relative indifference because of the associated lack of destructive effects.

With respect to both lower extrapyramidal levels *A* and *B*, most indifferent is turning telereceptors off, which has strong destabilizing effects on movements controlled by higher levels. For the level of synergies, similarly low danger, in the sense of disrupting effects, is produced by complications of the time-varying external force field and emergence of novel external forces and resistances, which are all handled successfully by mechanisms of dynamic stability. Finally, for the two lower levels, safe external effects include distraction of attention of the subject, which plays no noticeable role in involuntary, unconscious background components from *A* and *B*.

If we add *time* to the category of perturbations, this factor produces very small deautomating effects at the level of synergies. There are also only minor effects on stability of backgrounds from that level and changes in the tactile field: dry or wet skin, wearing gloves, wearing different clothes, etc.

Motor actions at the level of spatial field *C* tolerate easily exogenous, forced changes in the tempo, scale, and angular amplitude of movements – stimuli that are very morbid for extrapyramidal levels. As demonstrated by observations of the "nervous system plasticity," these actions handle very easily, in an adaptive manner, such exogenous stimuli and changes as, for example, tying-up an extremity, wearing an orthotic device, putting on (a healthy extremity) a prosthesis, using a different tool, etc. These facts, very close to phenomena of switching, have already been discussed (see Chapter 5). The range of external stimuli, which are safe with respect to their perturbing effect on actions at level *D*, is very wide and does not include any specific features.

The more informative last part of our description of specific features – the review of external stimuli showing selective, maximally destabilizing effects on functions of specific levels – is of a broad practical interest. Careful studies in this direction, in addition to enrichment of the list of specific features, should point at a whole range of very important information related to conditions, which contribute to maximal stability of motor skills with different structure and also to methods of most rational and effective *stabilization* of these skills. So far, not much has been known with respect to this issue.

Motor backgrounds from the level of paleoregulation *A* display maximal lability with respect to all stimuli causing *excitation of the vestibular (otolith) apparatus* and secondary tonic neck-trunk reactions. Pushes and shakes, passive turns, asymmetrical forces (for example, of the wind or of a static load), single-sided temperature stimuli to the skin of the face and neck, strongly different levels of luminance acting at the two eyes, etc., as a rule, lead to strong deautomating effects on all movements with components from the level *A*.

The next group of particularly effective stimuli for level A is related to changes in signals from *tangoreceptors of the extremities*.

It is common knowledge that changing the habitual handle of a tool destabilizes movements. Every expert would rather change the working part of a familiar tool than its parts grasped by the hand: The shaft of a knife if more precious than its blade. Comparably strong destabilizing effects are produced by changes in the tactile field, which have just been enumerated and addressed as indifferent with respect to the level of spatial field: putting on gloves, skin dryness, effects of antalgic corrections (see earlier) from local painful spots on the effector: callouses, felons, burns, etc. Finally, there is a clear destabilizing effect of *time* on backgrounds at level *A* in contrast to its very small effects on components from the level of synergies.

The level of synergies *B* also has a comparably characteristic group of dangerous stimuli. First of all, this group includes *changes in working posture*, shape and height of the seat, height of the workbench or work field, size and scale of the tool (work tool, musical instrument, gymnastic instrument, etc.), which influence relative location of body parts and joint angle configuration. In addition, stimuli of the same type include *changes in the working organ* – even a partial change in its participation in the holistic synergic action. Further, selectively destabilizing effects are likely to be produced by exogenously forced *rhythm*; movements from sublevel *C1* handle such changes well, whereas synergies clearly prefer their own, autochthonous rhythm ad libitum. Finally, destabilizing effects at the level of synergies are produced by simultaneous *recruitment of another synergy*; such double-tasking is clearly beyond the scope and distribution of "attention" at the level of synergies.

Let us mention one more factor. It is highly characteristic of movements (and motor elements of actions) with backgrounds from the level of synergies *B*, but is not typical of backgrounds from *C2*, that a well-learned skill proceeds without visual control, whereas *recruitment* of sublevel *C2* exerts a *strong destabilizing effect*.

The striatal sublevel *C2* is characterized primarily by two sources of destabilizing stimuli: First, when its movements are forced to follow cyclical *exogenous metrics*, and second, *weakening* of the function of the background rubro-spinal *level A* caused by corresponding destabilizing factors. The first group can be illustrated by the typical, mentioned a few times, example of walking over railway sleepers or stripes drawn on the floor. As far as the second group of destabilizing factors is concerned, every person has definitely noticed the strong and selective effects of an arm "falling asleep," or its overcooling, or hypesthesia caused by some factor on the two main motor features of sublevel *C1*: quickness and process precision. Every performing musician and draftsman knows that it is impossible to start working immediately after coming into the room from the cold weather. However, this destabilizing factor is due not to the direct importance of tactile and proprioceptive afferent signals for the control of movements at level *C1*, but primarily to deautomation of these movements caused by problems at the rubro-spinal level *A* intimately linked to sublevel *C1*. We have already had the opportunity to witness the degree of intimacy between these two levels in pathology, in particular in syndromes of functional losses; the same intimate link is seen in the discussed example illustrating how much coordination at sublevel *C1* needs foundations from level *A*.

Among stimuli beyond the range of normal everyday life, we have to mention the very selective destabilizing effects on sublevel *C1* of two groups of intoxication. First, these are effects of alcohol (disorders of standing and walking and of accurate movements from those controlled by sublevel *C1*), and second, effects of veronal poisoning,[10] which produce even stronger and even more selective disorders of walking, writing, speech, and various spatial manipulations.

External factors that are most effective in destabilizing movements and components at the upper sublevel of spatial field *C2* also form a very impressive group. We will place first stimuli *distracting subject's attention*: noise, conversation, attracting attention to something else, etc. Such stimuli act at motor actions at the lower sublevel *C1* very weakly and they, as it is well known, easily proceed in a machine-like fashion (walking, well-learned musical passage, habitual work automatism, etc.). With respect to lower levels *B* and *A*, we could easily place the factor of attention distraction into the group of most harmless stimuli with respect to destabilization. As far as the discussed movements at the *C2* level are concerned, distraction of attention has a destabilizing effect not only on actions independently led at that sublevel but also on background automatisms controlled by that level for the level of actions despite the fact that these automatisms proceed outside the field of consciousness. Sometimes, this fact allows distinguishing between motor actions at sublevels *C1* and *C2*.

The second fact, which has strong disorganizing effects at sublevel *C2*, is any kind of a *change in visual control*, which always has crucial importance for actions at that sublevel. Such destabilizing effects are seen when one tries to control movements while looking in the mirror or through prisms, etc. However, the broad plasticity inherent to both sublevels of spatial field very quickly corrects the emerging movement disorders and, following brief exercise, they regain good automaticity. Experiments with destabilizing effects of this kind should involve manipulations that are unexpected and have an element of surprise. Effects stronger than those produced by modified visual feedback can be seen under *complete removal of visual control*. Movements at the lower sublevel of spatial field show only minor destabilization after closing the eyes: for example, walking (assuming that there is no fear of running into objects), drawing lines, work of both hands during piano playing, and of the left hand during playing the violin, etc. Coordination of actions at the pyramidal sublevel in many cases becomes impossible, and this refers not only to actions that, by their essence, cannot be performed by blind.

There are strong negative effects, if not complete destruction, of movements at sublevel *C2* during deep fatigue, sleepiness, migraine, etc. Among directly toxic effects we could mention here those that induce, in some way, brain ischemia, for example carbon monoxide poisoning, altitude hypoxemia, etc.

Finally, the level of actions *D* has not allowed so far to identify a single characteristic of stimuli, which would have selective destabilizing effects on the controlled actions. Similar to other levels, movements at the level of actions show deautomation when they are switched to another, unusual, level. We have already discussed this form of deautomation earlier (Chapter 8), but we have agreed not to discuss this form in depth. Of course, violations of background automatisms by perturbing stimuli deautomate the actions themselves, but these disruptions have already been reflected in the discussion of disruptions of the corresponding lower levels. So far, no disrupting effects of specific external factors on leading corrections at the level of actions have been discerned.

These are the observations that could be collected at this time relevant to description and systematization of level-specific features of human motor actions.

Notes

1 Most commonly encountered forms of *stuttering* represent hypodynamic perseverations with the purely functional genesis. Their relation to specific levels is unclear; most likely, they are related to the lower sub-level of spatial field *C1*, which is why we mention them here. The commonly observed disappearance of stuttering during singing supports this conclusion. The reason is that singing, based primarily on striatal backgrounds, mobilizes the striatal sub-level and relieves its functional hypodynamia.

2 The patient is asked to write the word "revolution." If he writes "rution," this is an example of simple sensory agraphia. If the task is performed in the form of "rereretion" or "revvvvv," or as multiple repetitions of the correct writing of the word – these are all different examples of perseveration, defined by specific features of the injury localization and structure. Finally, writing the word "revolution" in response to other various tasks is "jamming." It is instructive to recollect the wonderful episode in N.S. Leskov (in the story "Kotin-Donlets and Platonida") about a neuropathic child who systematically reveals his scare of the teacher in the form of stereotypical perseveration during writing his own name: "Konsstantintintintintin tin ..."

3 Such an exceptional expert in child speech as K.I. Chukovsky emphasizes this repeatedly:

> Younger children use rhyme not for play, not for embellishment of their speech, but to facilitate it. Given the underdeveloped articular apparatus of a baby, *it is much easier for the baby to pronounce similar sounds than different ones*. For example, it is easier to say 'pokochinochi' than 'pokoynoy nochi' (good night). This is the reason that smaller babies with less developed speech tend more toward using rhymes.
>
> *('From two till five,' p. 238, 1937)*

> Babies are abnormally developing or sick when they do not perform such language exercises. These are indeed exercises, and it is hard to imagine a more rational system of exercise for mastering the art of speech than such *numerous repetitions of all possible sound variations*. ... To learn how to control (language sounds) at will, he repeats them again and again and, *to save the efforts*, each new sound combination differs by only one sound while all others stay unchanged.
>
> *(ibid., p. 240)*

A similar word perseveration in his four-year-old son is mentioned by A.N. Tolstoy: "Nikita looked at me sternly and said: Listen, listen (he has the habit of repeating some words twice), this is silly ... Give me back the paper, and you can write, write, write a very short story." (A.N. Tolstoy, "Adventures of Nikita Roschin," Preface)

4 A few examples useful for speech movement analysis: "Clara stole corals from Karl, Karl stole a clarinet from Clara"; simple but not easy tongue-twisters: "King – eagle – king – eagle," "Sasha walked along the highway," "cinq sangsues a cent cinq sous" etc.

5 Later, we will show that this question has to be split into three more specific ones.

6 *Ultimate precision* is typical of the upper sub-level of spatial field *C2*, although it does not monopolize this characteristic; similarly, process precision is typical of sub-level *C1*.

7 In Chapter 5, we mentioned *adaptive variability*, which distinguishes action at the level of spatial field from patterns at the level of synergies, as a necessary factor to adjust these actions for optimal solutions of the motor tasks. This statement might create an impression that large variability, i.e. *low* grouping, in some cases is essential for target-related precision and marksmanship. Indeed, movements, such as accurate shooting at flying objects, demonstrate a clear combination of both indices: large adaptive variability of the aiming movement and very high marksmanship, i.e. small mean error of the shot. However, more detailed analysis shows that, in this case, as well as in similar tasks, the corrections that ensure high *marksmanship* have to ensure similarly high grouping, otherwise shots would not be consistently accurate; variability of the aiming action is served by corrections of a completely different origin and composition. This is seen clearly in examples where target-related adaptive variability is regulated by a *different level* compared to target-related precision: For example, object chain actions by a toreador, with large variability of auxiliary movements, and the ultimate gesture of pricking the bull's medulla (*C2*), which is maximally precise

and standardized, because any error with respect to target-related precision is mortally dangerous. To summarize, at the *point of target-related precision*, marksmanship is essentially inseparable from grouping.

8 The exchange should be at the level itself, not in the performing organ, which can show in this sublevel easy switching. Of course, for such an experiment, one has to select actions with asymmetrical shapes.

9 A distinguishing feature from level *B*, which also does not allow transfer from the right to the left hand, is that actions at level *D* can withstand any other variations, whereas level *B* is grossly perturbed by them.

10 Similar effects can be seen under other barbiturates, but veronal produces the most clear picture.

Commentaries

C1

NIKOLAI BERNSTEIN IN 1947

From Summarizing to Planning

V.L. Talis

INSTITUTE FOR INFORMATION TRANSMISSION PROBLEMS (KHARKEVICH INSTITUTE) RAS
MOSCOW, RUSSIA

Nikolai Aleksandrovich Bernstein was born into a family of physicians. His grandfather – Nathan (Nikolai) Osipovich Bernstein (1836–1891) (Figure C1.1) was born in Odessa where he returned after defending his thesis at the Moscow University (because of being Jewish, he was not allowed to live in Moscow). In 1865, N.O. Bernstein became professor of anatomy and physiology at the Imperial Novorossiya University,[1] but in 1871 he taught only anatomy while sharing his teaching assignments with Ivan Sechenov.[2] Additionally, for 14 years, N.O. Bernstein served as the Chairman of the Odessa Society of Physicians and was in charge of a ward in the Red Cross Field Hospital during the Russian-Turkish war of 1877–1878.

N.O. Bernstein's older son Aleksander Nikolaevich Bernstein (1870–1922) (Figure C1.2), Nikolai Bernstein's father, a well-known Moscow psychiatrist, was a student of Korsakov[3] and founded Tsentral'nyi Preemnii Pokoi, (The Central Hospital Ward). This was the first clinic in Moscow where mentally ill people were kept without restraints. A.N. Bernstein's work was inspired by the psychiatry of those times, when in both Europe and Russia experimental scientific approaches to diagnostics were being developed and, at the same time, there was progress toward treating mentally ill people more humanely. In the obituary of A.N. Bernstein, Gilyarovsky[4] wrote:

> The majority of mentally ill patients were not able to obtain appropriate placements and, because of the inter-agency frictions, would often end up in police custody in unbearable conditions. The Central Hospital Ward was founded to remediate the situation. Aleksander Nikolaevich became the head of this ward.
>
> *(Gilyarovsky 1922)*

The close relationship and the influence of his father on Nikolai Bernstein is evident in their correspondence of 1914, when the 18-year-old son, entering the Faculty of History and Philology of the Moscow State University, wrote: "I found an essay about Gogol[5] in Korolenko,[6] where he expresses the same proposition about his cyclothymia[7] as you do, and refers to Bazhenov's 1902 essay,

FIGURE C1.1 Grandfather of N.A.Bernstein, Nathan (Nikolai) Osipovich Bernstein (approx.1869). (Source: courtesy of A.S. Bernstein).

FIGURE C1.2 Father of N.A.Bernstein, Aleksander Nikolaevich Bernstein (1913). (Source: courtesy of A.S. Bernstein).

which you perhaps know.[8] He points out that Gogol's father suffered from specific features of the very illness that you have described; the alternating periods of feverish activity, the inventiveness, which coincides with manifestations of his comic talent, and times of dark depression, apathy and horror of death, from which, it would appear, he died" (Talis 2020). It is surprising but the ideas expressed in the quoted 1914 letter persisted through Bernstein's life, and were reflected in one of his final notes in 1965 in the popular journal *Science and Life* entitled "Death from the fear of death" (Bernstein 1965). After Bernstein entered the Faculty of History and Philology of the Moscow State University, he transferred to the Medical Faculty, moved by the patriotic desire to take part in the First World War, which had just begun. After graduation from the Medical Faculty, Bernstein served as an army doctor in the Red Army from December 1919 to March 1921. After demobilization, he worked in the fields of psychiatry and psychology at several Moscow institutes. At the same time, he

FIGURE C1.3 N.A.Bernstein (1929).
(Source: courtesy of A.S. Bernstein).

completed several courses in the Department of Mathematics. Bernstein wrote in his autobiography that in 1922 he "started research in the physiology of movements at the Central Institute of Labor, which eventually became my main specialty ..."[9] Bernstein kept up to date with the emergent scientific ideas of his time both in physiology and psychology, having obtained medical education and mastered the use of mathematical methods (see for instance Bernstein 1922, 1962). The beginning of his research on motor activity occurred in close proximity to Vygotsky,[10] Ukhtomsky,[11] Gastev,[12] and Meyerhold.[13] Bernstein's first publication appeared in the first issue of the journal founded by his father, who had died aged 52 in May 1922, at the peak of his creative abilities. The first book by Nikolai Bernstein *General Biomechanics* (Bernstein 1926a), dedicated to the memory of his father, was a 432-page volume containing 237 illustrations. Along with this book, Bernstein publishes its popular adaptation – a 183-page-long *Biomechanics for Instructors* (Bernstein 1926b), which contained 84 illustrations. He described the latter book as a "significantly simplified and shortened version" of the first book that "is appropriate for any student or semi-professional," so that "the most the author can expect from his readers is knowledge of four arithmetic operations and the basics of mechanics and technical drawing" (Bernstein 2020). This publication, translated into English for the first time in 2020, became required reading for the Biomechanics course at the request of the professor who taught the course and who was one of the reviewers of this book for the publisher. In about 100 years since it has been written, it has not become outdated!

On the Construction of Movements was the third book published during Bernstein's lifetime. It was also supposed to be accompanied by a popular adaptation, *On Dexterity and Its Development*. The proofs of the book *On Dexterity and Its Development* with the corrections made by the author in 1948 are kept in the Russian State Library in Moscow.[14] In the proofs, feeling that a storm was gathering, N.A. Bernstein wrote on the photograph depicting a foreign ski jumper: "It would be desirable to replace this photo with a similar one of a Soviet skier! Prof. N.A. Bernstein." Unfortunately, the aggressive persecution of Bernstein in 1948 stood in the way of this publication during Bernstein's lifetime. Its proofs awaited their time for more than 40 years and finally were used for the first publication initiated by Bernstein's student and biographer, I.M. Feigenberg in 1991 (Bernstein 1991). Soon thereafter, in 1996, the book was translated into English by Mark Latash (Latash & Turvey 1996). With the book that you see in front of you, Professor Mark Latash logically completed the feat of heroism of translating both – the shorter popular book followed by the extended volume for professionals!

The last book by N.A. Bernstein, the 340-page-long *Essays on the Physiology of Movements and Physiology of Activity*, was published in 1966, the last year of his life. Unlike *On the Construction of Movements*, in the first two sections of the "*Essays* ...," only the second essay (10 pp. long) is original whereas, for instance, the fifth one (16 pp. long) is a revised first chapter from *On the Construction of Movements*. The concluding third part of the last book consists of five essays written in the 1950s and 60s and only the conclusion itself (16 pp. long) was written specially for this book. Thus, Bernstein's last book can be considered rather his "scientific testament," where he collected his selective papers, introduced his students and, most importantly, outlined, as he had done in *On the Construction of Movements*, the future paths of development for biological science. Many of these paths have not been trodden to this day ...

Between 1926 and 1947 three more books were published with N.A.Bernstein as the editor: *The Methods of Movement Research* (Popova and Mogilyanskaya 1934), *Research in Biodynamics of Locomotion* (Bernstein 1935), and *Research on Biodynamics of Walking, Running and Jumping* (Bernstein 1940). The 1945 list of "scientific works of Professor Bernstein"[15] contains the title of a book *Contemporary Search into Physiology of Neural Processes* that is listed as a 1936 publication without a reference to the number of pages (indicating that it was not published), and the only 1931 article entitled "Determining mass, density and location of the centers of gravity of body links in humans" is described as "written for a biophysics journal but was not published due to the unexpected closure of the journal." Judging by the list of Bernstein's works, there were regularly published articles for various encyclopedias and popular journals. As reported by B.I. Khodorov,[16] Bernstein had the ability to write "60 pages in one sitting without a single correction, added by a large dictionary of terms that he himself created"[17] – a clear sign of a genius: A genius understands everything, all schemes and logic are built internally, but there is a problem – all of this needs to be written out for the public and even twice if a version for the general public is required (plus, it is hard to know in advance the level of understanding of either profession and general public considering the interdisciplinary nature of Bernstein's work).

In the foreword to the book *On the Construction of Movements* Bernstein indicates that he started working on it as early as 1939. However, in the list of aforementioned "scientific works of Professor Bernstein," there is a book entitled *On the Construction of Movement and Their Systematization Based on the Neurophysiological Characteristics*, which is listed as submitted to *The Physical Culture and Sport* publishing house. It seems that Bernstein supposed that, if you have something to say, you need to put it on paper and send it to publishers (similar to how within some scientific disciplines it is customary now to save written works in electronic archives); once it's done – move on to new writings. If nobody wants to publish – then so be it – those who come after you will sort it out ... As it happened, in 2003 I.M. Feigenberg with his co-author I.E. Sirotkina published the book *Contemporary Search into Physiology of the Neural Process*, which was not published in 1936. In 2019, Springer Publisher published the *Atlas of Human Walking*, which Bernstein gave as a gift to the Dortmund University in 1929 during his last trip to Europe, when he did not come with Springer Publisher in agreement to publish it (Figure C1.3).

Bernstein called the book *On the Construction of Movements* in a way the culmination of the 25-year-long scientific career of the author and his colleagues that started at the Central Institute of Labor in 1922. After leaving this Institute in January of 1925, Bernstein started his research on piano strokes at the Gosurdarsvennyi Institut Muzykalnykh Nauk (The State Institute for Musical Sciences), which has the evocative acronym "GIMN."[18] Bernstein himself played piano and oboe; he had perfect pitch and composed music as well. His expert musical knowledge is obvious from the examples that he uses in the book *On the Construction of Movements*. He writes: "if each of the degrees of freedom of the arm and hand of a pianist produces an error of one degree, these errors may sum up to produce 5–6 cm deviations of the fingertip leading to a third or quart error (although in various individual links, e.g., finger phalanges, the contributing errors may be under 0.05 cm)" (Chapter 2, p. 14); "the leading level clearly controls all the lower, background levels participating in any given movement such that each of the background levels performs its part in the movement orchestra, while, to a degree, losing its individuality and only following the baton of the conductor" (Chapter 3 p. 46); "this level has major difficulties in creating new, unexpected motor combinations similar to a large orchestra asked to improvise" (Chapter 5, p. 82) and on pp. 75–77 of Chapter 5 we can see a carefully analyzed violin player's motion.

Starting in the mid-1930s, Bernstein studied "normal walking, walking when carrying a load, changes in walking due to fatigue, during rehabilitation, and in the course of a lifetime (development of walking in children, walking in elderly), walking of amputees using prosthetic limbs, the central nervous pathology of walking, characteristics of running in famous athletes, development of running in children, the technique of running jump, etc."[19] Let us note here that Bernstein had a sincere respect for "the dexterity" of professional athletes and once, in response to being asked why he gave an A to one of the student-athletes who did not perform well on the exam, he said: "Can you make a three-meter jump? Well, he can!"[20]

In the Fall of 1941, after the Great Patriotic War in Russia started in June (known as World War II), Bernstein together with his wife, his three-year-old son and eleven-year-old stepdaughter evacuated to Buryatia (the city of Ulan-Ude), and a year later, in the Fall of 1942 he moved to Uzbekistan (the city of Tashkent) where he practiced occupational therapy at the military hospital. At that time, Luria,[21] Zaporozhets,[22] Leontiev,[23] and other followers of Vygotsky, who fully incorporated Bernstein's ideas into their approach, started to apply Bernstein's multi-level concept of movement construction to rehabilitation of wounded soldiers. In the summer of 1943, Bernstein returned to Moscow with his family and resumed the work, interrupted by the war, at the Central Institute of Physical Culture. He dedicated his book *On the Construction of Movements* to the "Eternal, blessed memory of those who gave up their lives in the fight for the Soviet motherland." In May 1948, this book receives the USSR highest scientific award at the time – the Stalin Award. The book was very popular with young researchers who had interrupted their studies to fight in the war and were now returning home ready to resume their research. B.I. Khodorov recalls "even at the front, during the battle of the Oder, when we would literally dig ourselves in the soil while Nazis attacked us from the air, I would sit there and write down ideas on electrotone and accommodation that would come into my head." As R.S. Person[24] recollects these post-war years: "at that time, gradually, young scientists – physiologists started to show up and began to read the literature. Before, we did not have access to foreign publications. We only had an incomplete translation of Sherrington's works into Russian and that is what we had used for studying. Very few of us knew foreign languages."[25]

On the occasion of receiving the Stalin award, Bernstein wrote a letter he sent to Tbilisi to his student and first biographer Chkhaidze[26] dated June 2, 1948:

> Well, Lev Vladimirovich, the big celebration has, reached our street, indeed. Do you understand how this will change the suspicious, hostile attitude toward physical movement of all those who did not want or could not take it seriously but used hearsay to look down on it? "It is this biomechanics, mechanisms, Bernstein's antics" etc ... And, again, nothing is said about conditional inhibition and dominance; no "alliluevshina"[27] etc. Now the air will clear and only serious, worthy opponents will be left; the ones who know how to use scientific

> argumentation and who can engage in interesting debates. In particular, I am sure that it will become easier for you to work there, in Tbilisi.
>
> *(cited after Kandelaki 1991)*

The term "conditional inhibition" probably hints at the works of Pavlov,[28] and the term "dominance" – at the works of Ukhtomsky. However, very soon, the critique of Bernstein that had always been ongoing, partly due to the difficulties scientific community experienced understanding his views and partly due to his own reluctance to join the discussions, began to escalate soon after the notorious session of the Academy of Agricultural Science in the summer of 1948. Faultfinding during party meetings, public criticism from his graduate students, and the publication of accusatory articles in the special and central press ... Only the highest state award likely protected the scientist from the tragic outcome (for more details see Feigenberg & Latash 1996).

However, no logical explanation applies here. For example, the directive nature of the persecution similar to the "witch hunts" of the Middle Ages is evident from the protocols of the meetings held at Bernstein's places of employment where the criticisms of sport scientists appeared due to "an anti-party group of theatre critics"![29] On November 4, 1950 Bernstein writes to Chkhaidze:

> You know yourself that the conditions under which I currently work are far from being peaceful: the entire staff at my current Institute of Neurosurgery[30] treats me very friendly and with utmost respect, but routinely more and more arrows are launched at me, from the same ambush. And that creates the constant tension. So far, more than one agency has heard my sincere and complete acknowledgments of my shortcomings – in OPD[31] and my other publications. But so far, I have not been able to discover any mistakes and, consequently, had not had an opportunity to acknowledge those, especially because others not only failed to convince me of the presence of any mistakes but were unable to even point them to me with any clarity. I can only make exception for the famously known item in OPD about the priority of I.P. Pavlov, which I publicly have acknowledged everywhere as my error (but the error of a political, not a factual, nature).
>
> *(Kandelaki 1991)*

The other letter to Chkhaidze written on May 8, 1951 describes the on-going persecution:

> The meeting at the Burdenko Institute[32] was awful because of the absolutely crushing, indecent spectacle by Zhukov.[33] He created panic at the local headquarters and the impression with the taken aback (and very friendly to me) staff that I will perish irrevocably. Fortunately, the Board of the Academy took the diametrically opposite view (probably, someone's head got bitten off) and decided to the contrary: expeditiously organize a tiny laboratory for me at the Institute of Neurosurgery, give me space, about 12.000 rubles of financial support and one staff member; also, to stop verbally badgering me incessantly as punishment for the past and give me the opportunity to show through my resumed work my real face and methodological loyalty. This happened in December; the meeting I have described now (dedicated to destroying me) was, it appears, the highest point of the barometer. Since then, so far, it has rather quietly kept rising.
>
> *(Kandelaki 1991).*

Confident of the help from his friends and other scientists – physicians, members of the Board of the Academy of Medical Sciences of USSR (AMS USSR), Bernstein sends letters to the "higher-ups" to defend himself. For example, on December 21, 1951 he writes to Myasnikov:[34]

> Dear Aleksander Leonidovich, having discussed, as you have suggested, my issues with Sergei Yevgenievich Severin,[35] I … submit … my petition. Sergei Yevgenievich initially mistook it for a personal issue, but I explained to him that my request pertains in the first place to the just started and much needed laboratory of movement physiology, which I must and can launch and get on the right track …[36]

I would like to point out that the scrupulous approach to citations, the exceptional sensitivity to the written word, the wide scope of interests that an attentive reader will undoubtedly notice in this book – these very qualities of the brilliant scientist were considered as shortcomings, when Bernstein was labeled a "cosmopolitan" because he was citing foreign authors or when he was accused of "idealism" that was detrimental to "materialism" for the figurativeness of his writings. As the addressee of Bernstein's letter above, Myasnikov remembers it:

> We were not supposed to even cite foreign authors. The editors were supposed to cross them off anyway as it was the Soviet science that was the progressive one and only someone who was against everything Soviet could kowtow to the West.

Even the names of certain diagnostic approaches or methods were "russified" – changed to Russian. The Erb's point for listening to aortal diastolic sounds was renamed Botkin's point (Botkin pointed out its significance citing Erb).[37] Bittorf's sign[38] became Tushinsky's symptom, even though Michail Dmitrievich Tushinsky himself never attempted to take credit for Bittorf's discovery. The well-known Rumpel-Leede phenomenon became Konchalovsky phenomenon[39] (cited after Mlechin 2019).

In January 1952, Bernstein collects all his equipment and leaves the Burdenko Institute of Neurosurgery to join the Institute of Neurology.[40] Feigenberg, who was employed at the Institute of Neurology at that time, remembered that, whenever Bernstein had presented in front of his colleagues, he never tried to simplify things, thus often making it difficult to understand him:

> I remember his presentation in one of the meetings when a professor of neurology, an intelligent middle-aged man said "Nikolai Aleksandrovich, I've heard your presentation on this topic several times already and each time I have not been able to understand your main point." Nikolai Aleksandrovich's response was rather harsh: "Yes, I agree, you have heard my presentation several times and never were able to understand it."[41]

On March 19, 1953 the Laboratory of Movement Physiology and Pathology in the Institute of Neurology was also closed and Bernstein left, after submitting his resignation, to go on disability. As his students remember him at that time, Bernstein was an "idealist," lonely, sincere, somewhat naïve, who, for example wore a pin "Peace to the world" out of his desire to contribute to world peace which they considered odd.[42]

Bernstein, while unemployed at that time, had to support his teenage son and his wife who was on disability. Written translations for reference journals *Biology* and *Mathematics* became his main source of income. He did not stop his intense scientific research, however, although it became mainly theoretical. For example, his notes from 1949 are entitled "'W-coordinates' of time" and

FIGURE C1.4. N.A.Bernstein (1948).
(Source: courtesy of A.S. Bernstein).

"'U-coordinates' of space" and have the subtitles: "I see my impression"; "Image on the Retina"; "Definition of Time. Borderline action of 'present'" and "Natural Philosophy of Two Signs of Time" (Talis 2020). It comes as no surprise that, in the beginning, Bernstein could not imagine that he had become a target of persecution for one reason – because his last name sounded Jewish (his father had abandoned Judaism and become Russian Orthodox) and in response to a remark from one of his graduate student that he was a "rootless cosmopolitan" during a staff meeting, Bernstein responded: "Why do you call me 'a rootless cosmopolitan'? I know my father very well and I even knew my grandfather. My father was a prominent Moscow physician. I simply can't be 'a rootless cosmopolitan.'"[43]

I believe that Nikolai Bernstein was so far removed from what was happening around him that the suddenly organized persecution came as a total shock. He wanted to work and was sincere in his desire to be useful for his country that was, among other things, in need of prosthetic limbs in those post-war years. Two of his letters to the famous writer Chukovsky,[44] archived at the Russian State Library in Moscow,[45] demonstrate his state of mind. For example, his letter from February 20, 1950 is the letter of a man deeply hurt, asking Chukovsky to let him know "who and what were they like – all these 'torturing, pencil-sharpening' Kiskins, Rapoports, and others and what were the vicissitudes of your struggles with pedagogical illiteracy and men in cases?" We can see a man who is broken, regretful and feels very lonely ("I have no right to suggest the annotation for your memoirs"). Bernstein apologizes "for the unsolicited but sincere request from one of your readers." He concludes his letter with the following statement, which seems to be the entire reason for writing it: "I named – just to give you an idea – only a few of the things I would have asked you have I had a pleasure of knowing you personally." In his second letter dated November 6, 1959 Bernstein again mentions "'Kiskins' whose vicious and vulgar arguments you mentioned already in your book 'From 2 to 5'," but here Bernstein already shows no doubt that "people like them are a pitifully small and dying out minority." The change in his mood by 1959 was prompted by the renewed interest in Bernstein from the group of young scientists because of the emergence of the new science – cybernetics that studied management principles in technical and biological systems. The expertise in mathematics allowed Bernstein, on the one hand, to publish strictly applied math studies on isolating frequencies in biological records using encephalograms for this analysis (Bernstein 1962), while on the other hand – along with the prominent mathematicians of his time – to develop theoretical foundations of modeling in biology. V.L. Naidin[46] remembers that at that time Bernstein already was not willing to make any concessions to editors or critics and, in response to the objections of one of the editors at the encyclopedia where they co-authored a publication (Bernstein & Naidin 1966), he said: "Never, under any circumstances, agree to change your fundamental positions. If she does not agree, just withdraw your article."[47]

The reader of this book will find here results of the outstanding experiments of Nikolai Aleksandrovich Bernstein and their summary in their entirety. However, one will be pressed to find explanation of his views on human mental processes which he might not have wanted to outline directly so as not to listen to the "cry of Boeotians," as this was an ideologically loaded topic. In his foreword to the book, Bernstein writes:

> At least, as of now, a number of views expressed in the book have already attracted attention and even helped in the practical work of colleagues working in such diverse fields as physical education, neuropathological semiotics and diagnostics, psychophysiology of labor and defense programs, pedagogy of artistic performance, etc.

Unfortunately, these fields are yet to implement Bernstein's ideas, and the publication of this book potentially can advance their wide utilization in these fields and beyond. Nikolai Aleksandrovich Bernstein was a genius in such diverse areas as medicine, mathematics, and music; he used his free time to make railway models belonged to the so-called generation Z of the great surge of enthusiasm in the post-revolutionary Russia (Smirnov 2013). His perfect pitch allowed him to use it to determine the frequency of the obturator rotation during cyclographic filming, and the topic of his first article (Bernstein 1922) was perception of sound musical intervals by subjects with standard hearing.

Abundant citations of Bernstein's works by many well-known foreign authors (such as Karl Pribram 1971) and Bernstein's fruitful collaboration within the famous Gelfand seminar are just a few examples of Bernstein's scientific contacts up to the end of his life. Bernstein life work was dedicated to the movements research with the goal of "taking a documented actual live movement, which represents a ready integral solved by nature" and gradually "peel[ing] off it all its external mechanical components – which can always be achieved through analytical means." Bernstein believed that

> after the peeling whatever is left in our hands represents the central innervation nucleus of movement, which we need to study further from the purely physiological point of view, identifying what originates from what and how the process unfolds. At the moment, we are at the very beginning of such study.
>
> *(Bernstein 1966, p. 208)*

Notes

1 Presently "Odesa I. I. Mechnikov National University."
2 Ivan Michailovich Sechenov (1829–1905) was a Russian scientist, physiologist, professor and educator, founder of the first Russian school of physiology and natural scientific materialistic approach in psychology.
3 Sergei Sergeevich Korsakov (1854–1900) was a famous neuro-psychiatrists, author of works in neuropathology, psychiatry, and forensic medicine.

4 Vasily Alekseevich Gilyarovsky (1875–1959) was a Russian and Soviet psychiatrist, member of USSR Academy of Medical sciences.
5 Nikolai Vasilievich Gogol (1809–1852) was a great Russian writer famous for works such as the novel *Mertvyie Dushi (Dead Souls)* and the play *Revizor* (*The Government Inspector*).
6 Vladimir Galaktionovich Korolenko (1853–1921) was a Russian writer, journalist and human rights activist. The essay referred to is Korolenko (1909).
7 Cyclothymia is now referred to as bipolar disorder.
8 Nikolai Nikolaevich Bazhenov (1857–1923) was a Russian psychiatrist, who worked for the humane treatment of mentally ill patients. The essay referred to is Bazhenov NN (1902).
9 Personal files of N.A. Bernstein in the Archive of the Russian Academy of Medical Science, Moscow (F. 9120.V. 8/3. Unit 19).
10 Lev Semenovich Vygotsky (1896–1934) was a famous Soviet psychologist.
11 Aleksei Alekseevich Ukhtomsky, count, monkhood name Alipiy (1875–1942) was Russian and Soviet physiologist, member of the USSR Academy of Sciences (1935), the founder of the theory of dominance. The Bishop of Okhtinsk (1931–1942).
12 Aleksei Kapitonovich Gastev (1882–1939) was metal worker, revolutionary, poet, founder of the famous Tsentralnyi Institut Truda (TsIT), the Central Institute of Labor.
13 Vsevolod Emilyevich Meyerhold (1874–1940) was a Russian and Soviet theatre director, actor and theatrical producer.
14 The proofs of the book *On Dexterity and Its Development* in the Manuscript Department of the Russian State Library, Moscow (F. 743, V 95, Unit 1).
15 See n. ix.
16 Boris Israelevich Khodorov (1922–2014) was one of the founders of the Russian school of research on membrane biophysics and neurophysiology.
17 Interview with B.I. Khodorov 2010 (Talis 2020).
18 This acronym in Russian spells the word "hymn."
19 See n. ix.
20 Interview with B.I. Khodorov 2010 (Talis 2020).
21 Alexander Romanovich Luria (1902–1977) was a Russian neuropsychologist, often credited as a father of modern neuropsychological assessment. He developed an extensive and original battery of neuropsychological tests during his clinical work with brain-injured victims of World War II, which are still used in various forms.
22 Alexander Vladimirovich Zaporozhets (1905–1981) was a Soviet developmental psychologist. He studied the psychological mechanisms of voluntary movements, perception and action. He was one of the major representatives of the Kharkov School of Psychology.
23 Alexei Nikolaevich Leontiev (1903–1979) was a Soviet developmental psychologist, and the founder of activity theory.
24 Raisa Samuilovna Person (1918–2016) was a Soviet scientist, the founder of the contemporary electromyography.
25 Interview with R.S. Person 2010 (Talis 2020).
26 Levan Vladimirovich Chkhaidze (1911–2001) was the founder of biomechanics science in Georgia, he and his co-athor Chumakov have written the forst biography by NA Bernstein in Russian (Chkhaidze, Chumakov 1972).
27 "Alliluevshina," from the word "Hallelujah," here "no 'alliluevshina'" means "no praising."
28 Ivan Petrovich Pavlov (1849–1936) was the Russian and Soviet scientist, physiologist, vivisectionist, the founder of the science of the higher nervous activity, the school of physiology, the 1904 Nobel laureate in physiology and medicine "in recognition of his work on the physiology of digestion."
29 Personal files of D.D. Donskoi in the Central State Archive of the Institute of Physical Culture (F.9, № 1, Unit 260–292).
30 Presently "Federal State Autonomous Institution 'N. N. Burdenko National Medical Research Center of Neurosurgery' of the Ministry of Health of the Russian Federation."
31 Abbreviation stands for the title of Bernstein's book *O Postroenii Dvizeni* – OPD (*On the Construction of Movements*).

32 See n. xxix.
33 P.I. Zukov was at that time docent of Russian State University of Physical Education, the co-author of anti-Bernstein article (Zhukov & Kozhin 1950) "who, as it turned out later, obtained PhD without completing college studies" (Zatsiorsky 1991).
34 Aleksander Leonidovich Myasnikov (1899–1965) was a Soviet physician, member AMS USSR (1948).
35 Sergei Yevgenievich Severin (1901–1993) was a Soviet biochemist, member of the Academy of Sciences of USSR, AS (1968), member of the AMS (1948), the Lenin award laureate (1982).
36 See n. ix.
37 Sergey Petrovich Botkin(1832–1889) was Russian, W.H. Erb (1840–1921) was a German physician.
38 Bittorf-Tushinsky test (A. Bittorf (1876–1949) was a German physician and M.D. Tushinsky (1882–1962) was a Soviet physician) – receiving capillary blood for the purpose of detection of cells of a reticuloendothelium in it for differential diagnosis of a rheumatic and septic endocarditis (*Big Medical Encyclopedia*).
39 Carl August Theodor Rumpel (1862–1923), Carl Stockbridge Leede (1882–?) were German physicians; Konchalovsky M.P. (1875–1942) was a Russian Soviet physician.
40 Presently "Research Center of Neurology" in Moscow.
41 Interview with I.M. Feigenberg 2010 (Talis 2020).
42 Interview with V.L. Naidin 2009 (Talis 2020).
43 Interview with D.A. Ginzburg 2009 (Talis 2020).
44 Korney Ivanovich Chukovsky (1882–1969) was one of the most popular children's poets in the Russian language.
45 Personal files of K.I. Chukovsky in the Manuscript Department of the Russian State Library, Moscow (F. 620, V 61, Unit 7).
46 Vladimir Lvovich Naidin (1933–2010) was a physician specializing in neurorehabilitation, one of the founders of the Russian school of neurorehabilitation.
47 Interview with V.L. Naidin 2009 (Talis 2020).

References

Bazhenov NN (1902) Bolezn' i smert' Gogolya' (The Illness and Death of Gogol). Public reading at the annual meeting of The Moscow Society of Neuropathologists and *Psychiatrists. Kushnerev@K* 1: 38 [In Russian].

Bernstein NA (1922) K voprosu o vospriyatii velichin (o roli pokazatel'noi funktsii v prozessakh vospriyatiya velichin) (On the question of the perception of quantities (about the role of exponential function in the processes of the perception of qualities)). *Zhurnal Psykhologii, Nevrologii i Psykhiatrii (J of Psychology, Neurology and Psychiatry)* 1: 21–54 [In Russian, abstract in German].

Bernstein NA (1923) Issledovania po biodynamike udara s pomoshju svetovoi sapisi (Study on stike biodynamics by means of light registration) *Issledovanija TsIT (Studies of TsIT)* 1, 1: 19–79.

Bernstein NA (1926a) Obshchaya Biomekhanika (General Biomechanics) *Moscow TsIT-VTsSPS* [In Russian] 416 pp.

Bernstein NA (1926b) Biomekhanika dlja Instructorov (Biomechanics for Instructors) *Novaya Moskva (New Moscow)* [In Russian] 184 pp.

Bernstein NA (Ed.) (1935) Issledovanija po Biodinamike Locomozii (Research in Biodynamic of Locomotion) *Moscow-Leningrad, Isdatel'stvo VIEM (VIEM Publishing)* 544 pp.

Bernstein NA (Ed.) (1940) Issledovanija po Biodinamike Chod'bi, Bega, Prizka (Research on Biodynamic of Walking, Running and Jumping) *Moscow, Fizkultura i Sport (Physical Culture and Sport)* [In Russian] 312 pp.

Bernstein NA (1962) K analizu neperiodicheskich kolebatel'nykh sum s permenennymi spetrami po metody vzveshennykh reshetok (Towards an analysis of aperiodic oscillatory sum with the variable spectra using the methods of weighted lattices) *Biofizika (Biophysics)* 7, 4: 376–381 [In Russian].

Bernstein NA (1965) Smert' ot stracha smerti [The death from the fear of death]. *Nauka I Zhizn' (Science and Life)* 2: 149 [In Russian].

Bernstein NA (1966) Ocherki po Fiziologii Dvizenii i Fiziologii Aktivnosti (Outline of the Physiology of Movement and the Physiology of Activity*) Moscow Medizina (Medicine)* [In Russian] 350 pp.

Bernstein NA (1991) O Lovkosti i ee Razvitii (On Dexterity and its Development) *Moscow Fiskultura I Sport (Physical Culture and Sport)* [In Russian] 280 pp.

Bernstein NA (2020) Biomechanics for Instructors. Translation by Rose Whymen, *Springer Publisher*, 155 pp.

Bernstein NA, Naidin VL (1966) Movement, *Encyclopedia for Children, 7, Moscow, Pedagogika (Pedagogy)* [In Russian].

Chkhaidze LV, Chumakov SV (1972) Formula Shaga (Formula of the Step), *Moscow Fizkultura i Sport (Physical Culture and Sport)* [In Russian] 118pp.

Kandelaki E (1991) Letters of Nikolai Aleksandrovich Bernstein to Levan Vladimirovich Chkaidse, *Teoria i Practica Fizicheskoi Kulturi (Theory and Practice of Physical Culture) 3*: 41–47 (1991) [In Russian].

Feigenberg IM, Latash LP (1996) N. A. Bernstein: The Reformer of Neuroscience In: Latash, ML, Turvey, MT (Eds.) *Dexterity and Its Development. Mahwah, NJ, LEA Publishers*

Gilyarovsky VA (1922) Pamjati A.N. Bernsteina (A.N. Bernstein Obituary). *Zhurnal Psykhologii, Nevrologiii Psykhiatrii (Journal of Psychology, Neurology and Psychiatry)* 1: 5–9 [In Russian].

Korolenko VG (1909) Tragediya velokogo yumorista: Nieskol'komyslei o Gogol'e (The tragedy of great humanist: several thoughts about Gogol) *Russkoye Bogatstvo (Russian heritage)* 4, 6: 27 [In Russian].

Latash ML, Turvey MT (Eds.) (1996) *Dexterity and Its Development.* Mahwah, NJ, LEA Publishers.

Mlechin LM (2019) 'V rukach etich tipov proniknovenie k nav svobodnoi misli'. Isoljatsija ot vneshnego mira nanesla usherb scem sferam zisni SSSR ('These rascals are the gatekeepers for the free thought to reach us'. Isolation from the rest of the world was damaging for all aspects of life in USSR) *Novaya Gaseta #100*, September 10 [In Russian].

Popova TS, Mogilyanskaya ZV (1934) Technika Isuchenija Dvizenii (The Techniques of Movement Research (Ed. by NA Bernstein) *Standartisatcija I razionalisatcia (Standardization and rationalization)* [In Russian] 560 pp.

Pribram KH (1971) *Languages of the Brain: Experimental Paradoxes and Principles in Neuropsychology*, Prentice-Hall Publishers.

Smirnov AI (2013) *Sound in Z: Experiments in Sound and Electronic Music in Early 20th Century Russia.* Koenig Books, London, 281 pp.

Talis VL (2020) *Doktor Kotoryi Lyubul Parovoziki (The Physician who loved steam engines)* (manuscript) [In Russian].

Zatsiorsky VM (1991) *Introduction to* Bernstein NA (1991) O Lovkosti i ee Razvitii (On Dexterity and its Development) *Moscow Fiskultura I Sport (Physical Culture and Sport)* [In Russian].

Zhukov P, Kozhin A (1950) Protiv vul'garisatzii teorii fizicheskogo vospitanija (Against the vulgarization of the physical culture education) *Pravda (Truth)*, August 21: 233[In Russian].

C2

AT THE FEET OF A QUIET GIANT

Micha Mirsky

FORMERLY OF LOMONOSOV NATIONAL UNIVERSITY, MOSCOW, USSR, DEPARTMENTS OF NEUROLOGY, HADASSA MEDICAL CENTER, AND OF PHYSIOLOGY, HEBREW UNIVERSITY, JERUSALEM, ISRAEL, & SEATTLE UNIVERSITY, SEATTLE, WA, USA

My family moved to Moscow, USSR, from Kaunas, Lithuania – at that time, totally *bourgeoise* little Republic – in 1934, when I was already full three years old. My father, Lev Moiseyevich Mirsky, who was then a European authority in Movement Science, especially in the theory and methodology of physical education, was invited by the Soviet government and given an apartment on campus of the National Central Order of Lenin Institute for Physical Culture (ГЦОЛИФК) that became much later the Russian Academy of Physical Education. The research part of it, called the Central Scientific Research Institute of Physical Culture (ЦНИИФК), was located close to where we lived on campus, some hundred yards or less from our windows.

My memories of Nikolay Aleksandrovich Bernstein (NAB for the rest of this text), whom I called "Uncle Kolya" then, were being formed earlier and remained brighter than those of other acquaintances of my parents. I believe his first visits to our home happened in 1936 or 1937 when I was about six. My mother was a good piano player and, more importantly, an expert professional typist. She brought with her from Kaunas a fine typewriter with a Latin keyboard, and she amazed NAB with her enthusiastic and artistic performance (his words), as she was entering French, German, and English names and titles into his reference lists in Russian texts he prepared for publication. He could not believe her ability to combine high speed, fluidity, *and precision of movements*, while paying attention to the smallest details and never making a mistake. He said she was so successful in her typing because of her perceiving it as an intriguing play rather than a hard and tedious labor.

His image of those times is still bright in my memory, an amazingly beautiful and kind human.

My more deliberate impressions of NAB are dated to the fall of 1943, when we returned to wartime Moscow, father from serving at the Navy Rehabilitation Medical Center for patients with war injuries of the central nervous system in the far east of Russia, the rest of us from evacuation to Siberia since the autumn of 1941. The following summer, when I was twelve and a half and it was the summer break at my junior high school, I was free for the next two months and came from school to my father's office at ЦНИИФК, and it was locked! I knocked at the next door, and NAB opened it. "Uncle Kolya, have you seen my Old Man?" (Friends called my father Old Man.) NAB looked at me closely, tilting his head, waited a bit, and said, slowly, "I have not had the honor of seeing your esteemed Parent today."

He did speak like that to me, a kid of 12, looking at me with close and full attention, then glanced at his watch, so that I realized he was in a hurry, and added, "Might it be possible for the Doctor to find time tomorrow? For I am in a real time shortage now." I had even turned around to see who he saw in the hall behind me but there was no one there, and I then knew that he spoke about me that way, in the old-fashioned style, in the third person. And I learned later that it was his way of addressing anyone, with full and equal respect, not playing that role but feeling that way about anyone, of any age or social standing. I told him,

"Uncle Kolya, I am on summer break now!" He responded, "Very busy now, come to see me tomorrow, and it is better to call me by my full first name and middle name here. I'll be here at one o'clock tomorrow."

"I'll be running here at that time to arrive promptly," I said, and surely I came on the full run next day, for all of it was deeply and acutely interesting to me. It was the first time for me to be inside his lab, and I was so amazed by the amount and diversity of strange mechanical contraptions inside – a lot of intricately connected levers and hinges that did not look like medical instruments to me – that I used my "Uncle Kolya" address again, asking him, "Aren't you a physician?!" and he smiled at me kindly and said, "I most certainly am; I am a neurologist."

"Why do ***you*** need all these iron thingies?" and he said, "That ***is*** a very good question!" and smiled at me, kindly, again.

I want to emphasize here that thanks to my ability to memorize accurately the smallest details of my life experiences I do remember clearly the words he said to me then, and can see the nuances of his facial expression. One example: back in 2005, I recalled the precise words of what the great Israel M. Gelfand said at the memorial for M.L.Tsetlin, despite having heard them thirty-nine very long years earlier, and only after having compared the text of my recollection with the old stenographic record that Lena Glagoleva (the wife of Misha Berkinblit) still had in Moscow, Yura Arshavsky and I saw that my recollection was totally accurate, word for word, and it was, namely,

> In order to see, and even more so, to create something new, one needs to have the ability to look at the world through the eyes of a child who first opened them in this world; it seems to me, Misha Tsetlin possessed this ability.

With the same accuracy I remember the words NAB used then, and I was visiting with him every day later that summer. Summer break, nothing to do. Soon he was giving me, the boy who was into his only third year of studying the German language, hard assignments, like a recap of chapters in Bethe's physiology course book. Later, thanks to NAB, I knew German much better. He was letting me take books home and was expecting me to bring recaps later. That was the beginning of my learning from him, before I started studying anatomy and physiology at school. Once, having read one of my favorite pages of OPD "the mnestic component of coordination," I asked my physiology teacher at school what it meant (I was a naughty teenager indeed). The poor lady did not have a clue … Many years later, I met NAB during a conference at the Institute for Neurophysiology in Moscow and asked him, "How did you know then that I would not buckle under such a load?" He was smoking, coughed quietly couple of times, and said, "Well, I did *not* know that. And I am very glad that you did not buckle." That is how he was, a man impossible not to fall in love with all your heart and soul, he was an incredible human, a *Mensch*. Much later yet, in Dortmund, I spoke at length about NAB with the Germans working in the local Institute for Work Physiology, where NAB worked in 1929, they still remembered and admired him. He was a man of enormous power, warmth, and insight, he *was a Visionary*, as a matter of fact.

So, back then, a 12+ years old kid, when I came to see him next day, he said, "What is it that interests you?" – "Would you be laughing at me?" – "No." I told him then, "I cannot get it – *how* can a human or an animal move so effortlessly?" [From my earliest age, I was afraid of heights and fast moving objects, could not make myself jump down from a very low support base (e.g., couple of big dictionaries), was clumsy, cross-eyed from birth, had really no depth perception and no natural coordination, not even a passable one (was born ridiculously and drastically premature).] And NAB told me, "You know, our interests coincide." I said, "I am so really pleased to hear it, but I do understand how awfully little free time you have." He replied: "I have questions, and you may be able to help me answer them." – "Why me?" – "Because it seems to me that you and me are challenged by the same curiosity, and similar problems are fascinating each of us. So, pick up this glass from the shelf." I took that glass, quite carefully.

"Very well done," he said. "Why so?" – "Because you do not rush it. You think how to do it best. You try not to break it. That's exactly what interests me. While starting to move toward it, what do you see?" "I see the glass." "And your hand?" "No good for me in looking at my hand – I would not see the glass then."

Another typical NAB episode was described to me by my father. His younger friend and colleague, A.D. Novikov, was serving as the Director of ЦНИИФК at that time. A group of very serious people, among them a couple of Full Members of the Academy of Sciences, leading Soviet experts in the field of bridge construction, had an important problem to solve, in order to complete an important project. They could not solve it themselves no matter how they tried. So they came to see the Director, told him of their predicament – they heard that only NAB might be able to help them in finding the solution. They asked him to summon NAB for exactly that reason, and Novikov replied that he would not even dare to think of *summoning* Professor NAB for anyone, but he could walk to NAB's lab and *ask* him to help. It was about 1948, after NAB was awarded the Stalin Prize for this very OPD monograph (later on, this prize was renamed as the State Prize). The problem was in the fields of bridge and road construction, and the specifically crucial question was how to design a bridge for a certain combination of conditions and requirements. NAB agreed to think about it, and next morning gave them *two* solutions of that problem. On a piece of paper, he wrote what he called the "standard" and the "non-standard" solutions. My father evidently repeated the story Novikov told him – the Academicians fell speechless. It seemed to them utterly impossible, albeit absolutely clear and obvious *post factum*, but no one could come up with these solutions ever, before NAB wrote them down! All about NAB is in that episode – a physician, a neurologist, who was able to solve such problems in such a way! The great mathematician Israel M. Gelfand said about him, "If at the end of my career someone would have told me that I know and understand mathematics as well as NAB felt and understood it, I would be proud and happy."

At NAB's funeral in 1966, it struck me to read on one of the wreaths near his casket, "*To dear and beloved Kolya from Uncle Seryozha.*" It was from the 86 years old Academician Sergey Natanovich Bernstein, a famous mathematician, who was often amazed and surprised to see that his then quite young nephew had deep instinctive insights into areas of mathematics of which his uncle did not know (I heard this story from either Mika Bongard or Misha Smirnov). And one more episode during that funeral at the Neurophysiology Institute in Moscow. There was a closed-chain wall around the casket that many members of our group (Alik Kots, Mark Shik, Vadim Safronov, Grisha Orlovsky, Fedya Severin, Tolya Feldman, Yura Arshavsky, Misha Berkinblit, Volodya Smolyaninov, me, and many others) formed by linking our arms firmly and holding fast, to honor the wish that NAB expressed to his family not long before his death: "*I do not want Donskoy to come close to my coffin.*" That man, D.D.Donskoy, worked for many years at ЦНИИФК and served as its deputy director for research when I had worked there. He was known as "DD." One of the first students of NAB, he publicly betrayed him during the disgraceful years of persecution of "cosmopolitans without motherland" in Soviet science (it was enough for you to be Jewish, or disagree with Pavlov or Lysenko, or cite too many foreign authors in your articles to be demoted, fired, black-listed, and quite officially ostracized). [Strangely enough, my heart goes out to DD now – he was very bright and gifted, knowledgeable and interesting, and many did love and respect him, but *this has happened* to him, and he has killed a part of himself by having done that to NAB.] So, the very moment he has entered the hall, we have built our living closed-chain wall around the gasket, and DD could not come near it, for we were unmovable.

When I graduated high school, I wanted to be a physiologist and went to study biology and chemistry at the Potyomkin Pedagogical Institute in Moscow, for at the famed Lomonosov National State University and at the First and Second Medical Institutes I was not allowed to sit for the entrance exams – welcome to the mean and dishonorable year 1950. Misha Berkinblit was already studying there, at the Department of Physics and Mathematics. In 1956, back in Moscow after having worked as a high school teacher and a coach and college instructor on the Sakhalin Island, I managed, to everyone's and my own immense surprise, to reenter my Potyomkin Institute, this time as a PhD student at the Chair of Human and Animal Physiology, where Yura Arshavsky has already started his second year of PhD studies. From time to time, Mark Shik used to join Yura and me there, and Misha Berkinblit would come and teach the three of us those parts of mathematics that we did not know well enough then. Only later did I learn that, in his younger years, NAB had studied both medicine among other life sciences and, quite seriously, mathematics.

The following I heard both from one of my oldest dear friends, Yakov Mikhaylovich Kots (we called him Alik) and from my Moscow University Psychology Department elder and famous colleague, A.M. Luria. In 1957, when the famous "Father of Cybernetics," great American mathematician, Norbert Wiener, visited Moscow, he was greeted at the Belorusskiy Railway Station by a very high-caliber delegation of government-approved members of the Soviet Academy of Sciences, who were ready to take him to a well-planned banquet reception. Wiener just frowned at them, did not shake hands with them, and announced,

> No way am I going with you! I am going to get me a taxi and go visit with the very Nicholas Bernstein, who is held under house arrest by your damn authorities! Rosenblit and I were hailed in 1943 for our little story describing how insulin production is controlled in the body based on the feedback of the blood sugar amount changes, big deal! Bernstein has built the whole Physiological Cybernetics theory way before us, starting as far back as 1929!

And he went to visit NAB, leaving the poor officials on the platform …

Next morning, the house arrest was lifted. A rather short time later, at the monthly gathering of the Moscow Physiological Society, of which I was a member, the old University hall across from the Kremlin was full to its capacity, with people standing along the old walls and sitting on stairs in all isles. I came maybe two hours before the formal time and got a seat near the podium. NAB looked pale and almost frail but one could feel his quiet noble dignity and sense his total concentration on the hugely valuable contribution to knowledge that he was going to reveal. He brought notes but I do not remember seeing him look at them even once while he gave his lecture. He spoke clearly and evenly, moving only to indicate "Next slide, please!" to the projector operator in the middle of that amphitheater. Everyone was simply mesmerized. It felt to me as if I were an amateur musician suddenly present as Mozart played his new amazing masterpiece. NAB studied and knew many languages, so did I and others, but even his incredibly high level of mastery of the

Russian *language* was by far surpassed by his great and unique talent of *speech composition and delivery*. His eyes almost closed at times, he spoke smoothly, as if reading a perfectly edited text from a list held before his inner eyes. Many a question from the audience seemed insultingly mundane as compared to the greatness and richness of his vision and the flight of his insights. He wrote poetry and plays since childhood, he was a musician, and it seems to me he was quite a poet when he had to formulate his insights.

Another episode from my later interactions with NAB is connected to what I perceive and describe as my "tragic mistake" as a photographer. It had happened in 1965, during the "Cybernetics and Sport" conference held in the Grand Hall of what was once Razumovsky's Palace on Gorokhovskaya Street in Moscow of Tsarist times, in ГЦОЛИФК. The conference was inspired and organized by another of my oldest and dearest friends, Volodya Zatsiorsky, who managed to invite NAB as the keynote speaker. When NAB appeared on the podium, the whole audience in the full Great Hall rose to their feet and started to applaud, not so loudly but as one person whose whole soul is being expressed in the kindest way. He waited for a long time, for more than a minute or so, silently receiving this powerful flow of love and acceptance that ended with a *crescendo* before he finally nodded once and started speaking. It was there that he was being crucified a decade and a half earlier, by a pack of murderous bastards. One could hear a pin drop while he presented his key note paper, his every word was absorbed.

During the long intermission, Alik and I went out of the Hall next to NAB, each of us holding one palm carefully under his elbow, he was quite frail and weakened by his illness already, but refused to go to the buffet to eat something, nor did he want us to bring him a cup of tea. Clear memory – NAB and Alik deep in conversation, sitting in the lobby next to each other on the wooden sill of the window connecting the lobby and the Hall, and I am on some very old low ottoman, facing them. Soft, even lighting, I put a borrowed good old Russian portrait lens (the then legendary Helios-40, 90 mm/F1.2) on my camera, and am delightedly taking series of close-ups of them talking quietly to each other, feeling the uniqueness of these moments and sensing that many of my shots would indeed turn out to be really classy. Alik started to ask NAB something again, at mid-sentence, and NAB suddenly stopped him, saying sternly, "Y.M., wait please, let me finish what I am saying, and *listen while I am still alive*." Alik cried out, pleading with him, "My dear N.A., please do not say things like that!," and NAB told him, quite firmly, "*One must say everything*." It was only later, after his death, that we learned that NAB knew how and when he would die, long time ahead and in detail.

So, having just finished two full rolls of rare and fortuitous portraits of NAB, and in my anxious and childish hurry to see them printed ASAP, I ran across the campus to the ЦНИИФК photo lab and asked my old friend Nina (the veteran who started working there maybe twenty plus years earlier as a teenager, and knew, adored, and loved NAB with her whole heart) to please develop both rolls, ASAP! And she did start right away, and damaged them mortally, because in her honest hurry she rolled them by mistake into the container *emulsion layer to emulsion layer* … After having finished developing, rinsing, and fixing the films, and rinsing them again, she produced two totally clear rolls. Seventy-two good NAB shots were gone forever! She almost collapsed there with shock and horror, and was shaking with inconsolable crying and sobbing but there was now nothing we could do to bring them back to us. She remembered how in the forties, during the typical Moscow winter frosts, someone had stolen her overcoat from the common cloakroom, and when she had to leave work in the evening without it, facing the long way to her home in the suburbs, NAB made her put his own on, and he ran out into the freezing cold wearing just his rather flimsy suit. "He was always like a father to me," she sobbed. Even today, fifty-five years later, this remains a deeply painful recollection to me.

One knowledge is very strong in me – one of NAB's most amazing strengths consisted in his ability to find such a way to formulate the idea, such uniquely precise words that do describe accurately what is going on. His immortal "Image of the Desired Future" as the necessary precursor of initiating the current action does explain fully *what* you must have *before* you even started moving, up to the moment, e.g., when as a kid, I *carefully* took that glass off the shelf. That image "of fingers carefully taking the glass" *sets the tone to everything* that is about to occur because the very formulation of the motor task dictates the composition of the performing ensemble and the needed accents in the performance.

Later in life, forty or so years ago, when I started treating patients in the United States as a Physiotherapist, I was both encountering and using that NAB Principle while working in the clinics. My most challenging patient, whom we called Scotty, was a young man with a heavy ischemic brain injury syndrome. I worked with him for many years in a row. When he stood inside the parallel bars in our PT gym one day, his arms, like those of a child with cerebral palsy, were forced upward, above and behind his head, face thrusted far forward, its features distorted scarily into a Greek tragedy masque, and the more he tried to bring his arms down to reach the bars with his hands, the more *and more* forcibly they were thrusted upward and back. Sitting in front of and close to him on one of those low four-wheeled square stools that PTs used to use, spotting him to prevent equilibrium disturbance or loss, my ears next to the bars, I got quite suddenly inspired and did *whisper* to him, a former good jazz guitar player, "Scotty, could you try and touch these here damn bars *so carefully and quietly* that even I would not be able to hear it when you did?" And suddenly, for the first ever time in so many long months, his upper arms, elbows, and hands started moving forward and downward gently, lower and lower, with infinite care, until his fingers almost tenderly, and soundlessly, were wrapped around the bars. The masque was gone, a great gentle smile spread on his face, and he hummed, slowly and triumphantly, something that sounded like "Aha!," he could never speak after his injury. I could not move, nor could I dare to believe my eyes. No sound in our usually noisy Gym, everyone frozen and looking at us. That was, maybe, the greatest moment in the whole history of my life and my work, maybe the brightest illustration of the great Bernsteinian insight. Important is not the successful mechanical movement by and in itself, but rather its *carefulness*, which sets and controls its successful completion.

Now, more than eighty years since I first met NAB, I feel profoundly fortunate and grateful that I was given the rare and early chance to be close to one of the greatest scientists of our time. It seems to me now that what and how he taught me had shaped my life, that I "*had studied at his feet, dreamt to know the way he thought*" (I am using here the words of a fine young writer, Daniel Silva). Nowadays, more and more people of the Big World are learning about his legacy. In 1977, I was doing research and taught at the Hebrew University in Jerusalem. In the library, I was standing near the shelves leafing busily through the first English language book by NAB, the *Co-ordination and Regulation of Movements* (1967, Pergamon Press). He died in January 1966, when that book was being prepared for publication. A young student took a book from another shelf nearby and was looking at books on my shelf, and I noticed that the book she took was one by Dick Schmidt. She looked at me, smiled, and said "Shalom!" I did reply in kind and asked, smartly, "Psychology exam?" She shook her head emphatically and told me, "*No!*", explaining that she was interested in *movement studies*. I was still a naughty boy indeed, so I said, naughtily, "Place that beautiful book back on its shelf, and take this one." So she did, and we parted.

About forty if not more years later, Tamar Flash had been my host at her Applied Mathematics Department, Weizmann Institute of Science in Rehovot, Israel, where she introduced me to her friend, Eilat Almagor, the world-renowned Feldenkrais Method practitioner and instructor. And Eilat said, "Micha, do you remember me?" – "Vaguely, yes, but where have we met?" – "At the library, in Hadassa, and you have really changed my life then by introducing me to Bernstein!" Myself, I never met Moshe Feldenkrais; I believe he was a pupil of one of my father's students. His method is based on using movements in certain patterns to increase self-awareness and well-being, and is practiced all over the world today. Eilat was one of his best students, and when he was dying, he said she will continue to keep his work going. On her invitation, I gave seminars on Bernsteiniana within her Feldenkrais Method workshops near Jerusalem, and years later, Master Classes for German Feldenkrais Method instructors not far from Cologne. Discussions with my audiences made me feel certain that Moshe and NAB, who have never met each other, to my knowledge, taught us similar things in differing contexts. *Moshe:* as kids, we learn by moving and playing, and to learn more and more, *we use the mistakes we make. NAB*: playing with the image of desired future gives us the key to hone and polish movements *using sensory corrections* in the process of developing skills by doing *repetitions without repetition*. Everyone in those and my other audiences loved this elegant masterpiece by NAB from Chapter 8: "Correctly performed exercise repeats time after time *not the means used to solve* a particular motor task but rather *the process of solving that task*, changing and refining the means from time to time."

No idea how far ahead my path in life will continue. I have lived for a mere 88 years and a bit, but for about 82 of them I knew NAB. *I do miss you always, my dear Nikolay Aleksandrovich, and I do thank you very much indeed.*

C3

WE ARE OUR ACTIONS

Would N.A. Bernstein Research BCI?

Mihai Nadin

INSTITUTE FOR RESEARCH IN ANTICIPATORY SYSTEMS UNIVERSITY OF TEXAS AT DALLAS

Finally, the translation: Years ago, and more than once, Mark Latash, who did a lot to present the work of N.A. Bernstein to the scientific community, deplored the absence of a translation of his most important book, *On the Construction of Movements*, published in Moscow in 1947 (1947a). Before him, David Adams – who in the 1970s went to study in Moscow (and worked with the Georgian physiologist, Tengiz Oniani) – translated a first chapter (Adams 1978). Micha Mirsky, who knew Bernstein quite well (he wrote about his conversation with "Uncle Kolya," as he called him) and interacted with him (his father's office was next to Bernstein's) going back to the fall of 1944, also tried his hand at translating Bernstein's famous book. The famous mathematician Israel Moyseyovich Gel'fand (Kyoto Prize, MacArthur Fellowship), usually cited as Gelfand, was supportive of Mirsky's attempt. In his endorsement of Mirsky's effort (intended to help secure funding), he correctly pointed out that the book gave "rise to a new way of understanding how complex systems work" (Gelfand 2009). I shall borrow from Gelfand the moniker NAB, "standing for Nikolay Aleksandrovich Bernstein (1896–1966)" fully aware that given Bernstein's inclination to a formalism meant to express respect, he might have objected to the unintentional slight.

Maybe others have also tried to translate the book. (At this time, a translation into the French is in the works.) This is significant for understanding the role played by those who dedicate themselves to the task of facilitating access to contributions written in languages other than the dominant English. Within the culture of the Soviet Union, which for a long time locked itself in a condition of autarchy contrary to the free exchange of ideas, other publications had a similar destiny. For anyone aware of the major questions that science addresses, it becomes clear that NAB's influence in the Soviet Union was maybe not very deep, but of lasting effect. During a time when to even acknowledge him was considered risky, his reputation deservedly increased. Bernstein's activity gained respect abroad also, in particular in Germany, and eventually in the USA, after several eminent Soviet scholars made America (the USA and Canada) their home of choice – quite a change from their home of birth.

Recognition is good, but influence, in the sense of affecting the progress of science, is better. On this account, NAB's impact is rather under-, not over-, whelming. In a very thorough presentation in the *Journal of Motor Behavior*, Bongaardt and Meijer (2000) placed NAB's "Theory of Movement Behavior" in a historic perspective They took note of the fact that from ca. 140 original contributions (Feigenberg 1988), only 24 had been translated (into German and English).

The situation has changed since. While his work is constantly referenced, there were rather few examples of being carried forward. Almost 20 years later, in *Human Movement Science*, Profeta and Turvey (2018) acknowledged the progress made in "the science of the control and coordination of movement," indirectly confirming that it is not really triggered by NAB's understanding of movement. Asking whether "Bernstein's levels" could accommodate "Complementary Conceptions of Synergy" (Bruton and Dwyer 2018), take note of NAB's "Elegant solution to the problem of coordinated movement and the development of skilled action (p. 2766). He has followers, scientists who deem his "construction of movement" as foundational; but in the new context of science, his most significant ideas are not found to be inspiring, or remain ignored. I had the opportunity to review an article in which Bernstein's fundamental position vis-à-vis Pavlov is fully ignored (Pool et al. 2019), despite the fact that he addressed exactly the aspects discussed in that more than deficient publication. The "gatekeepers" at the *Nature Human Behavior* journal would not agree that the omission of NAB's fundamental work undermined the article in question. Actually – and this is extremely important – the Bernstein "revolution" starts with his exceptional analysis of the shortcomings of Pavlov's reflex theory. While acknowledging Bernstein, the neuroscience of motoric expression either avoids his fundamental ideas or seeks details that justify more fashionable, i.e., computational, models. Is the explanation to be found in the language barrier? This translation will provide an answer. Or should we try to identify what else explains the very timid integration in current research of his breakthrough ideas? This is the subject I shall try to address in this study.

Marginalia

NAB was an autodidact (as were, by the way, none other than Galileo Galilei, for example, George Boole, and Charles Darwin), trained in a medical school, but not in physiology in particular. You could succeed without a diploma in the early stages of the Soviet Union because the revolution promoted opportunity and downplayed the role of Academia, where the regime suspected less than full support. Home-schooled Konstantin Tsiolkovsky – a pioneer of rocket science – is yet another example mentioned by those who view talent and dedication more important than a diploma.

From various sources of testimonies (by friend and foe), it is apparent that NAB was a difficult character. He expected to always be right, while never failing to be extremely polite. A Soviet intellectual characterized as a "loyal dissident" (according to Meijer and Bruijn 2007), dedicated to the idealistic image of a new society – but not unaware of its limitations. It was easier for him, as it was for other intellectuals, writers, and artists, to be a devoted Soviet citizen, dedicated to the ideals of socialism, than to question any of its shortcomings. Out of conviction, he served as a physician in the Red Army. To acknowledge prejudice, opportunism, dogmatism, and the authoritarian nature of an ideology of intransigence and intolerance that eventually affected his life was not a feature of his profile. The winner of a Stalin Prize (associated with the name of a dictator, yet on a par with the Nobel Prize, associated, as we know, with the name of the inventor of dynamite), for the book we comment upon in these lines, was unceremoniously kicked out of his work once he was denounced as an "uprooted bourgeois." Ideals died the day when some of his comrades (some friends included) played their well-rehearsed role of declaring him and his ideas as dangerous. The way Bernstein was treated in the context of a campaign of anti-Semitism that affected many (some Jews lost their lives) was interpreted in a manner that NAB would have rejected. His parents converted under Tsarist Russia in order to advance their careers. An atheist himself, he never felt any affection for an identity that implied religion. To spell it out clearly: NAB was rejected not as the person he never was, but rather as an author of ideas that the ideology of the day considered

FIGURE C3.1 Gheorghe Marinescu (1863–1938) Early attempts at describing the motoric system using cinema recordings.
(Source: Courtesy of the Romanian Film Archive and provided by film historian Bujor Ripeanu).

inadmissible. This fact deserves attention because, at a later time, in a society that does not punish nonconformity to dogma, ideas as revolutionary as Bernstein's were killed by simply being ignored. I refer here to the scientific destiny of Robert Rosen, and of Walter Elsasser. We shall return to this in our discussion of NAB's most controversial views.

Andres Kurismaa (at the time a graduate student at the University of Osnabrück's Institute for Cognitive Science), who connected the dots between anticipatory systems (as I presented the subject in 2011 at the University) and NAB's model of the future underlying motoric expression, insisted that I pay attention to Bernstein's writings. I knew about his work, but had not begun to study his various contributions in detail. Somewhere in my past was the connection to another scientist who studied movement – Gheorghe Marinescu (1863–1938), known in the French-speaking world as Marinesco. He was an early user of film recording for evaluating conditions such as hemiplegia, paraplegy, and locomotion ataxia (Figure C3.1). Incidentally, he was also fascinated by the electric aspects of the physiology of movement (transmitting electric waves associated with motoric expression over telephone lines).

But NAB, as I soon learned from reading some of his books, was on a path that transcended the focus on neurology that characterized Marinesco's distinguished work. Questions regarding the role of the future articulated by NAB also tempted Leon Abgarovich Orbeli, Pyotr Kuzmich Anokhin, Konstantin Andreyvich Ukhtomsky, Ivan Solomonovich Beritashvili, and a few others who noticed how purpose affects action. And all of this in the 1930s and 1940s, in a country where the official line prohibited such views.

Discovering the Pioneers

In the state of mind of a researcher who on his own discovered anticipatory expression related to creativity (Nadin 1972, 1990), I was encouraged to find out that, almost at the same time, Robert Rosen worked on defining anticipation. Ignorance could be bliss: as I understood, after the abovementioned discussion at the University of Osnabrück, NAB came into the picture as the "giant" on whose shoulders I was not standing, because I was not aware of his broader perspective.

Neither were his distinguished colleagues (mentioned above) the source of inspiration for my own questions. Nevertheless, I could ignore his views only if I were willing to pay the price that those dedicated to research ultimately pay, willy-nilly, for their ignorance when the state of bliss is smashed. At this juncture, only one thought: although Bernstein's work eventually made it across the borders between the Soviet Union and the rest of the world, his work, as well as that of his colleagues interested in anticipation, remained, as I mentioned in the introductory lines, of limited impact. The 2014 Nobel Prize in Physiology and Medicine ignored the work of Beritashvili (Nadin 2014).

The absence of an English translation of Bernstein's most important work on motion – published in 1947 – partially explains the situation. Reading the various books and articles on motoric expression published after NAB's foundational contribution, the readers will realize that what is missing is not the formal acknowledgment of his role, but rather the opportunity to benefit from a new angle, a new epistemological horizon. Science advances through discovery and dialog. NAB submitted his original views – discovery, based on a long-term commitment to experimental work and testing of hypotheses – but no dialog ensued. It was a conceptual revolution that engaged few others – his followers, mainly. In the Soviet Union, until his late rehabilitation, his ideas were not acceptable even as a subject of criticism. The book on Pavlov's view remained unpublished, not because NAB avoided offending his memory (the nice moralizing tale that since Pavlov could not counter, Bernstein refrained from attacking his theory). It was not published because it did not receive the "seal of approval." The Germans speak of *Totschweigen*: "death through silence" or "wall of silence," i.e., not even talking about something destined to be ignored. In the West, some scientists heard about the manuscript, but the Nobel laureate Pavlov was already the scientist in whose footprints so many stepped. The prolonged absence of translations has undermined communication: NAB's work was acknowledged but rarely analyzed in depth. Some exceptions: Pickenhain and Schnabel (1975), Sporns and Edelman (1993), Feldman and Meijer (1999), Bongaardt and Meijer (2000) and very often Latash (e.g., Latash and Latash 1994).

Returning to context: His book On *the Construction of Movements* (1947a, in Russian) and a subsequent book, *On Dexterity and Its Development* (1947b/1991) – which Iosif Moiseevich Feigenberg, also known as Josef, a scientist I shall return to, published – allowed me not only the realization of NAB's pioneering work, but also of many aspects I had failed to realize in the absence of healthy scientific dialog (Figure C3.2). Many other scientists interested in NAB's work would (or should) probably say the same. Cross-pollination is as vital in our interaction as it is in agriculture.

No time for regrets. We never live in the past of errors and missed opportunities, but rather in a *sui generis* present of new opportunities. NAB's book, in its English translation by Mark Latash – tirelessly ascertaining the role NAB played in the establishment of a solid scientific foundation for a science of motion – should help us all (not only historians of science) to catch up with his original ideas. The fact that a scientist who himself has contributed much to the field made it his priority to translate the book deserves to be highlighted. If Bernstein sometimes encountered less than favorable circumstances, he is now the beneficiary of probably the most competent translation he could have hoped for.

"Activity Constructs a Motor Skill"

In his introductory remarks, Bernstein wrote candidly about the rather incomplete factual material, the hypothetical nature of some of his interpretations, and even about some subjectivity that

FIGURE C3.2 NAB's books *О построении движений*. M.: Медгиз, 1947 (*On the construction of movement*, 1947 edition), and *О ловкости и её развитии*. M.: Физкультура и спорт, 1991. *Публикация подготовлена профессором И. М. Фейгенбергом (On Dexterity and Its Development,* Professor I.M. Feigenberg, Ed., 1991) found their way onto my desk and into my library.

infiltrated conclusions. It is in connection to his own words that some comparative thoughts are justified, and examples juxtaposed. It struck me that in my research of anticipatory expression (since 1995 and ongoing), at times I pursued, unknowingly, means and methods that belong to the "family" of experimental work that NAB carried out. The fact that I had the advantage of eventually living in a time and a place of more scientific opportunities – though not necessarily of less dogmatism – is not my merit.

These images, of documentary interest, conjure little discussion – you cannot really compare the technology of the 1930s to that of the twenty-first century. They juxtapose NAB's experiments (late 1930s) and the data collection on the AnticipationScope (ca. 2007–2008). In some ways, they testify more to the similarity of perspective: how to describe human action as it unfolds naturally, i.e., avoiding the danger of disturbing in the process of measuring. His focus on activity (the title of this section is a quote from the book) and ours on action are very close. As we shall see, there is no agreement on what experimental data afford, and no agreement on how to interpret what we observed. NAB conceived of the kymocyclographic camera for making visible the pattern of the kinds of movement he focused on. I conceived of the AnticipationScope, combining the time-indexed digital recording of movement with physiological sensors capturing data significant to the movement (Figure C3.3). He analyzed goal-oriented motoric expression: athletic performance, work, piano playing. We at the Institute for Research in Anticipatory Systems analyzed similar subjects, but with additional means. Not limiting ourselves to the motoric system, but considering also the cardiovascular profile, the neural profile, and the hormonal profile, we came close to a holistic view. The result is the Anticipatory Profile (Nadin 2012), not the description of particular movements.

NAB's major realization, in short, is the variability principle: "Repetition without repetition," behind which the relation between a large number of degrees of freedom and the possibility of control processes looms large. Incidentally, this idea was documented in my contribution to the volume *Anticipation – Learning from the Past. Early Soviet/Russian Contributions to a Science of Anticipation* (Nadin 2015). This volume is the result of the discovery that before Nadin (1972, 1987) and Rosen (1985), there was an entire school, in what used to be the Soviet Union, that advanced hypotheses on anticipatory expression. Well, before them were Leonardo (to whom we shall return) and a number of others who noticed anticipation expression (Nadin 2010). NAB was probably the most significant, given his dedication to nothing less than how life is expressed: in motion. He did not specifically seek to generalize his findings (which integrate evolutionary aspects as well) to nature. In regard to the human being, he definitely proved that the mechanistic view, as he described the dominant reflex theory of the time, was not the answer to understanding motion.

Activity Means Purposeful Action

Recent spectacular accomplishments associated with an understanding of the human being contrary to what Bernstein ascertained in his book have cast a shadow on the celebration of its translation. Does NAB's path in exploring one of the most difficult aspects of how life unfolds lead into the nowhere of inconsequential speculation? Or do we pursue directions in defiance of what the "compass" of informed experiments and generalization indicate? The question of whether NAB would pursue brain-computer interface (BCI), featured in the title of this study, is only illustrative of the concern with the condition of science: the theology of the machine model vs. the understanding of the unfolding of life as an expression of anticipatory processes.

If Bernstein's book, in the author's own words, is more a "program for studies to be conducted in the immediate future than a dogmatic report ..." let's imagine a NAB asked by his employer (a Soviet Union Institute) to dedicate his work to how the brain guides limb control. This is hypothetical. In reality, he was asked to work in prosthetics (a subject to be revisited soon). In other words: to address, in this hypothetical situation, the hot issues of brain–computer interface that dominate the agenda of motoric expression research. Let's not forget: The Soviet Union tried to compete with the West (i.e., capitalist science) – yet another aspect to be soon recalled. This "what-if" scenario needs to be properly framed. As a Soviet scientist recognized for his contributions, NAB – instigated by an idealistic Alexei Gastev, who organized the Biomechanics Laboratory at the Central Institute of Labor (CIL) – studied motoric aspects of work: hitting a chisel with a hammer, the performance of gymnasts, and piano playing as part of a broader interest in all aspects of such activities. That scientists are often opportunistic is an observation that is neither new nor equivalent to a value judgment on their character. As much as we wish to idealize science – in principle it does not carry a party membership card – we all realize that society, in some form or another, provides the means for facilitating research. NAB studied motoric control associated with labor because the Soviet society wanted to extract more efficiency from its laborers, the living machines of the time. He himself mentions the Stakhanov model – a competitive scheme (subject to many jokes) meant to reward the most productive. The Soviet gymnasts were trained to be better than those in capitalist societies, and so were the pianists. Ideology penetrated the scientific agenda. It goes to NAB's credit that he internalized the motivation, but never compromised the integrity of the scientific endeavor to which he dedicated his life. During my own experience

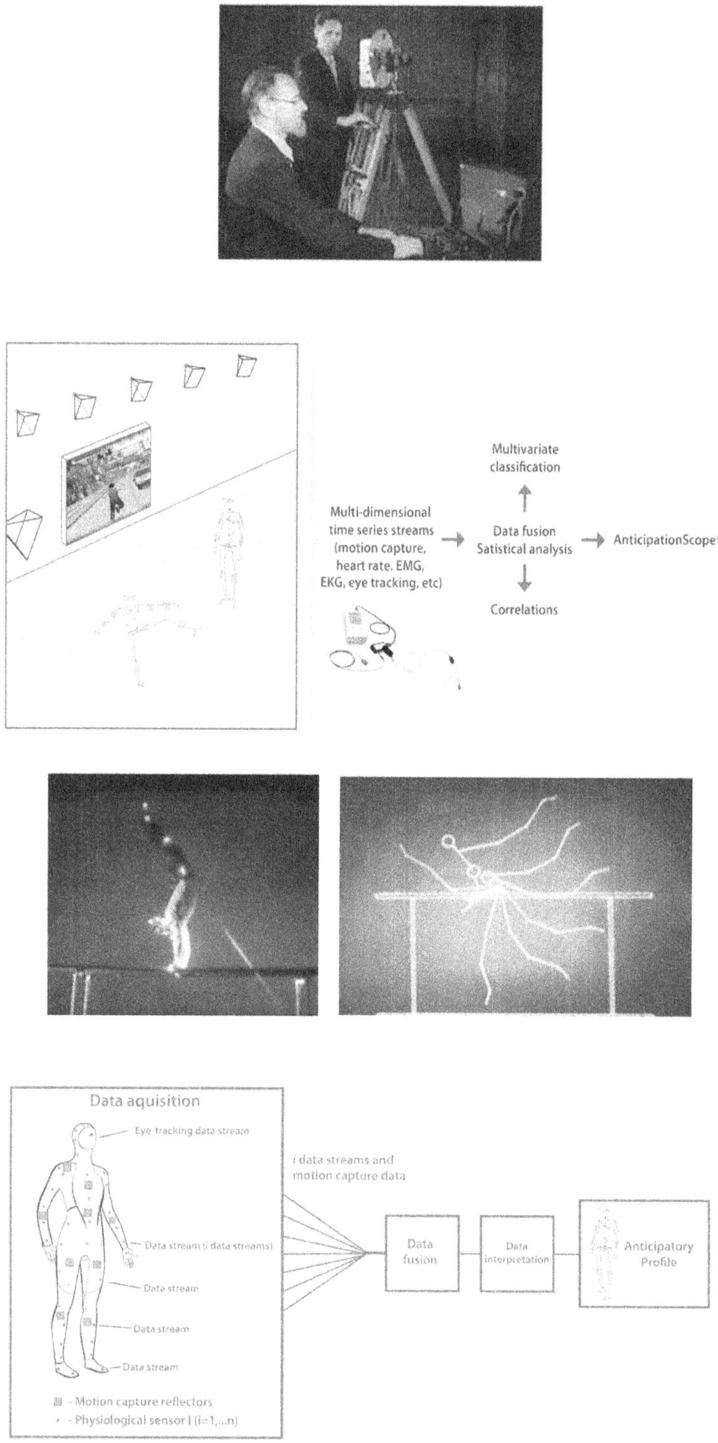

FIGURE C3.3 The kymocyclographic method and the AnticipationScope.

Unaware of NAB's experiments, we tried to describe the anticipatory expression of piano playing, and of gymnastics, which were also of interest to Bernstein. His outcome: visual representations. The outcome of the AnticipationScope: a combination of data defining the Anticipatory Profile and dynamic representations of purposeful actions.

under a communist regime, I learned from many of my distinguished colleagues that this was not easy. Science could land you in jail. Gastev, the enthusiastic Bolshevik, who instigated NAB to research labor, was killed by his intransigent comrades.

The book we discuss synthesized research carried out over 25 years. It appeared after the end of the brutal World War II. It is dedicated "to the memory that burns bright forever of comrades who gave their lives fighting for the Soviet Motherland." These words deserve our attention. Deep down, the superb researcher was respectful of the lives of his comrades. Eventually the idealistic NAB himself, with convictions he never gave up, suffered. His scientific arguments – in particular those that criticized Pavlov's view – were deemed contrary to the official line. Consequently, after having received the Stalin Prize (also known as the State Prize) for this book, he was ostracized from the scientific community after a meeting called by those who wanted to clean Soviet science of any foreign influence. He was eventually rehabilitated. Still, if the Soviet Union had said, "Nikolai Alexandrovich, we need your help for our war heroes ..." (millions of men lost limbs or were paralyzed). Or: "Look, we need a powerful military, brain control of artificial limbs could help us defend the Motherland," NAB probably would have obliged. After all, among the many places where he worked – Institute for experimental psychology (1925–1927), Institute for the Protection of Labor (1927–1933), Institute for Music Studies (1926–1940), Central Institute for Handicapped work (1932–1941); the Institute for Experimental Medicine (AIER) was also the Institute for Prosthetic Appliances (from which he was fired in March 1949). However, and with this hypothesis we end the "what-if" scenario, a Bernstein approach would have been grounded on his fundamental view of purpose-driven motoric control or, better yet, of goal-driven action.

Let us recall that NAB's work – I make reference to many of his articles that eventually fused into the book published in English translation – the brain does not, and as a matter of fact cannot, "prescribe peripheral mechanics (or muscle activation)," as Mark Latash noticed (private communication of September 25, 2019).

At the Institute for Prosthetic Appliances, NAB contributed original ideas to designing and constructing devices (lower limb, arm) based on the "construction of movement," his foundational work (Figure C3.4).

Without providing here an exhaustive overview of his original contributions, let us mention the article "Normalization of Walking on Prosthesis with an External Power Supply" (Shishmarev et al. 1974), and an even more specific application of NAB's ideas (Farber and Jacobson 1995), "The theoretical aspects of our development are based on the work of Bernstein" – with specific reference to his articles "On the Construction of Movements" (1947, in Russian), and "Essays on the Physiology of Movements and Physiology of Activity" (1966, in Russian). Of course, at that time, BCI had not entered the picture; but there is the documentary evidence that the passionate physiologist of movement did not shy away from technical innovations.

Fundamental Knowledge as Scaffolding for Applications

The purpose of insisting on such specific aspects is self-explanatory: there is a mechanistic-reductionist path, which he rejected, and there is a Bernsteinian path to addressing impairments, based on a foundational view of motoric expression as part of purposeful actions through which the human being experiences the world. The question of whether NAB would have gotten on the BCI bandwagon – as many reputed researchers (e.g., Georgopoulos, Schwartz, Hochberg, among others) have, can be answered in the negative. To implant microelectrodes in order to activate a prosthetic arm of impressive capabilities is a spectacular feat. But it was in

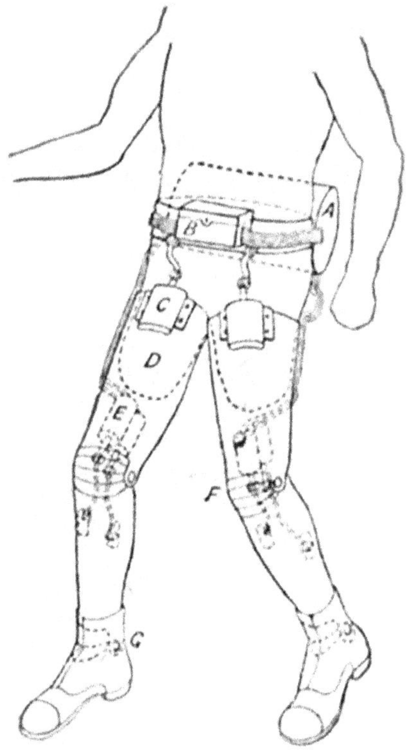

FIGURE C3.4 Prosthetics with an external power supply for activating small electric engines, inspired by NAB's views on the motoric.
(Source: from a private communication with Vera Talis).

contradistinction to NAB's view of the integrated motoric system and its relation to the brain or to the mind – or to both. It is a bit disappointing to find out that NAB's fundamental contributions were ignored by those who got involved, and as they still are, in the many attempts to emulate human motion that we experience in the age of AI and robotics.

But let us proceed step-by-step: the machine model of the living – the human being, in particular – going back to Descartes (but not only, see de la Mettrie, *L'homme machine* 1748) – implied the brain control of motoric expression (although Descartes, agnostic of the brain, focused on the pineal gland). This model was eventually adopted once the computational paradigm took over. Indeed, in its current embodiment, the machine model is understood as "We are computers" (as Popper 1966, saw coming) or, more recently, "deep learning" neural networks. Brain control of motoric components (limbs, in particular) translated as the path of control mechanisms for artificial replacements. The prostheses of the Industrial Age, based on the mechanics of the machines in factories, were successfully replaced by artificial "spare parts" (arms, legs, knees, etc.). Bernstein, who was learning about the brain by examining motoric expression, deplored the impossibility of examining the brain directly. With the emergence of neuroscience, monkeys became the substitute, and in 1970, brain signals recorded in the motor cortex of monkeys made available data on some forms of controlling motoric expression (Humphrey et al. 1970). Shortly afterwards (Georgopoulos et al. 1983), identified correlations to movement, and even signals before the onset of action – a real Bernstein moment, but never acknowledged as such. Wasn't this, one wonders, what Bernstein described as "the model of the required future"? Unaware of his pioneering work, exceptional scientists have been reinventing the wheel. But not really, since they failed to understand the depth of motoric expression, its epistemological meaning. Not even Benjamin Libet (1985), whose work inspired the early stages of my own research on anticipatory expression, was aware of NAB. He showed what "readiness potential" is; but while having access to real brains via EEG data, Libet did not understand motoric expression as the conduit to the brain. The fact that mind processes took place in advance of actions (the mind "guessing") suggested to him some selection procedures in the brain.

Of course, the case can be made that we are comparing apples and oranges. Implanted recording microelectrodes are based on a science quite different from that underlying the understanding of life from the perspective of goal-oriented actions. Neither Libet (himself only marginally acknowledged by the superb technologists of brain–computer interface), nor Georgopoulos, nor Schwartz (2004), nor Jiping He and colleagues (2005) ever considered the type of questions that NAB formulated in researching the motoric system and its relation to the brain and the nervous system. The reciprocal statement is also worth considering: Did NAB ever consider the challenges implicit in conceiving prosthetic devices interfaced to the brain or the nervous system? The Institute for Prosthetic Appliances turned out to be one of those places where competition with capitalist science led to developments worth remembering, even if the technology deployed at that time was quite rudimentary.

As an anecdotal parenthesis to our focus on Bernstein's theoretic contributions and their relevance to practical aspects, it is worth recalling that almost at the same time NAB, correspondent member of the Academy of Medicine, got his Stalin Prize, a one-armed, uneducated blue-collar worker received the same distinction. He conceived and manufactured (literally) a prosthetic arm, in the context in which millions of disabled veterans of the World War II desperately sought help in order to get on with their lives. The Kononov arm (Ruka Kononova) was featured in contrast to the American "Wonder Arm" of that time (indeed functional, but pretty ugly). In the Soviet Union, the Kononov prosthesis was even compared to the Tupolev bomber (an impressive aviation accomplishment). The rudimentary prosthetic arm was yet another "Potemkin Village" promise: i.e., the illusion that diverts attention from the real deception. Frances Bernstein, 2015, wrote about this in "Prosthetic Manhood in the Soviet Union at the End of World War II." NAB was working on a prosthesis that reflected his understanding not only of how arms functioned, but also of how the motoric is an *integrated* system within which control processes appropriate to the degrees of freedom (of the arm in this case) afford the appropriate performance. The question of whether Bernstein would have done BCI, if he had lived in the computer age as his late colleagues did, in order to make use of a technologically advanced prosthesis is relevant to the Kononov moment in the sense that fundamental knowledge should provide the scaffolding for applications even during the time when technology bursts ahead of science.

Within a short time at the Prosthetic Institute, NAB came up with concrete answers to practical questions pertaining to conceiving, designing, and manufacturing prosthetics (as we have seen, Figure C3.4). The interface was not to computers – the digital age was far off – but to the technology available at that time. Theoretic discoveries, tested in the laboratory using "original" recording methods – mirror recordings of movements, cyclographic techniques to track movements followed by the use of his original cyclocamera – constituted not only a perspective – the prosthesis must reflect the nature of human activity, its purposefulness – but also the foundation. At present (and this is true also for the last 50 years) such a foundation is missing most of the time, although the results of this "seeking in the dark" cannot be ignored. On a website dedicated to movement and its relevance to sports (Myszka 2018), NAB's ideas are applied to football: "how the masterful movers' ability would show itself in the skillful organization of movement." The entry is titled "Respecting the movement problem." One wishes that the stars of BCI technology had realized the same. Although ignoring NAB's fundamental contributions, good reviews of how BCI support prosthetic control were made by Taylor (2002), Schwartz (2004, 2007), and Hochberg et al. (2006), proving the point I am making: there is a lot of show-and-tell in spectacular neuroprosthetic control (a journal is even dedicated to the subject), but little on the underlying science of such control mechanisms.

The "One Problem" Scientist

NBA would have had a tough time getting funded by DARPA, but he would have continued on his path. I am not the first to notice that Luria (1987) was right in his assessment: Bernstein "was a rare case of a scientist who practically devoted his entire life to one problem: The physiological mechanisms of human movements and motor actions" (p. 85). I wish that Luria had chosen the word "processes," instead of "mechanisms."

The fact that the brain is an effective mapping processor became known once the neuronal theory of the brain was advanced. But the path of using the brain (in particular the motor and parietal cortices) for the purpose of motoric control in disregard of its characteristics is rather dangerous. Unrelated signals can be used in order to accomplish tasks that in our days are described as typical of the "transhumanist" age, i.e., transcending biology. For example: no longer associating data to motoric activity as an expression of purposeful life, but using the biological substratum (the brain, in particular) for controlling the flight of not just one drone (or whatever), but of several. Implanted arrays in both hemispheres of the brain could indeed serve as conduits for synchronous operations that will result in enhanced soldiers and multi-drones, as well as in behaviors that negate the human nature of the species. NAB probably would have shared in the enthusiasm for scientific and technological progress, but not to the extent of joining in the "show of bells and whistles" to which, in our days, even the most brilliant among us too often succumb.

Kononov's arm prosthesis worked fine, within the constraints of a mechanical contraption useful in one case and only one situation: when it was custom-designed and individually produced. That over time it failed, as the manufactured arm aged and the individual using it underwent biological changes, goes almost without additional explanation. Evidently, BCI-driven technology can also be custom-made, and even endowed with adaptive features: it can "learn." And it is, not unlike Kononov's artifact, successful – to the extent that it inspired a great amount of media reporting, including a relatively large article in the *New Yorker* (Khatchadourian 2018). But the experiment, at a scale of science and technology deployment that no one else can even dream about (never mind the huge cost), had to come to an end. There were too many parameters to even consider: picture a person having an array of sensors (the extremely impressive Utah arrays) implanted in the brain and the possible interaction between the Utah arrays and the brain. As one reporter put it, it was "a Frankenstein" image: two cylinders were protruding from the subject's head. Never mind the "hard-wired" solution defying the awareness of the adaptive nature of the human being, in particular that of the motoric system. "Give me a human with a hundred electrodes in his head," as the leading scientist of the Revolutionizing Prosthetics Project said, "and I can do all sorts of things," (Khatchadourian 2018, p. 63). Interfacing to the Applied Physics Lab (APL) arm, yet another amazing technological accomplishment (receiving millions of dollars in subventions), was one of them. The spinocerebellar degeneration of the subject willing heroically (or desperately) to link her brain to the robotic arm was of no concern. For the sake of preserving the integrity of this report, it is only fair to acknowledge that one major player, Schwartz et al., (2006) was aware of some of NAB's work (but not of *The Construction of Movement*). He even understood variability and highlighted NAB's contribution to rejecting the localizationist dogma:

> muscles must act to generate force that moves the body in context of the environment. The environment is constantly changing, so that in order to make the same movement, different muscle activation patterns would be needed.

Further down, the explanation:

> we arrive at the conclusion that the hypothesis of cellular localization of muscles necessarily leads to a denial of cellular localization of conditioned reflexes.

Unfortunately, this correct understanding of NAB's position is abandoned when the author delves into neuroprosthetic engineering. Moreover, the experiments (in which not only Schwartz, but also many others were involved) on monkeys (*Macaca mulatta*) entirely ignore NAB's very detailed description of the levels of construction of movement, corresponding to a deep understanding of how motoric expression testifies to evolution: "No other physiological function has shown such an intense and, moreover, accelerating phylogenetic progression" (Bernstein 1947). As readers will notice, NAB was quick to correct himself: "The evolutionary progression of the central processing systems." It is worth (and justified) contrasting the consequences of the reductionist view (expressed in BCI-driven prosthetics technology) to the promises of an integrating foundational perspective, such as the one NAB advanced.

This understanding was expressed before computational reductionism became dominant: more precisely at the NCMRR [National Center for Medical Rehabilitation Research] Conference on Prosthetic-Orthotic Research for the 21st Century (Bethesda, Maryland, July 23–25, 1992). Dr. Dudley Childress, recognized for contributions in rehabilitation engineering, spelled out a view of what motoric expression is that came close to NAB's fundamental perspective:

> I view motion analysis data as being similar to the observational data of the planets some 500 years ago. Much data was collected by many observers with varying degrees of accuracy, and it was difficult to put the measurements together in a coherent and unified fashion. Kepler, through mathematical description, and Newton, with the theory of gravity, were ultimately able to bring the information together in an understandable and relatively simple way. I think the same thing will be done with human locomotion.

He finished the editorial with a cautious assessment: "Other paradigms may already exist for the field; for example, I wouldn't be surprised if some of the concepts of Nikolai Bernstein (1967) might be of this nature." They are – in particular his views on anticipation.

Bernstein and Anticipation

There are good reasons, as we shall see, for considering *On the Construction of Movements* in the category of contributions made by Kepler, Newton, Einstein, and the quantum mechanics physicists. The focus on NAB's fundamental perspective invites us to consider the broader implications of his long-term dedication to understanding movement as an expression of life. At this juncture, I would mention Latash's book *Fundamentals of Motor Control* (2012), in which he describes motor control as "an area of science exploring natural laws." While the focus is on motoric expression, Latash builds upon NAB's views ("father of contemporary motor control"), delivering not only a solid historic review, but also a guide for future work in understanding motor control. After describing NAB's multi-level system for the constructions o movement, Latash makes it explicit that "nearly all human movements are built on several levels." Bergson (1911) quite suggestively framed the larger problem in almost Bernstein terms: If I raise my arm, the position which it successively occupies in space explain its movement, or does the movement explain the positions which the arm successively occupies in space? Let us take note that, without explicitly acknowledging each other, NAB and Bergson also shared in their views on the nature of the living.

The meaning of what NAB contributed (I make reference to Veresov 2006) is evinced in two lines of thought that stand out: "from reactiveness to activeness" and "from mechanism to organism." We should add: from action to purposeful action. Behind this formulation one finds "Contemporary Issues on the Physiology of Neural Processes" (finished in 1936, published in 2003) and *The Physiology of Human Movement and Activity* (1987). Luria's above-mentioned description of NAB's research agenda, puts the entire endeavor into focus. While it is true that the motoric is, as NAB stated, a window to the brain, the aim of his work was not to form a theory of the brain, but rather to suggest a broad view of what life is and how it is expressed in action (human action, in particular). Goethe's thought – *We know ourselves through action* – goes hand-in-hand with *We know the world through our actions*. Indeed, in writings that reflect the questions NAB entertained after publishing his two foundational texts (1965/1988, 1969), he attempted a difficult synthesis. The living, at all levels, is identified through activity: it "pursues goals." Activity itself unfolds against the background of the variation and selection processes through which evolution is expressed. Every living entity is "making itself" (an idea that many years later necessarily emerged in my own views, Nadin 1991), in what he described as the "anti-entropic structuring of the self." Reporting on this, Bongaardt and Meijer (2000, p. 666) summarize NAB's position: "activity is not restricted to movement behavior, but can be found at all levels of life, from molecular complexes to brains." NAB was even more precise: activity is evinced "in all possible manifestations of vital activity both in onto- and morphogenesis, and in all forms of interaction of the living organism with its surroundings" (Bernstein 1969, p. 120).

The slime mold (without a brain, of course), and also plants and all kinds of living entities (from the mono-cell to the human being) display behaviors – which means they are active – informed by goals that lie in the future. The unfolding of life, in all its forms known so far (archaea, fungi, protists, bacteria, plants, animals), is by necessity *anticipatory*.

To be alive is to act in order to maintain life. Hence, the possible future – a new entity – becomes part of the complex process through which the living constitutes itself. This process integrates reactive components (interaction of matter) and anticipatory actions – from defensive behavior patterns to creative endeavors. When, based on its still rudimentarily acquired data, Bernstein advanced the variability principle – no motion is ever identical to a previous motoric expression – he was probably not aware of the very broad implications of his observations. In my opinion, repetition without repetition actually qualifies as a universal law of the living. Let us shortly explain why I came to this conclusion.

Aristotle is credited with a formula (recalled by Elsasser 1958): No two blades of grass are the same. Likewise, Leibniz referred to leaves on a tree ("never the same"). Today we know more: there are no two identical cells in an organism, not to mention neurons, or identically folded proteins (Nadin 2018a). The mechanical view, exemplified by the self-duplicating machines that von Neumann (inspired by the Turing machine) advanced, originated in the theology of the machine: Humans made machines of all kinds in order to become more effective. To ascertain that humans are machines – for example, life as an internal combustion engine, an idea shared by many reputable scientists in the last century – is to hypostatize: ascribe real existence to our own constructs. The making of an idol (a piece of wood, after all, as good for burning as for veneration) leads to theology. Evidently, the theology of the machine ascertains what logic defines as identity: perfect repetition. The reason is an expression of the expectation that machines embody physical laws that actually capture the identical in the reality. The law predicates monotonic performance: each time an action is triggered, the machine delivers the same. A tourist from Sirius seeing an athlete might have seen "an impressive machine, unlike any other" (Bernstein 1991, translated into English, Latash and Latash 1994). But as we know, no machine has the degrees of freedom (static and dynamic) that the living has. The real subject is actually the distinction between living and non-living: machines are repetitive, the living never repeats itself, the living re-creates nature as being constituted by *the different*, not by *the same*. NAB was direct: "manifestations of vital activity" means interaction of the living with the ever-changing world to which it belongs. But vitalism was declared a reactionary view, as was, in the same sense, the Aristotelian *final cause*, i.e., the purposefulness of actions. Yet NAB wrote about the pursuit of goals as they relate to preserving life, and produced the proof of this in his experiments.

On account of ideological reasons, the Soviet Union censors blocked any thought not aligned with the official dialectic materialism, even when authors – Bernstein, in particular – openly subscribed to it, or claimed to do so. Science in the West was not subject to censorship in the sense of making it impossible to disseminate ideas not aligned with the ideology of those in power – i.e., the scientific establishment. Nevertheless, a healthy scientific debate on the nature of life was for all practical purposes made impossible. Rosen's *Life Itself* (1991) was published, but met with what the Germans call *Stillschweigen* (tacit silence), i.e., the silence adopted by a community of researchers and academics who know who butters their bread. The same happened to the work of Walter Elsasser, a distinguished physicist who would have received his Nobel Prize were it not for his writings of a new foundation of biology. The research of anticipation, definitory of the living, was and still is affected by the same attitude. In view of our justified expectation that this translation of Bernstein's book will be conducive to the reevaluation of many views on the motoric, it is justified to dwell, shortly, on how science is advanced in our society.

The expectation of profitability is implicit in the specific determinism of capitalism. The so-called socialist society (the Soviet Union and the countries aligned with it) cared more for ideological correctness than for an effective science. Capitalist society sees no value in ideology. But for all practical purposes it turns science into a marketable commodity: it monetizes its concepts. With this in mind, it becomes clear why a science that contributes to ever higher profits is preferred, and accordingly supported (by government as well as by business). The quick fix and ever faster machinery have a higher return on investment than living processes do. Treating a chronic migraine with analgesics is a form of reactive medicine. The pill is like the hammer to those who see every problem as a nail. But there are migraines that just won't go away – hammer (all kinds of pills) or not. Incremental medicine, as Atul Gawande labeled it (2017) yields no return on investment, even if it is the only effective medicine. "We devote vast resources to intensive, one-off procedures, while starving the kind of steady, intimate care that often helps people more. Natural processes unfold at the rhythm of life. Evolution, grounded in the anticipatory nature of the living, is steady, but slow. A knee or hip or shoulder replacement will return the person subject to surgery to productive life in a short time. There is a lot of profit in investing in the technology of replacements – monetizing pain and opioid addiction prove the point. In accord with (bio) physics, an anticipation-based healing procedure (such as genetic repair) will take as long as the deterioration process took (Nadin 2018b). But the outcome – motion, mobility – will be natural and sustainable (a replacement part will not be needed, as is the case with the prosthesis). These are, of course, only marginal thoughts along the line of highlighting NAB's perspective and why it is not so certain that he would have succeeded better in the USA of today than he did in the Soviet Union of pretty dismal memories. It might be the case that what the Soviet regime was not prepared to accept as a breakthrough in science will face similar obstacles in the "free world," but for reasons different from those that made NAB's activity so difficult. However, the real challenge is to continue in the spirit of his broad views in order to avoid the drawbacks of reactive medicine, and of reactive science.

In the Spirit of Bernstein's Perspective

We are a bit ahead of ourselves. In reality, the thought just articulated is the outcome of pursuing NAB's research agenda, but within a context of many new ways to formulate questions, articulate hypotheses, experiment, and eventually advance new ideas. Even if the dominant forces at work in our society are rather focused on more lucrative avenues (such as BCI research already discussed). It is in this context that in 2004, yet another "repetition without repetition" occurred: The Institute for Research in Anticipatory Systems (initiated in 2002) became part of the University of Texas at Dallas. From the broad research agenda, human action took center place. The project *Seneludens* (Nadin 2005) focused on how aging and anticipation expression through action are affected as we get older. Instead of documenting – as did NAB, and Marinescu tried before him – various forms of movement deterioration in the aging, the Institute's researchers asked questions that involved more than anatomy and physiology. We considered the muscular, skeletal, nervous, and even gastrointestinal systems. As a result, we came to associate actions and performance.

The AnticipationScope™ (Figure C3.5) can be considered a development in the spirit of NAB's kymocyclographic method. But that is as far as the connection goes. The focus on quantifying motion changed from aging to performance in general, and what makes other connections possible. NAB worked with athletes; we hosted members of the Olympic gymnastics team, but also golfers and tennis players. He famously analyzed piano playing (1926–1937; Bernstein and Popova 1930) of distinguished soloists (although the experiments were quite crude); we made our own attempt at the same, trying to capture the motion of the entire body playing the piano, and later when a person is engaged in activity of an aesthetic intent (Figure C3.6).

As long as the positivist premise, which informed NAB's experiments as well as ours, was disclosed within the deterministic view in firm control in today's science, there was sufficient funding and institutional support. But the broader agenda of understanding anticipatory processes as definitory of the living could not overcome what NAB and, later on in history, Robert Rosen or Walter Elsasser encountered. Indeed, the revolution in Bernstein's thought is in the anticipatory perspective. There is, of course, no vitalism to detect. The understanding of life as different from the non-living (physico-chemical reality) does not imply such forces, but rather accounts for the fact that biology and physics are not reducible to each other. Take only NAB's "principle of equal simplicity." It is derived from an understanding of motion that is the outcome not of mechanistic control forces, but rather of self-organization driven by the goal of the action. In our research, we

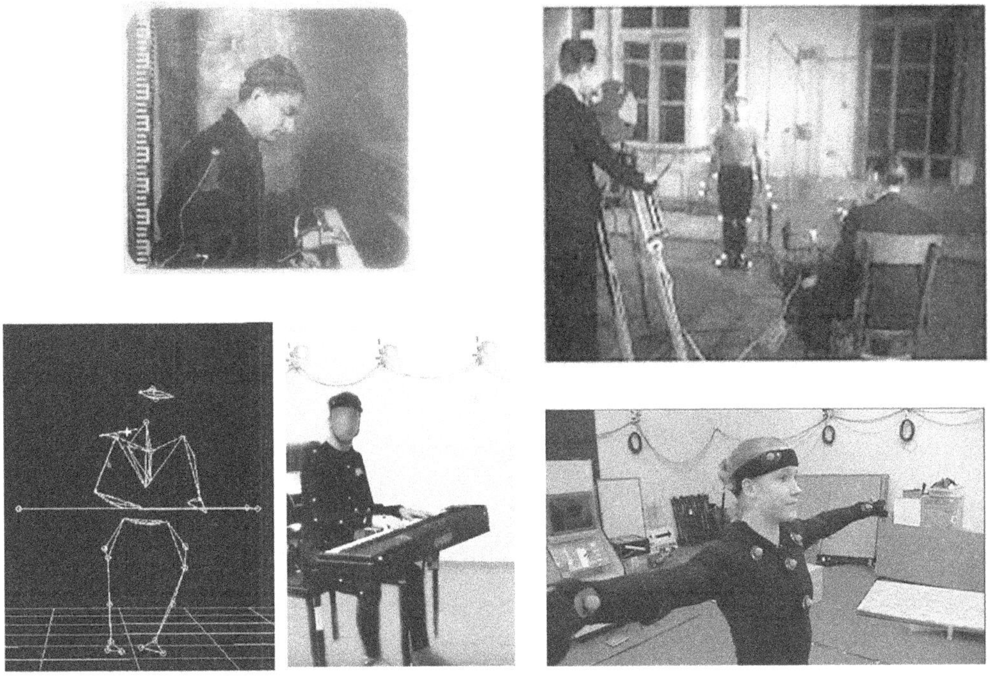

FIGURE C3.5 From Bernstein's kynocyclographic method to the AnticipationScope.

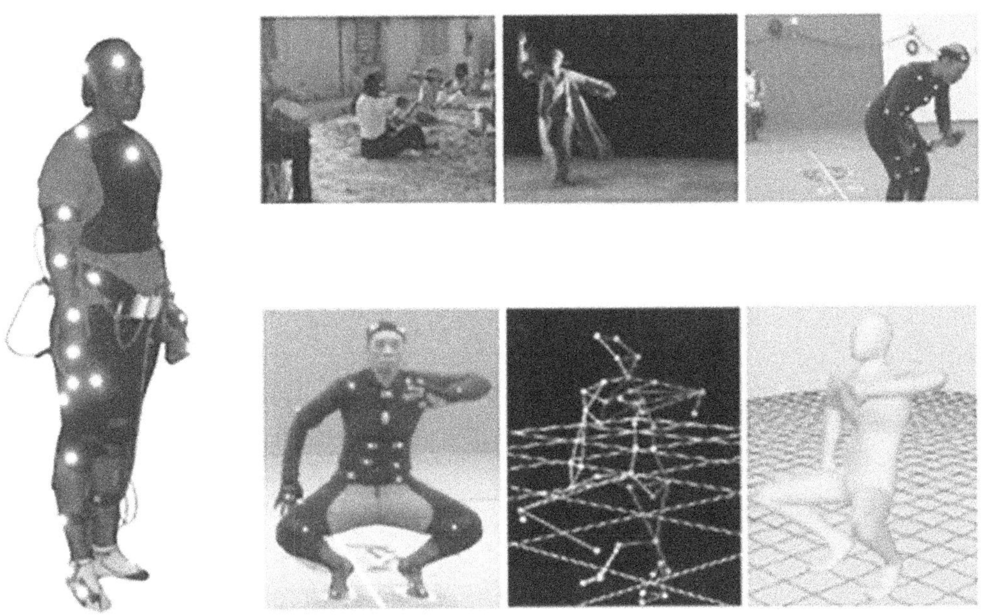

FIGURE C3.6 Capturing movement of aesthetic intent. Germaine Acogny, the "mother" of modern African dance, suggesting movement of therapeutic nature.

documented how the possible future – let's say pregnancy in its various phases – affects the current state of a pregnant woman's anatomy. For example: her skeletal system changes ahead of the added weight and modified body geometry.

Bernstein had success in advancing the idea of a model of the future. Rosen (1985) worked on a definition that is Bernsteinian through and through:

> An anticipatory system is a natural system that contains an internal predictive model of itself and of its environment, which allows it to change state at an instant in accord with the model's predictions pertaining to a later instant.

What neither afforded was 1) to deconfound anticipation from probability; 2) to advance the understanding that anticipatory processes are holistic: they engage the entire organism. Thus they are non-deterministic by necessity. Understanding motion on account of the awareness of anticipation is different from the mechanistic perspective that still dominates research.

The Journey of Discovery Continues

The book, *The Construction of Movement*, finally translated into English was only a preliminary report on a very ambitious scientific endeavor. NAB continued his dedicated inquiry, insisting on the need to find better ways to describe movement and associate it with the purposes triggering

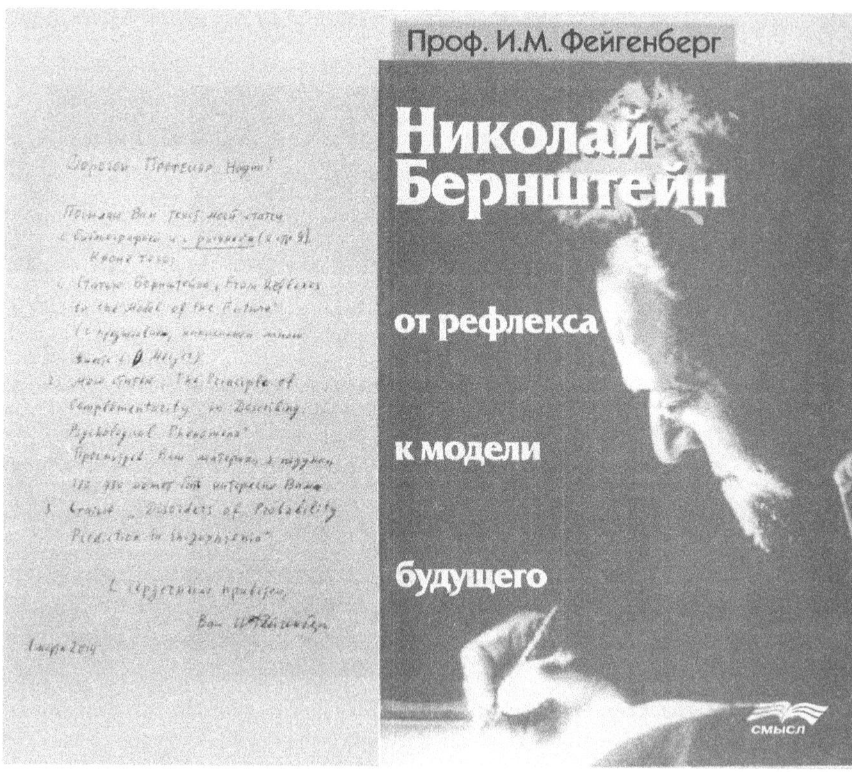

FIGURE C3.7 Learning with and from I.M. Feigenberg: Николай Бернштейн: от рефлекса к модели будущего» – М.: Смысл, 2004. His monograph about NAB.

it. Iosif M. Feigenberg, whose dedication to NAB cannot be emphasized enough, inspired him to address the future in terms of probability. I benefited from interactions with Feigenberg and from reading his book on Bernstein's physiology of movement (Figure C3.7).

We were supposed to meet at the workshop I dedicated to the Soviet/Russian forerunners of anticipation, but it was not to be. In some of my messages to him, I tried to explain why the probability view he suggested to NAB is inadequate for describing the possible future, quite different from the probable future. After all, nothing is probable unless it is possible, but not everything possible is probable.

Probabilities reflect what we don't know about a process: "10% chance of rain" actually means 90% lack of knowledge of those processes that result in precipitation, or lack of it. While deterministic processes have a well-defined phase space (i.e., constant number of variables at work), living processes continuously generate new phase spaces. Let us recall only the example of pregnancy mentioned above. Skeletal changes (spine configuration) are part of the larger phase space. Anticipation, for instance, is expressed in the secretion of oxytocin (which partakes in triggering birth contractions, but also eases the pain of the nipple once the newborn starts to suckle). Prolactin is secreted in advance of lactation, stimulating the mammary glands. On account of such observations, a very simple principle was stated and experimentally confirmed in studies of motion in general: *Anticipatory processes are always expressed in action.* Motoric expression is a realization not within a well-defined path to which quantitative descriptions can be associated, but rather along choices continuously redefined, along a multitude of possible paths. The physics (or chemistry) of life – subject to gravity, we still fall down; we need to touch (apply pressure) to the piano key in order to generate a sound; the physical properties of our ears or eyes are prerequisite to hearing and seeing – is quite well described and explained by physics (or chemistry) using quantitative descriptions. Elsasser thought that Niels Bohr's concept of complementarity: "two different sets of order for inanimate and for living things" would embody a "generalized complementarity" (1998, p. 144). The "two types of order" that "exist in the world" are such that "they never contradict each other." And he drew the following conclusion:

> conventionally designed experiments cannot teach us anything beyond the fact that organisms are physio-chemical systems. *If the holistic properties are to be verified experimentally*, a different type of experiment from that conventionally used *by physicists and chemists* are required.
>
> *(1998, p. 148)*

Chances are that NAB would have agreed with him and worked toward facilitating such experiments. The biology of life is about quantities, but also about *meaning*. Feigenberg was genuinely interested in the dialog regarding such thoughts, and indicated to me that NAB was aware of the need to find better ways to scientifically describe life (Figure C3.8).

NAB's "essential" and "nonessential" variable reflected his focus on "future requirements," and "of that which is not yet, but which must be the case" (Bernstein 1967 p. 187). He, again famously, described a reality we still face in respect to understanding that some goal-oriented actions are successful, while others will fail. This is the non-deterministic aspect, which for those who are given to the theology of the machine (never failing!) remain inscrutable: "Biologists understand the problem but lack the mathematical skills, and mathematicians have the skills, but don't understand the problem," (Bernstein 1965/1988, p. 246). It turns out that Gelfand, among others, argued for a mathematics of life different from the mathematics within which he became famous for his contributions. In an attempt to describe Leonardo da Vinci's intuitive description of anticipation, I made use of "a mathematics of generatively structured entities" advanced by Lev Goldfarb (and some of his collaborators; Goldfarb and Golubitsky 2001). Within this mathematics, the "formative

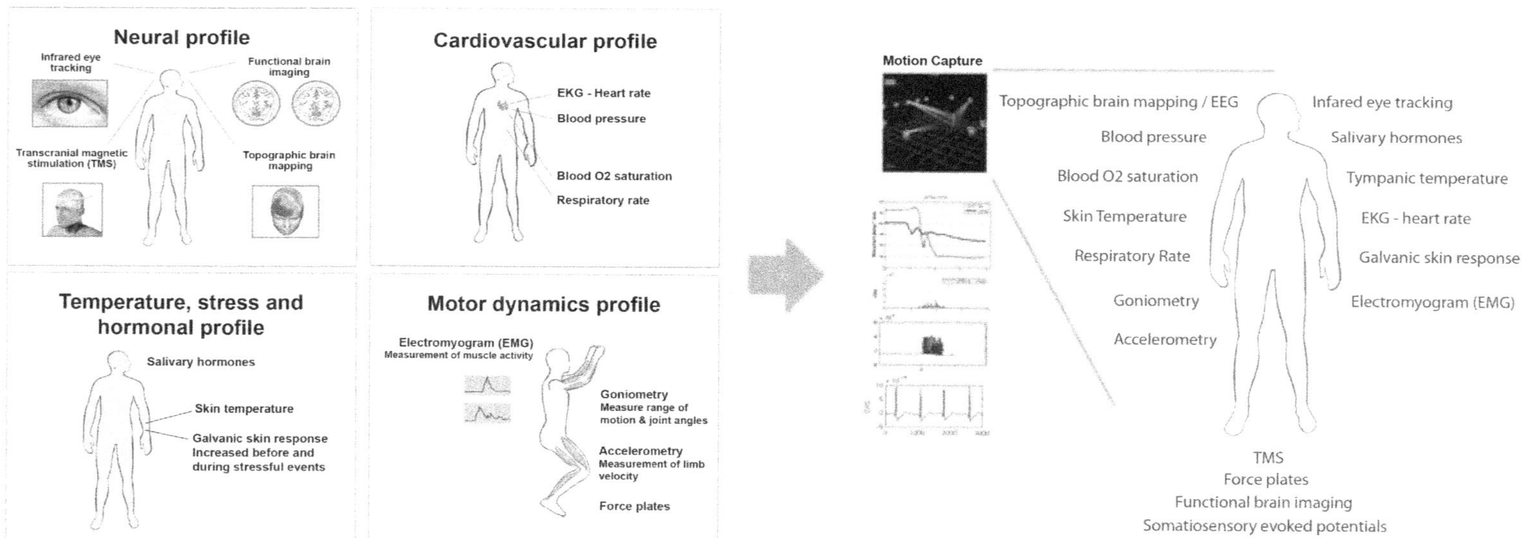

FIGURE C3.8 The Anticipatory Profile reflects the holistic nature of living processes.

evolution, i.e., the constructive path," is of central focus. A structural measurement process – by necessity continuous – replaces the discrete method of associating numbers to positions in space and time. Out of respect for Bernstein's legacy, I shall provide some details of this attempt, very much in the spirit of his inquiries.

Leonardo da Vinci (*Trattato della Pittura*, 1498) took note of what we experience during our actions:

> when a man stands motionless upon his feet, if he extends his arm in front of his chest, he must move backwards a natural weight equal to that both natural and accidental which he moves towards the front.

Over time, this observation caused many to wonder how the act of readjusting takes place. It is, like many other actions, an expression of autonomous processes. Addressing postural adjustment (Gahery 1987) proved that the compensation that Leonardo da Vinci noticed – the muscles from the gluteus to the soleus tighten as a person raises his arm – slightly *precede* the beginning of the arm's motion. In short, the compensation occurred in anticipation of the action. Gahery further took note of the fact makes it clear that every action (movements, tasks performed, etc.) involves "purposive" and "non-purposive" movement. This applies to postural adjustments (involving postural preparations, accompaniments and reactions), or to gait dynamics, for example. Without making reference to NAB's work, and, obviously, not having read *The Construction of Movements*, Gahery confirmed Bernstein's experimental results. With the AnticipationScope, we went further, addressing actions (such as catching a ball, moving on an unstable path, motions associated with stepping into a car, among others) that in the absence of anticipatory behavior could harm those carrying them out. The simple action of taking a seat – e.g., how high, how soft or rigid, how warm or cold, how slippery – involves a multitude of adjustments, all autonomous in nature.

The Anticipatory Profile (Nadin 2012) we derived, using mathematics of generatively structures entities, account for the anticipatory components of our actions – even when they are not purposive. Synergy comes into expression, as well as agonist (supportive) and antagonist (opposing)

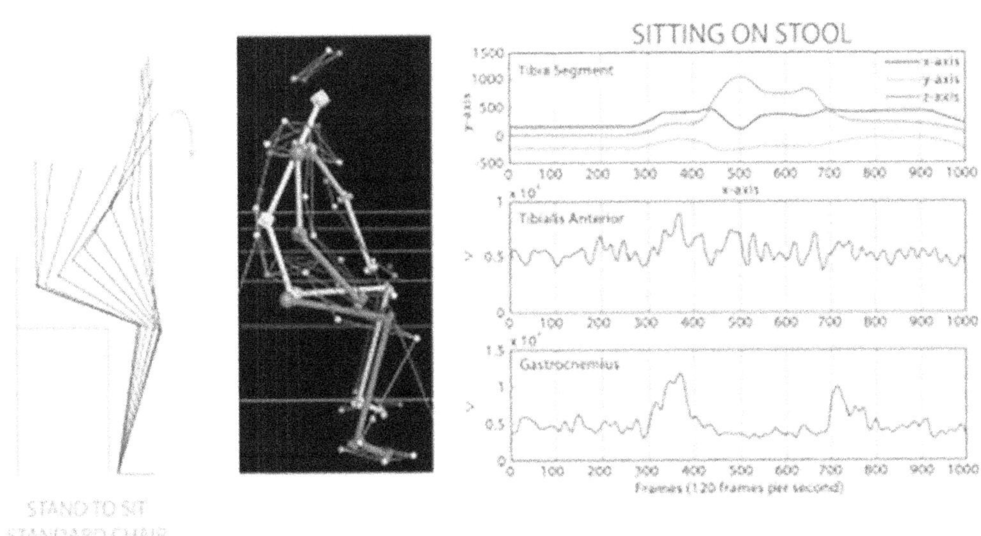

FIGURE C3.9 Anticipation expression in stand-to-sit motoric expression

factors. Without anticipation, we would fall when we raise our arms. Or we would either destroy our chairs (imagine a weight equal to that of a human being falling on a chair) or hurt our backs and buttocks (Figure C3.9).

Where is NAB when one needs him, or someone like him, most? *The Construction of Movements* in English places Bernstein among those who are dedicated to a better understanding not only of how the motoric system functions, but actually of how life is lived. Will they put in the effort needed to understand him?

References

Adams D (1978) http://culture-of-peace.info/vita/1978/bernstein.pdf. 46 pp.

Bergson H (1911) The Cinematographical Mechanism of Thought and the Mechanistic Illusion – A Glance at the History of Systems – Real Becoming and False Evolution. In: (Mitchell A, Trans.) *Creative Evolution* (Chapter 4). New York: Henry Holt and Company. pp. 272–370.

Bernstein F (2015) Prosthetic Manhood in the Soviet Union at the End of World War II, *Osiris* 30 (1): 113–133.

Bernstein NA (1936/2003) Contemporary Issues on the Physiology of Neural Processes, *Motor Control* 10 (1): 1–6.

Bernstein NA (1947) *On the Construction of Movements*. Moscow: Medghiz.

Bernstein NA (1965/1988) Auf den Weg zu einer Biologie der Aktivitat. In: Pickehain & Schnabel (Eds.) *Bewegungsphysiologie von N.A. Bernstein*. Leipzig.

Бернштейн А. Н. Очерки по физиологии движений и физиологии активности. М.: «Медицина," 1966.

Bernstein NA (1967) *The Co-ordination and Regulation of Movements*. Oxford: Pergamon Press. (Original work in Russian published 1940.)

Bernstein NA (1969) The Immediate Task of Neuro-physiology in the Light of the Modern Theory of Biological Activity, *Proceedings, 18th International Congress of Psychology: Symposium on Cybernetic Aspects of Integrative Brain Activities*, 119–125.

Bernstein NA (1987) *The Physiology of Human Movement and Activity*. Moscow: Nauka.

Bernstein NA (1991) *On Dexterity and Its Development*. Moscow: Physical Culture and Sport Press.

Bernstein NA, Popova TD (1930) Studies on the Biodynamics of the Piano Strike (English translation 2003). In: Kay BA, Turvey MT, Meijer OM (Eds) (2003) An Early Oscillator Model. Studies on the Biodynamics of the Piano Strike, *Motor Control* 7: 1–45.

Bongaardt R, Meijer OG (2000) Bernstein's Theory of Movement Behavior: Historical Development and Contemporary Relevance, *Journal of Motor Behavior* (April), 32 (1): 57–71.

Bruton M, O'Dwyer N (2018) Synergies in Coordination: A Comprehensive Overview of Neural, Computational, and Behavioral Approaches, *Journal of Neurophysiology* 120: 2761–2774.

da Vinci L (1498) Trattato della Pittura. (First published in France as *Trattato della Pittura di Lionardo da Vinci*, 1632, Paris: Presse Giacomo Langlois.)

Elsasser WM (1958) *The Physical Foundation of Biology*. London: Pergamon Press.

Elsasser WM (1998) *Reflections on a Theory of Organisms*. Baltimore: Johns Hopkins University Press. (Originally published in 1987 by ORBIS Publishing, Frelighsburg, Quebec.)

Farber BS, Jacobson JS (1995). An Above-knee Prosthesis with a System of Energy Recovery: A Technical Note, *Journal of Rehabilitation Research and Development* (November) 32 (4): 337–348.

Фейгенберг И. М. Ф 36 «Николай Бернштейн: от рефлекса к модели будущего» – М.: Смысл, 2004.

Feigenberg J (1988) Chronologisches Verzeichnis aller Publikationen N.A. Bernsteins. In: Pickenhain L and Schnabel G (Eds.) *Bewegungsphysiologie von N.A. Bernstein*, 255–263. Leipzig: Barth.

Feldman A, Meijer OJ (1999) Discovering the Right Questions in Motor Control: Movements (1929) Bernstein's Heritage, *Motor Control* 3 (June): 105–134.

Gahery Y (1987) Associated Movements, Postural Adjustments and Synergy: Some Comments about the History and Significance of Three Motor Concepts, *Arch Ital Biol*, 125 (1): 345–360.

Gawande A (2017) The Heroism of Incremental Medicine, *The New Yorker*, January 23 www.newyorker.com/magazine/2017/01/23/the-heroism-of-incremental-care.

Gelfand I (2009) Letter to Professor Tamar Flash, Wiezmann Institute of Science, Rehovot, Israel (Provided by Professor Dr. Mike Mirsky, October 11, 2019).

Georgopoulos AP, Caminiti R, Kalaska JF, Massey JT (1983) Spatial Coding of Movement: A Hypothesis Concerning the Coding of Movement Direction by Motor Cortical Populations, *Experimental Brain Research* (July) 51 (27): 247–260.

Goldfarb L, Golubitsky O (2001) "What Is a Structural Measurement Process?" Faculty of Computer Science, University of New Brunswick Technical Report TR01-147, October.

He J et al. (2005) Design of a Robotic Upper Extremity Repetitive Therapy Device, *Proceedings of the 2005 IEEE 9th International Conference on Rehabilitation Robotics*, 95–98.

Hochberg LR et al. (2006) Neuronal Ensemble Control of Prosthetic Devices by a Human with Tetraplegia, *Nature* (July 13) 442: 7099, 164–171.

Humphrey DR, Schmidt EM, Thompson WD (1970). Predicting Measures of Motor Performance from Multiple Cortical Spike Trains, *Science* (November 13) 170: 3959, 758–652.

Khatchadourian R (2018) Degrees of Freedom, *The New Yorker*, November 26, 56–71 www.newyorker.com/magazine/2018/11/26/how-to-control-a-machine-with-your-brain.

Latash LP, Latash ML (1994). A New Book by N.A. Bernstein: "On Dexterity and Its Development." *Journal of Motor Behavior*, 26 (1) (February 28): 56–62.

Latash ML (2012) *Fundamentals of Motor Control*. Cambridge MA: Academic Press/Elsevier.

Libet B (1985) Unconscious Cerebral Initiative and the Role of Conscious Will in Voluntary Action, *The Behavioral and Brain Sciences*. 8 (4): 529–566.

Luria AR (1987) Bernstein, Nicholas. In: Gregory RL (Ed.) *The Oxford Companion to the Mind*, 85–86. Oxford: Oxford University Press.

Mettrie de la JO (1748) *L'homme Machine*. Leiden: Elie Luzac, Fils.

Meijer OG, Bruijn SM (2007) The Loyal Dissident: N.A. Bernstein and the Double-Edged Sword of Stalinism, *Journal of the History of Neuroscience* (January–June) 16 (1–2): 206–224.

Myszka S (2018) www.football beyond the stats.wordpress.com/2018/08/30/respecting-the-movement-problem.

Nadin M (1972) *Determinations of Modern Art: Elements of Meta-Aesthetics* (in Romanian, doctoral dissertation). University of Bucharest.

Nadin M (1987) *Minds as Configurations: Intelligence is Process*. Graduate Lecture Series, Ohio State University, Columbus, January 29. Published 1991 *MIND-Anticipation and Chaos*. From the series Milestones in Thought and Discovery. Stuttgart/Zürich: Belser Presse.

Nadin M (1990) *Die Kunst der Kunst. Metaaesthetik* (The Art of Art. Meta-Aesthetics). Stuttgart: Belser Verlag.

Nadin M (1991) *The Civilization of Illiteracy*. Dresden: Dresden University Press.

Nadin M (2005) Seneludens: Maintaining Vital Anticipatory Characteristics in the Aging. www.anteinstitute.org/seneludens.html.

Nadin M (2010) Anticipation and Dynamics: Rosen's Anticipation in the Perspective of Time; and Annotated Bibliography: Anticipation. Special issue of *International Journal of General Systems*, 39 (1).

Nadin M (2012) The Anticipatory Profile. An Attempt to Describe Anticipation as Process. In: Nadin M (Ed.) Anticipation, *International Journal of General Systems* (January) 41 (1): 43–75.

Nadin M (2014) Navigation Nobel: Soviet Pioneer, *Nature* 515: 37, November. See also: www.nature.com/articles/515037c.

Nadin M (2015) Variability by Another Name: "Repetition Without Repetition." *Learning from the Past. Early Soviet/Russian Contributions to a Science of Anticipation*. Cognitive Science Monographs 25, pp. 329–340. Cham, CH: Springer.

Nadin M (2018a) *Redefining Medicine from an Anticipatory Perspective, Progress in Biophysics and Molecular Biology*. Amsterdam: Elsevier.

Nadin M (2018b) Machine Intelligence – A Chimera, *AI & Society*, pp. 1–28, London: Springer Verlag (Springer Nature).

Pickenhain L, Schnabel G. (Eds.) (1975) *Bewgungsphysiologie von N. A. Bernstein* [N.A. Bernstein's Physiology of Movement] (1st ed.). Leipzig: Barth.

Pool ER, Pauli WM, Kress CS, O'Doherty JP (2019) Behavioral Evidence for Parallel Outcome-Sensitive and Outcome-Insensitive Pavlovian Learning Systems in Humans, *Nature Human Behavior*, February 25. www.nature.com/articles/s41562-018-0527-9

Popper KR (1966) *Of Clouds and Clocks. An Approach to the Problem of Rationality and the Freedom of Man*. Seattle: Washington University Press.

Profeta VLS, Turvey MT (2018) Bernstein's Levels of Movement Construction: A Contemporary Perspective, *Human Movement Science*, 57 (February) 111–133.

Rosen R (1985) *Anticipatory Systems: Philosophical, Mathematical, and Methodological Foundations*. London: Pergamon Press.

Rosen R (1991) *Life Itself: A Comprehensive Inquiry Into the Nature, Origin, and Fabrication of Life* (Complexity in Ecological Systems). New York: Columbia University Press.

Schwartz AB (2004) Cortical Neural Prosthetics, *Annual Review of Neuroscience* (July 21) 27: 487–507.

Schwartz AB, Cui XT, Weber DJ, Moran DW (2006) Brain-Controlled Interfaces: Movement Restoration with Neural Prosthetics, *Neuron* (October 5) 52 (1): 205–220.

Schwartz AB (2007) Useful Signals from Motor Cortex, *Journal of Physiology*, Symposium on the Physiology of Brain-Computer Interfaces, March 14 (report commissioned by the Editorial Board) https://physoc.onlinelibrary.wiley.com/doi/abs/10.1113/jphysiol.2006.126698

Shishmarev VYu, Moreinis IS, Bogomolov AI, Koroktov AI (1974) Normalization of Walking on Prosthesis with an External Power Source. In: On Theory and Practice of Robots and Manipulators. International Centre for Mechanical Sciences (Courses and Lectures), vol 201. Springer, Vienna.

Sporns O, Edelman GM (1993) Solving Bernstein's Problem: A Proposal for the Development of Coordinated Movement by Selection, Child Development, 64:4, 960–981.

Taylor DM, Helms Tillery SI, Schwartz AB (2002) Direct Cortical Control of 3D Neuroprosthetic Devices, *Science* (June 7) 296: 5574, 1829–1832.

Veresov N (2006) Leading Activity in Developmental Psychology: Concept and Principle. *Journal of Russian and East European Psychology* (July) 44 (25): 7–25.

C4

WHILE APPRECIATING BERNSTEIN'S WORK, WE NEED TO CRITICALLY EVALUATE HIS IDEAS FROM THE VANTAGE POINT OF PRESENT SCIENTIFIC KNOWLEDGE

Anatol G. Feldman

UNIVERSITY OF MONTREAL, DEPARTMENT OF NEUROSCIENCE, AND THE CENTER FOR INTERDISCIPLINARY RESEARCH IN REHABILITATION (CRIR), CANADA

I was privileged to read Bernstein's book in Russian and understand the tremendous difficulties in translating this book into English – even in Russian, Bernstein's texts are often difficult to understand, and I congratulate Mark Latash for successfully overcoming numerous difficulties in clarifying Bernstein's thoughts in English.

While reading this book, now in English, I renew my feeling of how brilliant Nikolai Bernstein was in his scientific inquiries. According to Einstein (1922), the major purpose of such inquiries whether in natural sciences or psychology is to consider experimental facts ("our experiences") and bring them into a logical system, which is exemplified by scientific theories created by outstanding physicists and mathematicians. Not being a physicist or mathematician, Bernstein in his work revealed a clear tendency (also noted by a math giant, Israel Gelfand, personal communication) to create such a logical system in behavioral neuroscience. Although Bernstein did not succeed in achieving such a challenging goal, he gave a clear example of what researchers should desire to reach in their scientific endeavor. There were objective and subjective reasons that limited Bernstein's achievements. One of the reasons was indeed the ugly political system suppressing his publications in the atmosphere of fear of imprisonment and even death for many outstanding biologists and other innocent people in the dark period of KGB purges in the former Soviet Union.

One should also keep in mind that, while critically evaluating the ideas of his predecessors, and advancing new ideas, Bernstein was not free from sharing some mistaken general views or preconceptions in behavioral neuroscience established by them.

Apparently, it would be a disservice to science and disrespect to the memory of Bernstein to dogmatically consider his ideas and refer to his work each time we want to support our own ideas without a critical evaluation of them from the vantage point of present scientific knowledge.

For example, Bernstein (1967) emphasized that the nervous system often acts in an anticipatory and predictive manner, while considering physical properties of objects with which the body interacts. Bernstein expressed this idea or rather opinion in a very general, colloquial form as that the nervous system "creates a model of the desired future" (Feigenberg 2014; Latash 2014). There is no indication that Bernstein using this figurative expression, implying the existence of hypothetical internal models postulated in computational theories of motor control (Hollerbach 1982; Wolpert and Kawato 1998; Todorov and Jordan 2002). Nevertheless, proponents of computational theories eagerly refer to Bernstein to support their claim for the existence of such models to justify the idea of the computational nature of brain functioning (Fodor 1975; Shadmehr and Wise 2005), which appears to be misleading (Feldman 2019).

From the vantage point of present knowledge, one can identify some limitations of Bernstein's approach to actions and perception. One of them is related to the problem of the relationship between posture and motion. Intuitively, he understood that neural control processes for posture and movement are intrinsically related, being just different manifestations of the same neural mechanism. They "go hand in hand," as nicely said by Stuart (2005). However, Bernstein was unable to suggest how, specifically, posture and movement are related. The fact is that postures are associated with the balance of all forces or torques acting on body segments as well as with the stability provided, in particular, by proprioceptive reflexes that resist deflections from and tend to restore the initial posture. By producing movement, the system creates a deflection from the initial, stable posture and the problem is to explain why movements are not opposed by posture-stabilizing mechanisms. Although appreciated by many, the suggestion that stabilizing mechanisms are suppressed when movement is initiated (von Holst 1954) appears to be incorrect: stabilizing mechanisms are necessary not only for maintaining posture but also for producing movement trajectories (Feldman 2016). I want to suggest a common reason that prevented such prominent scholars as Von Holst and Bernstein from providing a physiologically feasible solution to the posture-movement as well as to other important problems in behavioral neuroscience. The reason was the dominant and very powerful philosophy accepted during earlier attempts to explain how animals and humans generate intentional movements. I will call this philosophy "mechanical reductionism." It became dominant long before Bernstein and von Holst, probably since Newton formulated laws of mechanics, when it became clear that these laws are universal and biological creatures obey such laws and behave according to them without making specific efforts to do so. Mechanical reductionism starts from the belief that biological creatures do make specific efforts and mechanisms to behave according to such laws. In an implicit form, such a belief was endorsed by the prominent physicist, Helmholtz (1866) who assumed, as if it were self-evident, that the brain directly specifies, or, in modern terms, pre-programs, motor commands to muscles to produce the required motor outcome. Unfortunately, this belief, reflecting an insufficient understanding of how physical laws manifest themselves was shared by von Holst and Bernstein (see his Figure 2.1 in Chapter 2 of the translated book), which prevented him from solving not only the posture-movement problem but also other problems he formulated, including the redundancy problem for movements involving multiple muscles and degrees of freedom of the body. To solve these problems, we need to overcome the tenets of mechanical reductionism, with the understanding that the brain can control movements only indirectly, by changing neurophysiological parameters that influence but can remain independent of, mechanical variables that describe the motor outcome (Feldman 2019; Feldman 2015; Latash 2014) One can only guess how far Bernstein could have progressed with his beautiful mind in the understanding of the neural control of action and perception if he had been able to overcome the tenets of mechanical reductionism.

References

Bernstein NA (1967) *The Co-ordination and Regulation of Movements*. Pergamon, Oxford

Einstein A. (1922). *The Meaning of Relativity*. Princeton, NJ: Princeton University Press.

Feigenberg JM (2014). *Nikolai Bernstein – from Reflex to the Model of the Future Theories in Behavioral Neuroscience*. Springer, New York

Feldman AG (2015) *Referent Control of Action and Perception: Challenging Conventional Theories in Behavioral Neuroscience*. Springer, New York

Feldman AG (2016) The Relationship between Postural and Movement Stability. In: *Progress in Motor Control. Theories and Translations*, Jozsef Laczko and Mark Latash (eds) pp. 105–113. Springer.

Feldman AG (2019). Indirect, Referent Control of Motor Actions Underlies Directional Tuning of Neurons. *J Neurophysiol* 121 (3): 823–841.

Fodor JA (1975) *The Language of Thought*. Thomas Crowell, New York, NY.

Helmholtz H (1866) *Handbuch der Physiologischen Optik*. Voss, Leipzig.

Hollerbach JM (1982) Computers, Brains, and the Control of Movement. *Trends Neurosci* 5: 189–192.

Latash ML (2014) Bernstein's "Desired Future" and Physics of Human Movement. *Cognitive Systems Monographs* 25: 287–299.

Shadmehr R, Wise SP (2005) *The Computational Neurobiology of Reaching and Pointing: A Foundation for Motor Learning*. Cambridge, MIT Press.

Stuart DG (2005). Integration of Posture and Movement: Contributions of Sherrington, Hess, and Bernstein. *Human Movement Science* 24 (5): 621–643.

Todorov E, Jordan MI (2002) Optimal Feedback Control as a Theory of Motor Coordination. *Nat Neurosci* 5 (11): 1226–1235.

von Holst H (1954) Relations between the Central Nervous System and the Peripheral Organs. *Br J Anim Behav* 2 (3): 89–94.

Wolpert D, Kawato M(1998) Multiple Paired Forward and Inverse Models for Motor Control. *Neural Netw.* 11 (7–8): 1317–1329.

C5

BERNSTEIN AND REHABILITATION

Mindy F. Levin

SCHOOL OF PHYSICAL AND OCCUPATIONAL THERAPY, MCGILL UNIVERSITY

When reading the English translation of Bernstein's book *On the Construction of Movements*, one cannot help but be impressed by the depth and encyclopedic breadth of Bernstein's command of anatomy and physiology from simple organisms to humans. His knowledge extended from basic functioning of physiological systems to comparative biology as well as to neuropathology. It is clear that he was a keen observer of motor behavior and appreciated the relationship between cognition and movement. The text is filled with detailed descriptions of behavior to substantiate his explanations of the construction of movements in addition to more objective "cyclogram" representations of movements that he and his students recorded.

One is also impressed with the quality of the translation of this book by Mark Latash. Bernstein's writing style is effusive, descriptive, and often quite convoluted. The text is rife with newly coined terminology, often borrowed from other languages, such as Greek, Latin, German, or French, included to better represent a particular idea. The writing style is in stark contrast to modern scientific writing that is sharply focused on being concise and direct. Bernstein's style is actually refreshing, allowing the reader to glimpse inside the processes underlying the thoughts – a luxury that we no longer allow in scientific literature.

Over the course of nine chapters (including numerous footnotes and figures – over 100!), starting from Chapter 2, Bernstein lays out the details of his five levels of the construction of movements in a roughly hierarchical manner from bottom to top – Levels A to E – rubro-spinal (paleokinetic), thalamo-pallidar (synergies), pyramidal-striatal (spatial field), parietal-premotor (action formation) and higher cortical (symbolic or intention). While this structure may be somewhat hypothetical, at least in some ways it is consistent with the present associations of functions with different brain areas. What follows are detailed descriptions of each level, largely based on Bernstein's own observations and experiments, but often referring to the work of other scientists of the day, including Lapicque, Magnus, Matthews, Sherrington, and Hughlings-Jackson, to name a few.

In the last two chapters, Bernstein puts many of the concepts together, dealing with motor skill acquisition (learning) and all its related features, as well as the consequences of central nervous system injury and finally a discussion on precision, variability, and stability.

Many of the ideas expressed by Bernstein go beyond the popular ideas of the period and are still relevant today, particularly with respect to the concept of coordination as a coordinative structure, representing a mastering of the ability of the nervous system to solve motor problems, and the concept of mutual adjustments of the coordinative elements of a skill. This is also true for his observations on aspects of movement dynamics, movement precision, variability, and stability. Bernstein himself did not try to work out the mechanisms underlying the actions and disordered actions that occur at each of the levels or of the system as a whole. This was left for later scientists to elaborate. Indeed, his teaching has created a basis for elaboration of many modern concepts and theories of motor control by other scientists. These include the dynamical systems theory (i.e., Miller, Galenter, and Pribram 1960; Schoner and Kelso 1988) and the equilibrium-point hypothesis (Feldman 1966, 1996) with its extension to the referent body configuration (Feldman 2015), implying that motor actions are controlled indirectly, by changing neurophysiological parameters that pre-determine, in a feedforward way, where, in the spatial frame of reference or system of coordinates, motor actions can emerge without being preprogrammed (Feldman and Levin 1995).

Bernstein's ideas form the basis for many descriptions of motor equivalence or alternative kinematic solutions for performing motor tasks such as compensatory synergies used for reaching when additional degrees of freedom are involved (Kaminski et al. 1995; Ma and Feldman 1995) as well as the eloquent uncontrolled manifold hypothesis (UCM) for the description of the structure of good and bad variability used for the stabilization of a given task variable (Scholz, Schoner, and Latash 2000).

Some of the ideas of Bernstein, later operationalized by the work of Feldman in the threshold control theory of motor control (Feldman and Levin 1995), go beyond purely structure-function explanations, popular at the time. In many sections of the book and in particular in Chapter 9, Bernstein provides numerous examples of the consequences of damage to specific brain structures or areas, illustrating his detailed knowledge of this relationship. At the same time, however, it is evident that he understood that motor deficits could not entirely be explained by damage or deficits to any one structure or pathway, although the details of the actual mechanisms underlying motor deficits are not elaborated.

Rather than focusing on the necessity to consider the intricate interaction between cognitive, sensory, and motor functions, the majority of clinical scientists today adhere to standard structure-function concepts and biomechanical explanations of movement to try to explain and remediate disordered movement. In the area of the rehabilitation of upper limb deficits in patients with stroke, for example, several recent trials have contrasted the effects of novel treatments based on increased training intensity compared to usual care. The concept of "more training is better" is based on findings of substantial functional reorganization in the forelimb area of the primary motor cortex of adult primates who were exposed to a greater amount of therapy than a control group (Nudo and Milliken 1996). However, in contrast to standard care that encompasses both evidence-based practice, clinical reasoning, and patient preferences, several upper limb training trials evaluating the effects of intensive practice have not resulted in substantially better functional outcomes (Daly et al. 2019; Lo et al. 2010; Rodgers et al. 2019).

One explanation for the lack of better outcomes is that the narrow view of motor recovery, while founded on solid principles of neuroplasticity, does not take into account broader concepts of movement production as outlined by Bernstein. Our current work addresses this problem. It is based on the premise that upper limb recovery in individuals who have sustained a stroke may be improved if the focus of treatment includes remediating an individual's specific motor impairment within the framework of the threshold control theory, elaborated by Feldman and Levin (1995) and Levin and Feldman (1994). This theory adheres to the Bernsteinian notion that movement is controlled by neural networks, including descending systems regulating parameters that define the limits of the range of possible movements (i.e., spatial thresholds for activation and inactivation of muscles in specific joint ranges), rather than the specific structure of the movement itself (i.e., directly specifying EMG patterns), as is usually proposed (Todorov and Jordan 2002). Our previous work in people with stroke and other neurological pathologies has shown that deficits in spatial threshold control lead to impairments such as spasticity, muscle weakness, and disordered voluntary muscle activation patterns that affect upper limb motor performance (Levin and Feldman 1994; Levin et al. 2000; Musampa et al. 2007) such as the ability to correct errors in a one-trial learning paradigm (Subramanian et al. 2018). However, current approaches to upper limb rehabilitation based on the premise that motor output can be improved by repetitive practice do not account for the individual's specific motor deficit. Importantly, since movements themselves are impaired, their repetition may likely lead to only small improvements in motor function, while reinforcing undesirable movement patterns (Takeuchi and Izumi 2012).

We are currently testing novel training programs to enhance the range of possible upper limb movements (i.e., Levin et al. 2017; Sror et al. 2019) that focus on individualizing treatment incorporating the identification and expansion of spatial threshold control at the elbow in patients with sub-acute stroke. The training is organized in order to increase the range of controlled elbow movement while exploiting the ability of the nervous system to adaptively choose movement patterns that meet the task demand. Effects of these training programs that exploit the natural variability and problem-solving ability of the nervous system may lead to better functional recovery of the upper limb and better rehabilitation outcomes.

References

Daly JJ, McCabe JP, Holcomb J, Monkiewicz M, Gansen J, Pundik S (2019) Long-Dose Intensive Therapy Is Necessary for Strong, Clinically Significant, Upper Limb Functional Gains and Retained Gains in Severe/Moderate Chronic Stroke. *Neurorehabilitation and Neural Repair* 33: 523–537.

Feldman AG (1966) Function Tuning of the Nervous System with Control of Movement or Maintenance of a Steady Posture. 2. Controllable Parameters of the Muscle. *Biophysics* 11: 565–578.

Feldman AG (2015) *Referent Control of Action and Perception: Challenging Conventional Theories in Behavioral Neuroscience*. Switzerland: Springer Nature.

Feldman AG (1996) Spatial Frames of Reference for Motor Control. In: Latash M. (ed). *Progress in Motor Control, Vol. I., Bernstein's Traditions in Movement Studies*. Champaign, IL: Human Kinetics, pp. 289–314.

Feldman AG, Levin MF (1995) The Origin and Use of Positional Frames of Reference in Motor Control *Behavioral and Brain Sciences* 18: 723–806.

Kaminski TR, Bock C, Gentile AM (1995) The Coordination between Trunk and Arm Motion during Pointing Movements. *Experimental Brain Research* 106 (3): 457–466.

Levin MF, Baniña MC, Frenkel-Toledo S, Berman S, Soroker N, Solomon JM, Liebermann DG (2017). Personalized Upper Limb Training Combined with tDCS for Sensorimotor Recovery in Spastic Hemiparesis: Design of a Randomized Clinical Trial. *Trials* 19 (1):7.

Levin ML, Feldman AG (1994) The Role of Stretch Reflex Threshold Regulation in Normal and Impaired Motor Control. *Brain Research* 657: 23–30.

Levin MF, Selles RW, Verheul MH, Meijer OG (2000) Deficits in the Coordination of Agonist and Antagonist Muscles in Stroke Patients: Implications for Normal Motor Control. *Brain Research* 853 (2): 352–369.

Lo AC, Guarino PD, Richards LG et al. (2010) Robot-Assisted Therapy for Long-Term Upper-Limb Impairment after Stroke. *New England Journal of Medicine* 13; 362 (19): 1772–1783.

Ma S, Feldman AG (1995) Two Functionally Different Synergies during Arm Reaching Movements Involving the Trunk. *Journal of Neurophysiology* 73 (5): 2120–2122.

Miller G, Galanter E, Pribram KH (1960) *Plans and the Structure of Behavior*. NY: Holt, Rinehart, Winston.

Musampa NK, Mathieu PA, Levin MF (2007) Relationship between Stretch Reflex Thresholds and Voluntary Arm Muscle Activation in Patients with Spasticity. *Experimental Brain Research* 181: 579–593.

Nudo RJ, Milliken GW (1996) Reorganization of Movement Representations in Primary Motor Cortex Following Focal Ischemic Infarcts in Adult Squirrel Monkeys. *Journal of Neurophysiology* 75 (5): 2144–2149.

Rodgers H, Bosomworth H, Krebs HI, et al. (2019) Robot Assisted Training for the Upper Limb after Stroke (RATULS): A Multicentre Randomised Controlled Trial. *Lancet* 6; 394 (10192): 51–62.

Scholz JP, Schöner G, Latash ML (2000) Identifying the Control Structure of Multijoint Coordination during Pistol Shooting. *Experimental Brain Research* 135 (3): 382–404.

Schoner G, Kelso JAS (1988) Dynamical Patterns of Biological Coordination: Theoretical Strategy and New Results. In: Kelso JAS, Mandell A, Shlesinger MF (eds). *Dynamic Patterns in Complex Systems.* Singapore: World Scientific, pp. 77–102.

Sror L, Vered M, Treger I, Levy-Tzedek S, Levin MF, Berman S. A Virtual Reality-Based Training System for Error Augmented Treatment in Patients with Stroke. *2019 International Conference on Virtual Rehabilitation (ICVR)*, Tel Aviv, Israel, 2019, pp. 1–2.

Subramanian SK, Feldman AG, Levin MF (2018) Spasticity May Obscure Motor Learning Ability after Stroke. *Journal of Neurophysiology* 119: 5–20.

Takeuchi N, Izumi S-I (2012) Maladaptive Plasticity for Motor Recovery after Stroke: Mechanisms and Approaches. *Neural Plasticity* 2012: 359728.

Todorov E, Jordan MI (2002) Optimal Feedback Control as a Theory of Motor Coordination. *Nature Neuroscience* 5: 1226–1235.

C6

NEURAL CONTROL PRINCIPLES

Bernstein's Insights from Biomechanics of Human Movement

Boris I. Prilutsky

SCHOOL OF BIOLOGICAL SCIENCES, GEORGIA INSTITUTE OF TECHNOLOGY, ATLANTA, GA, USA

Vladimir M. Zatsiorsky

DEPARTMENT OF KINESIOLOGY, THE PENNSYLVANIA STATE UNIVERSITY, STATE COLLEGE, PA, USA

Analysis of Biomechanics of Human Motion as a Tool to Understand Motor Control

Nikolai Aleksandrovich Bernstein was one of the first to use biomechanical analysis of human motion to infer neural control principles underlying movement control. It is remarkable that his deep insights into the neural control of movement were made in the 1930s–40s when no methods for in vivo recording of neuronal activity were available. Bernstein received his medical degree in neurology from the Moscow University. He also took courses in physics and mathematics. After working as a neurologist for several years following his graduation, he started his research in biomechanics of human movement as the director of biomechanics laboratory at the Institute of Labor in 1922. His educational background, the methods of movement recordings and analyses he developed, and a wide range of motor tasks he studied contributed to his discoveries of major motor control principles. Before commenting on Bernstein's insights into neural control of movement and their impact on current research, we briefly discuss strengths and weaknesses of his biomechanical analysis.

Experimental Methods and Data Analysis

As Muybridge (Muybridge 1887) and Marey (Marey 1879), Bernstein recorded movements in natural settings. He, however, was able to substantially improve spatial and temporal resolution of his recordings with developing the cyclographic recording method and methods of three-dimensional recordings with a mirror (Bernstein 1967). Both spatial and temporal resolutions of these methods were close to those of Braune and Fisher (Braune and Fischer 1895–1904). However, in contrast to Braune and Fischer who analyzed only several strides of walking in one subject, Bernstein's laboratory was able to analyze and publish an enormous amount of biomechanical data collected in hundreds of subjects performing a wide range of motor tasks. In the course of almost 20 years of experimental research, Bernstein investigated movements of highly skilled professional workers (Bernstein 1923), musicians (Bernstein and Popowa 1929) and athletes, as well as locomotion of children of different ages (Bernstein 1940). This was achieved partially owing to the developed methods of simplified marker digitizing and calculations of time derivatives and center of mass position.

Another factor enabling Bernstein to observe the rich complexity of recorded movements, largely unavailable to his predecessors and contemporaries, was his methods of calculating time derivatives. He did not smooth body marker trajectories as severely as was done by Braune and Fischer (Braune and Fischer 1895–1904). As a result, the computed body segment accelerations and forces contained high frequency components up to 20 Hz (Woltring 1984). Although many high frequency waves in accelerations and forces reported by Bernstein appeared repeatable across experiments, subjects, and similar motor tasks (e.g., Figure 4.8), some of them have not been reproduced by other researchers and were likely contaminated by high frequency noise. A detailed analysis of this issue was given by Woltring 1984. Nevertheless, major signal components contained in the calculated acceleration and forces were valid and later reproduced in other studies. These calculations permitted Bernstein quantification of movement variability and identification of the role of reactive forces in movement control.

Bernstein was also responsible for another advancement in biomechanics that contributed to his subsequent analysis of movement complexity – development of methods for measuring and calculating inertial properties of body segments in the same subjects participating in motion recordings. Although the obtained segmental inertial properties were not as accurate as those determined by next generation of gamma scanner, computer tomography and MRI methods (for review see Zatsiorsky 2002), they allowed Bernstein to conduct inverse dynamics analysis of freely moving upper and lower extremities, except in movement phases in which the most distal segment was in contact with an external environment, e.g. ground during the stance phase of locomotion. Methods of accurate recordings of ground reaction forces and incorporating them into inverse dynamics analyses became available in the late 1930s (Elftman 1939; Manter 1938), so Bernstein could not incorporate them in his experimental studies that ended at that time.

Mechanical Complexity of Musculoskeletal System

Bernstein's Degrees-of-Freedom Problem

Following the study of Braune and Fischer (Braune and Fischer 1895–1904), analysis of whole body dynamics during human movements led Bernstein to realize the complexity of the musculoskeletal system and consequences of this complexity for control of movement. He wrote:

> The mobility of the kinematic chains in the human body is enormous and reaches tens of degrees of freedom. The mobility of the wrist with respect to the scapula and of the tarsus with respect to the pelvis adds up to seven degrees of freedom, the mobility of a fingertip with respect to the thorax is 16 degrees of freedom. The presence of mobile fingers enriches the mobility and deformability of the hand compared, for example, with the forelimb of hoofed quadrupeds by 22 additional degrees of freedom.
>
> *(Chapter 2, p 13, Bernstein 1947)*

His analysis of human musculoskeletal morphology has been later confirmed and expanded by next generation of researchers in detailed quantitative analyses of the musculoskeletal morphology of the whole human body. For example, Morecki with colleagues (Morecki et al. 1971) have determined that the total number of moving bones in the human body is 206 and they are interconnected by a total of 147 joints. These consist of 85 joints with 1 degree of freedom

(DOF), 33 joints with 2 DOF, and 29 joints with 3 DOF. Thus, the total number of DOF in the human body is 244. Morecki and his colleagues have also determined the total number of skeletal muscle compartments (they called them "actons") producing moments with respect to each DOF; this number exceeds 640 (the exact number depends on how muscle compartments are defined). This means that on average each DOF is served by approximately 3 muscle compartments. It is clear from the above analysis that any whole-body motor task in a three-dimensional space can be performed with unlimited number of possible kinematic patterns in each DOF, which reflects the kinematic redundancy of the musculoskeletal system. Furthermore, specific kinematics of a motor task can be produced by an unlimited number of muscle activities and forces, reflecting muscle redundancy. There is a much greater neuronal redundancy in the central and peripheral nervous systems that makes it possible to produce similar motoneuronal and muscle activities with various contributions of the central and peripheral (sensory) neuronal inputs. A further complication for motor control, realized by Bernstein and prior researchers, e.g. Braune and Fischer 1895–1904, is that muscle-tendon mechanical and physiological properties (the force–length–velocity relationships) are highly nonlinear; see for example Zatsiorsky and Prilutsky (2012).

According to Bernstein, this complexity of the musculoskeletal system and, in particular, the kinematic and muscle redundancy, have critical impact on movement control. This follows from his formulation of the famous degrees-of-freedom problem of motor control and motor coordination: "The main problem of movement coordination is ... in overcoming the redundant degrees of freedom of a moving organ, i.e. in turning it into a controllable system." (Chapter 2, p. 15, Bernstein 1947).

Complex Dynamic Interactions between Multiple Body Segments

Another significant complication for movement control of the multi-segment musculoskeletal system, recognized by Bernstein, arises from dynamic interactions (interaction moments) between moving multiple interconnected body segments. Even in the simplest planar kinematic chain consisting of two segments, motion of each segment induces several interaction moments affecting motion of the adjacent segment and arising from gravitational, inertial, centripetal and Coriolis forces; e.g. (Zatsiorsky 2002). The interaction moments are motion-dependent and their influence on body segments also depend on the initial positions and velocities of each body segment in the kinematic chain before initiation of movement. A typical kinematic chain in the human body, i.e. the upper or lower extremity, consists of at least 3 to 4 major segments that are free to move in a three-dimensional space. This leads to much more complex dynamical interactions between body segments during motion. Based on his analysis of dynamics of human motion, Bernstein made a profound conclusion that significantly influenced the subsequent developments in motor control: There is no one-to-one correspondence between central motor commands to the muscles and the resulting motion. In other words, the same motor commands can cause different movements depending on initial position and velocity of body segments and on interactions of the body segments with changing reactive internal forces and external forces (e.g., effects of gravity on interaction moments depend on orientation and movement direction of body segments). The opposite is also true – the same movement may result from different motor commands. The unpredictable dynamics of the human musculoskeletal system led Bernstein to conclude that accurate performance of motor tasks is impossible without constant sensory corrections; see below.

It is worth mentioning that Bernstein arrived to the above conclusions by investigating movements of humans, a relatively large biological species with substantial body segment inertia, which is responsible for considerable motion-dependent gravitational and interaction moments

(Hollerbach and Flash 1982; Zernicke et al. 1991). In small animals like insects and rodents, but not cats (Wisleder et al. 1990), effects of gravitational and inertial forces are negligible (Hooper 2012), suggesting less complex dynamics and possibly different organization of motor control in these species.

Complexity of Incoming Somatosensory Information

Bernstein came close to realize that mechanical complexity of the musculoskeletal system and its kinematic and muscle redundancies results in analogous issues of uncertainty in mapping from many motion-dependent somatosensory signals from multiple sources (muscle length, stretch velocity and force afferents; cutaneous mechanoreceptors, etc.) and the few resulting perceptions of position and motion of joint angles and the end-effector of an extremity, as seen in the many-to-few mapping between the motor commands and the resulting movements (Reed 1984). As with motor commands and motion, there is no one-to-one mapping between somatosensory inputs and the resulting perception (Gibson 1966; Köhler 1929). This is because somatosensory perceptions depend also on previous experiences and behavioral goals of the organism in the surrounding environment. For instance, different perceptions and the corresponding different motor responses to similar somatosensory inputs may be produced at the different fusimotor sets of the nervous system with diverse gains in somatosensory proprioceptive feedback pathways established by activation of the static and dynamic gamma motoneurons (Prochazka et al. 1985).

The uncertainties in mappings between motor commands and the motor responses and between motion-dependent somatosensory inputs and the resulting somatosensory perceptions discussed by Bernstein suggest that movement and movement perception emerge in dynamic interactions among neuronal activities in the central nervous system, mechanical and physiological properties of the musculoskeletal system, external environment and motion-dependent sensory feedback.

Organization of Neural Control of Movement

Bernstein proposed several ways by which the nervous system could control the musculoskeletal system and deal with the complexities arising from kinematic and muscle redundancies and from unpredictability of resulting movements. He was one of the first to propose a major organizational principle of complex systems, in his case the motor control system – a hierarchical, modular organization.

Modularity of Sensorimotor Control

Bernstein proposed several levels of the movement control system each responsible for a limited number of functions. A top executive level, e.g., "D – the level of actions," formulates motor behavioral goals and issues relatively simple instructions to lower levels. Lower levels, "A – level of paleokinetic regulations," "B – the level of synergies" and "C – the level of spatial field," are responsible for own specific functions from a selection of particular motor responses, appropriate postural adjustments, modulation of gains in somatosensory pathways and motor reflexes, etc. Such a division of labor among the motor control hierarchical levels makes the control of the highly redundant neuromechanical system and processing of sensory inputs more manageable. Each level is capable of relatively independent functioning and receives multiple input signals from afferents of its own level ("leading afferentation") and from other levels of the hierarchy. Examples of independent autonomic functioning of various levels of the motor control system have been well documented and include coordinated locomotion of cats with deactivated motor cortex (Beloozerova and Sirota 1993), decerebrate cats (Lundberg and Phillips 1973; Shik et al. 1966; Waller 1940) and spinalized cats (Forssberg and Grillner 1973; Frigon et al. 2017). In the above examples of the cat model with several upper hierarchical control levels separated or removed, the animals are able not only to coordinate movements of their limbs, activities of multiple muscles and to appropriately change body posture, they also adjust their limb and muscle coordination to speed of a single treadmill belt or to distinct speeds of separate belts for the left and right limbs and select the appropriate locomotor gait for a given speed. The hierarchical organization of motor control with relatively autonomous control levels simplifies motor control and processing of motion-dependent sensory information by reducing the number of independently controlled parameters (Gelfand et al. 1971).

Examples of low levels of motor control hierarchy include relatively simple spinal reflexes – stretch reflex, flexion reflex, cross-extension reflex (Nichols and Houk 1976; Sherrington 1910a), and more complex reflexes – scratch reflex (Deliagina et al. 1975; Sherrington 1910b), paw shake response (Pearson and Rossignol 1991) and wiping reflex (Fukson et al. 1980; Giszter et al. 1989). Each level of motor control hierarchy is capable of leading autonomous functioning of the lower hierarchical levels by integrating inputs from them and sensory feedback, which is evident from well-coordinated reflex responses and complex automatic locomotor behavior mentioned above. The relatively low levels of motor control hierarchy are quite robust as their motor performance does not suffer substantially in the face of perturbations such as adding additional mass to body segments during a paw shake response, constraining selected joints during a wiping reflex or separating the lumbar spinal cord from supraspinal control and changing speed of treadmill in the locomoting cat. These examples indicate that the lower motor control levels even when separated from the higher ones are "aware" of their motor behavior goal and whether the goal is achieved; the latter is accomplished with the use of the relevant somatosensory feedback ("leading afferentation," Chapter 2, Bernstein 1947), without which the motor performance will suffer (see below).

Synergistic, Non-individualized Sensorimotor Control

Another important feature of the organization of motor control proposed by Bernstein in light of complexity of the musculoskeletal system is movement control by motor synergies. The meaning of Bernstein's definition of synergies is similar to that of stereotyped muscle (Ivanenko et al. 2004; Maier and Hepp-Reymond 1995; Tresch and Jarc 2009) and kinematic (Berkinblit et al. 1986; Bianchi et al. 1998; Giszter et al. 1989) synergies that are derived from experimental data using computational methods of dimensionality reductions in the contemporary research. According to Bernstein (Chapter 4, Bernstein 1947), Level B, "the level of synergies and patterns, alias the thalamo-pallidar level" in his movement construction hierarchy, is responsible for coordinating large groups of muscles in whole body stereotyped automatic movements like locomotion. The subcortical centers thalamus ("thalami optici") and globus pallidus ("globi pallidi"), a part of the extrapyramidal involuntary motor control system, are the major anatomical substrates responsible for forming motor synergies. The leading afferent information at this level is proprioception encoding joint-angle positions and velocities, as well as information from various mechanoreceptors in skin, ligaments and joint capsules, in the body-centered coordinate system without reference to the external environment. Bernstein considered this level of motor control well suited for counteracting reactive forces in multi-joint kinematic chains because of the fast, disynaptic afferent pathways signaling movement related information. Efferent pathways of this level regulate activity of spinal premotor interneurons projecting to motoneuronal pools of multiple muscles. This leads to a synergistic, non-individualized muscles control (Gelfand et al. 1971) that simplifies the degrees-of-freedom-problem by substituting controlling, for example, 30 muscles of an extremity by controlling three joints of a kinematic chain (Chapter 4, Bernstein 1947).

Bernstein's proposal of the location and function of the level of synergies and patterns was based on available at the time anatomical and physiological knowledge and biomechanical analysis of normal and pathological movements. It was already recognized, as noted by Bernstein, that local injuries to thalamus and/or globus pallidus or structures closely associated with them result in discoordination of whole-body movements as seen, for example, in Parkinson's patients.

Rigorous testing of Bernstein's proposed hierarchical levels of the motor control system, and specifically the level of synergies, started by his followers at the end of the 1950s – early 1960s (Gelfand et al. 1971; Gelfand and Tsetlin 1971). These studies have revealed in particular that electrical stimulation of the mesencephalic locomotor area of the midbrain of the cat decerebrated at the precollicular-postmammilar level, demonstrated well-coordinated locomotion (Shik et al. 1966). Subsequent studies of the synergistic motor control have involved computational decompositions of recorded EMG activities of multiple muscles into low-dimensional patterns (Tresch and Jarc 2009), cluster analyses of EMG bursts occurring at the same time (Krouchev et al. 2006) or correlation analyses of motor units from multiple muscles (Laine et al. 2015) in intact subjects and reduced animal preparations. These studies have supported the synergistic control of multiple muscles postulated by Bernstein.

Recent studies have aimed at investigating specific roles of spinal and brainstem circuits and somatosensory feedback in forming motor synergies in frogs, rodents and cats. It has been shown in particular that frog brainstem preparations shares 60–70% of muscle synergies (low-dimensional representations of EMG activities of multiple muscles) during jumps, swim, kicks and steps with intact frogs; while only 35–60% of the synergies are shared by the spinal frogs (Roh et al. 2011). A large number of EMG burst groups is likewise shared between fictive and real locomotion of the decerebrate and intact cat, respectively (Markin et al. 2012), whereas generally fewer EMG burst groups are observed during spinal cat locomotion (Desrochers et al. 2019; Higgin et al. 2020). Thus, Bernstein's proposal concerning the synergistic control of multiple muscles in whole-body locomotor movements has been supported, although the exact location of neural circuits controlling these synergies appears to be in the brainstem caudal to the thalamus and globus pallidus.

The next higher level of Bernstein's motor control hierarchy, Level C or "the cortico-striatal level of spatial field," is responsible for forming a different type of motor synergies that are goal-directed with respect to the external environment, as opposed to Level B stereotyped motor synergies that are organized without regards to the external environment. The leading afferentation at Level C originates in the visual and somatosensory cortices and provides integrative information about movement of the working point (end effector) of an extremity in the external space. The organization of movement synergies at this level ensures high precision (low variability) of the working point during aiming movements (Bernstein 1923; Scholz et al. 2000), tracing external visual targets (Parsa et al. 2017), precise or simple locomotion that requires foot placement in a specific location to maintain balance or satisfy movement goals (Auyang et al. 2009; Klishko et al. 2014; Krishnan et al. 2013). Accuracy and precision of movements of the working point in these motor tasks is achieved by motor synergies that take advantage of motor redundancy – principle of motor abundance (Gelfand and Latash 1998), by purposeful covariation of redundant non-essential kinematic degrees of freedom of the extremity for stabilization of the essential working point movement characteristics (Latash et al. 2007).

The Principle of Sensory Corrections

The principle of sensory corrections is another general principle of neural control of movements proposed by Bernstein. As discussed above, kinematic and muscle redundancy of the musculoskeletal system, nonlinear mechanical and physiological properties of muscles, motion-dependent reactive forces arising in multi-segment kinematic chains due to segmental inertia and accelerations, and external forces make prediction of movement outcome impossible. Thus, it is necessary to constantly monitor movement and make movement corrections to insure the desired movement outcome.

Bernstein has described a number of movement disorders related to interruption of somatosensory input at different levels of the movement control hierarchy, which are characterized by various types of movement discoordination and ataxias depending on the lesion location (vestibular, cerebellar, tabetic ataxias). In these patients, movements deteriorate dramatically in the dark or after closing the eyes, which indicates an important role of visual feedback in compensating for lack of proprioception in these patients. The observations described by Bernstein have been later confirmed with additional quantitative descriptions of movement deficits and their mechanisms (Chesler et al. 2016; Cole 1995; Sainburg et al. 1995). For instance, it has been shown that patients lacking proprioception straggle to control reactive forces (interaction intersegmental moments) during reaching arm movements, in complete agreement with Bernstein's predictions and with his principle of sensory corrections.

Bernstein's ideas continue to inspire further research of the role of somatosensory signals in movement control. In current animal studies aimed to understand the mechanisms of sensory movement corrections, motion-dependent somatosensory feedback is removed by pyridoxine injections (Stapley et al. 2002), genetic manipulation of muscle spindle afferents (Akay et al. 2014), muscle self-reinnervation that removes monosynaptic stretch reflex from self-reinnervated muscles (Abelew et al. 2000; Gregor et al. 2018), removal or anesthesia of foot cutaneous receptors (Bolton and Misiaszek 2009; Bouyer and Rossignol 2003), etc. These studies have demonstrated so far that effects of somatosensory feedback on movement are muscle- and task-dependent. For example, perturbation of muscle length feedback from ankle extensors in the cat affects ankle kinematics only during downslope locomotion but has no effect during level and upslope locomotion (Abelew et al. 2000; Gregor et al. 2018). Removal of muscle spindle sensory feedback from the mouse triceps surae significantly reduces activity modulation of this muscle as a function of locomotion speed; removal of spindle input from quadriceps femoris does not have this effect (Mayer et al. 2018).

Biomechanical Aspects of Motor Skill Aquisision

Bernstein devoted the whole of Chapter 8 in his 1947 book to the proposed mechanisms of motor skill acquisition or "movement construction." Biomechanics of the musculoskeletal system (the large number of degrees of freedom, nonlinear properties of muscles, motion-dependent reactive forces) make acquisition of a novel motor skill, involving multi-segment kinematic chains, an extremely difficult, time-consuming and multi-stage process. Bernstein proposed mechanisms of this process based on his biomechanical analysis of movements of highly skilled workers, musicians and athletes, children acquiring locomotion skills, and patients with various neurological disorders. According to Bernstein, during initial stages of motor skill aquisition, the novice attempts to execute a desired motor task by using the highest levels of the motor control hierarchy (level C2 – pyramidal, level D – the level of actions, and/or level E – symbolic coordinations). These levels of voluntary movement control attempt to reproduce a general movement "scaffold," i.e., trajectories of end-effectors of body extremities, resembling the intendent movement. The leading afferentations in these stages of movement construction are assumed to be visual and vestibular that allow the novice to evaluate the movement performance with respect to the motor task goal and external environment. The involved top levels of motor control system are not well suited for harnessing multiple degrees of freedom and selecting and/or organizing appropriate muscle synergies to counteract motion-dependent reactive forces. These tasks are better performed by the specialized lower motor control levels (C1 – the striatal level of spatial field, B – the level of synergies and A – the level of paleokinetic regulations), whose leading afferentation is proprioception.

A typical strategy in the initial stages of whole-body motor skill acquisition is to limit the number of independently controlled DOF and effects of unpredictable reactive forces by coactivating antagonistic muscles, "freezing degrees of freedom." Needles to say that such movement execution is tiring and not energy efficient. The role of the higher motor control levels in the initial stages of skill acquisition is not only a reproduction of the intendent movement scaffold. They also presumably affect the lower motor control levels through descending inputs and modulation of their activities and processing of proprioceptive motion-dependent signals. Bernstein thought that the latter process is largely involuntary but could possibly be guided by conscious perception of effort or ease with which the movement is performed. In each repetition of the practiced skill, the learner attempts to find a performance that, on the one hand, satisfies the movement goal and general movement scaffold and, on the other hand, makes the performance "easier." In each new motor task repetition during the initial stages of skill practice, the movement is different; Bernstein called this process "repetition without repetition." This process resembles the process of trials and errors.

With motor skill development, a greater role in movement construction is assumed by the lower involuntary motor control levels, while the contribution of voluntary levels becomes smaller. In later stages of motor skill acquisition, the learner is capable of counteracting reactive forces in multi-segment kinematic chains by short, appropriately timed activity bursts of appropriate muscles. This leads to reduction of muscle coactivation; the learner "releases degrees of freedom, one after another."

In the final stages of motor skill acquisition, the lower motor control levels (primarily the level of synergies B – the thalamo-pallidar level) are responsible for selecting, constructing and tuning complex muscle synergies that allow for efficient use of reactive forces in the body, external forces and unique features of individual's musculoskeletal system that now assist and stabilize the movement instead of perturbing it. This makes the movement more efficient and stable. Bernstein wrote about this stage of whole-body motor skill development (Chapter 4, p. 62):

> removing active forces that used to be spent on counteracting reactive forces, by itself, yields both energetic and inertial economy, because the movement starts to flow by itself without the necessity for steadfast control and corrections. In addition, such dynamically stable forms, obviously, represent forms that correspond most closely to design of the kinematic chain and, as such, are most economical. ... *Clearly, for each motor task, there cannot be many such solutions. Indeed, there is either a single solution or, in the best-case scenario, a few for each given task.*
>
> *[highlighted by the authors].*

The process and corresponding mechanisms of motor skill acquisition or, in general, movement construction proposed by Bernstein were mostly conjectures and were not rigorously tested at the time. Subsequent studies on biomechanics of highly skilled movements and neural mechanisms of motor skill acquisition appears to support a number of Bernstein's conjectures. In accordance with Bernstein's suggestion that highly skilled movements are unique solutions of motor control problems, skilled arm reaching (Flash and Hogan 1985), tracing and curved movements (Haruno and Wolpert 2005), maintenance of arm posture (Bottasso et al. 2006), exerting external forces by a leg (Prilutsky and Gregor 1997), human (Crowninshield and Brand 1981), and cat locomotion (Prilutsky et al. 1997) demonstrate kinematic and muscle activity invariant patterns that satisfy optimality criteria. The kinematic invariants of arm movements include hand straight line trajectory, bell-shaped velocity profile, and the dependence between the hand velocity and trajectory curvature. They all satisfy the related optimality criteria of maximum hand movement smoothness (Flash and Hogan 1985) and minimum trajectory variability (Harris and Wolpert 1998). The stereotyped muscle activity features of skilled tasks are characterized by reciprocal activation of one-joint antagonists, simultaneous activity of synergistic muscles (cosine tuning of muscle activity), and dependence of activity of muscles crossing two DOF on the desired moments at these DOF. These features of muscle activity satisfy optimization criteria of minimum fatigue, perceived effort, energy expenditure and motor noise; e.g. (Prilutsky 2000; Prilutsky and Zatsiorsky 2002).

The above kinematic and muscle activity features are not hard wired in the nervous system but emerge over practice of a given motor tasks, as predicted by Bernstein. For instance, long practice of arm reaching in complex viscous force fields changes jerky hand movements into smooth straight-line trajectories with bell-shaped velocity profiles with a parallel decrease in muscle coactivation and sense of effort (Shadmehr and Mussa-Ivaldi 1994; Thoroughman and Shadmehr 1999). The increase in hand movement smoothness during practice of maximum-speed reaching arm movements is accompanied by changes in muscle coordination that include decreasing muscle activity and coactivation and counterbalancing the reactive motion-dependent moments by muscle moments (Schneider et al. 1989).

How does the nervous system select the most appropriate movement patterns during the process of motor skill acquisition or movement construction? Exploring movement variability, inborn in the complexity of the musculoskeletal system, and selecting better and better motor solutions during a trial-and-error process seem in line with Bernstein ideas. This process can be formally described by reinforcement learning that evaluates the consequences of actions (practice trials) based on reward signals and maximizes a value function over the period of learning in a closed-loop manner; e.g. (Sutton and Barto 2018). The value functions that might guide selection of efficient and stable motor solutions during motor learning, suggested in the literature, include the minimum interactions between the motor control levels (Gelfand and Tsetlin 1971), maximum of movement smoothness (Flash and Hogan 1985), minimum of afferentation (Prilutsky et al. 1991), minimum of end-point variability (Harris and Wolpert 1998), etc. The number of studies exploring biomechanical and neural mechanisms of movement control and learning is rapidly growing, and the ideas of Bernstein published over 50 years ago are guiding many of these studies.

Bernstein Parables

One of the authors of these commentaries (VMZ) had the opportunity to have two rather lengthy (about 1 hour each) discussions with Prof. N.A. Bernstein. The meetings occurred at Dr. Bernstein's apartment in Moscow in September–October 1965, i.e. about 55 years ago. Prof. Bernstein passed away in January 1966, i.e. approximately 4–5 months after the meetings. The author at that time was a young scientist, who just recently defended his PhD dissertation.

Prof. Bernstein's talk was serious in message and at the same time rather "poetic," with nice parables and jokes.

First Parable

A cousin of God went to God and complained to him: "Everybody knows about you while I am absolutely unknown."

"So what do you want?" – asked God.

"I want to become a famous scientist. Like Newton and Einstein."

"OK" – said God.

> I am giving you an opportunity to make three great discoveries. When you make all of them, you can publish a paper. As everybody knows, I can do everything and I know everything. I am giving you the following power: from now on you are as almighty as I am. You can do everything; there will be no limitations in your experiments. And your personal knowledge and mental power will be equivalent to that of all of contemporary science. You will know as much as all the world's scientists together know. You can go and make your discoveries.

The first question was whether there is life on Mars. (It was in 1965 and then it was a popular topic of semi-scientific discussions.)

No problem. The cousin flew to Mars, looked around, and returned back with his report (he unfortunately was not allowed to publish it until he made all three discoveries).

The second research question was: how is an electron built? Once again, no problem. The cousin made himself very small, went inside the electron, looked around, made all the necessary recordings, and prepared his report.

The third question was about motor control: how does the brain control human movements? The cousin did the following; he recorded activity of each neural cell in the brain. At that time, the most popular estimate of the number of neural cells in the brain was about 15 billion (now it is closer to 100 billion). He also built a huge building with an atrium. A wall in the atrium contained 15 billion bulbs and each bulb represented a neural cell. The cousin ordered himself a very comfortable chair (in this case it was important), sat in the chair, and started looking at the wall. The bulbs were blinking, and the cousin started contemplating what this signified. What do these neural activities mean? It seems that he is still thinking to this day, and because of that, his research papers are yet to be published.

So, what is the moral of this parable? Its take-home message? The conclusion is that scientific problems are different. There exist problems where the question is easy to ask. For instance: is there life on other planets? Obtaining an answer is difficult, but the question by itself is simple. There also exist problems where formulating a precise question is by itself difficult. Brain control of movements is one such problem. I was greatly impressed by this parable.

Second Parable

Everybody knows that after the six days of creation, God took a day off. The reason was evident: after such difficult work he was tired. It is much less known that God also took a small break after the fifth day, prior to creating man. There were two reasons for that. The first was that he was already tired and needed some rest. The second reason was that he decided to refresh his knowledge. So, he sent angels down to Earth and asked them to bring him university handbooks on physiology and other biological subjects from different times.

Initially angels brought him textbooks from the eighteenth century. God looked at them and decided that these textbooks were too simple and that he would not be creating a man following these textbooks. So, he ordered the angels to bring him textbooks from the nineteenth century. Again, too simple. Then angels brought textbooks from the twentieth, twenty-first centuries and so forth. Finally, the angels brought the textbooks from an unknown future century. They were perfect and described the whole of human biology with absolute precision. God was satisfied and decided to create a man following these descriptions.

Contemporary science is very young. Starting from the experimental studies of Galileo, it is only about 400 years old. What will happen after 1400 years, or after 4000 years? At some point, will science know everything about human biology?

Some people believe that scientists who are not biologists already know something about life sciences that biologists do not. If these people are the lab directors, they invite mathematicians and computer engineers with the expectation that they will explain how the brain works, in particular how it controls movements. In other words, they think that this knowledge is already available, but it is described in other textbooks. I.e. in textbooks on mathematics and computer science, but not in textbooks on physiology and biomechanics.

I can assure you (these are Dr. Bernstein's words, not mine) that these lab directors are wrong. Use math, use computers, but do not expect that somebody from other fields of science will explain to you how the brain works or how it controls human movements. This knowledge can be obtained only in neurophysiological and motor control labs.

In general, Dr. Bernstein's language was very picturesque and figurative. Here is one more example that I remember. The topic of discussion switched to how to determine the quality of published research, to distinguish strong studies from the weak ones.

Dr. Bernstein offered the following classification. You may divide all new publications into three groups. The publications of the first group do not change our understanding of the complexity of the topic under discussion. For instance, we know that ligaments are elastic bodies. Somebody measures the elasticity of a given ligament and publishes the results. It could be a useful study. Thank you so much. This study does not change however our understanding of the complexity of the problem. We knew before the experiment that the ligaments possess elastic properties. Now we know the number and we know as previously that the ligaments are elastic.

The studies of the second type make the picture more complex. We know in advance that something is not simple. A new scientist steps in the field and tells: your knowledge is wrong; my research shows that the problem is even more complex than you were originally thinking. Dr. Bernstein added: very often such research is published by young and ambitious scientists. While saying that, he cunningly looked at me.

And finally, there is the third group of research. These investigations provide an explanation; make complex issues easier to understand. They are seldom and they are really great.

This is what I remember from the talk which happened 55 years ago.

References

Abelew TA, Miller MD, Cope TC, and Nichols TR. Local loss of proprioception results in disruption of interjoint coordination during locomotion in the cat. *J Neurophysiol* 84: 2709–2714, 2000.

Akay T, Tourtellotte WG, Arber S, and Jessell TM. Degradation of mouse locomotor pattern in the absence of proprioceptive sensory feedback. *Proceedings of the National Academy of Sciences of the United States of America* 111: 16877–16882, 2014.

Auyang AG, Yen JT, and Chang YH. Neuromechanical stabilization of leg length and orientation through interjoint compensation during human hopping. *Exp Brain Res* 192: 253–264, 2009.

Beloozerova IN, and Sirota MG. The role of the motor cortex in the control of accuracy of locomotor movements in the cat. *J Physiol* 461: 1–25, 1993.

Berkinblit MB, Feldman AG, and Fukson OI. Adaptability of innate motor patterns and motor control mechanisms. *Behavioral and Brain Science* 9: 585–638, 1986.

Bernstein NA. Studies on biomechanics of hitting using optical recordings. *Annals of Central Institute of Labor* 1: 19–79, 1923.

Bernstein NA. *Studies of the Biodynamics of Walking, Running and Jumping* Moscow: Central Research Institute of Physical Culture, 1940.

Bernstein NA. *The Co-ordination and Regulation of Movements*. Oxford; New York: Pergamon Press, 1967.

Bernstein N, and Popowa T. Untersuchung über die Biodynamik des Klavieranschlags. *Arbeitsphysiologie* 1: 396–432, 1929.

Bianchi L, Angelini D, Orani GP, and Lacquaniti F. Kinematic coordination in human gait: relation to mechanical energy cost. *J Neurophysiol* 79: 2155–2170, 1998.

Bolton DA, and Misiaszek JE. Contribution of hindpaw cutaneous inputs to the control of lateral stability during walking in the cat. *J Neurophysiol* 102: 1711–1724, 2009.

Bottasso CL, Prilutsky BI, Croce A, Imberti E, and Sartirana S. A numerical procedure for inferring from experimental data the optimization cost functions using a multibody model of the neuro-musculoskeletal system. *Multibody System Dynamics* 16: 123–154, 2006.

Bouyer LJ, and Rossignol S. Contribution of cutaneous inputs from the hindpaw to the control of locomotion. I. Intact cats. *J Neurophysiol* 90: 3625–3639, 2003.

Braune W, and Fischer O. *Der Gang des Menschen*. Leipzig: B. G. Teubner, 1895–1904.

Chesler AT, Szczot M, Bharucha-Goebel D, Ceko M, Donkervoort S, Laubacher C, Hayes LH, Alter K, Zampieri C, Stanley C, Innes AM, Mah JK, Grosmann CM, Bradley N, Nguyen D, Foley AR, Le Pichon CE, and Bonnemann CG. The role of PIEZO2 in human mechanosensation. *The New England Journal of Medicine* 375: 1355–1364, 2016.

Cole J. *Pride and a Daily Marathon*. Cambridge, MA: MIT Press, 1995.

Crowninshield RD, and Brand RA. A physiologically based criterion of muscle force prediction in locomotion. *J Biomech* 14: 793–801, 1981.

Deliagina TG, Feldman AG, Gelfand IM, and Orlovsky GN. On the role of central program and afferent inflow in the control of scratching movements in the cat. *Brain Res* 100: 297–313, 1975.

Desrochers E, Harnie J, Doelman A, Hurteau MF, and Frigon A. Spinal control of muscle synergies for adult mammalian locomotion. *J Physiol* 597: 333–350, 2019.

Elftman H. Forces and energy changes in the leg during walking. *American Journal of Applied Physiology* 125: 339–356, 1939.

Feldman AG. Once more on the equilibrium-point hypothesis (lambda model) for motor control. *J Mot Behav* 18: 17–54, 1986.

Feldman AG. Active sensing without efference copy: referent control of perception. *J Neurophysiol* 116: 960–976, 2016.

Flash T, and Hogan N. The coordination of arm movements: an experimentally confirmed mathematical model. *The Journal of Neuroscience: The Official Journal of the Society for Neuroscience* 5: 1688–1703, 1985.

Forssberg H, and Grillner S. The locomotion of the acute spinal cat injected with clonidine i.v. *Brain Res* 50: 184–186, 1973.

Frigon A, Desrochers E, Thibaudier Y, Hurteau MF, and Dambreville C. Left-right coordination from simple to extreme conditions during split-belt locomotion in the chronic spinal adult cat. *J Physiol* 595: 341–361, 2017.

Fukson OI, Berkinblit MB, and Feldman AG. The spinal frog takes into account the scheme of its body during the wiping reflex. *Science* 209: 1261–1263, 1980.

Gelfand IM, and Latash ML. On the problem of adequate language in motor control. *Motor control* 2: 306–313, 1998.

Gelfand IM, and Tsetlin ML. On mathematical modeling of the mechanisms of the central nervous system. In: *Models of the Structural-Functional Organization of Certain Biological Systems*, edited by Gelfand, I.M., Gurfinkel VS, Fomin SV, and M.L.T. Cambridge, Massachusetts: MIT Press, 1971, p. 1–22.

Gelfand IM, Gurfinkel VS, Tsetlin ML, and Shik ML. Some problems in the analysis of movements. In: *Models of the Structural Functional Organization of Certain Biological Systems*, edited by Gelfand IM, Gurfinkel VS, Fomin SV, and Tsetlin ML. Cambridge, Massachusetts: MIT Press, 1971, p. 329–345.

Gibson J. *The Senses Considered as Perceptual Systems*. Boston, MA: Hougton-Mifflin, 1966.

Giszter SF, McIntyre J, and Bizzi E. Kinematic strategies and sensorimotor transformations in the wiping movements of frogs. *Journal of Neurophysiology* 62: 750–767, 1989.

Gregor RJ, Maas H, Bulgakova MA, Oliver A, English AW, and Prilutsky BI. Time course of functional recovery during the first 3 mo after surgical transection and repair of nerves to the feline soleus and lateral gastrocnemius muscles. *J Neurophysiol* 119: 1166–1185, 2018.

Harris CM, and Wolpert DM. Signal-dependent noise determines motor planning. *Nature* 394: 780–784, 1998.

Haruno M, and Wolpert DM. Optimal control of redundant muscles in step-tracking wrist movements. *J Neurophysiol* 94: 4244–4255, 2005.

Higgin D, Krupka A, Maghsoudi OH, Klishko AN, Nichols TR, Lyle MA, Prilutsky BI, and Lemay MA. Adaptation to slope in locomotor-trained spinal cats with intact and self-reinnervated lateral gastrocnemius and soleus muscles. *J Neurophysiol* 123: 70–89, 2020.

Hollerbach MJ, and Flash T. Dynamic interactions between limb segments during planar arm movement. *Biol Cybern* 44: 67–77, 1982.

Hooper SL. Body size and the neural control of movement. *Current Biology: CB* 22: R318–322, 2012.

Ivanenko YP, Poppele RE, and Lacquaniti F. Five basic muscle activation patterns account for muscle activity during human locomotion. *J Physiol* 556: 267–282, 2004.

Klishko AN, Farrell BJ, Beloozerova IN, Latash ML, and Prilutsky BI. Stabilization of cat paw trajectory during locomotion. *J Neurophysiol* 112: 1376–1391, 2014.

Köhler W. *Gestalt Psychology*. New York, NY: Liveright, 1929.

Krishnan V, Rosenblatt NJ, Latash ML, and Grabiner MD. The effects of age on stabilization of the mediolateral trajectory of the swing foot. *Gait Posture* 38: 923–928, 2013.

Krouchev N, Kalaska JF, and Drew T. Sequential activation of muscle synergies during locomotion in the intact cat as revealed by cluster analysis and direct decomposition. *J Neurophysiol* 96: 1991–2010, 2006.

Laine CM, Martinez-Valdes E, Falla D, Mayer F, and Farina D. Motor Neuron Pools of Synergistic Thigh Muscles Share Most of Their Synaptic Input. *The Journal of Neuroscience: The Official Journal of the Society for Neuroscience* 35: 12207–12216, 2015.

Latash ML, Scholz JP, and Schoner G. Toward a new theory of motor synergies. *Motor Control* 11: 276–308, 2007.

Lundberg A, and Phillips CG. T. Graham Brown's film on locomotion in the decerebrate cat. *J Physiol* 231: 90P–91P, 1973.

Maier MA, and Hepp-Reymond MC. EMG activation patterns during force production in precision grip. II. Muscular synergies in the spatial and temporal domain. *Exp Brain Res* 103: 123–136, 1995.

Manter JT. The dynamics of quadrupedal walking. *J Exp Biol* 15: 522–540, 1938.

Marey ÉJ. *Animal Mechanism: A Treatise on Terrestrial and Aërial Locomotion*. New York: Appleton, 1879.

Markin SN, Lemay MA, Prilutsky BI, and Rybak IA. Motoneuronal and muscle synergies involved in cat hindlimb control during fictive and real locomotion: a comparison study. *J Neurophysiol* 107: 2057–2071, 2012.

Mayer WP, Murray AJ, Brenner-Morton S, Jessell TM, Tourtellotte WG, and Akay T. Role of muscle spindle feedback in regulating muscle activity strength during walking at different speed in mice. *J Neurophysiol* 120: 2484–2497, 2018.

Morecki A, Ekiel J, and Fidelus K. *Bionika Ruchu [Bionics of Movement] (in Polish)*. Warszawa, Poland: Panstwowe Wydawnictwo Naukowe, 1971.

Muybridge E. *Animal Locomotion: An Electro-Photographic Investigation of Consecutive Phases of Animal Movements, 1872–1885*. Philadelphia University of Pennsylvania, 1887.

Nichols TR, and Houk JC. Improvement in linearity and regulation of stiffness that results from actions of stretch reflex. *Journal of Neurophysiology* 39: 119–142, 1976.

Parsa B, Zatsiorsky VM, and Latash ML. Optimality and stability of intentional and unintentional actions: II. Motor equivalence and structure of variance. *Exp Brain Res* 235: 457–470, 2017.

Pearson KG, and Rossignol S. Fictive motor patterns in chronic spinal cats. *J Neurophysiol* 66: 1874–1887, 1991.

Prilutsky BI. Coordination of two- and one-joint muscles: functional consequences and implications for motor control. *Motor Control* 4: 1–44, 2000.

Prilutsky BI, and Gregor RJ. Strategy of muscle coordination of two- and one-joint leg muscles in controlling an external force. *Motor | Control* 1: 92–116, 1997.

Prilutsky BI, and Zatsiorsky VM. Optimization-based models of muscle coordination. *Exercise and Sport Sciences Reviews* 30: 32–38, 2002.

Prilutsky BI, Herzog W, and Allinger TL. Forces of individual cat ankle extensor muscles during locomotion predicted using static optimization. *J Biomech* 30: 1025–1033, 1997.

Prilutsky BI, Raitsin LM, and Poltorapavlov NV. Biomechanical aspects of automation of a motor skill. *Theory and Practice of Physical Culture* 13–19, 1991.

Prochazka A, Hulliger M, Zangger P, and Appenteng K. 'Fusimotor set': new evidence for alpha-independent control of gamma-motoneurones during movement in the awake cat. *Brain Res* 339: 136–140, 1985.

Reed ES. From action gestalts to direct action. In: *Human Motor Actions – Bernstein Reassessed*, edited by Whiting HTA. North-Hilland: Elsevier Science Publishers B.V., 1984, p. 157–168.

Roh J, Cheung VC, and Bizzi E. Modules in the brain stem and spinal cord underlying motor behaviors. *J Neurophysiol* 106: 1363–1378, 2011.

Sainburg RL, Ghilardi MF, Poizner H, and Ghez C. Control of limb dynamics in normal subjects and patients without proprioception. *J Neurophysiol* 73: 820–835, 1995.

Schneider K, Zernicke RF, Schmidt RA, and Hart TJ. Changes in limb dynamics during the practice of rapid arm movements. *J Biomech* 22: 805–817, 1989.

Scholz JP, Schoner G, and Latash ML. Identifying the control structure of multijoint coordination during pistol shooting. *Exp Brain Res* 135: 382–404, 2000.

Schoner G, and Kelso JA. Dynamic pattern generation in behavioral and neural systems. *Science* 239: 1513–1520, 1988.

Shadmehr R, and Mussa-Ivaldi FA. Adaptive representation of dynamics during learning of a motor task. *The Journal of Neuroscience: The Official Journal of the Society for Neuroscience* 14: 3208–3224, 1994.

Sherrington CS. Flexion-reflex of the limb, crossed extension-reflex, and reflex stepping and standing. *J Physiol* 40: 28–121, 1910a.

Sherrington CS. Notes on the scratch-reflex of the cat. *Quart J Physiol* 3: 213–220, 1910b.

Shik ML, Severin FV, and Orlovsky GN. Control of walking and running by means of electric stimulation of the midbrain. *Biophysics* 11: 659–666, 1966.

Stapley PJ, Ting LH, Hulliger M, and Macpherson JM. Automatic postural responses are delayed by pyridoxine-induced somatosensory loss. *The Journal of Neuroscience: The Official Journal of the Society for Neuroscience* 22: 5803–5807, 2002.

Sutton RS, and Barto AG. *Reinforcement Learning: An Introduction*. Cambridge, Massachusetts; London, England: The MIT Press, 2018.

Thoroughman KA, and Shadmehr R. Electromyographic correlates of learning an internal model of reaching movements. *The Journal of Neuroscience: The Official Journal of the Society for Neuroscience* 19: 8573–8588, 1999.

Tresch MC, and Jarc A. The case for and against muscle synergies. *Current Opinion in Neurobiology* 19: 601–607, 2009.

Turvey MT. Coordination. *Am Psychol* 45: 938–953, 1990.

Waller WH. Progression movements elicited by subthalamic stimulation. *J Neurophysiol* 3: 300–307, 1940.

Wisleder D, Zernicke RF, and Smith JL. Speed-related changes in hindlimb intersegmental dynamics during the swing phase of cat locomotion. *Exp Brain Res* 79: 651–660, 1990.

Woltring HJ. On methodology in the study of human movement. In: *Human Motor Actions – Bernstein Reassessed*, edited by Whiting HTA. North-Holland: Elsevier Science Publishers B. V., 1984, p. 35–73.

Zatsiorsky VM. *Kinetics of Human Motion*. Champaign, IL: Human Kinetics, 2002.

Zatsiorsky VM, and Prilutsky BI. *Biomechanics of Skeletal Muscles*. Champaign, IL: Human Kinetics, 2012.

Zernicke RF, Schneider K, and Buford JA. Intersegmental dynamics during gait: Implications for control. In: *Adaptability of Human Gait*, edited by Patla AE. North-Holland: Elsevier Science Publishers, 1991.

C7

MOVEMENT IN SPACE

Following Bernstein's Thread

Apostolos P. Georgopoulos

DEPARTMENT OF NEUROSCIENCE, UNIVERSITY OF MINNESOTA MEDICAL SCHOOL, MINNEAPOLIS, MINNESOTA

Multilinked limb movements are immensely complex due to multiple actuators (muscles) acting on multiple joints on the background of a gravitational field and changing posture; add the dynamic interactions of various sorts as the multilinked limb moves in various speeds and directions, and you have an almost intractable challenge to compute practically instantaneously (millisecond by millisecond) in a closed analytic form all the simultaneous components of the moving limb to guide it to reach your goal. (Bernstein correctly points out that artificial systems typically keep their "limbs" constructed with minimal links and actuators to facilitate their control.) Yet, your goal is routinely reached in a quick, effective, elegant and seemingly effortless way, somehow obviating the need for tortuous, numerous, and time-consuming calculations. How does the brain achieve this? In my opinion, the brain has evolved over a long evolutionary period as a massively interacting neural network that handles (and solves) this problem as a "smart" device (Georgopoulos et al. 1987): the "solution" lies in the hardwired anatomical connections and physiological, dynamic interactions among the vast number of neurons dedicated to this function. Modern neuroscience has been largely devoted to study perception (mainly visual perception) and associated brain systems, and yet the vast majority of the brain is concerned with movement. As Bernstein correctly points out in Chapter 1, "Nowhere in the phylogenesis, watching the world is the ultimate goal, something self-sufficient. Receptor systems happen to be either signal systems – we have already discussed this role – and then any degree of their sophistication cannot by itself ensure biological advantage for the animal if the effector apparatus is defective, or they are part of the process ensuring adequate, coordinated work of the effectors – we will discuss this role later – and here the secondary character of their activity follows directly from the semantic content of performed tasks. Hence, both the signal and corrective roles of receptors is ensured only by effectors of the animal or species." No other statement could be closer to the truth!

The nature of a smart device is in achieving a result without solving the problem analytically. Two examples would suffice to illustrate the point. The first concerns the estimation of the area of an irregular shape. The usual way is to measure it using a grid of regular shapes (e.g. small squares) and count the number of squares and their fractions within the shape, a process that can be time-consuming and potentially not very accurate, depending on the irregularity of the shape. Here, the smart device is the planimeter by which you just trace the shape and you get its area! The second, and simpler example, and one closer to (my) home since it was solved by a Greek, concerns estimating the volume of an irregular solid, a practically impossible problem to solve analytically. Yet, Archimedes discovered that you simply have to immerse the solid in a liquid (e.g. water) and measure the volume of the liquid displaced – it was then that he exclaimed Eureka! Although the problem of instantaneous, accurate, dynamic control of multilinked limb movements is obviously different from measuring areas and volumes, the examples above illustrate ways by which analytically time consuming or intractable tasks can be easily solved by a smart device or procedure.

In the motor system, the prime example of a smart device is the spinal cord. The main observation that speaks to that was the discovery that electrical stimulation of a locus in the midbrain (the "Mesencephalic Locomotor Center") can elicit locomotion in the cat (Shik et al. 1966). Although this discovery has been well acclaimed on its own merit and further extensively investigated (see Musienko et al. 2012 for a recent assessment), its importance for the idea of the spinal cord as an integrated "smart" motor device has not been appreciated enough. Here you have an enormously complicated pattern of multilinked movements of four limbs (in the cat) that can be elicited by a gross electrical midbrain stimulation. This observation demonstrates that the spinal locomotory circuits are integrated as a whole and are controlled as a whole: exactly how a smart device would be operated. The smart motor spinal device has all the features necessary for local adjustments, as a need arises, such as avoiding obstacles. As we move from locomotion to directed limb movements, it becomes apparent that spinal circuitry is well organized and adapted to subserve such movements (Georgopoulos and Grillner 1989). Studies in the frog spinal cord have documented the existence of an intricate motor vector field organization (Mussa-Ivaldi et al. 1994) that can be vectorially summed to produce a potentially large repertoire of motor behaviors. In the cat, the spinal cord's propriospinal system has been shown to mediate the implementation of central motor commands (Illert et al. 1978), and a scheme by which motor cortical commands might engage spinal circuitry has been proposed (Georgopoulos 1996). In summary, the spinal motor apparatus fits ideally the concept of a smart motor device which allows for integrated, fast and accurate movements to be executed without the need to compute from first principles all elemental details of activations of individual muscles. This smart spinal device is used for various functions, including antigravity support (through the vestibular system), reaching, grasping, and object manipulations, among others. Of these, reaching to targets in extrapersonal space is a good example of telereceptor-movement matching, a topic I discuss next.

As Bernstein points out in Chapter 1, the appearance of telereceptors made a huge difference in enabling the animal to detect stimuli and objects at a distance from its body and to act accordingly: "Telereceptors turned out to be a powerful centralizing factor by giving the animal possibilities of reacting to stimuli from infinitely far away, compared to the body size." He goes on to contend that, in evolutionary time,

> there is ongoing development of the system of telereceptors, which assume the leading role and contribute to the evolution of central processing systems and the brain and to the whole qualitative revolution, which has been described before and which follows the gradual dominance of receptors of this type.

He also correctly points out that a crucial survival advantage was conferred by the evolution, at the same time, of a well-developed motor system that could be guided by the telereceptor information: "An increase in the leading role of telereceptors in the evolution of motor functions is also due to the fact that telereceptors brought about complex integrative movement forms ..." In this regard, the most effective telereceptor is the retina and the visual information it supplies is of the utmost usefulness for everyday life. However, the effective use of that information crucially depends on the possibility that arm movements (and whole body movements, when needed) be guided according to the visual information received. In most real-life situations, such visuo-motor guidance needs to be quick and effective, for the animal to achieve its goal. An important aspect of this matter is that visuomotor coordination takes place in extrapersonal space. In fact, as Bernstein points out in Chapter 5, this space is totally pervasive in action and underlies all telereceptor-motor interactions.

Several areas of the frontal and parietal lobes are involved in visual control of reaching. Although frontal motor and premotor areas are expected to play a central role in the initiation and generation of arm movements, the participation of posterior parietal cortex in reaching (Mountcastle et al. 1975) was an unexpected discovery that linked the well-known involvement of the parietal cortex in spatial functions (Critchley 1969) to arm movements operating in extrapersonal space. The intricate interactions between parietal and frontal areas involved in visuomotor coordination of the arm have been investigated extensively by Caminiti and colleagues (see Battaglia-Mayer et al. 2003; Caminiti et al. 2010). Ultimately, the motor command is generated in the motor cortex as the result of integrated processing of cortical (frontal and parietal) and subcortical (cerebellar and basal ganglia) signals for transmission to the spinal cord. In the early 1980s, we investigated reaching movements and their neural mechanisms by training monkeys to move an articulated manipulandum to stimuli presented on a drawing board (Georgopoulos et al. 1981). We then changed the location of the stimulus at various times after its initial onset and observed how the pointing movement changed its course. We found that the arm movement was closely coupled to the visual stimulus, becoming practically its "slave," following the visual command faithfully (Figure C7.1). We then investigated the neural mechanisms underlying this tight dependence of the motor action on the visual stimulation by recording the activity of single neurons in the motor cortex during performance of this task (Georgopoulos et al. 1983a). We found an extremely close coupling of the change in cell activity in the motor cortex to the visual stimulus with a latency as short as 40 ms (Figure C7.2), barely accounting for neural signal transmission delays from the retina to the motor cortex through known anatomical pathways. No wonder that primates are masters in visuomotor coordination, reaping all the advantages conferred by it for survival. Obviously, such an effective coordination is possible only if the multi-linked arm movement is controlled effectively as a whole, with its end-point as the focus. The massive subcortical structures of the cerebellum and basal ganglia are intimately involved in this control, both feeding back processed signals to the arm area of the motor cortex (for reaching). Remarkably, the basic local neural mechanisms of cell activity in relation to the direction of arm movement in space are very similar in all cortical and subcortical areas investigated, consisting of broad directional tuning and local inhibitory loops (see Mahan and Georgopoulos 2013 for a review).

In Chapter 5 of this book, Bernstein discusses complex movements in space with defined trajectories, such as imitation of a gesture, tracing or copying a shape, etc. In discussing the difference between tracing and copying a shape, he states: "we should note that, in all the described examples, the moving extremity (more exactly, its working point) leads to a spatial contour of the movement at sub-level C1 – as a directly prescribed contour (a typical movement – tracing a shape), and at sub-level C2 – as a prescribed pattern presented as a geometric shape without limitations on its location and scale (a typical movement – copying a shape). The lowest variability and highest accuracy are seen at this level exactly with respect to the working point. Particularly remarkable examples are cases when the working point does not correspond to a point on the extremity itself but represents a point on the hand-held tool: the tip of the pen, pencil, or knife, the sweet spot of the tennis racket, the hammer's head, etc.." Here, Bernstein draws attention to four remarkable

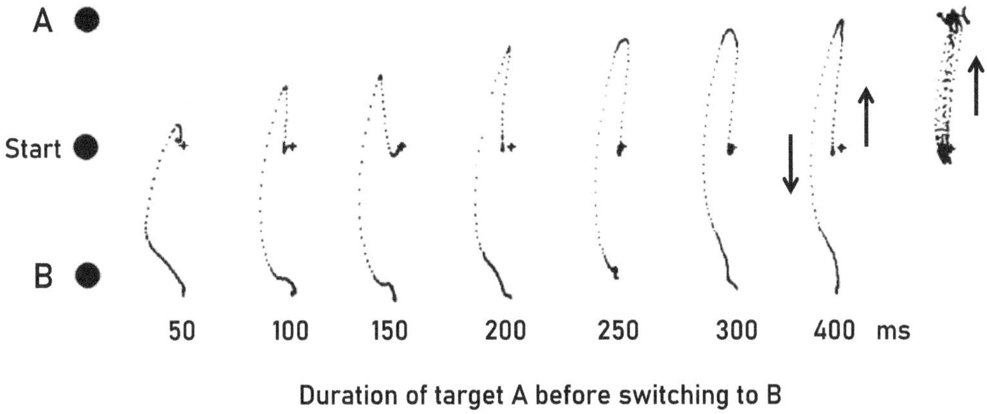

FIGURE C7.1 Hand trajectories when switching targets at various intervals. Dots are hand position every 10 ms.
(Source: Adapted from Figure 6 in Georgopoulos et al. 1981.)

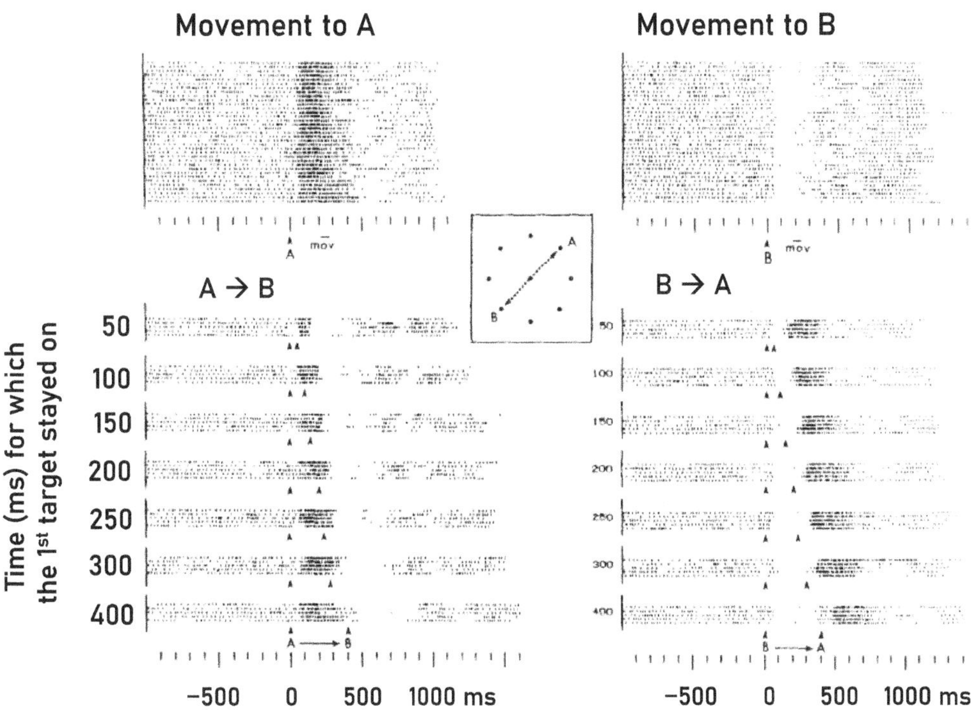

FIGURE C7.2 Repeated trials (spike trains) for movements to target A, target B, and target shifts. Short vertical lines indicate the occurrence of a neural spike. Recordings from a single cell in monkey motor cortex.
(Source: Adapted from Figure 7 in Georgopoulos et al. 1983a.)

points, namely (1) the exact spatial constraint in tracing as shape ("as a directly prescribed contour"); (2) the simply spatially congruent reconstruction in copying a shape ("as a prescribed pattern presented as a geometric shape without limitations on its location and scale"); (3) the reference to the "working point" as the controlled output of the effectors; and (4) the extension of the "working point" beyond the body extremity itself to a point on a hand-held tool. Indeed, we discovered in the early 1980s that, indeed, moving a manipulandum ("extended working point") in different directions in extrapersonal space (on a drafting board) was associated with orderly changes in cell activity in the motor (Georgopoulos et al. 1982) and parietal cortex (Kalaska et al. 1983) of behaving monkeys, such that the activity of a single cell was highest for movements in a particular direction (the cell's "preferred direction") and decreased gradually with movements made farther and farther away from the preferred direction; we called this "directional tuning." Such an orderly directional tuning has been observed for movements in 3D space (Schwartz et al. 1988) and in all cortical and subcortical areas studied (see Mahan and Georgopoulos 2013 for an extensive review) and also for effectors other than the arm, including the tongue (Murray and Sessle 1992). Recently, evidence for directional tuning in various frontal and parietal areas was obtained in human subjects using functional magnetic resonance imaging (fMRI) (Fabbri et al. 2010).

Epilogue

Our discovery of directional tuning in space (Georgopoulos et al. 1982) is perhaps the most direct link to Bernstein's insight that space is an integral background in the construction of movements, especially multilinked limb movements directed to objects of interest. Remarkably, a very similar experiment and dataset from the motor cortex did not lead to any new insights. More specifically, the joint-focused interpretation of motor cortical cell activity during reaching culminated in a paper by Murphy et al. (1982) titled "Precentral unit activity correlated with angular components of a compound arm movement" in which 3D arm motions at various joints were recorded simultaneously with single activity in the motor cortex of monkeys reaching to targets in 3D space. The neuronal discharge rates were analyzed with respect to angular joint motions and muscle activity, with essentially negative results. The main conclusion was that

> No simple relation between the EMG [electromyogram] and single precentral neuron activities was observed in these experiments, even in those cases where the muscles recorded were activated by ICMS [intracortical microelectrode stimulation] at the unit recording site. A similar lack of a simple relationship has been noted for presumed single joint movements . . . Although the motions occurring at the two joints were produced quite differently, shoulder and elbow units tended to behave very similarly in this task.
>
> *(Murphy et al. 1982, p. 144)*

Obviously, an analysis of motor cortical cell activity during multijoint (i.e. multilinked) movements using motion/muscle components as a reference did not lead to any meaningful insights. Ironically, it was that same year that our paper was published, reporting on the orderly relations between motor cortical activity and direction of arm movement in 2-D space (Georgopoulos et al. 1982).

An unambiguous measure predicting accurately the direction of movement in space (the "Neuronal Population Vector") was extracted from the activity of the neuronal ensemble (Georgopoulos et al. 1983b, 1986, 1988), and movement trajectories for reaching in 3D space were constructed (Georgopoulos et al. 1988), and extended to free 3D drawing movements (Schwartz 1994) by monkeys. A similar analysis was used to construct successfully movement trajectories

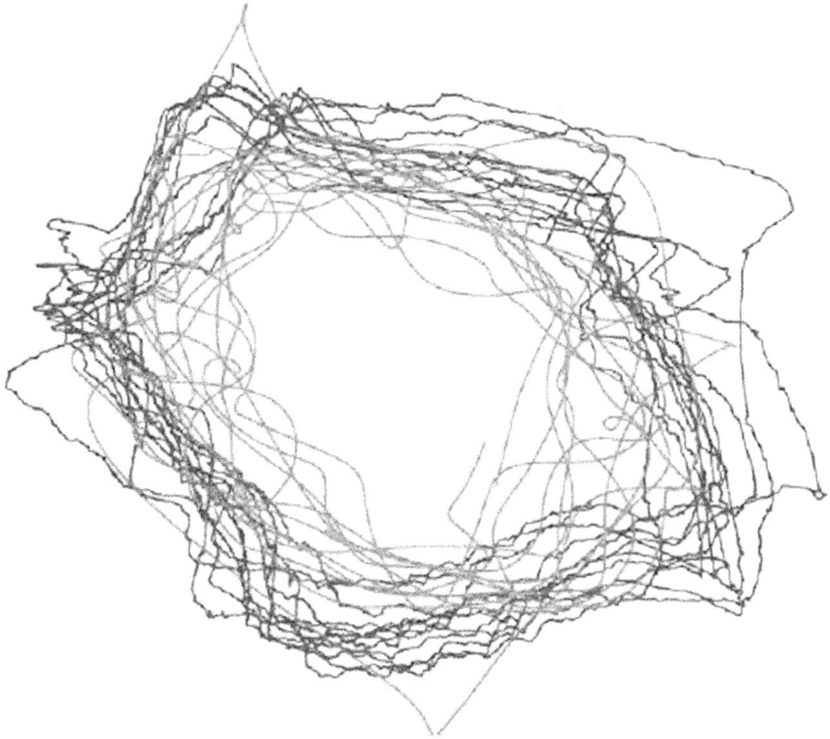

FIGURE C7.3 Actual (blue) and predicted by MEG (magenta) hand trajectories during copying a pentagon.
(Source: Adapted from Figure 1 in Georgopoulos et al. 2005.)

during copying a pentagon from a set of magnetoencephalographic (MEG) recordings in humans (Figure C7.3) (Georgopoulos et al. 2005). This general neural - -> movement-in-space translation method is being used in neuroprosthetics to control trajectories in space of prosthetic arms in real time for various activities (Collinger et al. 2013). A succinct and lucid account of this application was given by Courtine et al. (2013). Altogether, these findings provide a concrete neural substantiation of Bernstein's ideas on the construction of movements in space, as explicated in this book – and still guiding us!

Acknowledgments

I dedicate this Commentary to the late Gerald M. Edelman who introduced me to Bernstein's work in the early 1980s. Gerry was a devotee of Bernstein whom he considered one of the greatest motor scientists and thinkers of the twentieth century. This book makes this assertion all too clear.

References

Battaglia-Mayer A, Caminiti R, Lacquaniti F, Zago M (2003) Multiple levels of representation of reaching in the parieto-frontal network. *Cerebral Cortex* 13: 1009–1022.

Caminiti R, Chafee MV, Battaglia-Mayer A, Averbeck BB, Crowe DA. Georgopoulos AP (2010) Understanding the parietal lobe syndrome from a neurophysiological and evolutionary perspective. *Eur J Neurosci* 31: 2320–2340.

Collinger JL, Wodlinger B, Downey JE, Wang W, Tyler-Kabara EC, Weber DJ, McMorland AJ, Velliste M, Boninger ML, Schwartz AB (2013) High-performance neuroprosthetic control by an individual with tetraplegia. *Lancet* 381: 557–564.

Courtine G, Micera S, DiGiovanna J, del R, Milla´ n J (2013) Brain–machine interface: closer to therapeutic reality? *Lancet* 453: 516–517.

Critchley J (1969) *The Parietal Lobes*. New York: Hafner.

Fabbri, S, Caramazza A, Lingnau A (2010) Tuning curves for movement direction in the human visuomotor system. *J Neurosci* 30: 13488–13498.

Georgopoulos AP (1996) On the translation of directional motor cortical commands to activation of muscles via spinal interneuronal systems. *Cognitive Brain Res* 3: 151–155.

Georgopoulos AP, Grillner S (1989) Visuomotor coordination in reaching and locomotion. *Science* 245: 1209–1210.

Georgopoulos AP, Kalaska JF, Massey JT (1981) Spatial trajectories and reaction times of aimed movements: effects of practice, uncertainty and change in target location. *J Neurophysiol* 46: 725–743.

Georgopoulos AP, Schwartz AB, Kettner RE (1986) Neuronal population coding of movement direction. *Science* 233: 1416–1419.

Georgopoulos AP, Schwartz AB, Kettner RE (1987) Neuronal coding and robotics. *Science* 237: 301.

Georgopoulos AP, Kettner RE, Schwartz AB (1988) Primate motor cortex and free arm movements to visual targets in three-dimensional space. II. Coding of the direction of movement by a neuronal population. *J Neurosci* 8: 2928–2937.

Georgopoulos AP, Kalaska JF, Caminiti R, Massey JT (1982) On the relations between the direction of two-dimensional arm movements and cell discharge in primate motor cortex. *J Neurosci* 2: 1527–1537.

Georgopoulos AP, Kalaska JF, Caminiti R, Massey JT (1983a) Interruption of motor cortical discharge subserving aimed arm movements. *Exp Brain Res* 49: 327–340.

Georgopoulos AP, Caminiti R, Kalaska JF, Massey JT (1983b) Spatial coding of movement: A hypothesis concerning the coding of movement direction by motor cortical populations. *Exp Brain Res Suppl* 7: 327–336.

Georgopoulos AP, Langheim F, Leuthold A, Merkle AN (2005) Magnetoencephalographic signals predict movement trajectory in space. *Exp Brain Res* 167: 132–135.

Illert M, Lundberg A, Padel Y, Tanaka R (1978) Integration in descending motor pathways controlling the forelimb in the cat. 5. Properties of and monosynaptic excitatory convergence on C3–C4 propriospinal neurones. *Exp Brain Res* 33: 101–130.

Kalaska JF, Caminiti R, Georgopoulos AP (1983) Cortical mechanisms related to the direction of two-dimensional arm movements: Relations in parietal area 5 and comparison with motor cortex. *Exp Brain Res* 51: 247–260.

Mahan MY, Georgopoulos AP (2013) Motor directional tuning across brain areas: Directional resonance and the role of inhibition for directional accuracy. *Front Neural Circuits* 7:92.

Mountcastle VB, Lynch JC, Georgopoulos AP, Sakata H, Acuna C (1975) Posterior parietal association cortex of the monkey: command functions for operations within extrapersonal space. *J Neurophysiol* 38: 871–908.

Murphy JT, Kwan HC, MacKay WA, Wong YC (1982) Precentral unit activity correlated with angular components of a compound arm movement. *Brain Res* 246:141–145.

Murray GM1, Sessle BJ (1992) Functional properties of single neurons in the face primary motor cortex of the primate. III. Relations with different directions of trained tongue protrusion. *J Neurophysiol* 67: 775–785.

Musienko PE, Zelenin PV, Lyalka VF, Gerasimenko YP, Orlovsky GN, Deliagina TG (2012) Spinal and supraspinal control of the direction of stepping during locomotion. *J Neurosci* 32(48): 17442–17453.

Mussa-Ivaldi FA, Giszter SF, Bizzi E (1994) Linear combinations of primitives in vertebrate motor control. *Proc Natl Acad Sci U S A* 91: 7534–7538.

Schwartz AB (1994) Direct cortical representation of drawing. *Science* 265: 540–542.

Schwartz AB, Kettner RE, Georgopoulos AP (1988) Primate motor cortex and free arm movements to visual targets in three-dimensional space. I. Relations between single cell discharge and direction of movement. *J Neurosci* 8: 2913–2927.

Shik ML, Severin FV, Orlovsky GN (1966) Control of walking and running by means of electrical stimulation of the midbrain. *Biophysics* 11: 756–765.

C8

BERNSTEIN'S ANATOMICAL LEVELS OF CONSTRUCTION

John Rothwell

DEPARTMENT OF CLINICAL AND MOVEMENT NEUROSCIENCE, UCL QUEEN SQUARE INSTITUTE OF NEUROLOGY, LONDON, UK

Like me, I expect that many people's understanding of Bernstein's ideas about motor control is limited to the notions of synergies and levels of movement construction. But this classic volume demonstrates that Bernstein's interests ranged far wider than that. Bernstein takes his theoretical ideas about motor control and attempts to map them onto specific anatomical/physiological levels of the central nervous system. He then describes how these levels evolved and how they develop in the human infant, and ends with a description of how his ideas account for the effect on movement of pathologies at different points in the nervous system. It is an extraordinary intellectual achievement with detailed descriptions ranging from biomechanics and the physiology of nerve conduction to discussion of the different types of apraxia. He integrates ideas over so many different fields that it can be difficult in parts to keep up with the train of thought.

However, like many attempts at a grand unification theory, it suffers from the problem of trying to shoehorn a theoretical concept into anatomical and physiological facts on the ground. In some parts it simply does not work.

The book as a whole is an attempt to describe how the nervous system controls muscle contractions in order to move a multijointed body in space and interact meaningfully with objects in that space. Bernstein defines four main levels of movement construction, in which appropriately transformed afferent input ("sensory fields") guides and corrects movements controlled by that level. It is a hierarchical system, not unlike that proposed by Hughlings-Jackson in the late nineteenth century in which higher levels do not replace the function of lower levels, but rather control the operation of the lower levels and add new control features.

The lowest level, A, controls muscle tone and some simple spinal reflex arcs. It is mapped anatomically onto the "rubro-spinal" level consisting principally of the spinal cord, red nucleus, substantia nigra, vestibular cerebellum and the vestibular nuclei. Descending commands from the supraspinal centers switch the excitability of reflex pathways and adjust the level of muscle tone. This type of arrangement accounts for the effects of decerebration on muscle tone and the effect of vestibular and neck reflexes on axial and proximal muscle contraction. According to Bernstein, this level controls very few independent movements in humans, limited to repetitive alternating movements such as a tremulous vibrato on a violin or the pathological tremor of Parkinson's disease that are due to activity in the spinal reciprocal inhibitory pathway.

The next level, B, is the thalamo-pallidal level, which uses more highly processed afferent input from proprioceptors and deep pressure receptors to control muscle synergies that move the limbs and trunk along desired trajectories. These movements, because they do not receive any afferent input from vision, are made in a body-centered coordinate system, and are not designed to interact directly with the environment (Bernstein gives an example of the stepping movements of a newborn infant). Afferent input is processed in the thalamus and then passed to the pallidum. The only motor output of this system is from the pallidum to the red nucleus.

Level C controls movements in space using processed inputs from visual and tactile receptors (and to a lesser extent vestibular and auditory receptors). This level governs movements that interact with objects in the environment. Anatomically it corresponds to a cortico-striatal anatomy. The main areas of cortex are the visual, somatosensory and motor cortices plus the neocerebellum, which can influence lower levels either indirectly, through the extrapyramidal system via projections to the striatum (caudate and putamen), or directly through the pyramidal pathway. Bernstein subdivides this level into C1 (extrapyramidal) and C2 (pyramidal) with the main difference being increased accuracy/precision of movements made by C2.

Levels D and E utilize progressively larger areas of frontal and parietal cortex to control motor output. They generate movement sequences that allow humans to operate and interact with tools and to formulate theoretical concepts of tool use. A disorder of these levels leads to the apraxias.

Bernstein gives an example of how the same movement (a circle) can be made with different objectives, and that these correspond to his levels of control. Level A may control a rapid piano vibrato in which a finger makes rapid alternating movements, the tip of which roughly describes a circle. This results purely from activity in the local spinal reciprocal inhibitory pathway. The next level of complexity can be seen in the circular arm movement made by a gymnast or dancer. In this case, the motion is made relative to the body and is controlled by processed afferent input that describes the joint angles needed to perform the task. At level C, a circular movement can be drawn from imagination with a pencil on paper or copied from another drawing. The movement is made in relation to external space. Level C2 is needed for the more complex and accurate replication of the drawn circle. In level D, a circular arm movement may occur when making a stitch with a needle or untying a knot. The movement requires a complex sequence of actions and is made in relation to the shape and meaning of objects in space. Finally a professor can draw a circle on a blackboard to illustrate a geometrical theorem. In this case, the exact form of the circle is not relevant since it stands in for a concept representing the relationship between elements of the mathematical scheme. This is the uppermost level E. In each case, the movement requires different levels of precision and control that relate directly to the objective of the movement itself.

The theory is elegant but the problem lies in its transition to anatomical and physiological reality. The basic concept of brainstem control of spinal circuits together with direct and indirect cortical control of each is not unlike that of the present day. However the anatomical parcellation of each level is rather distant (almost unrecognizable) from modern day ideas. This is not surprising. A lot has happened in 70+ years to change the way we understand the connectivity and physiology of the nervous system. Indeed application of genetic fingerprinting of individual populations of neurones is even now changing our notions of connectivity within cortical and spinal circuits.

Perhaps more surprising is that the details do not match up to what was known in the early 1940s and available to Bernstein when the book was written. For example, it seems unreasonable for Bernstein to ascribe to the spinal cord only the regulation of muscle tone and the reflexes of reciprocal inhibition and tendon jerks. In the early years of the century Sherrington (1906) had described the scratch reflex in spinalized dogs, which persisted even when deafferented. Similarly, Graham Brown had studied stepping movements elicited in spinalized cats that again persisted after deafferentation (see description in Jones et al. 2011). Clearly the spinal cord was capable of rather more than Bernstein gives it credit. It may be that Bernstein thought that in humans these processes had migrated to the brainstem in the process of encephalization. However, he gives no descriptions of humans with complete spinal section to back this up and probably such patients were not available since most cases at the time died within a year of injury from medical complications and infections.

The notion of a thalamo-striatal connection, which is the basis of level B, is also odd. Most scientists at the time (see description in Jones 1985) thought that the main basal ganglia projection from thalamus to basal ganglia was to the striatum rather than the pallidum (in fact there are very few connections from thalamus to pallidum, although there are dense projections in the opposite direction from pallidum to thalamus). So why assign the striatum to level C and pallidum to level B?

So should the modern reader simply dismiss Bernstein's attempt to map his theoretical approach onto basic anatamo-physiological circuits of the CNS? After all, present day concepts of the organization of the nervous system are very different from those of 70+ years ago. At one level the answer must surely be "yes" since the specific model presented here is wrong and incompatible with modern views. However, the problem itself has not disappeared. We still struggle to understand how theoretical concepts of control are instantiated within the nervous system.

For example, in each of his levels of control, Bernstein requires a different level of integration of afferent input. Level B requires interaction of proprioceptive information to control movements at the level of joint angle, and level C require integration of visual and other inputs to guide movements in relation to objects in space. The text is strangely silent on how these operations might occur or even where this happens. However, we are gradually beginning to answer this type of question at a neural level. There is excellent work in the vestibular system showing how, for example, convergence of inputs from otolith and canal signals allows many neurones in the vestibular nuclei to selectively encode translational motion while being unaffected by changes in head orientation. It is perhaps a long way from understanding the basic operations that fail and lead to apraxia, but at least it is a step in the right direction that I think Bernstein might appreciate.

References

Jones EG (1985). *The Thalamus*, Plenum Press New York, p 625.

Jones JG, Tansey EM, Stuart DG (2011). Thomas Graham Brown (1882–1965): Behind the scenes at the Cardiff Institute of Physiology. *Journal of the History of Neurosciences* 20: 188–209.

Sherrington CS (1906). Observations on the scratch reflex in the spinal dog. *Journal of Physiology* 34: 1–50.

C9

LEARNING FROM NIKOLAI BERNSTEIN

A Personal Account

Francesco Lacquaniti

DEPARTMENT OF SYSTEMS MEDICINE AND CENTER OF SPACE BIOMEDICINE, UNIVERSITY OF ROME TOR VERGATA, ROME, ITALY
LABORATORY OF NEUROMOTOR PHYSIOLOGY, IRCCS FONDAZIONE SANTA LUCIA, ROME, ITALY

Yury Ivanenko

LABORATORY OF NEUROMOTOR PHYSIOLOGY, IRCCS FONDAZIONE SANTA LUCIA, ROME, ITALY

Introduction

Both authors of this commentary met Bernstein's ideas very early in their career. Here, we sketch our previous encounters with Bernstein, how they influenced our work, and how the issues presented in the book *On the Construction of Movements* (*OCM* in the following) are relevant to our research directions. Lacquaniti was formed in the West (Italy and United States) and Ivanenko in the Soviet Union (later Russia). Thus, their stories provide one perspective of how Bernstein influenced some work on motor control both within and outside his country, and how these two different paths converged when Lacquaniti and Ivanenko started collaborating together in Rome.

Lacquaniti's Initial Path under Bernstein's Influence

When Lacquaniti began his postdoc at the University of Minnesota in Minneapolis in 1979, he hurried to read the book *The Coordination and Regulation of Movements* by Nikolai Bernstein. This book – a collection of papers originally published by Bernstein separately between 1934 and 1962 – had gained immediate popularity in the States since its publication in English translation in 1967. There, one finds several key ideas that are also elaborated in *OCM*. In particular, Bernstein's concepts of hierarchical organization and movement coordination proved to be highly influential for our work, as it has been for many other scientists (see other commentaries in *OCM*).

In Minneapolis, Carlo Terzuolo (Lacquaniti's mentor) together with Paolo Viviani tackled the problem of the motor patterns used for key pressing (1980), a related task having been studied more than 50 years before by Bernstein and Popova (1929). Terzuolo and Viviani argued for a parallel organization of the motor engrams for typing, in line with Bernstein's proposal of a parallel arrangement of motor patterns. Next, Terzuolo and Viviani, with Lacquaniti joining forces, studied drawing and handwriting in search of movement invariances. They discovered a lawful relationship between the instantaneous velocity of the pen's tip and the curvature of the traced figure, taking the form of a two-thirds power (Lacquaniti et al. 1983). Still in Minneapolis, Soechting and Lacquaniti (1981) focused on another classical problem enunciated by Bernstein, the coordination of multiple degrees of freedom (see Chapter 2, *OCM*). They found an invariant relation between the rate of shoulder flexion and elbow extension during planar pointing movements. This research line was later extended to 3D pointing, confirming a tight coupling of shoulder and elbow motions (Lacquaniti et al. 1986), and to the coordination between distal (wrist and hand) and proximal (elbow and shoulder) degrees of freedom (Lacquaniti and Soechting 1982; Lacquaniti et al. 1987).

When Lacquaniti (1989) reviewed the experimental work on arm movements carried out first in Minneapolis and later in Milan, he cast it in the context of the hierarchical organization of the neural control of movements proposed by Bernstein. In this model, the higher organization levels are only concerned with the general movement properties, i.e. those that remain invariant under translation, rotation, amplitude or time scaling of the movement. This hypothesis accounts for the well-known motor equivalence phenomenon: the form of a given letter or word tends to be preserved through wide changes in execution size, speed and even effector (right or left hand, foot, mouth, see the examples of writing abilities of amputees in Chapter 5 of *OCM*). These shape invariants are therefore independent of the specific set of muscles that are used in each instantiation of a movement, as in the famous example provided by Bernstein of a circle drawn with the straight arm in front or to the side of the person (other examples in Chapter 2, *OCM*). One central representational element used for the construction of arbitrary drawing or writing movements is given by the segment of geometrical curve corresponding to an elliptic or parabolic arc, and the whole shape could be constructed by joining together several such arcs. Thus, the invariants describing the movement of the end effector, the hand in the case of drawing, would be prescribed at higher levels, while the coupling between limb segment angular motion and the resulting activations of the muscles would depend on lower levels of control. At lower levels of organization, the specific parameters of the movement (location, amplitude and speed of execution) would be defined on the basis of current task demands. At still lower levels, the required patterns of activation of muscles, including muscle synergies, would be specified.

However, this hypothesis of information flow leaves the problem of inter-level compatibility unresolved: how do high-level, abstract instructions get translated in the low-level language of the commands sent to the effector? Bernstein hypothesized that different levels of the neural organization share a common language (see also Profeta and Turvey 2018). A tentative answer to this issue can be provided for movement invariances. Thus, the power law relating hand speed and curvature appears to be compatible with both high-level instructions and low-level commands. Indeed, the law can be decoded from cortical processes, since activity vectors of neural populations in motor cortex obey the power law during drawing (Schwartz 1994). At the same time, however, the law can originate from simple harmonic motions (Lacquaniti et al. 1983) such as those output by spinal central pattern generators (CPGs), coupled with the viscoelastic properties of muscles (Gribble and Ostry 1996). Since also linear covariations between segment angular motion can be explained in terms of coupled oscillators with appropriate intersegmental phase (Soechting et al. 1986; Barliya et al 2009), the mechanisms underlying the power law and the segment covariation would be very similar. Notice that Bernstein and Popova (1929), investigating piano strikes movements, had already emphasized the role of intrinsic oscillations. In Chapters 4 and 8 of *OCM*, Bernstein discusses different levels underlying handwriting, an automatic lower-level involving "the oscillator synergy of the forearm, hand, and fingers, which typically proceeds with a clear rhythm" and higher levels that "impose meaningful modulation on these monotonous rhythmical oscillations, thus turning them into non-cyclical letters and words."

Strikingly, it has recently been shown that also the locomotor trajectories of *Drosophila* larvae, which depend on peristaltic oscillatory waves of muscle contractions along the body axis, follow the power law (Zago et al. 2016). Therefore, even rather simple nervous systems are able to produce movement trajectories complying with the power law (Loveless et al. 2019).

In Milan, Lacquaniti and Claudio Maioli worked on another problem related to hierarchical motor control, the problem of coordinate transformations in the control of posture in cats standing on a tilting platform (Lacquaniti and Maioli 1994a, b). They found that cats maintain equilibrium in the face of the postural perturbations by controlling the global geometrical variables of limb length and orientation relative to the vertical. These global variables are mapped in the effector space by means of a planar covariation of the limb joint angles, which constrains the degrees of freedom of limb and body posture. However, the mapping is not one-to-one but one-to-many, since dynamic perturbations (platform tilts) yielding almost identical endpoint trajectories (i.e., identical changes in limb length and orientation) involve trajectories in the plane of joint angles covariation that can vary conspicuously from trial to trial. This idea of motor equivalence is closely related to the uncontrolled manifold hypothesis (Scholz and Schöner 1999), and is reminiscent of the famous observation made by Bernstein that professional blacksmiths hitting a chisel with a hammer showed stereotypical endpoint (hammer) trajectories but variable trajectories of the arm joints (Bernstein 1967). According to Bernstein, they showed "repetition without repetition," that is, repetitive solutions of the task with variable means (see Latash 2012). In Chapter 5 of *OCM*, Bernstein writes "naturally, control levels allow higher variability with respect to indifferent aspects and movement components, compared to those aspects that are of primary importance for the level." This concept is a clear precursor of the uncontrolled manifold hypothesis.

However, not all mappings between endpoint and limb segments admit redundancy of solutions. For instance, a stereotypical planar covariation of the angular motion of lower limb segments has been described for human locomotion (Borghese et al. 1996), allowing a one-to-one mapping between changes in limb length and orientation, on the one hand, and changes in limb segment angular motion, on the other hand (Ivanenko, Cappellini et al. 2007).

Another problem envisioned in the pioneering work of Bernstein is how the nervous system deals with the interaction dynamics in multi-jointed limbs. In Chapter 2 of *OCM*, he states:

> Any force emerging in one of the links immediately brings about a large system of reactive or reflective forces, which are transmitted to all other links. This mutual influence of all the links on each other in all feasible combinations creates altogether a huge number of force interactions. This makes the overall dynamic pattern of movement of the kinematic chain extremely complicated and, most importantly, practically unpredictable because of its extreme mechanical entanglement. Nevertheless, it is possible to make movement of a kinematic chain accurate, at least theoretically, which requires producing incredibly accurate mutual adjustment of its parts.

Later in the same chapter, he states: "it is beyond doubt … that movements are more economical and, therefore, more rational when the organism uses reactive and external force to implement them and, hence, it can add less active muscle forces." These great intuitions have found extensive experimental confirmation in several studies of voluntary movements (e.g., Soechting and Lacquaniti 1981; Hollerbach and Flash 1982; Bastian et al. 1996). Soechting and Lacquaniti also studied the reflex responses of several arm muscles to perturbations applied to the multi-jointed limb (Lacquaniti and Soechting 1984, 1986; Soechting and Lacquaniti 1988). They found that these responses depend on the angular motion and torques at all limb segments, independently of whether the perturbation directly affects a specific muscle. A striking example is provided by the reflex response of anterior deltoid muscle, a shoulder flexor, elicited by perturbations that pronate or supinate the forearm, even though these perturbations clearly do not stretch or shorten this muscle. Importantly, the responses are functional, in so far as they tend to compensate for the interaction dynamics, as hypothesized by Bernstein. These observations have been corroborated and expanded by later work (e.g. Maeda et al. 2017).

Ivanenko's Initial Path under Bernstein's Influence

Yury Ivanenko started his scientific career in 1979 under the supervision of Victor Gurfinkel at the Institute of Information Transmission Problems of the Russian Academy of Sciences, as a student of the Moscow Institute of Physics and Technology. Around that time, in late 1970s – early 1980s, he started to participate in the legendary motor control seminars directed by Mark Shik, one of the collaborators of Victor Gurfinkel, and many people who frequently (every two weeks) attended these seminars were pupils or fellows of Nikolai Bernstein (R. Pearson, Y. Arshavskiy, M. Berkinblit, A. Feldman, G. Orlovski, Y. Levik, and many many others). Victor Gurfinkel was a pupil of Nikolai Bernstein (Gurfinkel 1996) and had the biggest influence for the scientific career of Ivanenko. The works and concepts of Bernstein, such as the hierarchical organization and movement coordination, the notion of motor synergies, motor learning, the concept of muscle tone, etc, were often cited and discussed during these seminars, which were a great source of inspiration and fruitful ideas.

After entering the laboratory of Victor Gurfinkel, in 1980s Ivanenko performed his PhD studies, under supervision of Gurfinkel (in a sense, being a pupil of the pupil of Bernstein), related to the mechanisms of muscle contraction and muscle tone. These studies were significantly inspired by the ideas of Bernstein concerning the definition and concept of tonus. Traditionally, tonus had been linked to the activity level of muscle, and evaluated by the extent to which muscles resist to stretch. Bernstein considered tonus from a novel perspective, not so much as a condition of resistance, but as a condition of readiness. From his point of view, tonus is related to coordination, "as a state is related to an action or as a precondition to an effect" (Bernstein 1940). Bernstein also often refers to the control of tonus as an important component of any movement (e.g., see Chapter 3 of *OCM*). Accordingly, he formulates the definition of tonus: "muscle tonus is not only the current state of muscle tissue stiffness and viscosity and changes in this state, but the whole set of flexible and plastic changes of excitability of muscles during their functioning in an intact organism. Tonus is the flowing state of preparation of the neuro-muscular periphery to selective acceptance of an effector process and its realization" (Chapter 3 of *OCM*). These ideas were further developed by Ivanenko in his PhD studies on the muscle tonus at rest (e.g., Ivanenko 1986; Gurfinkel and Ivanenko 1987), and later when investigating interactions between the tonic state of the neural network and postural and locomotor control (Gurfinkel et al. 1995a; Ivanenko and Gurfinkel 2018; Ivanenko et al. 2013c; Selionov et al. 2009). It was further suggested that any system of motor control resembles a lock that needs two keys for its opening: one key for the rather diffuse activation of tonogenic structures, and the other key to activate a motor program that produces a required movement (and possibly including more specific tuning of this motor program, Gurfinkel et al. 1999). This concept of "*a lock with two keys*" (Gurfinkel et al. 1999) may prove useful in future studies on the neurobiology of motor programs. During the last decades, the importance of the proper tonic physiological state of the spinal networks has received great attention in the context of the mechanisms of recovery of locomotor function (Courtine et al. 2009; Edgerton and Roy 2012; Ivanenko et al. 2017). While the term "*tonus*" unites many phenomena, Bernstein's original interpretation (Chapter 3 of *OCM*) and the further developments in modern studies may be fruitful for various implications. However, it still needs to be better understood and investigated.

The concept of a multilevel hierarchical system for movement coordination, as a complete theory, was published in Bernstein's book *OCM* and represents a major scientific heritage of

Bernstein. Following these ideas, Ivanenko performed various studies with Victor Gurfinkel on human postural regulation and different levels of its control: one level sets a distribution of tonic muscle activity ("*posture*") and the other is assigned to compensate for internal or external perturbations ("*equilibrium*," Gurfinkel et al. 1995b; Ivanenko and Gurfinkel 2018). In this framework, Gurfinkel also devoted a large amount of work to the study of the system of the internal representation in the human brain, in order to comprehend how the brain forms the models on the basis of which a motor behavior is realized, in particular by studying spatially oriented postural automatic responses (Gurfinkel 1994). Ivanenko continued these works with Berthoz on the perception of movement (Ivanenko et al. 1997) and later with Lacquaniti on internal representations in motor control by investigating postural (Ivanenko et al. 1999) and locomotor (Ivanenko et al. 2011) responses in humans. For instance, these studies suggested the existence of a dynamic locomotor body schema used for controlling step length and path estimation based on a priori knowledge about intrinsic dynamics of limb segment motion and inherent relationships between gait parameters and body proportions (Ivanenko et al. 2011).

It is also worth noting that, in seventies and eighties, many people visited the Institute of Information Transmission Problems in Moscow, e.g., Sten Grillner, Shigemi Mori, Reggie Edgerton, Alain Berthoz, Jean Massion, and others. Many of them became leaders in their fields, the founders of their own scientific areas, but Nikolai Bernstein's and Victor Gurfinkel's concepts continue to serve as a source of fruitful ideas for many scientists. It is worth mentioning that, despite belonging to the scientific school of Nikolai Bernstein and his followers, only after leaving Russia and coming to the laboratory of Alain Berthoz at the Collége de France in Paris in 1995 and then to the laboratory of Francesco Lacquaniti at the Santa Lucia Foundation in Rome in 1998, did Ivanenko realize how well known and influential were Bernstein's ideas abroad, and what influence his ideas had on the progress in motor control studies.

East and West Meet Together

Lacquaniti and Ivanenko collaborated in several projects over the years. Here, we only consider a few research lines on locomotion directly related to Bernstein's work: identification of locomotor muscle synergies, developmental and evolutionary modularity. Among Bernstein's many interests, the study of human locomotion attracted him as an example of a phylogenetically old, automatic, stable behavior that involves extensive synergies of multiple limb and body muscles. The concept of synergy recurs frequently in *OCM*. It is formulated in rather general, abstract terms, mostly related to the hypothesis of hierarchical control of movements. In Chapter 4, he specifically assigns the organization of multi-muscle synergies and patterns to the structural-functional thalamo-pallidal level of the hierarchy. In Chapter 5, he contrasts the high variability of indifferent movement components (see above, with regards to motor equivalence) with the small variability and flexibility of movements at the level of synergies. He specifically mentions that "cyclogrammetric measurements of many highly automatized cyclical movements, such as walking, running, filing, etc., have shown that variability of consecutive synergic cycles usually does not exceed a few millimeters along spatial coordinates of the trajectory and a few milliseconds in their duration."

Muscle synergies. There are different formulations of the synergy hypothesis (Grillner 1985; Bizzi et al. 2008; Latash 2012; Profeta and Turvey 2018). Synergies are also at the basis of the principle of abundance (Gelfand and Latash 1998) and Uncontrolled Manifold (UCM) hypothesis (Scholz and Schöner 1999). In different versions, synergies are key to the solution of the degrees of freedom problem proposed by Bernstein. Following the framework of coordination and muscle synergies, Lacquaniti and Ivanenko performed several experiments at the Santa Lucia Foundation in Rome,

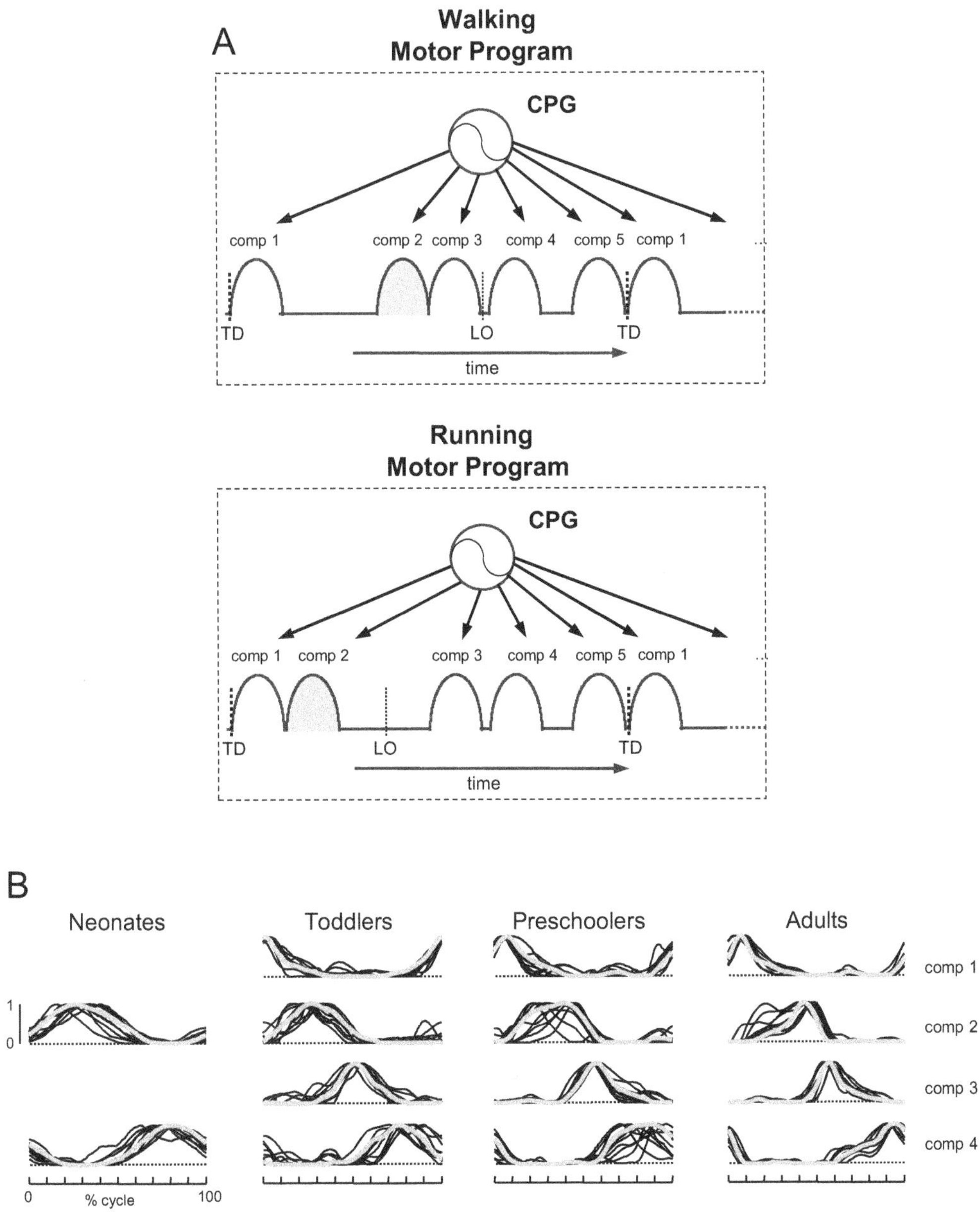

FIGURE C9.1A Motor synergies (modular organization of the muscle control) in human locomotion. A – locomotion motor program as a sequence of activation pulses for walking and running in humans. (Source: Cappellini et al. 2006). B – basic patterns from EMG profiles in 10 subjects of each group (neonates, toddlers, preschoolers and adults) obtained by decomposing bilateral multimuscle activity. During development the two basic pattern of stepping neonates are retained, augmented by two new patterns first revealed in toddlers. (Source: Dominici et al. 2011).

starting from 2003 with a very fruitful collaboration with Richard Poppele visiting our laboratory from Minneapolis. This set of studies was focused on muscle synergies and multimuscle control of human locomotion. Locomotion requires the activity of many muscles and the question of how the central nervous system coordinates this activity is central to an understanding of motor control. The so-called muscle modules (basic temporal patterns of lumbosacral motoneurone pool activity and corresponding muscle synergies) can be identified from multimuscle electromyographic recordings using factorization algorithms. We performed experimental studies during various conditions (walking at different speeds, with body weight unloading, running, hopping, skipping, etc.). These findings suggested that locomotor movements are controlled using a small set of basic temporal components or activation patterns (Figure C9.1A), shared by several different muscles and reflecting global kinematic and kinetic goals (Ivanenko et al. 2004, 2005, 2013a; Lacquaniti et al. 2012a). Spatiotemporal maps of spinal cord motoneuron activation also show discrete periods of motor pool activity linked to specific kinematic and kinetic events in the gait cycle (Ivanenko et al. 2006).

This framework proved to be very useful to study coordination of locomotor movements and multiple degrees of freedom at the muscle level, although there are still open questions about modular strategies for locomotion (Zelik et al. 2014) and neural substrates for muscle modules. For instance, the coordination of locomotion with voluntary tasks might be accomplished through a superposition of motor programs or activation timings that are separately associated with each task (Ivanenko et al. 2005). Furthermore, this approach also demonstrated plasticity in muscle synergies and basic activation patterns in neurological disorders involving a motor deficit (Cappellini et al. 2016; Grasso et al. 2004; Ivanenko et al. 2013a; Martino et al. 2015).

Development. The approach outlined above also revealed some principles of functional organization of the spinal locomotor output during early development in neonates and infants (Dominici et al. 2011; Lacquaniti et al. 2012b). In particular, the development of human locomotion from the neonate to the adult starts from a rostrocaudal excitability gradient and involves an increase in the complexity of the spinal locomotor output (Dominici et al. 2011; Ivanenko et al. 2013b). During development, the two basic patterns of stepping neonates are retained, augmented by two new patterns first revealed in toddlers (Figure C9.1B).

Bernstein (1967) proposed that, when people start learning a skill, they may restrict the number of degrees of freedom to reduce the size of the search space and simplify the coordination problem. We experimentally explored this topic by investigating the kinematic locomotor synergy including its development and evolutionary considerations. Figure C9.2B illustrates the planar covariation of limb segment elevation angles assessed by the principal component analysis. In adults, the full limb behavior in all gaits we considered can be expressed as the 2 degrees of freedom (dof) of planar covariation for each gait, plus the rotation of the covariation plane about a specific axis (1 dof), rotation that depends on the condition (Ivanenko, Cappellini et al. 2007). This extends the analysis to a full 3 dofs spatial control of locomotion, in which the third dimension may determine the gait pattern or its adaptation to different support surface conditions. In toddlers, the maintenance of a roughly constant planar covariation (Figure C9.2A) reduces the flexibility of the kinematic pattern, and thus restricts the manifold of angular segment motion. This coordination strategy is consistent with the exploratory learning hypothesis of Bernstein (1967) according to which children learn a task by first reducing the degrees of freedom.

Bernstein emphasizes that "during the second year of life, toddler's locomotion is neither walking, nor running, but something not yet differentiated" (Chapter 7 of *OCM*). Indeed, our extensive observations carried out on children over the years are consistent with this idea. Locomotor patterns in children are immature and lack adaptive features. For instance, despite millions of years of bipedal walking evolution, the gravity-related pendulum mechanism of walking does not seem to be implemented at the onset of independent walking, requiring each toddler to develop

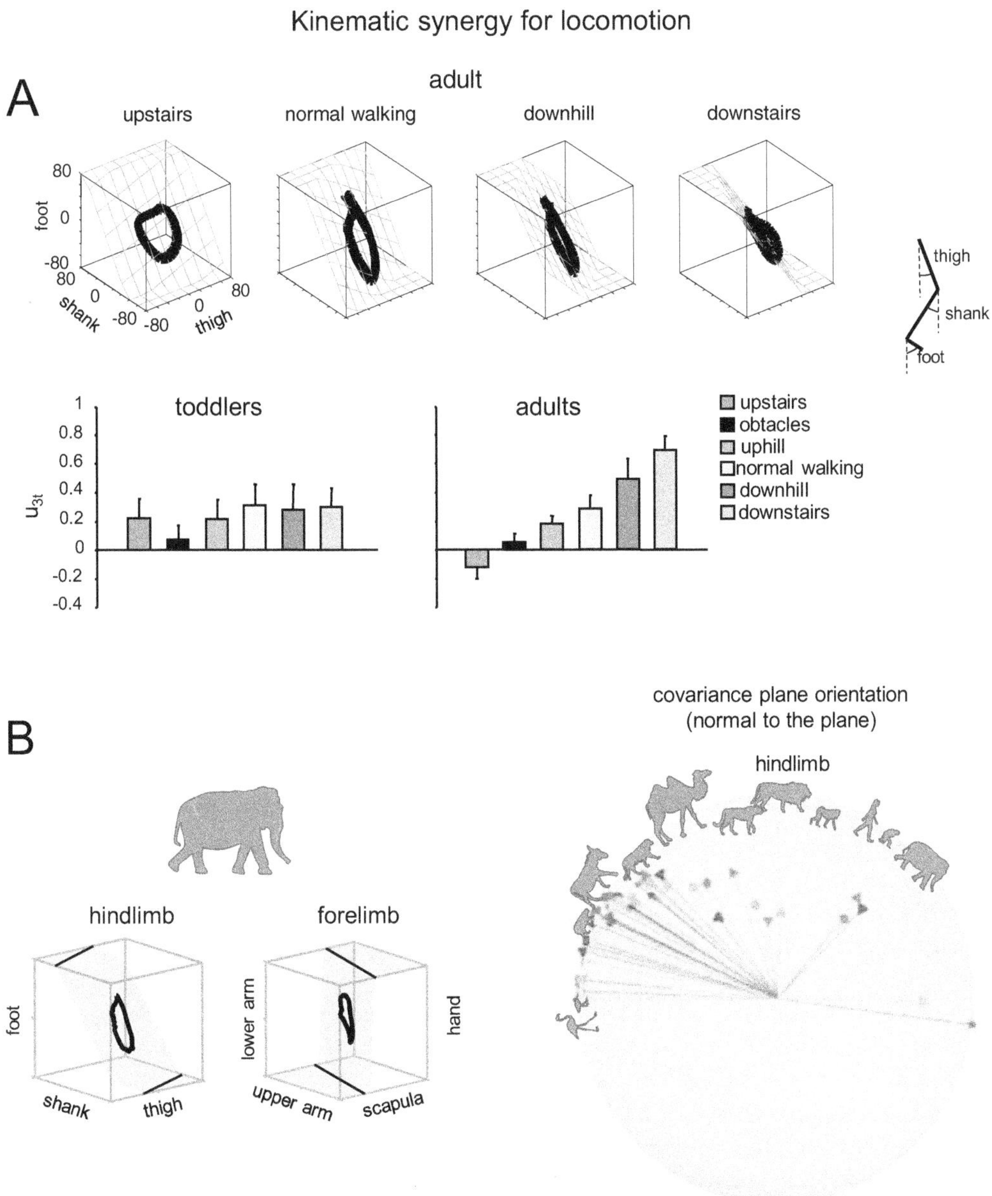

FIGURE C9.2A Kinematic synergies for locomotion.

A – intersegmental coordination assessed by principal component analysis of limb segment (thigh, shank, foot) elevation angles during walking over different surfaces in adults and toddlers.

Upper panels illustrate examples of 3D trajectories in segment angle space along with the interpolated planes in one adult. Lower panels illustrate changes in the orientation of covariation plane (u_{3t}, projection of the normal to the plane onto the thigh axis, mean+SD). Note a lack of adaptability to different locomotor conditions in toddlers. (Source: Dominici et al. 2010).

B – kinematic synergy involving the planar covariation oflimb segment motion holds in 54 different animal species, despite large diff erences in body size, mass, limb confi guration, and amplitude of movements.The left panel illustrates an example of gait loops with corresponding interpolation planes for one mammal (elephant). Note diff erent plane orientation for the hindlimb and forelimb. For the posterior limb, the right panel illustrates the characteristic of this kinematic synergy (specific covariation plane orientation) adopted by different species. (Source: Catavitello et al. 2018).

it progressively starting from about two years of age (Ivanenko et al. 2007b). Bernstein refers to active synthetic exploration by the baby, first of the spatial contours of its own body and then of the external spatial field (Chapter 7 of *OCM*). Our studies corroborate this hypothesis (Dominici et al. 2010) and show that in infants the maintenance of a roughly constant planar covariance – irrespective of the surface inclination and height (Figure C9.2A) – may be functional to the exploratory behavior. Also, the toddlers often tend to place the foot onto the obstacle or across the edges of the stairs (Dominici et al. 2010), and such foot placements can be interpreted as part of a haptic exploratory repertoire.

Bernstein also gave special attention to comparative and evolutionary findings (Chapter 1 of *OCM*). Related approaches have been recently resurrected in view of the idea that comparative analyses may reveal common principles of limb movement control. In particular, we discovered that the kinematic synergy involving the planar covariation of limb segment motion (Figure C9.2B) holds in 54 different animal species (44 species of mammals and 10 species of birds), despite large differences in body size, mass, limb configuration, and amplitude of movements (Catavitello et al. 2018). This kinematic synergy may represent one neuromechanical principle conserved in evolution to save mechanical energy and optimize the multisegment limb control.

Conclusion

Bernstein wrote in the Preface of *OCM*: "this book is, to a larger degree, a program for future urgently needed studies rather than a dogmatic summary of firmly established facts" and later "the author would be satisfied by appreciation that this book played the role of one of the modest bricks in the foundation of future physiology." Nothing could have been more prescient! When speaking about the scientific heritage of Bernstein, one should consider what influence his ideas had on subsequent progress in motor control studies and that his works still continue to be highly cited despite the long time elapsed since they were published (see also other commentaries to *OCM*). It should be stressed that, in addition to his great insights, Bernstein left an enduring legacy on experimentalists such as ourselves for his emphasis on careful, often highly innovative methodology, constant iteration between hypothesis, experimental design, interpretation of results, and revision of the hypothesis. Although Bernstein's ideas are perhaps better read today in a metaphorical rather than literal sense, his outside-in approach has been inspiring especially to counterbalance the inside-out approach of Sherrington, Adrian and others, which has long dominated Western neurophysiology (Stuart 2005). Another important legacy of Bernstein is represented by his optimism about future developments of the field. He had the dream that all aspects of motor coordination and planning would finally merge into one comprehensive science of a mathematical nature. Current computational efforts are inspired by Bernstein's optimism, as shown by the return to attention of his ideas about the hierarchical organization of the nervous system (Profeta and Turvey 2018; Merel et al. 2019).

Bernstein was a man of science with a great intuition, independence of opinion and extraordinary personality. The history that has shaped the interdisciplinary field of motor control may also guide us into the next era of this fascinating field of research. In this commentary, we have reported how his ideas influenced our own scientific work and how they are relevant to our current research directions. We are aware that our work only represents an intermediate step and will likely be superseded by future work. Nevertheless, it provides one testimony of the enduring influence of Bernstein's general program of research.

References

Barliya A, Omlor L, Giese MA, Flash T (2009) An analytical formulation of the law of intersegmental coordination during human locomotion. Exp Brain Res 193: 371–385.

Bastian AJ, Martin TA, Keating JG, Thach WT (1996) Cerebellar ataxia: abnormal control of interaction torques across multiple joints. *J Neurophysiol* 76: 492–509.

Bernstein N (1967) *The Co-ordination and Regulation of Movements*. Oxford: Pergamon Press.

Bernstein N, Popova TS (1929) Untersuchung der Biodynamik des Klavieranschlags. *Arbeitsophysiologie* 1: 396–432.

Bernstein NA (1940) Studies of the Biodynamics of Walking, Running and Jumping. Moscow, Researches of the Central Scientific Institute of Physical Culture. (In Russian). English translation In H.T.A. Whiting (Ed.) (1984), *Human Motor Actions. Bernstein Reassessed*, Amsterdam, North-Holland, pp. 171–222.

Bizzi E, Cheung VC, d'Avella A, Saltiel P, Tresch M. (2008) Combining modules for movement. *Brain Res Rev*. 57: 125–133.

Borghese NA, Bianchi L and Lacquaniti F (1996) Kinematic determinants of human locomotion. *J Physiol* 494: 863–879.

Cappellini G, Ivanenko YP, Martino G, MacLellan MJ, Sacco A, Morelli D, Lacquaniti F (2016) Immature spinal locomotor output in children with cerebral palsy. *Front Physiol* 7: 478.

Cappellini G, Ivanenko YP, Poppele RE, Lacquaniti F (2006) Motor patterns in human walking and running. *J Neurophysiol* 95: 3426–3437.

Catavitello G, Ivanenko Y, Lacquaniti F (2018) A kinematic synergy for terrestrial locomotion shared by mammals and birds. *eLife* 7: e38190.

Courtine G, Gerasimenko Y, van den Brand R, Yew A, Musienko P, Zhong H, Song B, Ao Y, Ichiyama RM, Lavrov I, Roy RR, Sofroniew MV, Edgerton VR (2009) Transformation of nonfunctional spinal circuits into functional states after the loss of brain input. *Nat Neurosci* 12: 1333–1342.

Dominici N, Ivanenko YP, Cappellini G, d'Avella A, Mondì V, Cicchese M, Fabiano A, Silei T, Di Paolo A, Giannini C, Poppele RE, Lacquaniti F (2011) Locomotor primitives in newborn babies and their development. *Science* 334: 997–999.

Dominici N, Ivanenko YP, Cappellini G, Zampagni ML, Lacquaniti F (2010) Kinematic strategies in newly walking toddlers stepping over different support surfaces. *J Neurophysiol* 103: 1673–1684.

Edgerton VR, Roy RR (2012) A new age for rehabilitation. *Eur J Phys Rehabil Med* 48: 99–109.

Gelfand IM, Latash ML (1998) On the problem of adequate language in movement science. *Motor Control* 2: 306–313.

Grasso R, Ivanenko YP, Zago M, Molinari M, Scivoletto G, Castellano V, Macellari V, Lacquaniti F (2004) Distributed plasticity of locomotor pattern generators in spinal cord injured patients. *Brain* 127: 1019–1034.

Gribble PL, Ostry DJ (1996) Origins of the power law relation between movement velocity and curvature: modeling the effects of muscle mechanics and limb dynamics. *J Neurophysiol* 76: 2853–2860.

Grillner S (1985) Neurobiological bases of rhythmic motor acts in vertebrates. *Science* 228: 143–149.

Gurfinkel VS (1994) The mechanisms of postural regulation in man. In: *Physiology and General Biology Reviews*. T. M. Turpaev & N. K. Koltsov (Eds.). USA: Harwood Academic Publishers GmbH, pp. 59–87.

Gurfinkel VS (1996) Bernstein and modern physiology of motor control (On his 100th birthday). *Human Physiology* 22(6): 751–756 (translated from Fiziologiya Cheloveka, vol. 22, N6, pp. 124–130).

Gurfinkel VS, Ivanenko YP (1987) Low threshold reactions in human muscles at rest. *Human Physiology* 13(2): 317–323.

Gurfinkel VS, Ivanenko YP, Levik YS (1995a) The influence of head rotation on human upright posture during balanced bilateral vibration. *Neuroreport* 7: 137–140.

Gurfinkel VS, Ivanenko YP, Levik YS, Babakova IA (1995b) Kinesthetic reference for human orthograde posture. *Neuroscience* 68: 229–243.

Gurfinkel VS, Ivanenko YP, Levik YS, Kazennikov OV, Selionov VA (1999) The neural control of posture and locomotion: a lock with two keys. In: *Motor Control Today and Tomorrow*. Gantchev GN, Mori S, Massion J (eds), pp. 113–121.

Hollerbach MJ, Flash T (1982) Dynamic interactions between limb segments during planar arm movement. *Biol Cybern* 44: 67–77.

Ivanenko YP (1986) Friction and thixotropy of passive skeletal muscles. *Human Physiology* 12(2): 229–235.

Ivanenko YP, Cappellini G, Dominici N, Poppele RE, Lacquaniti F (2007) Modular control of limb movements during human locomotion. *J Neurosci* 27: 11149–11161.

Ivanenko Y, Gurfinkel VS (2018) Human postural control. *Front Neurosci* 12: 171.

Ivanenko YP, Cappellini G, Dominici N, Poppele RE, Lacquaniti F (2005) Coordination of locomotion with voluntary movements in humans. *J Neurosci* 25: 7238–7253.

Ivanenko YP, Cappellini G, Solopova IA, Grishin AA, Maclellan MJ, Poppele RE, Lacquaniti F (2013a) Plasticity and modular control of locomotor patterns in neurological disorders with motor deficits. *Front Comput Neurosci* 7: 123.

Ivanenko YP, Dominici N, Cappellini G, Di Paolo A, Giannini C, Poppele RE, Lacquaniti F (2013b) Changes in the spinal segmental motor output for stepping during development from infant to adult. *J Neurosci* 33: 3025–3036a.

Ivanenko YP, Wright WG, St George RJ, Gurfinkel VS (2013c) Trunk orientation, stability, and quadrupedalism. *Front Neurol* 4: 20.

Ivanenko YP, Dominici N, Daprati E, Nico D, Cappellini G, Lacquaniti F (2011) Locomotor body scheme. *Hum Mov Sci* 30: 341–351.

Ivanenko YP, Dominici N, Lacquaniti F (2007b) Development of independent walking in toddlers. *Exerc Sport Sci Rev* 35: 67–73.

Ivanenko YP, Grasso R, Israël I, Berthoz A (1997) The contribution of otoliths and semicircular canals to the perception of two-dimensional passive whole-body motion in humans. *J Physiol (Lond)* 502: 223–233.

Ivanenko YP, Grasso R, Lacquaniti F (1999) Effect of gaze on postural responses to neck proprioceptive and vestibular stimulation in humans. *J Physiol (Lond)* 519 Pt 1: 301–314.

Ivanenko YP, Gurfinkel VS, Selionov VA, Solopova IA, Sylos-Labini F, Guertin PA, Lacquaniti F (2017) Tonic and rhythmic spinal activity underlying locomotion. *Curr Pharm Des* 23: 1753–1763.

Ivanenko YP, Poppele RE, Lacquaniti F (2004) Five basic muscle activation patterns account for muscle activity during human locomotion. *J Physiol (Lond)* 556: 267–282.

Ivanenko YP, Poppele RE, Lacquaniti F (2006) Spinal cord maps of spatiotemporal alpha-motoneuron activation in humans walking at different speeds. *J Neurophysiol* 95: 602–618.Lacquaniti F (1989) Central representations of human limb movement as revealed by studies of drawing and handwriting. *Trends in Neurosciences* 12: 287–291.

Lacquaniti F, Ferrigno G, Pedotti A, Soechting J F, Terzuolo C (1987) Changes in spatial scale in drawing and handwriting: kinematic contributions by proximal and distal joints. *J Neurosci* 7: 819–828.

Lacquaniti F, Ivanenko YP, Zago M (2012a) Patterned control of human locomotion. *J Physiol (Lond)* 590: 2189–2199.

Lacquaniti F, Ivanenko YP, Zago M (2012b) Development of human locomotion. *Curr Opin Neurobiol* 22: 822–828.

Lacquaniti F, Maioli C (1994a) Coordinate transformations in the control of cat posture. *J Neurophysiol* 72: 1496–1515.

Lacquaniti F, Maioli C (1994b) Independent control of limb position and contact forces in cat posture. *J Neurophysiol* 72: 1476–1495.

Lacquaniti F, Soechting JF (1984) Behavior of the stretch reflex in a multi-jointed limb. *Brain Res.* 311:161–166.

Lacquaniti F, Soechting JF (1982) Coordination of arm and wrist motion during a reaching task. *J Neurosci* 2: 399–408.

Lacquaniti F, Soechting JF (1986) EMG responses to load perturbations of the upper limb: Effect of dynamic coupling between shoulder and elbow motion. *Exp Brain Res* 61:482–496.

Lacquaniti F, Soechting JF, Terzuolo SA (1986) Path constraints on point-to-point arm movements in three-dimensional space. *Neuroscience* 17: 313–324.

Lacquaniti F, Terzuolo C, Viviani P (1983) The law relating the kinematic and figural aspects of drawing movements. *Acta psychologica* 54: 115–130.

Latash M (2012) The bliss (not the problem) of motor abundance (not redundancy). *Exp Brain Res* 217: 1–5.

Loveless J, Lagogiannis K, Webb B (2019) Modelling the mechanics of exploration in larval Drosophila. *PLoS Comput Biol* 15(7): e1006635.

Maeda RS, Cluff T, Gribble PL, Pruszynski JA (2017) Compensating for intersegmental dynamics across the shoulder, elbow, and wrist joints during feedforward and feedback control. *J Neurophysiol* 118: 1984–1997.

Martino G, Ivanenko YP, d'Avella A, Serrao M, Ranavolo A, Draicchio F, Cappellini G, Casali C, Lacquaniti F (2015) Neuromuscular adjustments of gait associated with unstable conditions. *J Neurophysiol* 114: 2867–2882.

Merel J, Botvinick M, Wayne G (2019 Hierarchical motor control in mammals and machines. Nat Commun. 10(1): 5489.

Profeta VLS, Turvey MT (2018) Bernstein's levels of movement construction: A contemporary perspective. *Hum Mov Sci* 57: 111–133.

Scholz JP, Schöner G (1999) The uncontrolled manifold concept: identifying control variables for a functional task. *Exp Brain Res* 126: 289–306.

Schwartz AB (1994) Direct cortical representation of drawing. *Science* 265: 540–542.

Selionov VA, Ivanenko YP, Solopova IA, Gurfinkel VS (2009) Tonic central and sensory stimuli facilitate involuntary air-stepping in humans. *J Neurophysiol* 101: 2847–2858.

Soechting JF, Lacquaniti F (1981) Invariant characteristics of a pointing movement in man. *J Neurosci* 1: 710–720.

Soechting JF, Lacquaniti F (1988) Quantitative evaluation of the electromyographic responses to multidirectional load perturbations of the human arm. *J Neurophysiol* 59: 1296–1313.

Soechting JF, Lacquaniti F, Terzuolo CA (1986) Coordination of arm movements in three-dimensional space. Sensorimotor mapping during drawing movement. *Neuroscience* 17: 295–311.

Stuart DG (2005) Integration of posture and movement: contributions of Sherrington, Hess, and Bernstein. *Hum Mov Sci* 24: 621–643.

Terzuolo CA, Viviani O (1980) Determinants and characteristics of motor patterns used for typing. *Neuroscience* 5: 1085–1103.

Zago M, Lacquaniti F, Gomez-Marin A (2016) The speed–curvature power law in Drosophila larval locomotion. *Biology letters* 12: pii: 20160597.

Zelik KE, La Scaleia V, Ivanenko YP, Lacquaniti F (2014) Can modular strategies simplify neural control of multidirectional human locomotion? *J Neurophysiol* 111: 1686–1702.

C10

GOAL-DIRECTED ACTION AND THE ARCHITECTURE OF MOVEMENT ORGANIZATION

Vitor L. S. Profeta

SCHOOL OF PHYSICAL EDUCATION, PHYSIOTHERAPY AND OCCUPATIONAL THERAPY, UNIVERSIDADE FEDERAL DE MINAS GERAIS

Michael T. Turvey

CENTER FOR THE ECOLOGICAL STUDY OF PERCEPTION AND ACTION
UNIVERSITY OF CONNECTICUT

Claudia Carello

CENTER FOR THE ECOLOGICAL STUDY OF PERCEPTION AND ACTION
UNIVERSITY OF CONNECTICUT

Introduction

The ecological approach to perception championed by James Gibson (1966, 1979/1986) was long notable in psychology for being oriented around activity. Even at the early stages in the development of Gibson's theory, as he tried to come to terms with the challenges that a theory of perception would have to accommodate, he provided analyses of actions: driving a car (Gibson and Crooks 1938), landing a plane (Gibson 1947), and controlling locomotion (Gibson 1958). Importantly, he appreciated that the perception-action relationship should be characterized as one of mutuality, not dominance by either achievement: "Spatial behavior and spatial perception are coordinate with one another. We have neither to see space before we can behave nor to make spatial responses before we can see" (Gibson 1950, p. 225). Reflecting his concern for how organisms make their way in the environment, his theory of perception is unique for being directed not at low-level sensory elements but at rich optical structure specific to the organism-environment relation about which it informs.

Psychology during his lifetime – and for quite some time thereafter – conspicuously lacked *any* theory of action, let alone one that was comparably rich (Rosenbaum 2005; Turvey 1974). Even theories that purported to foreground motor activity (e.g., Liberman, Cooper, Shankweiler, and Studdert-Kennedy 1967) did so with little regard for how acts are produced: "Curiously, action-based theories of perception and of mind in general have been advanced on a nonexistent theory of action" (Turvey 1977, p. 212). Consequently, when the English translation of Bernstein's *The Coordination and Regulation of Movements* (1967) began to circulate, it was mined by ecologically oriented psychologists with enthusiasm. Particularly appealing features included Bernstein's highlighting the computational obstacle presented by the very many degrees of freedom that are coordinated in implementing an action and, relatedly, his distinguishing an intentional action from the incidental muscle activity that accompanies it. Indeed, these issues are not only at the heart of what was missing from treatments of behavior in psychology but provided guideposts on how to build an appropriate theory (cf. Turvey 1977, 2013). As that theory has been forged, it has become notorious in favoring self-organizing physics over the computational machinery of nervous system in order to address how goal-directed actions emerge as degrees of freedom are harnessed.

With this new translation of *On the Construction of Movements,* we consider how to position Bernstein's (new) insights with respect to how the theory of action has developed within ecological psychology over the past 45 years, particularly in light of how that development was influenced in no small measure by his insights (Turvey 1990). We have chosen to focus on the architecture of movement organization qua action. The rationale for our focus is two-fold. First, this issue emerged as an explicit concern for understanding not only coordination but the relationship between perception and action as well as organism and environment generally (Turvey, Shaw, and Mace 1978; Michaels and Carello 1981; Shaw and Turvey 1981). Second, Bernstein's favored hierarchical architecture, which essentially dictates the organization of his book, seems to overvalue the role played by the nervous system in coordination. As noted, such an emphasis has been eschewed by those inspired by Bernstein (e.g., Kugler and Turvey 1987/2015; Profeta and Turvey 2018; Turvey 1990).

Bernstein's Hierarchy

Bernstein's attempt to relate structure and function in a way that suits a meaningful environment is kindred with the ecological emphasis on describing organisms relative to their environments and environments relative to their organisms (Gibson 1979). For the latter, this emphasis is apparent in the complementary concepts of affordance (opportunity for action offered by the environment; Gibson 1979/1986) and effectivity (capability to perform an action that is offered by the environment; Turvey, Shaw, Reed, and Mace 1981). In essence, the ecological approach considers the organism-environment relation, rather than just the organism or just the nervous system, as the unit of analysis (Turvey and Carello 1981, 2012). Against the backdrop of the evolution of the nervous system, however, Bernstein (1947/2020) proposed that what we would call *perception-action systems*[1] organize their actions hierarchically.

Bernstein's hierarchy provides a logical structure to understand how the large number of degrees of freedom of the motor apparatus can be constrained to support goal-directed movements (i.e., actions) in animals. In the proposed hierarchy, the central nervous system (CNS) is responsible for movement construction. Bernstein suggested that the CNS evolved in stages, with newer levels emerging from older levels. The hierarchy is composed of five levels, each accounting for a different class of motor problems. According to the action challenge that one is facing, the newest level involved in movement production takes the role of leading level, while older levels play the role of background levels.

Each level in Bernstein's hierarchy performs a specific function to allow a requisite action. Level A is responsible for regulating the state of the motor apparatus in order to facilitate muscular control. By modulating the state of muscles, Level A assures the responsiveness of the motor apparatus to influences from the higher levels. Level B establishes relations among parts of the motor apparatus. It is the level directly responsible for dealing with the degrees of freedom problem by forming functional units of interdependent elements (i.e., synergies) that, due to sensory inflow,[2] adapt to their fluctuations in order to produce consistent and repeatable movements. Level C refers to goal-directed movements happening relative to the surroundings. It reflects awareness of an objective environment through the detection of information about the surroundings. Level D refers to chains of movements oriented by a semantic relation with objects. It deals with problems that require many steps to be solved. Finally, Level E is similar to Level D in regards to chains of movements, but its semantics is more abstract (for example, rather than being constrained by an object to be used it might be constrained by the expression of emotion in dance). Cognizant of the importance of perceptual anchoring of actions, Bernstein ascribed to each level of movement construction specific sensory systems that would be responsible for adaptive corrections. These corrections are particularly relevant at Level B due to its essential role in coordination.

Bernstein's hierarchy is about dealing with the multitudinous degrees of freedom of perception-action systems. Or, as Bernstein put it, "movement coordination is overcoming excessive degrees of freedom of moving organs or its transformation into a controllable system" (this volume, p. 28). From an ecological standpoint, the problem of degrees of freedom cannot be considered independently from the problem of the degrees of constraints (Turvey et al. 1978). It is in this dual consideration that an architecture of movement organization must be understood. In what follows, we try to show how hierarchies (as logical types) may support the emergence of soft-assembled synergies, that is, functional units that arise flexibly.[3] We then expand the discussion to include heterarchies and coalitions, organizational styles that favor the requisite flexibility. Finally, we present examples that illustrate the limits for any strictly neural-based explanation of goal-directed movements.

Hierarchies

Hierarchies like the one proposed by Bernstein are common in many areas of science. Their relevance led to the rise of a *theory of hierarchies* (see, for example, Salthe 1985; Simon 1962), whose main purpose is to describe the features that hierarchical systems share. In general, hierarchies are considered coherent wholes composed of different levels, characterized by processes and events that have different characteristic times (Simon 1962). For the sake of simplicity, consider four constitutive, ordered levels, with activity at Level 1 the lowest, fastest and most noisy and activity at Level 4 the highest, slowest and least noisy. The changes of a lower level relative to a higher level are noisy fluctuations and its effects are observed only in the long term due to aggregation (Salthe 1985). The changes of a higher level relative to a lower level are constants, that is, they are structural constraints on the lower level.

The relative independence between levels implies that a hierarchical system has relatively weak vertical coupling between levels (Simon 1962, 1973). The implication is that any given level can be analyzed relatively independently of the other levels. It is of note that Bernstein's hierarchy sits on a single time scale: *It is not defined in terms of differences in temporal-spatial scales, and its levels are subsystems populating a common level (the level of the nervous system).*

A subsystem (as in Bernstein's hierarchy) is encapsulated, meaning interactions within it are stronger than its interactions with other subsystems. Thus, horizontal coupling is also weak. Consequently, the processes that happen in a particular subsystem occur relatively independently of the processes that occur in other subsystems at the same level, ascribing a certain degree of autonomy to subsystems to perform their functions (Allen and Starr 1982; Simon 1962; see also Gelfand, Gurfinkel, Tsetlin, and Shik 1971). Each subsystem is composed of other subsystems,

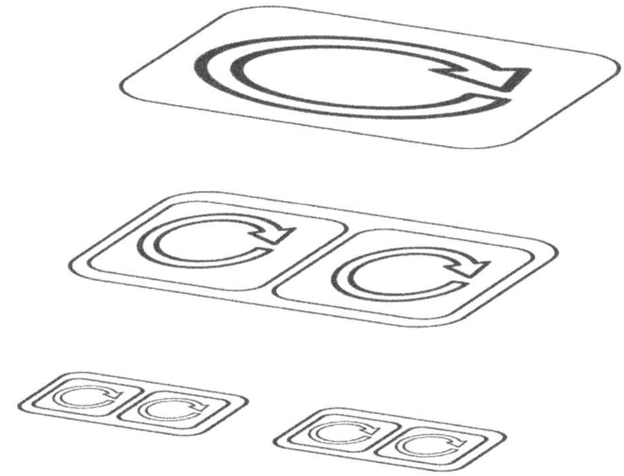

FIGURE C10.1 A nested hierarchical organization.
The top level of the hierarchy constrains the middle level, which is composed of two subsystems (rounded rectangles). Each subsystem at the middle level constrains and encompasses specific subsystems at the bottom level. The arrows represent the process inside each subsystem.

characterizing what Allen and Starr (1982) defined as a nested hierarchy: Elements of a lower level are encapsulated by elements of the level above (Figure C10.1).

Soft-assembled, Task-specific Hierarchies

Hierarchies are organizations of control; they define the direction of influence in a perception-action system and, consequently, how degrees of freedom are constrained at each level. Note, however, that the same subsystems may be combined in different synergies (i.e., may be softly assembled) to bring about different behaviors. That is, despite the autonomy of each subsystem, it is the interaction among subsystems that contributes the most to observed behavior, not the mechanisms and processes within each subsystem (Turvey and Carello 1996). In other words, no processes of a subsystem dominate the emergence of behavior.

To some extent, this idea is in Bernstein's hierarchy when he establishes the need for a leader-background level relationship. From an ecological standpoint, this concept needs to be radicalized in terms of a full-blown interaction-dominant approach to behavior (Van Orden, Holden, and Turvey 2003). That is, subsystems at different temporal and spatial scales may interact in a perception-action system in order to bring about a given behavior (cf. Gottlieb 1998). Consequently, the synergies that emerge from the interaction of subsystems are temporary, and their components are recruited explicitly to perform a particular task (Wagman and Hajnal 2016). That is, synergies are soft-assembled and function-specific rather than anatomy-specific (Turvey 1986). They do not exist off-line in a dormant state, waiting to be activated in a particular task (Kloos and Van Orden 2009).

Non-anatomical specification ascribes to soft-assembled synergies the capability of adapting when perturbed, that is, the relationship among the elements of a structure are adjusted to achieve a goal (Kelso 2009). For example, introducing a mechanical perturbation to the jaw during a participant's utterances of *bab* and *baz* induces an immediate remote adjustment (Kelso, Tuller, Vatikiotis-Bateson, and Fowler 1984). That is, depending on which utterance is being produced when the jaw is disturbed, adaptations can occur at the tongue, the upper lip, or the lower lip. The implication is that in order to preserve the functional integrity of an action, each element applies its own interactive capacity within the embedding context (Shaiman and Gracco 2002), and adapts according to the circumstances (Kelso 2009; Turvey 2007).

Soft-assembled synergies such as the one described by Kelso et al., with the context-dependence of their elements' functions and their immediate remote responses to perturbations, are difficult to explain by considering the nervous system exclusively. The mechanical properties of the muscle-tendon-fascia-skeleton system that contribute to movement stabilization and adaptation have to be considered as well (Loeb 1995, Valero-Cuevas, Yi, Brown, McNamara, Paul, and Lipson 2007). This premise is encompassed in the concept of biotensegrity (Levin 2006; Turvey and Fonseca 2014), which has been suggested to be the material resource for the emergence of synergies (Profeta and Turvey 2018; Turvey and Fonseca 2014).

A biotensegrity system is characterized by the capacity to sustain its mechanical integrity by continuous tension over discontinuous compression (Levin 2006). The stability of a biotensegrity system is obtained by the distribution of mechanical forces over the entire system, assuring responsiveness even at rest due to its prestressed condition (Turvey and Fonseca 2014). The organization of biotensegrity systems is nested, and interactions occur within and across levels. For instance, a biotensegrity model for the temporomandibular joint has been proposed by Scarr and Harrison (2017). The authors suggested that their model of the temporomandibular joint is a biotensengrity subsystem that interacts with other biotensegrity subsystems across the body. If other subsystems involved in the production of an utterance are also mechanically defined in terms of biotensegrity, then speech is characterized by the soft-assemblage of biotensegrity subsystems nested in a more extensive biotensegrity system that involves the whole body (thereby providing the basis for the remote adjustments seen in the perturbation studies).

Soft-assembled structures can also be put in place to seek information. Exploratory behavior by moving perceptual systems is key in ecological psychology (Gibson 1966). For instance, one can perceive the length of an occluded object by (a) wielding it in one hand, supporting it on one hand while pushing on the end with the other hand, or (b) supporting it on one knee while pushing on the end with a hand (Carello and Turvey 2015). That is, a multitude of possible organizations can achieve the same perceptual goal, with different parts of the body being coordinated to pick up the same information. This is permitted by biotensegrity, whereby tissue deformation of the entire musculoskeletal allows the detection of information throughout the system. Interestingly, the whole body can be engaged in such an exploratory behavior and implement different organizations to perceive different object properties. For instance, Palatinus, Kelty-Stephen, Kinsella-Shaw, Carello, and Turvey (2014) asked participants to perceive either the whole or the partial length of a rod attached to their back. Multifractal analysis of the center of pressure at the mm scale demonstrated differences between the two conditions. It has been suggested that multifractality can be used to describe biotensegrity systems as well (Turvey and Fonseca 2014). That is, multiple temporal and spatial scales interact in a nested biotensegrity structure forming soft-assembled, self-organizing synergies. This points to a need for an organizational architecture more flexible than the hierarchy.

Heterarchies and Coalitions

The solution for the degrees of freedom problem is a set of soft-assembled, self-organizing synergies. The capacity that perception-action systems have to switch between synergies suggests that they have available countless degrees of freedom, barely constrained (Fultot, Frazier, Turvey, and Carello 2019). Consequently, degrees of freedom may be harnessed in different ways, depending on the contingencies to achieve desired outcomes. Furthermore, the same organization may bring about different consequences depending on contextual contingencies in an environment that changes regularly. Together, these facts raise the question *What kind of organizational style allows the coexistence of many potentials for behaving in perception-action systems?* It is important to highlight that we are not discussing *how* an intended behavior is chosen; this would need a study of the dynamics of intention (Shaw 2001), which is beyond the scope of this chapter. Instead, our focus is on the kind of organizational style that empowers perception-action systems with the potential of generating uncountable soft-assembled, self-organizing hierarchies. The answer seems not to be hierarchical organization itself since hierarchies support a choice that was made, not the process of choosing (von Goldammer, Paul, and Newbury 2003).

The foregoing question was first addressed by McCulloch (1945), who was one of the fathers of neural network theory and cybernetics. McCulloch was interested in understanding how the nervous system brings about a given behavior considering the complexity in the relationships among its subsystems (or circuits in McCulloch's terms). He introduced the notion of *diallel*, which means a logical circle or contradiction in a neural circuit (von Goldeman, Paul, and Newbury 2003). To illustrate the concept, consider a system composed of three subsystems A, B, and C. Starting with a hierarchical organization, where influence flows from A to C, a *diallel* is created by connecting C back to A (Figure C10.2). In this new organization, A controls B, B controls C, and C controls A. Therefore, influence does not flow in one single direction as it happens in hierarchies, and no subsystem possesses absolute control over others. A *diallel*, therefore, implies that values in an organization are not hierarchically defined; they are context-dependent. For this organizational style, McCulloch coined the term *heterarchy*, from the Greek words *heteros*, "the other," and *archein*, "to rule").[4]

Unlike hierarchies, the theory of heterarchies is not well established. The term is used in different areas to describe different types of organizations. For instance, in analyzing the distribution of power in human societies, Crumley (1995) considers that different societies may present different types of organizations. Governments may present either a strict hierarchical (e.g., dictatorship regimes) or be a more heterarchical organization (e.g., democracies, in which who is in power changes according to the context). To Crumley (1995), heterarchies are unranked relations of elements to one another (i.e., different values may have an influence on others

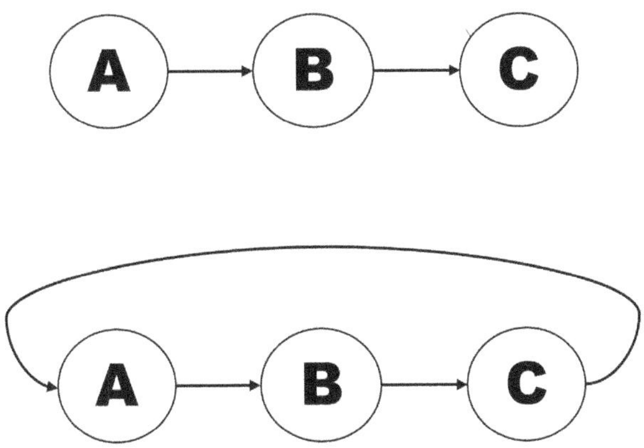

FIGURE C10.2 A *diallel* as proposed by McCulloch (1945). (top) A hierarchical flow of influences: Control flows from A to C. (bottom) A similar system with the addition of the influence of C on A (a *diallel*), which removes the status of absolute controller from A.

depending on the context). These elements, therefore, possess the potential of being ranked in many different ways.

Other authors have considered heterarchies as an organizational style that encompasses hierarchies. For instance, in discussing the need to provide a framework to discuss the functioning and resilience of ecological-social systems, Cumming (2016) pointed out that ecological-social systems can be described either as a hierarchy or a network. He also conveyed the possibility of hybrid variations and changes from one type of organization to another according to the circumstances. He proposed that heterarchies are a broader concept that encompasses both types of system organization.

In the ecological literature on perception-action systems, a taxonomy of control system organizations has been provided based on their connectedness dimension and control distribution (Turvey et al. 1978). The taxonomy was based on nets, and different types of control systems were defined based on combinations of properties of nets. Based on four control primitives, nets are defined by (1) states *S*, (2) a set of dominance relations *R*, (3) a function $\overline{DOM}$ whose domain is *S* and range is contained in *R*, and (4) a function DOM whose domain is *R* and range contains *S*. In this scheme, hierarchies are defined as a special case of heterarchies. More specifically, a hierarchy with interdependence of its subsystems is characterized by heterarchical principles superimposed over a hierarchical structure.

Some similar and complementary features may be highlighted among all the definitions of heterarchy provided above: (1) Heterarchies possess the potential of having their elements ranked in many different ways; (2) heterarchies are a broader class of organizational style that includes hierarchies; (3) hierarchies emerge from heterarchies due to constraints; (4) hierarchies may preserve properties of the heterarchy that they are derived from. Heterarchies, therefore, seem to provide a logical structure with uncountable potentials to form different soft-assembled, self-organizing hierarchies. That is, heterarchies express the abundance (Latash 1996) of degrees of freedom of perception-action systems, while hierarchies express situation-specific actualization of potentials due to constraints imposed upon a heterarchical logic. More specifically, hierarchies emerge due to constraints set upon heterarchies since constraints reduce the extension of interactivity within a heterarchy and create asymmetry in influences between elements and relationships (Turvey et al. 1978).

It is particularly noteworthy that the environment surrounding perception-action systems is also a heterarchy. The interaction between these two heterarchies is characterized by the perception-action system acting upon the environment and the environment constraining the perception-action system. Consequently, the degrees of freedom problem cannot be seen independently from the degrees of constraints problem (Turvey et al. 1978). Recognizing the relationship between the two problems is a matter of understanding a perception-action system and its environment in terms of their duality (not dualism). That is to say, perception-action systems and their environments are mutual and reciprocal (Shaw and Turvey 1981; Turvey and Shaw 1995, 1999), meaning the environment is as thoroughly organized as its organism as well as specific to it (Turvey et al. 1978). They are symmetrically related heterarchies, an organizational style of coordination known as *coalition* (Turvey et al. 1978; Shaw and Turvey 1981). A direct implication of coalitions is one of the principles of the ecological psychology style of inquiry: The minimal unit of analysis is the relation perception-action system-environment (Turvey and Carello 1981, 2012).

Considering the multitude of temporal and spatial scales interacting as well as the implications of coalitions for how perception-action systems bring about an action, the role ascribed by Bernstein to the CNS as ultimate movement constructor seems to be overvalued. This point is brought into bold relief by the rich behaviors performed by non-human organisms, to which we now turn.

Perception and Action without a Nervous System

Understanding the neural basis of behavior is almost universally treated as the proper target for understanding behavior itself. The unanimity of this conviction has long been undermined by scholars within ecological psychology, a voting bloc whose skepticism is summarized succinctly by the admonition: "Ask not what's inside your head but what your head's inside of" (Mace 1977, p. 43). This ecological predilection has promoted a focus on actions rather than the specifics of their embodiment in an effort to uncover generic principles by which the realization of those actions might be understood. Consequently, it has been argued that an understanding of behavior requires an understanding of behavior across all kingdoms, not just *animalia* (Carello, Vaz, Blau, and Petrusz 2012; Fultot, Nie, and Carello 2016a, b; Fultot et al. 2019; Turvey 2013; Turvey and Carello 2012). A principled understanding must accommodate behavior by all organisms, regardless of the sophistication of their neural machinery – indeed, regardless of whether they have neural machinery at all.

Examples of sophisticated behavioral achievements by unsophisticated organisms are abundant. *Cuscuta pentagona*, or Dodder, is a parasitic leafless plant in the morning glory family. This twining plant will decline to wrap around a potential host branch, even one it is tethered to, if it deems the host insufficiently nutritious (Kelly 1992). Its distinction between edible and non-edible is made without first "trying out" the host and guides its behavior even if no other alternative hosts are available. Only when dodder deems the host suitable does it produce haustoria, tiny structures that penetrate the host. The haustoria are continually reattached as dodder grows. The coiling and attaching behaviors are perceptually guided, presumably by an air-borne gradient of chemicals exuded by the host plant. Alternatively, if Dodder is placed in the middle of an empty arena, it is as likely to grow in one direction as any other. The introduction of viable hosts structures the chemical gradient in a way that guides Dodder in the direction of a preferred plant – tomato is superior to wheat. If only wheat is available, however, Dodder will grow toward it but not as often as it would if only tomato were available (Runyon, Mescher, and DeMoraes 2006). Just as with the coiling behavior, Dodder's ramifying behavior is perceptually guided by ecologically relevant information in the chemical gradient.

The realm of single-celled organisms also provides challenges for a generic account of behavior. The genus of testate, or shelled, amoeba has become a favorite example to illustrate that organisms without nervous systems engage in complicated sequencing and timing to accomplish end-directed behaviors (Turvey 2013). *Difflugia coronata, difflugia oblonga, difflugia tuberspinifera,* and so on, build shells with species-specific shapes. More precisely, they assemble intricate, sculptural structures (Figure C10.3) from minute grains of sand they have collected, distributing the different sized grains appropriately for the role they play in the structure (Armynot Du Chatelet, Noiriel, and Delaine 2013). The shelled amoeba locomotes by creating extended pseudopods that attach onto surfaces (Wohlman and Allen 1968). In turn, when *d. tuberspinifera* encounters organisms of the size of prey, it sends six pseudopods to embrace and explore the surface of the organism. Then, it retracts its pseudopods, and sends two longer membrane projections to probe the prey. In the course of the latter behavior, the amoeba reorients the prey "feet first" relative to the sharp opening of the test which it uses as a means to scrape the tough covering off the captured prey so that it can be ingested. During ingestion, the pseudopods continue to wave in a manner that maintains the position of the prey (Han, Wang, Lin, and Dumont 2008).

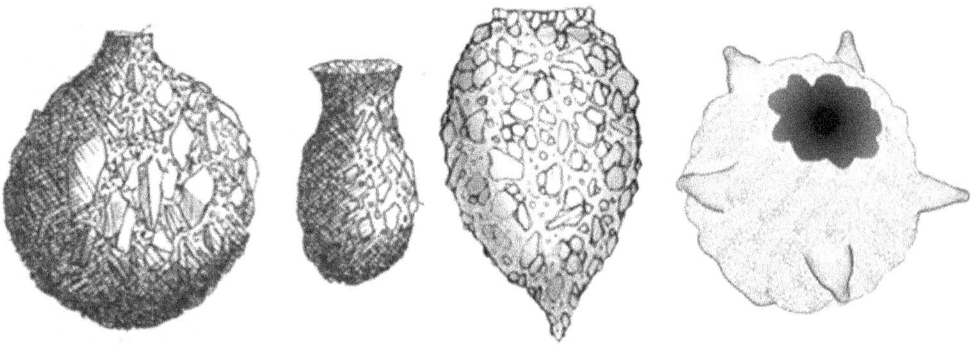

FIGURE C10.3 Four kinds of *Difflugia*, distinguishable by their shells. (Rightmost) *D. tuberspinifera* uses its sharp collar to peel prey.
(Source: Adapted from Meunier (1910, pl. 13) licensed under CC BY-NC-SA 4.0).

The foregoing behaviors by organisms without nervous systems are purposeful, and the degrees of freedom, be they tendrils or pseudopods, are constrained by properties of the surroundings. Coiling and attaching happen or they don't, growth is directed or not, pseudopods change in number and length, sand grains of particular sizes are collected, all according to the task. Therefore, coalitions seem to be a feasible type of coordination in these domains as well. Furthermore, behaviors of these kinds are also excellent examples of Bernsteinian intentional actions implemented through appropriately constrained degrees of freedom.

Concluding Remarks

In the present work, Bernstein developed a neural, hierarchical model of movement coordination. Despite the sophistication of the model, however, hierarchy seems to be a logical type that supports the actualization of actions but seems insufficient to deal with many simultaneous potentials to act. As we have tried to show, the latter is better accommodated by heterarchies/coalitions, logical types that are independent of their material instantiation, and seem to fit organisms across phyla. In regards to the neural foundation of his model, Bernstein essentially argues that the evolution of perception-action demands drives the evolution of the CNS. A similar argument has, in fact, been made in the service of ecological psychology. Whereas achievements by organisms such as Dodder and *Difflugia* essentially depend on morphodynamic changes wrought by growth processes, multicellular organisms require something else. They have more behavior modes and they need to switch between them with a speed not possible with growth processes:

> available macroscopic affordance structures can be exploited or realized by organisms that developed the capability to coordinate the timing of their functional transitions with the spatiotemporal constraints of those affordance structures … That kind of functional transition is the main contribution of their nervous systems.
>
> *(Fultot et al. 2019, p. 226)*

The benefit of considering the commonalities of behavior across phyla is that it encourages the uncovering of generic principles that respect the connectedness between human behavior and nature more broadly (Turvey and Fitzpatrick 1993; Turvey and Swenson 1991).

Acknowledgments

The authors thank the National Science Foundation, Grant BCS-1344725 (INSPIRE Track 1), for supporting discussions of perception-action without nervous systems.

Notes

1 Our use of perception-action systems is intended to acknowledge the variety of organisms whose behaviors are guided by information about their environments. Even though Bernstein's work addresses only the kingdom *Animalia,* we believe he would have endorsed the notion that the problems he identified were more general.
2 Although our ecological lens would see this as perceptually guided, we are keeping Bernstein's use of *sensory* when describing his model.
3 While we endorse this view, it should be noted that there is ongoing discussion about the nature and even the existence of synergies at the level of muscular-articular links as originally proposed by Bernstein (e.g., Latash, Scholz, and Schöner 2007; Singh, Iqbal, White, and Hutchinson 2018; Tresch and Jarc 2009).
4 www.britannica.com/topic/heterarchy

References

Allen TFH, TB Starr (1982) *Hierarchy: Perspectives for Ecological Complexity.* Chicago, IL: University of Chicago Press.

Armynot Du Chatelet E, Noiriel C, Delaine M (2013) 3D morphological and mineralogical characterisation of testate amoebae. *Microscopy and Microanalysis* 19: 1511–1522.

Bernstein N (1967) *The Coordination and Regulation of Movements.* Oxford, UK: Pergamon Press.

Carello C, Turvey MT (2015) Dynamic (effortful) touch. *Scholarpedia* 10: 842.

Carello C, Vaz D, Blau JJC, Petrusz S (2012) Unnerving intelligence. *Ecological Psychology* 24: 241–264.

Crumley CL (1995) Heterarchy and the analysis of complex societies. *Archeological Papers of the American Anthropological Association* 6:1–5.

Cumming GS (2016). Heterarchies: reconciling networks and hierarchies. *Trends in Ecology & Evolution* 31: 622–632.

Fultot M, Frazier PA, Turvey MT, Carello C (2019) What are nervous systems *for*? *Ecological Psychology* 31: 218–234.

Fultot M, Nie L, Carello C (2016a) Complementarity of symmetry and asymmetry. *Constructivist Foundations* 11: 335–340.

Fultot M, Nie L, Carello C (2016b) Perception-action mutuality obviates mental construction. *Constructivist Foundations* 11: 298–307.

Gelfand I, Gurfinkel V, Tsetlin M, Shik M (1971) Some problems in the analysis of movements. In: Gelfand I, Gurfinkel V, Fomin S, Tsetlin M, eds. *Models of the Structural-Functional Organization of Certain Biological Systems,* 329–345, Cambridge, MA: MIT Press.

Gibson JJ (1947) *Motion Picture Testing and Research.* Aviation Psychology Research Report No. 7. Washington, DC: US Government Printing Office, 181–195.

Gibson JJ (1950) *The Perception of the Visual World.* Boston, MA: Houghton Mifflin.

Gibson JJ (1958) Visually controlled locomotion and visual orientation in animals. *British Journal of Psychology* 49: 182–194.

Gibson JJ (1966) *The Senses Considered as Perceptual Systems.* Boston, MA: Houghton Mifflin.

Gibson JJ (1979/1986) *The Ecological Approach to Visual Perception.* Boston, MA: Houghton Mifflin.

Gibson JJ, Crooks (1938) A theoretical field-analysis of automobile-driving. *The American Journal of Psychology* 51: 453–471.

Gottlieb G (1998) Normally occurring environmental and behavioral influences in gene activity: From central dogma to probabilistic epigenesist. *Psychological Review,* 105: 792–802.

Han B-P, Wang T, Lin Q-Q Dumont HJ (2008) Carnivory and active hunting by the planktonic testate amoeba *Difflugia tuberspinifera. Hyrdrobiologia* 596: 197–201.

Kelly CK (1992) Resource choice in Cuscuta europaea. Proceedings of the National Academy of Science 89: 12194–12197.

Kelso JS (2009) Synergies: atoms of brain and behavior. In Sternad D, ed. *Progress in Motor Control: A Multidisciplinary Perspective*, 83–91. New York, NY: Springer Verlag.

Kelso JS, Tuller B, Vatikiotis-Bateson E, Fowler CA (1984) Functionally specific articulatory cooperation following jaw perturbations during speech: Evidence for coordinative structures. *Journal of Experimental Psychology: Human Perception and Performance* 10: 812–832.

Kloos H, Van Orden GC (2009) Soft-assembled mechanisms for the unified theory. In: Spencer JP, Thomas M, McClelland J, eds. *Toward a New Grand Theory of Development: Connectionism and Dynamic Systems Theory Re-considered*, 253–267, Oxford: Oxford University Press.

Kugler PN, Turvey MT (1987) *Information, Natural Law, and the Self-assembly of Rhythmic Movement*. Hillsdale, NJ: Lawrence Erlbaum Associates (reprinted 2015 by Taylor & Francis).

Latash ML (1996) The Bernstein problem: How does the central nervous system make its choices? In: Latash ML, Turvey MT, eds. *Dexterity and Its Development*. Hillsdale, NJ: Lawrence Erlbaum Associates.

Latash ML, Scholz JP, Schöner G (2007) Toward a new theory of motor synergies. *Motor Control* 11: 275–307.

Levin SM (2006) Tensegrity: The new biomechanics. In: Hutson M, Ellis R, eds. *Textbook of Musculoskeletal Medicine*, 69–80. Oxford, England: Oxford University Press.

Liberman AM, Cooper FS, Shankweiler DP, Studdert-Kennedy M (1967) Perception of the speech code. *Psychological Review* 74: 431–461.

Loeb GE (1995) Control implications of musculoskeletal mechanics. *Annual International Conference of IEEE-EMBS* 17: 1393–1394.

Mace WM (1977) James J. Gibson's strategy for perceiving: Ask not what's inside your head but what your head's inside of. In: Shaw RE, Bransford J, eds. *Perceiving, Acting, and Knowing*, 43–65. Hillsdale, NJ: Lawrence Erlbaum Associates.

McCulloch WS (1945) A heterarchy of values determined by the topology of nervous nets. *Bulletin of Mathematical Biology* 7: 89–93.

Meunier A (1910) *Microplankton des Mers de Barents et de Kara. Duc d'Orléans. Campagne arctique de 1907*. Bruxelles, Belgique: Imprimerie scientifique Charles Bulens.

Michaels CF, Carello C (1981) *Direct Perception*. New York, NY: Prentice Hall.

Palatinus Z, Kelty-Stephen DG, Kinsella-Shaw J, Carello C, Turvey MT (2014) Haptic perceptual intent in quiet standing affects multifractal scaling of postural fluctuations. *Journal of Experimental Psychology: Human Perception and Performance* 40: 1808–1818.

Profeta VLS, Turvey MT (2018) Bernstein's levels of movement construction: A contemporary perspective. *Human Movement Science* 57: 111–133.

Rosenbaum DA (2005) The Cinderella of psychology: the neglect of motor control in the science of mental life and behavior. *American Psychologist* 60: 308–317.

Runyon JB, Mescher MC, De Moraes CM (2006) Volatile chemical cues guide host location and host selection by parasitic plants. Science 313: 1964–1967.

Salthe, S.N. (1985). *Evolving Hierarchical Systems*. New York: Columbia University Press.

Scarr G, Harrison H (2017) Examining the temporo-mandibular joint from a biotensegrity perspective: A change in thinking. *Journal of Applied Biomedicine* 15: 55–62.

Shaiman S, Gracco VL (2002) Task-specific sensorimotor interactions in speech production. *Experimental Brain Research* 146: 411–418.

Shaw RE (2001) Processes, acts, and experiences: Three stances on the problem of intentionality. *Ecological Psychology* 13: 275–314.

Shaw E, Turvey MT (1981) Coalitions as models for ecosystems: A realist perspective on perceptual organization. In: Kubovy M, Pomerantz J, eds. Perceptual Organization, 343–415. Hillsdale, NJ: Lawrence Erlbaum Associates.

Simon HA (1962) The architecture of complexity. Proceedings of the American Philosophical Society 106: 467–482.

Simon H (1969) *The Sciences of Artificial*. Cambridge, MA: MIT Press.

Simon H (1973) The organization of complex systems. In: Pattee HH, ed. *Hierarchy Theory*, 1–28, New York: Braziller.

Singh RE, Iqbal K, White G, Hutchinson TE (2018) A systematic review on muscle synergies: from building blocks of motor behavior to a neurorehabilitation tool. *Applied Bionics and Biomechanics* 2018: 1–15.

Swenson R, Turvey MT (1991) Thermodynamic reasons for perception-action cycles. *Ecological Psychology* 3: 317–348.

Tresch MC, Jarc A (2009) The case for and against muscle synergies. *Current Opinion in Neurobiology* 19: 601–607.

Turvey MT (1974) A note on the relation between action and perception. In: Wade MG, Martens R, eds. *Psychology of Motor Behavior and Sports*. Urbana, IL: Human Kinetics.

Turvey MT (1977) Preliminaries to a theory of action with reference to vision. In: Shaw R, Bransford J, eds. *Perceiving, Acting and Knowing: Toward an Ecological Psychology*. Hillsdale, NJ: Lawrence Erlbaum Associates.

Turvey MT (1986) Beyond anatomical specificity. *Behavioral and Brain Sciences* 9: 624–625.

Turvey MT (1990) Coordination. *American Psychologist* 45: 938–953.

Turvey MT (2007) Action and perception at the level of synergies. *Human Movement Science* 26: 657–697.

Turvey MT (2013) Ecological perspective on perception-action: What kind of science does it entail? In: Prinz W, Bessard M, Herwig A, eds. *Action Science: Foundations of an Emerging Discipline* (pp. 139–170). Cambridge MA: MIT Press.

Turvey MT, Carello C (1981) Cognition: The view from ecological realism. Cognition 10: 313–321.

Turvey MT, Carello C (1996) Dynamics of Bernstein's level of synergies. In: Latash M, Turvey, MT, eds. Bernstein's "On Dexterity and Its Development," 339–376, Hillsdale, NJ: Erlbaum.

Turvey MT, Carello C (2012) On intelligence from first principles: Guidelines for inquiry into the hypothesis of physical intelligence (PI), *Ecological Psychology 24:* 3–32.

Turvey MTT, Fitzpatrick P (1993) Commentary: Development of perception-action systems and general principles of pattern formation. *Child Development* 64: 1175–1190.

Turvey MT, Fonseca ST (2014) The medium of haptic perception: A tensegrity hypothesis. *Journal of Motor Behavior* 46: 143–187.

Turvey MT, Shaw RE (1995) Toward an ecological physics and a physical psychology. In: Solso R, Massaro D, eds. The Science of the Mind: 2001 and Beyond, 144–169, Oxford, UK: Oxford University Press.

Turvey MT, Shaw RE (1999) Ecological foundations of cognition: I. Symmetry and specificity of animal-environment systems. Journal of Consciousness Studies 6: 85–110.

Turvey MT, Shaw RE, Reed ES, Mace WM (1981) Ecological laws of perceiving and acting: In reply to Fodor and Pylyshyn (1981). Cognition 9: 237–304.

Turvey MT, Shaw RE, Mace WM (1978) Issues in the theory of action: Degrees of freedom, coordinative structures and coalitions. In: Requin J, ed. *Attention and Performance VII*, 557–595. Hillsdale, NJ: Lawrence Erlbaum Associates.

Valero-Cuevas F, Yi JW, Brown D, McNamara RV, Paul C, Lipson H (2007) The tendon network of the fingers performs anatomical computation at a macroscale. *IEEE Transactions in Biomedical Engineering* 54: 1161–1166.

Van Orden GC, Holden JG, Turvey MT (2003) Self-organization of cognitive performance. *Journal of Experimental Psychology: General* 132: 331–350.

von Goldammer E, Joachim P, Newbury J (2003) *Heterarchy e Hierarchy: Two Complementary Categories of Description*. www.vordenker. de/heterarchy/a_heterarchy e.pdf. accessed 01.13.2020.

Wagman JB, Hajnal A (2016) Use your head! Perception of action possibilities by means of an object attached to the head. *Experimental Brain Research* 234: 829–836.

Wohlman A, Allen RD (1968) Structural organization associated with pseudopod extension and contraction during cell locomotion in Difflugia. *Journal of Cell Science 3*: 105–114.

C11

BERNSTEIN'S CONSTRUCTION OF MOVEMENTS

Bringing Task Constraints into Action Theory

Karl M. Newell

UNIVERSITY OF GEORGIA

Yeou-Teh Liu

NATIONAL TAIWAN NORMAL UNIVERSITY

It is an honor for us to comment on *On the Construction of Movements* – the long-awaited translation of Nicholai Bernstein's (1947) original Russian language manuscript. It builds substantially on the set of now well-known papers from 1930–1960 that were published as a collective in the western world in 1967 under the title of *The Coordination and Regulation of Movements*. The original Russian language version of this new book *On the Construction of Movements* also predates, though only by a few years, the Russian language version of the second translated Bernstein book *On Dexterity and Its Development* (1996). *On the Construction of Movements* is then the third Bernstein book published in English and the fullest expression of his wide ranging but coherent theorizing. Undoubtedly, construction of movements represents the pinnacle of his thinking on a biophysical account of movement and physical activity.

The scope and depth of the scholarship is prodigious even by the inflated standards and expectations of today. In the text, this is reflected in many diverse but related ways – the collection of which creates the unique and coherent view of biological motion that scholars in the movement domain are intrigued by and searching for. The breadth of this new Bernstein book on movement and physical activity holds challenges for contemporary work on motor control of all theoretical persuasions.

The theorizing and thinking seem so fresh in places that one has to pinch oneself to remind that the original manuscript was written more than 70 years ago. Nevertheless, Bernstein recognized the limitations of his treatise: including, in places, a lack of experimental support leading to the need for speculation in the theorizing. Given the protracted and convoluted sequence of publishing outlined above, it is perhaps not surprising that there is also some overlap across the three English language Bernstein books. Furthermore, a lot of relevant science on the theme of this book has gone on in the last 70 years.

A prominent feature with Bernstein's writings is that the natural phenomena of human movement and biological motion more generally, are central to the agenda of a theory of motor learning and development (see Chapters 7 and 8). There is still a refreshing breadth to the focus on the phenomena of human movement and physical activity as a biophysical problem. This involves a broad consideration of the role of the task itself in the organization of the levels of the movement system (Newell 1986) and the changes that arise from practice over the different time scales of adaptation, learning, development and evolution (Newell, Liu, & Mayer-Kress 2001).

The theoretical agenda to motor control spans several disciplinary levels of living systems with both an evolutionary biology backdrop and a developmental multidisciplinary ontogenetic framework. The center piece from our viewpoint is, in effect, the integration of the levels of construction framework (Bernstein 1996) with the degrees of freedom biophysical problem (Bernstein 1967) and task constraints (Bernstein 1947) expressed in the context of motor skill development (Chapter 7) and learning (Chapter 8). The multi-level framework of Bernstein's construction of movements has the organization at all levels of the system influenced by the goal of the action in question.

The format for our commentary is that of selecting from Bernstein's "constructions" a critical principle or proposition within each of three major themes. We consider the propositions in the broad context of the book, related contemporary research, together with open issues still remaining. Finally, by way of general introduction, we note that Profeta and Turvey (2018) have recently provided, from an ecological and dynamical framework, a substantial relook and contemporary dynamical/ecological perspective on Bernstein's hierarchy of the levels of construction, that we try hard not to overlap with.

Task Constraints and Levels of Construction

- Proposition 1: Each level of the construction of movements is a key to solving a particular class of motor tasks

The key word in this proposition is the word key. In several sections of this book, Bernstein forges links between the task constraints of the motor skill and the level of control in the hierarchy that is leading or background to the movement in action. The bigger picture of this mapping of perceptual-motor skills to levels of construction is the potential development of a task taxonomy that is grounded in a functional and structural neuroscience of the central and peripheral nervous systems. Bernstein is, in effect, bringing the task into the foundation of his theory of motor learning and control.

This is achieved through classifying motor tasks into functional task categories and linking their unique control feature to different levels of system organization and movement construction`. There have been many efforts to classify motor skills but these have been largely behavioral, one dimensional and diverse in theoretical perspective. These include but are not limited to: (Fleishman & Quaintance 1984 – constructs of human performance; Gentile 1987 – two-dimensional classification; Poulton 1957 – open and closed skills; Warren 2006 – behavioral dynamics stability/instability).

We can generalize the role of the task by proposing that task constraints are the goals of the action together with rules (where present) that specify or constrain how the movement pattern and the outcome are realized (Newell 1986; Newell & Jordan 2007). The task constraints coalesce with those of the individual and environment to channel the emergence of the movement dynamics in pursuit of the task goal. These three sources of constraint can all be captured in the same language of dynamics. The constraints at each level of the system may be construed as functional or structural according to their timescale of change. The triad of constraints on the development of coordination was an effort to get task constraints as central as those of the organism and environment in movement coordination, control and skill.

Bernstein holds that for the individual actor solving the coordination problem affords the emergence of a controllable system. This is consistent with a task category of the behavioral dynamics taxonomy (Warren 2006) that hinges on the relative contribution of physical and informational constraints to the control of the task at hand. In Warren's framework there are three perceptual-motor task categories based on different forms of dynamic stability/instability. Category 1 is stabilizing a passively stable system, with the example of bouncing a ball on a racket. Here the physics of the context dominates control of the action and the coordination solution is found through perceptual-motor search strategies. Category 2 is active stabilization of an unstable system, such as in learning to balance a pole. In this task category there is strong contribution of both physical and informational constraints for a successful task outcome. Finally, in category 3 there is the active stabilization of a neutrally stable system, such as in braking while driving a car. Motor learning and control in this last task category is dominated by task relevant information and, in our view, has in effect been the task category of choice in the study of perceptual-motor skills.

In the Warren (2006) framework, category 2 is consonant with Bernstein's definition of turning the coordination pattern into a stable system. Though it should be noted that stability is characterized mechanically in Bernstein and dynamically in Warren. The inability to stabilize the system most typically arises in the very beginning phase of learning a new motor skill where the learner has not produced previously a relevant collective variable for the task. In infant motor development, for example, through prone progression and learning to sit, stand and walk initially there are typically many failed attempts to construct the relevant collective variable before eventually transitioning to stability and task success. In the same way, in adult motor learning where a task relevant collective variable has yet to be formed, there will be through practice a transition to a new stable state – an example in this category is learning the rollerball task (Liu, Chuang, & Newell 2019). In category 3 a transition is not required to meet the task demands as the subject movement pattern is already stable even though the scaling of it to the task outcome remains to be completed. The different stability categories for tasks in Warren (2006) will lead to different learning dynamics and multiple pathways to the change in the performance outcome and movement pattern over time (Newell et al. 2001; Saltzman & Kelso 1987).

Bernstein proposes that the key to understanding the organization of a class of tasks is to map them on a common dimension to a level in his hierarchy, Levels C, D and E, that can be leading levels. Level C Space captures many of the standard perceptual-motor skill tasks that have been investigated in motor learning and control and that are not of the sequential task category of Level D Action. This substantial body of experimental studies on these two task types, however, has not been couched in Bernstein's 1996 Levels of construction framework. Indeed, the impact of the 1996 Dexterity translation of the levels of construction on motor control has been negligible, from both behavioral and neuroscience perspectives.

In the construction of movements Bernstein outlines 5 levels (A–E) of control in the hierarchy. This is a revision from Bernstein (1996) where just the lower four levels A to D were presented. Table C11.1 provides details of the full five-level hierarchy. These levels are well defined and characterized in the book to require only a brief description here, but sufficient for us to meaningfully integrate with the task dynamics. The introduction of level E and its link to symbols relates to a definable category of tasks but is under developed. Indeed, it was from theorizing on the primacy of dynamics over symbols as modes of control that led to the introduction of the dynamics

TABLE C11.1 List of levels of construction characterized by Bernstein in the second part of his book

A – Tonus, the level of paleokinetic regulations, known also as the rubro-spinal level within the central nervous system.
B – the level of synergies, also known as the thalamo-pallidar level.
C – the level of spatial field. It subdivides into two sublevels:
C1 – the striatal level that belongs to the extrapyramidal system, and
C2 – the pyramidal level related to the group of cortical levels etc.
D – the level of actions (object actions, actions with meaningful serial order, sequence learning, etc.), also known as the parieto-premotor level.
E – the group of the highest cortical levels of symbolic coordinations (writing, speech).

of coordinative structures and the construct of self-organization to the movement domain (Kugler, Kelso, & Turvey 1980).

Bernstein's evolutionary biology framework holds that the new levels of the brain were qualitative changes to the selection pressures for the functions that they now support. The levels of construction are generally still consonant with standard neuroscience accounts of brain encephalization but there is now available 70 years on considerably more experimental detail from the different disciplines of neuroscience. Shmuelof and Krakauer (2011) have reiterated the theoretical and contextual relevance of couching movement neuroscience in the framework of evolutionary biology.

In the Bernstein theory, each motor task discovers for itself, depending on its content and semantic structure, one leading level of construction. Or, in Bernstein's words, the search for a stability leads to the discovery of one of the sensory syntheses, which is most adequate by its qualitative features, together with the contents of its afferent components, and their synthetic unification for the desired solution. This level is defined as the leading level for a given movement with respect to the execution of the main, crucial sensory corrections and realization of the required recoding processes. The mechanism(s) for discovering the relevant leading level is, however, not specified and the implied role of perceptual processes and task constraint information in the levels of construction is not developed. The Profeta and Turvey (2018) reinterpretation of the levels of construction from the viewpoint of information in ecological psychology and direct perception provides a framework to address these limitations of the Bernstein approach.

Bernstein holds the early stage of practice to be in essence a search process for task relevant solutions. How engagement in the motor task leads to the discovery of the relevant leading level and its efferent and afferent components is, however, not outlined. In general, Bernstein proposes the strong contribution of perceptual processes to motor control without providing as a part of the theory a framework for information and perceptual processes in action (Profeta & Turvey 2018). Newell (1991, 1996) has in line with Bernstein outlined a search strategy approach to motor learning where the relevance and effectiveness of the type of information available to the learner is individual skill level and task specific.

Bernstein gives a more restricted role to synergies (Level B) than outlined in more contemporary motor control that generalizes them as abstract relations between elements of the system (Kelso 2009; Latash 2008; Turvey 2007). Thus, current theorizing places synergies as more than established couplings between neuromuscular functional groups. For Bernstein, the level of synergies does not play a leading role in movement construction – a position on the surface that is in contrast to much contemporary experimental and theoretical work.

It would be instructive to examine experimentally Bernstein's link of levels of the central nervous system to particular classes of motor task to ascertain how broad and robust this relation is. Task classifications are sought and implemented to facilitate general statements on the phenomena at hand. A challenge is rationalizing from theory the choice of the movement properties on which to define the task classification. Finally, relying on a single movement dimension to classify the task may not be sufficient to draw robust insights from the task synthesis.

The Degrees of Freedom Problem

- Proposition 2: The main problem of movement coordination is indeed in overcoming the redundant degrees of freedom of a moving organ, i.e. in turning it into a controllable system

In this book, Bernstein reaffirms explicitly what most scholars have assumed implicitly from the 1967 book: namely, that his emphasis on degrees of freedom was at the level of joint configuration (Latash 1996; Newell & Vaillancourt 2001; Turvey, Shaw, & Mace 1978). Thus, every degree of freedom of mobility (as Bernstein calls it here in "Construction") more or less coincides with a separate, independent direction of mobility in a set of joints of a kinematic chain. A degrees of freedom perspective can be used at any level of analysis but driven by Bernstein's theoretical framework the behavioral joint configuration level has been the contemporary emphasis in motor learning and control. Unraveling the active degrees of freedom in executing a task is another way to decompose the influence of task constraints in actions (Kugler et al. 1980).

Bernstein develops the position that a single degree of freedom task represents a very simplified control problem from that of a multiple (even two) degree of freedom task. He notes that this distinction is most evident in artificial systems. This is in part due to the fact that the degrees of freedom of the task constraints contribute to driving the organization of central and peripheral control structures. In this context, it is relevant that a number of theoretical perspectives over the last century of research on motor control have rested on laboratory experimental protocols with just one joint space degree of freedom to be constrained (e.g., linear positioning movement).

The centrality and significance of the construct of degrees of freedom in Bernstein's (1967) theorizing has opened up, however, the rationale for the implementation of multiple degrees of freedom in so-called whole-body movement tasks. This view is further developed in the book through the mapping of task constraints to a particular level of the hierarchy and its anatomical/physiological structure. The general position in the construction of movements is that among other influences, multiple degree of freedom tasks "change the balance of the contributions of internal, external and reactive forces and they make skilled movement more difficult given that the individual only has control over the internal forces, and even that can be unpredictable." There have been few direct experimental examinations of the reactive forces in movement control. In sum, as identified by Bernstein, the challenges of the "indeterminacy" of the system in action are only magnified in multiple degree of freedom tasks. Nevertheless, human movement is a multiple degree of freedom problem and shying away from approaching it will only restrict progress on both Bernstein's problem and the field of motor control more generally. And, after all, the perceptual-motor system typically solves the coordination problem in an elegant and functional way given often only a modicum of practice.

The number of degrees of freedom. The original emphasis of Bernstein (1967) on degrees of freedom in motor learning and control and that is elaborated here in "Construction of movements" was on the *number* of degrees of freedom to be controlled and how this changed with practice and the acquisition of skill (Turvey et al. 1978). This emphasis was due to the general position that no matter the theoretical framework, the larger the number of degrees of freedom to be controlled the greater the organizational demands (Bernstein here invokes complexity) on system control. This opened up a parallel consideration of the degrees of freedom at several levels of the system: task, joint motions, muscles, neurons and so on. Generally, as the analysis level of the system becomes more micro the greater the number of degrees of freedom requiring constraint.

The construct of degrees of freedom with Bernstein is on the joints as physical components and their potential planes of motion or the number of independent coordinates required to unambiguously define the joint configuration of the system. Bernstein illustrates the richness of this approach in the act of prehension where the many degrees of freedom of the arm-hand complex afford high dimensional complex movement behavior of the digits of the hand in grasping and object manipulation. A more complete biomechanical account of the state space of the system would typically assume that a position and velocity coordinate is required for each joint space degree of freedom.

In potential joint configurations there are approximately 245 degrees of freedom (Zatsiorsky 1998) though as Bernstein indicates this number depends on the assumptions made. For example, how the rigidity of the thorasic region of the spine is treated substantially changes the number of degrees of freedom: namely, as a rigid single (1) degree of freedom or a multiple (66) degree of freedom entity. The attendance to this level of detail in boundary conditions reflects Bernstein integrating the constraints of the structure and function of the biological system to the problem of movement coordination and control. However, considering only the number of potential planes of motion from joint configurations does not take-into-account the functional and structural constraints from other levels on those joint motions (e.g., muscles) that can contribute to reducing the dimension of the movement system output (Newell, Challis, & Morrison 2000). The inference is that setting up the degrees of freedom problem on the *potential* dimension of the joint configuration space typically overestimates the actual scope and severity of the degrees of freedom problem.

The advancing of principles for the directional order in recruiting the joint motions into a coordinative structure has not moved forward in a substantial way from the original ideas of Bernstein (1967). Only isolated task dependent evidence has been found for the original freezing and freeing postulates and this has not yielded yet viable general principles for the acquisition of coordination (Newell & Vaillancourt 2001), though it is generally held that there is a reduction in dynamical degrees of freedom as a function of motor learning (Mitra, Am azeen, & Turvey 1998). The explicit link of the degrees of freedom problem to task categories and the levels of construction remains to be examined.

Redundancy/degeneracy. Bernstein (1967) and in this book, outlined the challenges of organization of the degrees of freedom in a redundant system. Indeed, mastering the redundant degrees of freedom was his very definition of skill. This has been recognized by the motor control community but in our view remains still an under developed construct theoretically and experimentally. And, even more problematic, perhaps, is that the redundancy construct has been muddied by the introduction of the related but different construct of degeneracy (Edelman & Gally 2001); and even questions about the appropriateness/usefulness of the redundancy interpretation (Latash 2000).

Edelman and Gally (2001) proposed "that in addition to environmental effects that lead to increasing complexity during evolution, the degeneracy of biological networks in animals within competing populations also makes a major contribution to complexity. Degeneracy is the ability of elements that are structurally different to perform the same function or yield the same output. Unlike redundancy, which occurs when the same function is performed by *identical* elements, degeneracy, which involves *structurally different* elements, may yield the same or different functions depending on the context in which it is expressed. Indeed, there is mounting evidence that degeneracy is a ubiquitous property of biological systems at all levels of organization."

The distinction between redundancy and degeneracy, however, offered by Edelman and Gally (2001) has not been taken up in a general way by the motor control community. Rather, the tendency has been to follow the standard engineering definition of redundancy from the artificial systems viewpoint as the inclusion of extra components in the system that are not strictly necessary for functioning, but available in case of failure in the respective active components. An everyday example is having a spare wheel/tire in the boot of the car. In human movement, however, redundancy is when there are typically more joint degrees of freedom available than are required to provide a task relevant solution. This situation also affords, however, the potential of multiple solutions to the forming of a stable and task relevant coordination mode, and advantages in adapting to perturbations of the system. In this context, Latash (2000) has challenged the prevailing assumption that redundant degrees of freedom are necessarily a problem and, moreover, indicates that the presence of abundant degrees of freedom can be a positive for system control, such as in adapting to perturbations.

It appears that the dominant engineering-oriented use in motor control of the term redundancy prevails to cover both redundancy *and* degeneracy as defined above. Indeed, our interpretation is that the term redundant is now predominantly used in the same sense that Edelman and Gally define degeneracy. The reality is though that unlike a car carrying a spare tire or an airplane having more engines that are minimally necessary to accommodate the flight plans, the human system does not carry spare body segments or joint space degrees of freedom (e.g., heads, legs, knee joints, etc.) as back-ups. In the same vein, surgery is not called on to reduce the degrees of freedom (Latash 2000).

The coordination and control of the joint space degrees of freedom has been a central issue in the field of motor control – it is known widely as Bernstein's problem after the publication of the 1967 book became recognized. Nevertheless, apart from Latash (2000) and the introduction of the abundance interpretation, no new principles on the organization of joint space degrees of freedom have emerged in regard to the acquisition of movement coordination and none are proposed in the "Construction of movements." Indeed, recent commentary and interpretation suggests that Bernstein and followers may have overestimated the influence of redundancy to the degrees of freedom problem (Newell et al. 2000). It would seem that this long-standing negative perspective of the redundancy/degrees of freedom issue has been fueled in part by the adherence to a construct from artificial systems for an account of movement in living systems.

Degrees of freedom and attractor dynamics. The construct of degrees of freedom is also present in a more contemporary dynamical systems analysis of the active or dynamical degrees of freedom (Eubank & Farmer 1997). This approach shifts the emphasis of degrees of freedom from the motions of the individual components or elements to that of the attractor dynamic that is supporting the coordination and control of the components' output. In attractor dynamics the state of the system is determined through the identification of a low dimensional order parameter (Haken, Kelso, & Bunz 1985) or collective variable (Schöner & Kelso 1988). For example, in the bimanual finger movement protocol the relative phase of the movements of the two fingers is the order parameter (Haken et al. 1985). The determination of the relevant order parameter for a given task situation is not necessarily straight forward and there has been limited progress on this open contemporary issue (though see Whitte, Bock, Storb, & Blaser 2003 for methods to determine the collective variable in tasks with more degrees of freedom than the bimanual case).

Kay (1988) was the first to conduct a dimensional analysis of a movement task: namely, finger oscillations. The technique of correlation dimension rests on the theoretical position that the system dynamics can be reconstructed in its state space from a single time series of the observed movement. This method provides a determination of the number of independent degrees of freedom (here equations of motion) required to reproduce the time evolutionary properties of the system dynamics. Significantly, there need be no relation between number of degrees of freedom at the component (e.g., joint level) and attractor dynamic levels of analysis (Eubank & Farmer 1997).

In light of this, we find it useful to reserve the phrase degrees of freedom for the motions of the components or elements and use dimension to capture the geometric organization of the attractor dynamic (Newell & Vaillancourt 2001). An advantage to using the distinguishing labels of degrees of freedom for the physical components and dimension for the attractor dynamics is that the dimension of an attractor dynamic can be non-integer or fractal. There has been a growing interest in the dimensional analysis of single time series both in the frequency (Spectral Analysis) and time (Detrended Fluctuation Analysis – DFA) domains and the theoretical link to complex systems, self-organization and 1/f-like processes (van Orden, Kloos, & Wallot 2011; West 2006). Indeed, this approach may be the most active contemporary emphasis of Bernstein's degrees of freedom problem though it is not usually interpreted in this light.

Theory of Motor Learning and Development

- Proposition 3: The next movement does not repeat the previous one, therefore, exercise, in essence, is repetition without repetition. The solution represents not repetition or forcing of movement but its construction. Exercise correctly designed repeats not the means used to solve a given motor problem but the process of solving this problem, which improves the means across repetitions (original text edited/shortened by KMN)

The above quotation was also presented in the 1967 Bernstein book. This proposition on practice is central to Bernstein formulating a general theory of motor learning and development. As with a number of Bernstein's (1996) postulations on the processes of practice the repetition phrase has received little direct experimental investigation although there are references in the literature to the catchy notion of "repetition without repetition." Nevertheless, we do know that even given a very small margin of measurement error, that it is very unlikely that the kinematics of one trial are repeated on the next trial even in rapid short duration (100 ms) discrete movements. And, if by chance the movements of successive trials were duplicated within measurement error, this repeatability would not also follow to the lower level of muscle activity (EMG) and other more micro levels of analysis.

The constructs of learning and development are both based upon change in behavior over time. The nature and degree of change in movement skill through practice has been shown to be influenced by a multitude of factors (Newell 1991, 1996). Typically, there is more absolute change in the movement pattern and outcome between the trials of early than late practice. In this context the learner is not trying to repeat the kinematics of the previous trial but actually intending to *change* them so as to derive an outcome that is closer to the goal of the task than was realized on previous trial(s). The relevance of the phrase "repetition without repetition" is perhaps more appropriate later in the practice of a sufficiently skilled performer where there is a focus on reducing variability, although even here one can argue that the performer is not trying to repeat the previous trial. The performer is intending to repeat the action and may or may not have an intention to alter the movement or its outcome. As an instructional constraint in movement experiments the task goal of repeat the movement pattern and outcome of the previous trial is rarely implemented (though see Woodworth (1899) in the movement speed accuracy context).

Bernstein drew on his understanding of motor control to rationalize the postulation of practice being repetition without repetition. Namely, that motor skill cannot represent a fixed effector formula of a sequence of neuromuscular impulses. This is because motor coordination is the continuous adjustment of effector impulses to external conditions, which continuously change and have to be continuously monitored by receptor systems. And, as noted previously, the external and reactive forces are also unpredictable and even the internal forces uncertain. Given these conditions, Bernstein argues that it is impossible to expect any kind of standard sensory corrections. Nevertheless, these sources of variation collectively challenge the efficacy of the "forcing" concepts of movement repetition, such as a motor program view, but we are left with an open question as to the mechanism for systematic change in the construction of the movement pattern and outcome.

The repetition without repetition practice issue has become part of a growing contemporary interest in the role of movement variability in motor learning and development (Newell & Corcos 1993) and more specifically relates to issues pertaining to the degeneracy of movement (Edelman & Gally 2001). This has spawned the decomposition of movement kinematics and outcome data to investigate the structure of the movement variability in the time and/or frequency domains (van Orden et al. 2011; West 2006). The one place where the literal interpretation of repetition without repetition may be implemented is in simple laboratory tasks where the learner is often trying to maintain the same level of variation around a mean target goal value that they can consistently meet and that 'suffices' for the task goal.

Arising from the practice issue, Bernstein reports several significant phenomena evident in motor learning over time as background to developing a two-stage theory of motor learning. Movement execution is viewed as a successive construction process that constrains, within- and between-trials of practice sessions, the collective of unpredictable internal, external and reactive forces. Unlike many behavioral theories of learning and development Bernstein does not propose formal dynamical pathways of change over time (functions of learning). Indeed, the implication is that there is not a single function of change for learning and development given the varying confluences of constraint on movement in action (individual, environment and task – Newell 1986).

Bernstein seemed intrigued by the observation of the sudden change (jump) in movement pattern and/or performance outcome that often was present in the early stage of practice and getting the idea of the task. The sudden jump tends to induce an "Aha" effect on the part of the learner and has been modeled as reflecting a non-equilibrium phase transition. Task category 2 in the Warren (2006) framework of stabilizing an unstable system would show these kinds of qualitative transition changes as a function of practice time and conditions (Newell et al. 2001). Other tasks in this dynamic stability category are likely to include: bicycle riding, pole balancing, roller ball, water skiing, but their learning has rarely been studied. Bernstein does not draw out the distinctive stability properties of these kind of tasks or whether they reflect Level C or D.

It is from his rich description of the phenomena of motor learning and development that Bernstein proposes a two-stage model of motor skill acquisition. He provides an explicit account of the specific functions accomplished in each stage. He also points out that the borders and transition between the two phases are fuzzy. Sometimes, even, there are overlaps between the edges of two stages or even coexistence of the two stages. Overlapping is present with most stage-like theories of motor skill acquisition and behavior more generally.

The first stage is characterized by Bernstein as an active search for more and more adequate, in all aspects, solutions for the exercised motor problem. More specifically, the time period of this stage involves: 1) establishing the leading level; 2) defining motor composition of the movement; 3) identifying adequate corrections; and 4) switching of the background corrections to corresponding lower levels – automation.

This first stage covers a lot of functional changes that require considerable practice time to realize. Only the most elementary movements are confined with all their corrections at a single leading level. Also, the first attempt at performing an unfamiliar movement are built as single-level actions, unless there are old previously elaborated and stored in memory backgrounds.

The second stage is called Stabilization – this is now a common proposal in contemporary dynamical accounts of motor learning. During this period: 1) background levels are switched to lower levels for automation; 2) standardization of motor composition and its components; and 3) stabilization of the motor action takes place. In this stage and considering the construction of levels we see the embedding of common functions in stage 1 into stage 2. Finding experimental tests of the effect of practice and learning on the role of the higher order levels, including switching of the leading level, seems essential to address the potential of the levels of construction framework for motor learning and development.

Most contemporary theories of motor learning have a final stage that is interpreted variously as reflecting processes that characterize the movement control as "automatic" (Fitts 1964). Bernstein proposes that this is accomplished by switching the supporting movement components in the higher leading level to a lower level of the hierarchy. The process of switching technical movement components to lower, background levels represents what is commonly addressed as movement automation. When the leading level of movement construction is changed to a lower (background) or higher level, the "de-automation" of the skill occurs. The latter coins the secondary task protocol in experimental psychology where shared attention disrupts the task performance, and the former has been demonstrated for the last 20 years or so with empirical studies under the themes of focus of attention (e.g., Wulf 2013).

In every movement, independently of its overall level of construction, only the leading level and only corrections executed at that level are consciously perceived. Again, only Levels C–E can operate as the leading level. In several places Bernstein goes beyond automatic and uses the term optimal solution which has been an attractive feature in a number of contemporary perspectives of human performance.

Closing Comments

In closing, it seems apropos to remind that Turvey (1990) drew attention to the central role of Bernstein's (1967) theorizing in what was called Round 1 of the new agenda for movement coordination and control (Kugler et al. 1980). It was proposed that the initial theorizing and experimentation on "Bernstein's problem" was conducted largely in terms of how a device of very many independent variables might be regulated without ascribing excessive responsibility to an executive subsystem. However, Turvey (1990) also indicated that a second round of theory and research on Bernstein's problem had been ongoing for some time. This second round is motivated by similarities between coordination and physical processes in which multiple components become collectively self-organized (van Orden et al. 2011; West 2006). It remains to be seen what lasting role Bernstein's (2020) treatise on the construction of movements will play in this evolving research program and more generally in motor learning and control.

It is typically the case that new theoretical developments in the broad domain of movement and action tend to emphasize the most compatible class of motor tasks for experimental work. Bernstein, by contrast is elaborating on the control distinctions of several of the most relevant types of motor tasks. The outcome is that he is bringing the construct of task constraints to a more central theoretical position than has been typical to date. He is not just *using* a motor task to test some aspect of theory or practice that relates to the individual and/or environment but rather outlining mutual influences of the task with these more standard constructs for motor learning and development.

References

Bernstein, N. (1967). *The Co-ordination and Regulation of Movements.* New York: Pergamon.

Bernstein, N.A. (1996). On dexterity and its development. In M. L. Latash & M.T. Turvey (Eds.), *Dexterity and Its Development* (pp. 3–244). Mahwah N.J.: Lawrence Erlbaum Associates.

Edelman, G. M., & Gally, J. (2001). Degeneracy and complexity in biological systems. *Proceedings of the National Academy of Sciences USA,* 98, 13763–13768.

Eubank, S. G., & Farmer, J. D. (1997). Probability, random processes, and the statistical description of dynamics. In L. Lam (Ed.) *Introduction to Nonlinear Physics* (pp. 106–178). New York: Springer.

Fitts, P. M. (1964). Perceptual-motor skills learning. In A. W. Melton (Ed.), Categories of Human Learning (pp. 243–285). New York: Academic Press.

Fleishman, E. A., & Quaintance, M. (1984). *Taxonomies of Human Performance: The Description of Human Tasks.* New York, NY: Academic Press.

Gentile, A. M. (1987). Skill acquisition: Action, movement, and the neuromotor processes. In J. H. Carr, R. B. Shepard, A. M, Gentile & J. M. Hind (Eds.), *Movement Science: Foundations for Physical Therapy in Rehabilitation* (pp. 93–154). Rockville, MD: Aspen

Haken, H., Kelso, J. A. S., & Bunz, H. (1985). A theoretical model of phase transitions in human hand movements. *Biological Cybernetics*, 51, 347–356.

Kay, B. (1988). The dimensionality of movement trajectories and the degrees of freedom problem: A tutorial. *Human Movement Science*, 7, 343–364.

Kelso, J. A. S. (2009). Synergies: atoms of brain and behavior. *Adv Exp Med Biol.* PMID 19227496.

Kugler, P. N., Kelso, J. A. S., & Turvey, M. T. (1980). On the concept of coordinative structures as dissipative structures: I. Theoretical lines of convergence. In G. E. Stelmach & J. Requin (Eds.), *Tutorials in Motor Behavior* (pp. 1–49). New York: North–Holland.

Latash, M. L. (1996). The Bernstein problem: how does the central nervous system make its choices? In M. L. Latash & M. T. Turvey (Eds.). *Dexterity and Its Development* (pp. 277–304). Mahwah, NJ.: Lawrence Erlbaum.

Latash, M. L. (2000). There is no motor redundancy in human movement. There is motor abundance. *Motor Control*, 4, 259–261.

Latash, M. L. (2008). *Synergy*. Oxford: Oxford University Press.

Liu, Y-T., Chuang, K-L., & Newell, K. M. (2019). Mapping collective variable and synergy dynamics to task outcome in a perceptual-motor skill. *PloS One*, 14(4): e0215460

Mitra, S., Am azeen, P. G., & Turvey, M. T. (1998). Motor learning as decreasing active (dynamical) degrees of freedom. *Human Movement Science*, 17, 17–65.

Newell, K. M. (1986). Constraints on the development of coordination. In M. G. Wade & H. T. A. Whiting (Eds.), *Motor Skill Acquisition in Children: Aspects of Coordination and Control* (pp. 341–360). Amsterdam: Martinies NIJHOS.

Newell, K. M. (1991). Motor skill acquisition. *Annual Review of Psychology*, 42, 213–237.

Newell, K. M. (1996). Change in movement and skill: Learning, retention, and transfer. In M. Latash & M. Turvey (Eds.), *Dexterity and Its Development* (pp. 393–429). Hillsdale, NJ: Erlbaum.

Newell, K. M., & Corcos, D. M. (1993). Issues in variability and motor control. In K. M. Newell & D. M. Corcos (Eds.), Variability and Motor Control (pp. 1–12). Champaign Il: Human Kinetics.

Newell, K. M., & Jordan, K. J. (2007). Task constraints and movement organization: A common language. In W. E. Davis (Ed.), *An Ecological Approach to Human Movement: Linking Theory, Research and Practice.* Champaign, Il: Human Kinetics.

Newell, K. M., & Vaillancourt, D. E. (2001). Dimensional change in motor learning. *Human Movement Science*, 4–5, 695–716.

Newell, K. M., Challis, S., & Morrison, S. (2000). Dimensional constraints on limb movements. *Human Movement Science*, 19, 175–201.

Newell, K. M., Liu, Y. T., & Mayer-Kress, G. (2001). Time scales in motor learning and development. *Psychological Review*, 108, 57–82.

Poulton. E. C. (1957). On prediction in skilled movements. *Psychological Bulletin*, 54, 467–478.

Profeta, V. L. S., & Turvey, M. T. (2018). Bernstein's levels of movement construction: A contemporary perspective. *Human Movement Science*, 57, 111–133.

Saltzman, E. L., & Kelso, J. A. S. (1987). Skilled actions: A task dynamic approach. *Psychological Review*, 94, 84–106.

Schöner, G., & Kelso, J. A. S. (1988). A synergetic theory of environmentally-specified and learned patterns of movement coordination. I. Relative phase dynamics. *Biological Cybernetics,* 58, 71–80.

Shmuelof, L., & Krakauer, J. W. (2011). Are we ready for a natural history of motor learning? *Neuron*, *72*, 469–476.

Turvey, M. T. (1990). Coordination. *American Psychologist*, 45, 938–953.

Turvey, M. T. (2007). Action and perception at the levels of synergies. *Human Movement Science*, 26, 657–697.

Turvey, M. T., Shaw, R. E. & Mace, W. (1978). Issues in the theory of action: degrees of freedom, coordinative structures and coalitions. In J. Requin (Ed.), *Attention and Performance VII*. Hillsdale, N.J.: Erlbaum.

van Orden, G. C., Kloos, H., & Wallot, S. (2011). Living in the pink: Intentionality, wellbeing, and complexity. In *Philosophy of Complex Systems* (pp. 629–672). North-Holland.

Warren, W. H. (2006). The dynamics of perception and action. *Psychological Review*, 113, 358–389.

West, B. J. (2006). *Where Medicine Went Wrong: Rediscovering the Path to Complexity*. New Jersey: World Scientific.

Witte, K., Bock, H., Storb, U., & Blaser, P. (2003). A synergetic approach to describe the stability and variability of motor behavior. In W. Tschacher & J-P. Dauwalder (Eds.), *The Dynamic Systems Approach to Cognition: Concepts and Empirical Pardigms Based on Self-organization, Embodiment, and Coordination Dynamics. Studies of Nonlinear Phenomena in Life Science (Vol. 10)* (pp. 133–144), Singapore: World Scientific.

Woodworth, R. S. (1899). The accuracy of voluntary movement. *Psychological Review*, 3, 1–114.

Wulf, G. (2013). Attentional focus and motor learning: A review of 15 years. *International Review of Sport and Exercise Psychology*, 6, 77–104.

Zatsiorsky, V. M. (1998). *The Kinematics of Human Motion*. Champaign, Ill: Human Kinetics.

C12

HOW CONTEMPORARY THEORETICAL THINKING ABOUT MOVEMENT GENERATION RESONATES WITH BERNSTEIN'S OUTLINE OF A NEURAL ARCHITECTURE OF MOVEMENT

Gregor Schöner

INSTITUTE FOR NEURAL COMPUTATION, RUHR-UNIVERSITY BOCHUM, GERMANY

How the Language and Outlook of the Book Is Still Relevant

Reading Nikolai Bernstein's 1947 book for the first time now, in 2020, in its English translation, was an amazing experience for me. It felt like a connection was established through time, a resonance taking place between my current scientific thinking about movement and the synthesis of a master of movement science that was articulated over 70 years ago. Why that resonance and what does it imply? I want to first briefly characterize the style in which Bernstein talks about movement generation in this book that is highly appealing to me, and to link this style to current lines of thinking in movement science. Then, I want to sketch how my current research program on a neural theory of movement generation aligns astoundingly well with Bernstein's thinking as laid out in his book. This will give substance to my conclusion, that this book, after so many years, still contains core ideas for a research program on movement generation.

The first thing that strikes me reading this translation is the language employed. I am not referring to linguistic style, such as the somewhat complicated phrasing or the neologisms, which shine through in the translation. I am referring to the strictly functional language, the careful, nuanced description of dimensions of motor behavior that draw out many different layers of function, and that help build a functional architecture of the motor system that the book outlines. In a modern context, that language would be highly interdisciplinary, drawing on fields as diverse as biomechanics, muscle physiology, reflex physiology, cortical physiology, neuropsychology, motor development and motor learning. I am not sure, these disciplines or areas existed as articulated fields then, but Bernstein clearly found concepts that we now associate with these fields necessary and useful to understand how all components of movement generation are integrated. More on that below.

What is most remarkable about this functional language is, however, how it remains valid, useful, and even innovative today. I will elaborate on the subtle use the concepts of "space" and "objects" below as examples.

The functional language and how it is used to describe movement generation is, in fact, close to what some of us today call "dynamical systems" thinking in movement science (see Schöner 2014 for a review). Bernstein intuitively articulates the notion that intact movements emerge from the couplings among neural and mechanical systems in what amounts to attractor states in modern language. The key concept is stability (sometimes framed as controllability). He discusses at length the sensory channels that establish such stability by providing reference to motor processes.

In contrast, his language does not resonate as much with the abstract style of control theory, which was being developed in the West at about the same time, brought to the attention of a broader audience through Wiener's 1948 book (but unlikely to have reached Bernstein at the time). Bernstein does not talk about "computation," "information," and does not differentiate between the "plant" and the "controller." He resists abstraction from the physiological and biomechanical grounding of movement generation and in all his descriptions he explicitly refers to the neural pathways and circuits as units of analysis rather than the abstract computations that these circuits implement.

The neurophysiological language reflects, of course, the methods of Bernstein's time, and the concrete neurophysiological facts referred to are probably the most dated aspect of the book. Conceptually, however, that language is surprisingly modern and even innovative. Perhaps this is helped by the fact that his writing predates the strong shift in neurophysiology toward the notions of encoding and decoding that Hubel and Wiesel's discoveries made so attractive. Decoding is today the dominant tool through which function is assigned to any part of the nervous system. In contrast, Bernstein uses co-variation of neural activity in different structures, a fine-grained analysis of the behavioral effects of lesions (neuropsychology), and intuitive (and sometimes speculative) "mental simulation" to understand how the brain's neural networks may coordinate to bring about particular functions. The concepts of coding and decoding ultimately refer to a feed-forward view of the central nervous system, in which the connectivity from the sensory surface or to the motor system are used to characterize the functional significance of activation in a part of the nervous system. Bernstein's discourse is much more about coupling, mutual interaction among components, than about forward information processing.

This style of thinking is well suited to understand integration, how different component processes are coordinated and work together. Such synthesis is not easy and is strongly neglected in much modern physiologically based work in movement science. At the same time, we all recognize that synthesis is much needed to understand how motor function emerges from our bodies and nervous systems (Lisman 2015). Bernstein frames his thinking about the integrated neural system of movement generation in an evolutionary perspective, exploiting rich work in comparative neurophysiology already available at his time. That is a thoroughly modern perspective (Grillner & Robertson 2016). And finally, he is addressing development and skill learning as learning from experience, again in a wholistic and integrated fashion, not as the narrow problem of building an internal model (e.g., Egger et al. 2019).

How My Theoretical Research Program Resonates with Bernstein's Functional Architecture of Movement Generation

Over the last three decades, I have made a systematic effort to develop a theoretical language in which all component process of movement generation can be described and their integration be understood. This began with studying timing and coordination of rhythmic (Schöner & Kelso 1988), and discrete movements (Schöner 1990), and their relation to control (Schöner 1994). That led me to the degree of freedom problem and the concept of the uncontrolled manifold (UCM) (Schöner 1995; Scholz & Schöner 1999; Scholz, Schöner, & Latash 2000). With my colleagues, I developed Dynamic Field Theory (DFT (Schöner, Spencer, & the DFT Group 2016), a theoretical framework grounded in the activation of neural populations in the higher nervous system, to understand sensory-motor decision making (Kopecz & Schöner 1995), and movement preparation (Erlhagen & Schöner 2002). The concepts of DFT made it possible to address scene perception (Zibner et al. 2011) and the attentional selection of objects toward which actions may be directed (Grieben et al. 2020). The coordinate transforms critical for determining motor commands from scene perception were captured as steerable maps within DFT (Schneegans & Schöner 2012). We developed a neural dynamic account for the sequential organization of movement (Sandamirskaya & Schöner 2010; Tekülve et al. 2019) and used obstacle avoidance to study how multiple component tasks are integrated (Grimme, Lipinski, & Schöner 2012).

Integrating these component models into a comprehensive, neurally grounded theory of movement generation is a current and ongoing research program (Martin, Scholz, & Schöner 2009; Zibner, Tekülve, & Schöner 2015; Schöner, Tekülve, & Zibner 2019; Martin, Reimann & Schöner 2017). It was amazing for me to discover then, that our neural dynamic account aligns closely with the physiological account of movement generation sketched by Bernstein in 1947! This is somewhat true in how functions are mapped to parts of the central nervous system, but that may not be surprising and is true for other, much more abstract computational proposals for integrative theories (Scott 2004; Shadmehr & Krakauer 2008). What surprised me was the conceptual alignment in how the different dimensions of movement generation are unfolded and characterized functionally, always framed in terms of coupling and stabilization. Here I want to briefly go through a sketch of such an integrative theory, and align it roughly with the levels that Bernstein introduces in his book. This helps me point out many questions that remain open after so many years.

Figure C12.1 illustrates our overall architecture. I will step through it from the bottom upwards, following a similar path as Bernstein does in his book, from the periphery to the higher and central processes of movement generation (with one notable exception).

Reflexes and Descending Activation

The fundamental unit of movement generation is the individual muscle or muscle groups with their reflex loops. In Bernstein's words (from Chapter 2 of the book): "The reflex loop … represents a fundamental form of the movement-related neural process …" The fact that the force and movement generating periphery is reflex controlled is an important constraint for the neural signals that are required to bring about movement (Feldman 1986). Consider the stretch reflex as the simplest example, illustrated in Figure C12.2, and focus only on the agonist for simplicity (assuming that the antagonist covaries, keeping co-contraction to zero). The motor neuron sums descending activation with the sensory signal from muscle spindles and generates muscle activation when that sum exceeds a threshold. Muscle activation generates force, which tends to shorten the muscle, unless resisted by other forces acting on the joint.

Movement of the joint occurs when descending activation increases the motor neuron's activation beyond threshold, leading to muscle activation, force generation, and contraction of the muscle. As the muscle shortens, the sensory signal from muscle spindles drops, so that ultimately the sum of descending activation and sensory signal falls below threshold. Muscle activation decays and the arm is at the new postural configuration with a silent muscle.

Muscle activation of the agonist makes, in this simple picture, a single positive excursion and generates the force that brings about the movement (the antagonist is activated as well and helps break the movement). The time course of the descending activation is qualitatively different than muscle activation, however! Descending activation changes monotonically, with a continued elevated level as the new postural state is approached. So the descending activation is not isomorphic

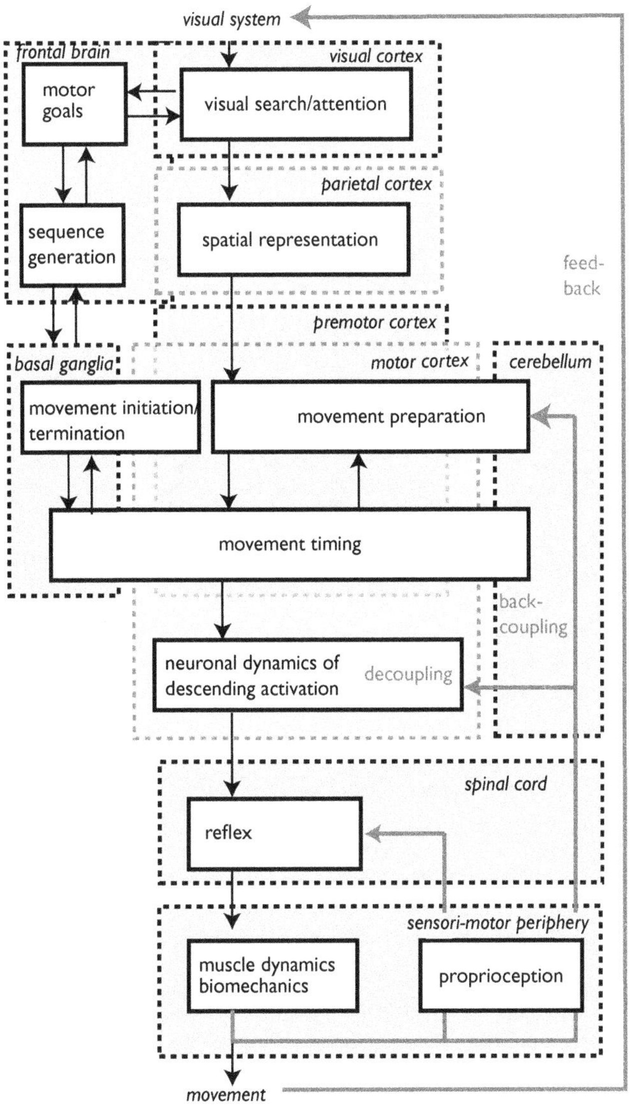

FIGURE C12.1 Overview over a neural processing architecture of object-oriented movements. The mapping onto neuroanatomy (gray boxes) is only approximate and tentative. Portions of the architecture have been realized based on neural dynamics in the cited models (see text for details).

to muscle activation nor to muscle force! Why is that? The descending signal needs to compensate for the reduced sensory signal from the muscle spindles, so that their sum is just below threshold at the end of the movement (again, we need to look at antagonists or external forces to understand why zero force is achieved with muscle activation just below threshold).

This is how the reflex solves the posture-movement problem (Feldman & Latash 2005). The implication is that descending activation is aligned with muscle activation only early during a movement. While muscle activation fades, descending activation evolves toward a new activation state that effectively represents the new postural state.

This is only the simplest manifestation of the larger theme evoked by Bernstein, that the motor periphery with its systems of reflexes determines the "movement problem," that is, constrains what

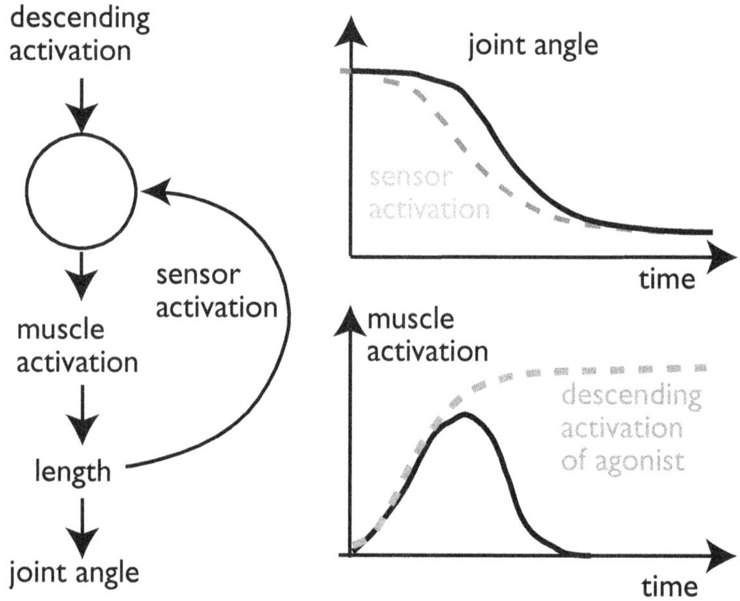

FIGURE C12.2 Left: Schematic of a model of the stretch reflex. Top right: Trajectory of the joint angle (solid) and the associated sensory activation from the muscle spindle of the agonist for that movement (dashed). Bottom right: Time course of muscle activation of the agonist (solid) and the associated time course of the descending activation that induces this muscle activation (dashed).
Note that the descending activation remains elevated after the movement to compensate for the smaller sensory activation, keeping their sum just below threshold.

activation patterns the central nervous system must generate and project onto the periphery to bring about the smooth and efficient movements that humans excel at.

In ongoing work with my students Cora Hummert and Rachid Ramadan, we are using simple models of the stretch reflex, of muscle dynamics and of the biomechanics to quantitatively estimate descending activation patterns from observed movement kinematics. We find that these activation patterns have the qualitative form illustrated in Figure C12.2, but that their temporal shape is modulated as movement speed and the region in work space vary. This modulation reflects the adaptation of descending activation patterns to physical interaction torques, keeping kinematics more invariant under variation of speed and movement parameters than without such modulation. Again, this is a theme anticipated by Bernstein (Chapters 2 and 3).

Generating Time Courses

To bring about a movement, time courses of neural activation patterns that ultimately deliver the descending activation, must be generated. One theme of our theoretical work is that such time courses are ultimately linked to space, that is, to the trajectories of an end-effector through space. Modern evidence for such a view comes from diverse sources, such as the relatively uniform patterns of tuning of neurons in motor and premotor cortex to the direction in space of the hand's trajectory (Georgopoulos, Schwartz, & Kettner 1986) and the correlation of their activity levels with the speed of the hand (Moran & Schwartz 1999).

Bernstein discusses movement timing in a manner that is closely tied to the degree of freedom problem and thus close to the actual articulator and muscle states, his level B (in Chapter 4 of the book). I will instead now first address Bernstein's level C (his Chapter 5) of movement generation that he characterizes by "The leading afferent system of this level is the synthetic spatial field." He elaborates:

> The completely formed spatial field is in full correspondence with the external world and is liberated from the unbreakable bonds to the body, which were so much constraining and devaluating for the spatial synthesis at the level of synergies. Further, the spatial field of level C is vast, immovable, and homogenous.

And further: "The next most important properties of the spatial field are its metrics and geometricity ..." and "Movements at the level of spatial field possess a clear goal-directed character: They move from somewhere, to somewhere, for a certain purpose."

These excerpts read like a description of the constraints we have made explicit in our theoretical modeling (Schöner, Tekülve, & Zibner 2019): Movements are prepared ahead of movement initiation as evidenced by the fact that movement parameters such as movement amplitude, movement direction, and movement time can be predicted from the first 20 ms of a trajectory. These movement parameters reflect the location in space to which the hand is moved, but depend on the initial position of the hand in space, requiring a coordinate transform from allocentric space to body space. And it is the metrics of movement parameters that determines how different sources of information about an upcoming movement are combined (Erlhagen & Schöner 2002). The neural maps that account for movement preparation can be modeled as dynamic neural fields, realized through population activity in motor and premotor cortex (Bastian, Riehle, & Schöner 2003).

We have then gone on to postulate that the time courses of neural activation patterns that drive the movement forward are also represented at this spatial level. Formally, timing amounts to creating a neural oscillation, an autonomous time course of neural activation, even if such an oscillation runs only for a single cycle in what is sometimes called an active transient (Schöner 2002). That timing references, at least in part, a spatial frame is plausible in light of the relative invariance of end-effector time courses, the fact that events in the world are what movement of the hand is effectively timed to, such as when time-to-contact is used to fine-tune movement generation (Schöner 1994b). When the movement of multiple end-effectors is conceived of as the activation of multiple such oscillators, then their synchronization through coupling is responsible for coordination (Schöner & Kelso 1988). Such coordination is tied to the spatial rather than the anatomical alignment of end-effectors (Mechsner et al. 2001), consistent with this Bernsteinian theme of a spatial field at level C.

In Chapter 8 on learning motor skills, Bernstein articulates an intuition that qualitatively different movements may be learned and become automatized, each tunable to accommodate the perceived current environmental situation. One way to think about this is to postulate different movement generators, each a bank of tunable neural oscillators. One such class of movements may be selected, but multiple elementary movements may also be superposed. Studying obstacle avoidance in reaching movements, we identified two movement components for transport of the hand to the goal and lift of the hand off the table surface and a potential obstacle there (Grimme, Lipinski, & Schöner 2012). The notion of movement primitives (Mussa-Ivaldi & Bizzi 2000) and its robotic variant (Schaal, Ijspeert, & Billard 2003) may be close in spirit to this conception although the way such primitives are combined has not been studied systematically.

Another argument of Bernstein's for a spatial representation of movements as "wholes" is the relative invariance of movement outcomes under change of effector, demonstrated in his studies of handwriting with unusual effector arrangements (his Figures 5.4 to 5.7). This supports the notion that movement generators would actually reside at his C (rather than B), which we model by neural populations of oscillators in movement parameter space.

A challenging and unsolved question is if the biomechanical dynamics of the articulator is likewise addressed at this level. In Bernstein's classification of movements kinematic tasks like tracing or reaching and tasks that are strongly constrained by body configuration, momenta, and gravity (his Figures 2.5 to 2.7, for instance) all reside at the same level. Our current theoretical work pursues the hypothesis that modulation of the timing signals at this spatial level may account for adaptation to force-fields and interaction torques. The vast majority of empirical work on such adaptation (in the tradition of Shadmehr & Mussa-Ivaldi 1994) has been performed in non-redundant articulators, making it undecidable in principle if the adaptation occurs at the level of the end-effector in space or at the level joint configurations.

The Degree of Freedom Problem

Which brings me to the problem on which Bernstein's influence is perhaps still the most remarkably felt. His level B (in his Chapter 4) addresses what we now call the degree of freedom problem. I have talked about Bernstein's level C first as it seems to me that his level B is the link between levels A and C. Without understanding the movement as a whole, its spatial movement parameters in the end-effector frame of reference, it is difficult to appreciate the degree of freedom problem, I submit. Again, our theoretical modeling deviates from Bernstein's conception at this point in that we postulate that the timing signals that pace the overall movement are generated at spatial level while Bernstein considered the capacity to "control movement in time gracefully" a property of the level B at which synergies are organized.

It is interesting that Bernstein addresses synergies and the abundance or redundancy of the degrees of freedom of the human movement apparatus before he talks about the higher-level goals and tasks (from levels C and up). He was definitely not viewing the degree of freedom problem as one of "computing" a solution given a task constraint. Instead, he arrived directly at the degree of freedom problem by comparing human movement to the single degree of freedom mechanisms of this time. It was obvious to him that moving beyond such mechanisms entails new problems. Even for two degrees of freedom there are continuously many movement paths that reach a goal. And he anticipated many of the dynamic repercussions of multiple degrees of freedom, including what we would now call interaction torques, that is, the mechanical couplings among different degrees of freedom in a kinematic chain. Their existence implies that motions cannot be planned and controlled independently for each degree of freedom, so that movement generation cannot be decomposed into multiple single-degree of freedom problems (which, ironically, is what is actually done for standard industry robots by making each joint so stiff that it can be controlled by simply rejecting the interaction torques as perturbations). Bernstein even intuits the problem of hysteresis in redundant articulators (Mussa-Ivaldi & Hogan 1991) when he says: "This means that movements of such a multi-link pendulum could be the same across cycles only if active muscle dynamical contributions would vary adaptively over successive cycles."

The way Bernstein talks about the degree of freedom problem is thus much broader and more comprehensive than the modern definition of the problem of redundancy. His discussion is framed in terms of interdependence, and the required coupling of neural control. That is where our current theoretical thinking coheres with Bernstein's picture. Like Bernstein, we do not think of synergies as few-to-many mappings that select a particular solution to the degree of freedom problem out of the continuum of possible ones. Instead, we think of it as a problem of coupling the degrees of freedom in such a way, that their neural control stays within a subset, a manifold, that is compatible with the current motor task. Our modeling (Martin, Scholz, & Schöner 2009; Reimann & Schöner 2017; Martin, Reimann, & Schöner 2019) has led us to three forms of such coupling. First, the neural signals (descending activation) to individual degrees of freedom may be reciprocally coupled. These couplings may be arranged so as to induce compensatory adjustments: a deviation of one signal then induces changes in other signals to keep the movement on track at the level of the spatial trajectory of the end-effector. In other words, these couplings may be set up to keep the arm on the solution manifold to the given state of the end-effector. We had earlier called this the "uncontrolled manifold" (UCM), based on the intuition that there was no need to control within the manifold, only a need to control *onto* the manifold, but that name might have been confusing (Schöner 1995; Scholz & Schöner 1999; Latash, Scholz, & Schöner 2007). I'll come back to this below.

The second form of coupling we named "back-coupling" (Martin, Scholz, & Schöner 2009). It was the idea that a deviation of the limb, its joint angles or muscle lengths, from the state specified as an equilibrium state by the descending activation patterns would lead to an update of the descending neural signals, that would yield to the deviation as long it is lay within that solution manifold. This would allow for the emergence of a new solution that kept invariant the spatial trajectory of the end-effector (or other types of task variables). A related idea was proposed by Mark Latash and colleagues (2005) in a slightly different context and has since been supported by empirical evidence (e.g. Ambike et al. 2016).

The third form of coupling occurs through outer feedback loops and is, therefore, effective only to the extent to which there is sufficient time for such loops to intervene. This is clearly the case for posture, for instance (Reimann & Schöner 2017). The idea is that such outer feedback occurs at the level at which movement is perceived centrally. That perception resides at the level of the spatial representation of the end-effector. A deviation caused by one particular descending neural signal induces a deviation at that spatial level. Signals to compensate for that deviation are sent to all degrees of freedom, effectively coupling them to keep the spatial trajectory of the end-effector invariant.

These three types of coupling explain motor equivalence (Martin, Scholz, & Schöner 2009) just as well as the structure of variance from repetition to repetition of a movement (Martin, Reimann, & Schöner 2019; Scholz & Schöner 2014). This ties in which Bernstein's key idea that the variation from cycle to cycle of a motor task is revealing of the subtle processes of control or stabilization. In fact, the structure in the variance of joint angles or muscle activation patterns that is sometimes referred to as the "UCM effect" is a signature of the organization of the neural control of movement toward particular motor goals that Bernstein considered to be the core problem of movement coordination (from his Chapter 2): "The main problem of movement coordination is indeed in overcoming the redundant degrees of freedom of a moving organ, i.e. in turning it into a controllable system."

Such a use of variation is very different from the rather widespread, but conceptually orthogonal notion that co-variation of degrees of freedom across time or across different movement tasks is a signature of synergies (e.g., d'Avella et al. 2006 and many others, see Kutch & Valero-Cuevas 2012, for review and Bruton & O'Dwyer 2018 for a very useful discussion). Such co-variation across time and movement tasks reflects essentially how trajectories in the space of the degrees of freedom are shaped, and how complex the dependence of these trajectories on varying movement parameters is. Those shapes do not, per se, speak to control and stabilization.

Scene Perception and Movement Sequences

Bernstein describes two higher levels of movement generation, his level D, (which he calls "parietal-premotor") and his group of levels E. Level D is the level of actions, which are typically directed at objects in the world.

That actions are oriented at objects is also a central tenet of our theoretical thinking. In fact, the very first mental act in any such action is to direct attention to the object, to bring it into the attentional foreground, often by shifting gaze. In real life, the visual scene is familiar already, so shifting attention to an object may be based on a scene representation (Hollingworth 2012). Only objects previously attended are part of a scene representation. So visual search for not previously inspected objects may also be necessary. Visual search is studied in the laboratory by presenting the visual scene at the same time as the description of the target object. Attentionally selecting the target object may thus be based on both scene perception and visual search. This can be understood in terms of neural processes using the notions of neural dynamic activation fields mentioned earlier (Grieben et al. 2020). This perceptual front-end is part of the overall neural dynamic architecture of movement generation I am outlining here (Figure C12.1). Consistent with Bernstein's classification, the neural substrate for these processes is partly located in parietal cortex, and the premotor areas, although frontal cortex and subcortical structures such as superior colliculus play a role as well.

The higher levels beyond the level D of actions play a role, ultimately, when sequences of actions are generated to achieve a particular goal. Most motor behavior of humans consists of such sequences, which are serially ordered in a way that makes functional sense. To make coffee, you put the filter in place first, then add the coffee, and only then turn on the machine, for instance. The sequential organization of such actions requires neural control that is not only capable of remembering and organizing actions in a particular serial order, but also of timing the initiation and termination of each element at just the right time (Sandamirskaya & Schöner 2010). Different actions may take different amounts of time, and the same action may take longer to complete in one instance, due to environmental conditions, than in another. So only controlling initiation and termination of each action enables the performance of sequences of actions in such daily life skills. The neural machinery for doing so is part of the overall architecture we outlined here (Figure C12.1, see also Tekülve et al. 2019, for a robotic implementation).

Bernstein discussed writing and speech as examples of higher level skills. Those are, of course, among the most impressive and the most cognitive of action sequences. Their fast rate and often somewhat rhythmic nature may have led many researchers to overlook the issue of controlling on- and offsets. But the larger theme, how action sequences tie into sequences of thought is fascinating and part of the research program I hinted at at the outset (see Richter, Lins, & Schöner 2017, for a first excursion into such a direction).

Motor Learning and Development

A last word on motor learning and development, central to Bernstein's thinking and the topic of the two last chapters of his book. Despite so much work on motor adaptation, theoretical accounts are almost exclusively limited to computational, normative models that define abstractly what the goal of motor learning is (Shadmehr & Mussa-Ivaldi 2012). A continuing challenge for theory is to provide accounts for learning autonomously from experience. This requires, first of all, an account for autonomous movement generation that integrates all aspects of the "construction of movement." And such accounts are still in their infancy as reviewed here. It then also requires an account for how motor learning is engaged and controlled. Demands arise on the processes of movement generation to enable autonomous learning. Motor commands must be kept neurally active long enough to update neural networks when the outcome of a movement has been perceived, for instance (Sandamirskaya & Storck 2015). The challenge of autonomous learning is also the reason why so little progress has been made toward a theoretical understanding of motor development.

There are many other topics that Bernstein touches, which continue to be stimulating for theoreticians. His notion of automatization and de-automatization is an example as is his analysis of different levels of motor learning. It seems really exceptional to me that a research monograph can outline research problems that remain programmatic 70 years later.

Conclusion

The functional, neurally grounded perspective Bernstein provides in his book can still be used as a rough road-map for neural process accounts of movement generation. In spite of many elements that are dated, especially in neurophysiology, and a fair amount of what we would now call speculation, Bernstein's style of analysis remains attractive and instructive. This is due, in large part, to his nuanced functional analysis of movement, his serious interest in understanding integration and how all processes and levels of movement generation interact, his emphasis on neural processes and their interaction or coupling, and his intuitive grasp of dynamics at all levels. It is tempting to contrast this with the sterility of the abstract, normative thinking that currently prevails in computational theories of motor control. I think, Bernstein may yet inspire theoreticians to stay closer to behavior, neural process, and integration.

References

Ambike, S., Mattos, D., Zatsiorsky, V. M., & Latash, M. L. (2016). Unsteady steady - states: central causes of unintentional force drift. *Experimental Brain Research, 234*(12), 3597–3611.

d'Avella, A., Portone, A., Fernandez, L., & Lacquaniti, F. (2006). Control of fast-reaching movements by muscle synergy combinations. *The Journal of Neuroscience, 26*(30), 7791–7810.

Bastian, A., Schöner, G., & Riehle, A. (2003). Preshaping and continuous evolution of motor cortical representations during movement preparation. *European Journal of Neuroscience, 18*(7).

Bruton, M., & O'dwyer, N. (2018). Synergies in coordination: A comprehensive overview of neural, computational, and behavioral approaches. *Journal of Neurophysiology, 120*(6), 2761–2774.

Egger, S. W., Remington, E. D., Chang, C. J., & Jazayeri, M. (2019). Internal models of sensorimotor integration regulate cortical dynamics. *Nature Neuroscience, 22*(11), 1871–1882.

Erlhagen, W., & Schöner, G. (2002). Dynamic field theory of movement preparation. *Psychological Review, 109*(3), 545–572.

Feldman, A. G. (1986). Once more on the equilibrium point hypothesis (λ-model) for motor control. *Journal of Motor Behavior, 18*, 15–54.

Feldman, A. G., & Latash, M. L. (2005). Testing hypotheses and the advancement of science: recent attempts to falsify the equilibrium point hypothesis. *Experimental Brain Research, 161*, 91–103.

Georgopoulos, A. P., Schwartz, A. B., & Kettner, R. E. (1986). Neural population coding of movement direction. *Science, 233*, 1416–1419.

Grieben, R., Tekülve, J., Zibner, S. K. U., Lins, J., Schneegans, S., & Schöner, G. (2020). Scene memory and spatial inhibition in visual search. *Attention, Perception, and Psychophysics*. (in press)

Grillner, S., & Robertson, B. (2016). The basal ganglia over 500 million years. *Current Biology, 26*(20), R1088–R1100.

Grimme, B., Lipinski, J., & Schöner, G. (2012). Naturalistic arm movements during obstacle avoidance in 3D and the identification of movement primitives. *Experimental Brain Research, 222*(3), 185–200.

Gurfinkel, V. S., & Cordo, P. J. (1998). The scientific legacy of Nikolai Bernstein. In M. L. Latash (Ed.), *Progress in Motor Control* (pp. 1–19). Champaign, Illinois, USA: Human Kinetics
Hollingworth, A. (2012). Guidance of visual search by memory and knowledge. In M. D. Dood & J. H. Flowers (Eds.), *The Influence of Attention, Learning, and Motivation on Visual Search, Nebraska Symposium on Motivation* (pp. 63–89). New York, NY, USA: Springer.
Kopecz, K., & Schöner, G. (1995). Saccadic motor planning by integrating visual information and pre-information on neural dynamic fields. *Biological Cybernetics, 73*, 49–60.
Kutch, J. J., & Valero-Cuevas, F. J. (2012). Challenges and new approaches to proving the existence of muscle synergies of neural origin. *PLoS Computational Biology, 8*(5).
Latash, M. L., Scholz, J. P., & Schöner, G. (2007). Toward a new theory of motor synergies. *Motor Control, 11*, 276–308.
Latash, M. L., Shim, J. K., Smilga, A. V, & Zatsiorsky, V. M. (2005). A central back-coupling hypothesis on the organization of motor synergies: a physical metaphor and a neural model. *Biological Cybernetics, 92*(3), 186–191.
Lisman, J. (2015). The challenge of understanding the brain: Where we stand in 2015. *Neuron, 86*(4), 864–882.
Martin, V., Scholz, J. P., & Schöner, G. (2009). Redundancy, self-motion and motor control. *Neural Computation, 21*(5), 1371–1414.
Martin, V., Reimann, H., & Schöner, G. (2019). A process account of the uncontrolled manifold structure of joint space variance in pointing movements. *Biological Cybernetics, 113*(3), 293–307.
Mechsner, F., Kerzel, D., Knoblich, G., & Prinz, W. (2001). Perceptual basis of bimanual coordination. *Nature, 414*, 69–73.
Moran, D. W., & Schwartz, A. B. (1999). Motor cortical representation of speed and direction during reaching movement. *Journal of Neurophysiology, 82*, 2676–2692.
Mussa-Ivaldi, F. a, & Bizzi, E. (2000). Motor learning through the combination of primitives. *Philosophical Transactions of the Royal Society of London. Series B, Biological Sciences, 355*(1404), 1755–1769.
Mussa-Ivaldi, F. A., & Hogan, N. (1991). Integrable solutions of kinematic redundancy via impedence control. *International Journal of Robotics Research, 10*(5), 481–491.
Reimann, H., & Schöner, G. (2017). A multi-joint model of quiet, upright stance accounts for the "uncontrolled manifold" structure of joint variance. *Biological Cybernetics, 111*(5–6), 389–403.
Richter, M., Lins, J., & Schöner, G. (2017). A neural dynamic model generates descriptions of object-oriented actions. *Topics in Cognitive Science, 9*, 35–47.
Sandamirskaya, Y., & Schöner, G. (2010). An embodied account of serial order: How instabilities drive sequence generation. *Neural Networks, 23*(10), 1164–1179.
Sandamirskaya, Y., & Storck, T. (2015). Learning to look and looking to remember: A neural-dynamic embodied model for generation of saccadic gaze shifts and memory formation. In P. Koprinkova-Hristova, V. Mladenov, & N. K. Kasabov (Eds.), *Artificial Neural Networks SE – 9* (Vol. 4, pp. 175–200). Springer, New York.
Schaal, S., Ijspeert, A., & Billard, A. (2003). Computational approaches to motor learning by imitation. *Philosophical Transactions of the Royal Society (London) Series B, 358*, 537–547.
Schneegans, S., & Schöner, G. (2012). A neural mechanism for coordinate transformation predicts pre-saccadic remapping. *Biological Cybernetics, 106*(2), 89–109.
Scholz, J. P., & Schöner, G. (1999). The uncontrolled manifold concept: Identifying control variables for a functional task. *Experimental Brain Research, 126*, 289–306.
Scholz, J. P., & Schöner, G. (2014). Use of the Uncontrolled Manifold (UCM) Approach to understand motorvariability, motor equivalence, and self-motion. In M. F. Levin (Ed.), *Progress in Motor Control* (Vol. 826, Chapter 7). Cham: Springer International Publishing.
Scholz, J. P., Schöner, G., & Latash, M. L. (2000). Identifying the control structure of multijoint coordination during pistol shooting. *Experimental Brain Research, 135*, 382–404.
Schöner, G. (1990). A dynamic theory of coordination of discrete movement. *Biological Cybernetics, 63*, 257–270.
Schöner, G. (1994a). From interlimb coordination to trajectory formation: common dynamical principles. In S. Swinnen, H. Heuer, J. Massion, & P. Casaer (Eds.), *Interlimb Coordination: Neural, Dynamical, and Cognitive Constraints* (pp. 339–368). Academic Press, San Diego.

Schöner, G. (1994b). Dynamic theory of action-perception patterns: The time-before-contact paradigm. *Human Movement Science, 3*, 415–439.

Schöner, G. (1995). Recent developments and problems in human movement science and their conceptual implications. *Ecological Psychology*, 7, 291–314.

Schöner, G. (2002). Timing, Clocks, and Dynamical Systems. *Brain and Cognition, 48*, 31–51.

Schöner, G. (2014). Dynamical systems thinking: From metaphor to neural theory. In P. C. M. Molenaar, R. M. Lerner, & K. M. Newell (Eds.), *Handbook of Developmental Systems Theory and Methodology* (pp. 188–219). New York, New York, USA: Guilford Publications

Schöner, G., & Kelso, J. A. S. (1988). Dynamic pattern generation in behavioral and neural systems. *Science, 239*(4847), 1513–1520.

Schöner, G., Spencer, J. P., & DFT Research Group, T. (2016). *Dynamic Thinking: A Primer on Dynamic Field Theory*. Oxford University Press.

Schöner, G., Tekülve, J., & Zibner, S. (2019). Reaching for objects : a neural process account in a developmental perspective. In D. Corbetta & M. Santello (Eds.), *Reach-to-Grasp Behavior: Brain, Behavior and Modelling across the Life Span* (pp. 281–318). Taylor & Francis.

Scott, S. H. (2004). Optimal feedback control and the neural basis of volitional motor control. *Nature Reviews Neuroscience, 5*(7), 534–546.

Shadmehr, R., & Mussa-Ivaldi, F. a. (1994). Adaptive representation of dynamics during learning of a motor task. *The Journal of Neuroscience, 14*(5), 3208–3224.

Shadmehr, R., & Krakauer, J. W. (2008). A computational neuroanatomy for motor control. *Experimental Brain Research, 185*, 359–381.

Shadmehr, R., & Mussa-Ivaldi, F. A. (2012). *Biological Learning and Control: How the Brain Builds Representations, Predicts Events, and Makes Decisions*. Cambridge, MA, USA: MIT Press.

Tekülve, J., Fois, A., Sandamirskaya, Y., & Schöner, G. (2019). Autonomous sequence generation for a neural dynamic robot: Scene perception, serial order, and object-oriented movement. *Frontiers in Neurorobotics, 13*, 1–27.

Wiener, N. (1948). *Cybernetics: or Control and Communication in the Animal and the Machine (2nd edition)*. Cambridge, MA, USA: The MIT Press.

Zibner, S. K. U., Faubel, C., Iossifidis, I., & Schöner, G. (2011). Dynamic neural fields as building blocks for a cortex-inspired architecture of robotic scene representation. *IEEE Transactions on Autonomous Mental Development, 3*(1), 74–91.

Zibner, S. K. U., Tekülve, J., & Schöner, G. (2015). The neural dynamics of goal-directed arm movements : a developmental perspective. In *International Conference on Learning and Development ICDL-EpiRob 2015* (pp. 154–161).

C13

WHAT DO SYNERGIES DO AT BERNSTEIN'S LEVEL OF SYNERGIES?

Mark L. Latash

DEPARTMENT OF KINESIOLOGY, THE PENNSYLVANIA STATE UNIVERSITY, UNIVERSITY PARK, PA, USA

Introduction

Nikolai Bernstein was not the first to use the word "synergy" with respect to the coordination of biological movements. Related terms, such as "asynergia," had been used by neurologists at least since the turn of the twentieth century (Babinski 1899). Nevertheless, the contribution of Bernstein to the refinement and development of this concept was so prominent that nearly all researchers refer to Bernstein when they discuss synergies (for recent reviews see Latash and Zatsiorsky 2016; Santello et al. 2016; Profeta and Turvey 2018). Frequently, however, references to Bernstein's ideas overlook the importance of the level of synergies in Bernstein's hierarchy, emphasize some of Bernstein's ideas at the expense of others, and lead to misrepresentation of the role of synergies, which is described in much detail in *On the Construction of Movements*.

Currently, the word "synergy" is used in the literature in three main meanings. In the clinical literature, this word is used in a meaning dramatically different from that of Bernstein's (e.g., DeWald et al. 1995): Synergies are pathological patterns of muscle activation following an injury to the central nervous system (CNS), which interfere with purposeful voluntary movements (e.g., "flexion synergy," "extension synergy," etc.). Bernstein did not use this term with respect to "pathological synergies"; hence, we will not discuss them here.

A more traditional meaning of this word emphasizes one of the two main features of synergies within Bernstein's scheme, namely reducing the number of degrees-of-freedom (DOFs) to make multi-joint kinematic chains controllable. This understanding is dominating the current literature, and synergies are commonly defined as co-varying patterns of involvement of variables produced by elements (elemental variables), such as joint angles, digit forces, and muscle activation levels (reviewed in Ting and McKay 2007; Tresch and Jarc 2009; Santello et al. 2016).

More recently, synergies have been defined as neural organizations ensuring dynamical stability of actions by multi-element systems (reviewed in Latash et al. 2007). This definition stresses the other main feature of synergies as described by Bernstein. Further, I will try to make arguments that only the third definition makes sense, whereas the traditional one is built on one of the very few erroneous conclusions drawn by Bernstein.

What Did Bernstein Imply under "Synergy"?

In his description of the level of synergies (level *B*), Bernstein mentions both aspects, reducing the number of DOFs and ensuring dynamical stability of action. Further, however, in particular in sections on movement disorders, Bernstein describes consequences of neurological injuries only with respect to the latter role of synergies, not the former one. Indeed, he emphasizes, on many occasions, loss of movement stability during disorders of the level of synergies but never mentions problems with "too many DOFs to handle." He does emphasize reduction of the number of DOFs ("freezing DOFs") at early stages of skill acquisition as the means of making multi-joint kinematic chains controllable in the absence of their dynamical stability. However, as discussed later, "freezing DOFs" is not a viable concept and, even if it were possible, it would not reduce the number of DOFs across levels of analysis. The seeming "freezing" of kinematic DOFs reflects lack of developed control patterns at a task-specific level.

Let me make a sacrilegious statement: Despite being a genius, Bernstein sometimes made misleading or even wrong statements. One of those is the (in)famous "problem of motor redundancy." After Bernstein had accepted this concept, it naturally led to the idea of elimination of DOFs by the CNS and synergies as the means of accomplishing this goal. It fit perfectly well the development of the control theory in the middle of the twentieth century and, unfortunately, got accepted as ultimate truth. To clarify this point, let me dwell a bit on the concepts of DOFs, redundancy, and abundance.

DOFs at the Level of Peripheral Variables, Redundancy, and Abundance

By definition, the number of DOFs is equal to the minimal number of potentially independent variables that allow describing behavior of a system of interest *at a selected level of analysis*. Most commonly, in movement studies, DOFs are associated with directly measured variables (or those computed based on directly measured ones), such as joint angles, muscle forces, effector (e.g., digit) forces, muscle activations, etc. Note, however, that the number of DOFs in such sets depends crucially on the selected level of analysis. Indeed, even activation level of a single muscle may be viewed as representing hundreds of DOFs corresponding to firing frequencies of individual motor units. Selecting level of analysis is done subjectively, commonly biased by views of the researcher and even by the available equipment.

Let us now consider whether "freezing" a joint in a kinematic chain (i.e., making its motion relatively small; it is never going to be zero) is going to help alleviate the DOF problem. This is highly unlikely due to the following reasons. First, the "non-frozen" joints still involve multiple muscles and a huge number of motor units. Second, given time-varying interaction torques, "freezing a joint" requires non-trivial control of muscles crossing that joint (cf. Koshland et al. 1991; Latash et al. 1995). Hence, at the motor unit level or muscle activation level, the number of DOFs remains unchanged or may even increase. Overall, attempts at counting DOFs without first accepting a theory on the neural control of movement that would explicitly specify control variables manipulated by the CNS are exercises in futility.

Note that all the mentioned variables recorded in typical experiments fail to qualify as neural control variables. Indeed, in one of his very first figures (Figure 2.1 in Bernstein's text), Bernstein makes a very important point that *no peripheral mechanical variable can be prescribed by the CNS* because of the inherent muscle force-length relationship and the unpredictable spread of interaction forces across elements of kinematic chains. Bernstein did not pay much attention to muscle reflexes; otherwise, he would be able to conclude that muscle activation levels (which are functions of the muscle mechanics due to the action of reflexes) also cannot be prescribed by the CNS.

We arrive at a set of intermediate conclusions. First, *any human movement is hopelessly redundant, at least at some levels of analysis. This means that the apparent redundancy in a motor system can never be eliminated.* Second, *attempts at addressing the problem of motor redundancy at arbitrarily selected levels reflecting peripheral variables (e.g., joint rotations, muscle forces, or muscle activations) may be justified in the fields of engineering and control theory but have no relevance to the field of the neural control of movement unless relations of those variables to hypothetical neural control variables are stated.* Third, *the notion of redundancy in biological movements is ill-defined and has to be either defined better (be my guest!) or replaced altogether.*

An alternative view rejecting the concept of redundancy was introduced not that long ago as the *principle of abundance* (Gelfand and Latash 1998; Latash 2012). According to this view, counting *peripheral* DOFs in a biological system is not very meaningful and the apparently redundant DOFs may be highly useful – more is better! – for ensuring dynamical stability of actions and possibility of performing multiple actions by a single set of effectors without negative interference (Zhang et al. 2008).

DOFs at Neural Control Levels

In the objective opinion of the author, there is only one theory of the neural control of movement that has at least a chance to be useful and productive, which is the equilibrium-point hypothesis (Feldman 1966, 1986) developed over the past years into the theory of hierarchical control with referent spatial coordinates (RCs; reviewed in Latash 2010; Feldman 2015). This theory follows the physical approach to motor control, i.e. search for laws of nature that define biological movements (reviewed in Latash 2016, 2017). It postulates that animals produce movements by changing parameters of relevant laws of nature for the involved effectors; these parameters are RCs. Setting a time profile of RC(t) leads to the emergence of muscle activations and mechanical variables, which are also affected by the (unpredictable!) external forces and spontaneous changes in the excitability of involved neuronal pools.

For a single animal (human), RCs can be introduced at different levels, from the whole body to individual limbs, joints, muscles, and motor units (Figure C13.1). The number of RCs for elements at each level of the hierarchy illustrated in Figure C13.1 increases in the top-to-bottom direction. This could be seen as another set of problems of motor redundancy. Indeed, one might ask a question: How is the CNS deciding which combinations of RCs to use at a lower level to satisfy constraints imposed by a lower-dimensional RC set at a higher level? The point is that the CNS does not do such things. It sets parameters of corresponding laws of nature and allows physics and physiology to generate variable solutions for such problems given the actual interaction between the body and the environment.

The back-coupling loops illustrated in Figure C13.1 allow the CNS to make some directions in the space of lower-level RCs more or less stable in a task-specific way (Latash et al. 2005; Martin et al. 2009) leading to selective stabilization of task-specific performance variables. A toolbox developed within the uncontrolled manifold (UCM) hypothesis (Schöner 1995; Scholz and Schöner 1999) has been used to explore stability of different performance variables across a variety of tasks, effector systems, spaces of variables, and populations (reviewed in Latash et al. 2007; Latash and Zatsiorsky 2016). This approach has recently been generalized for RCs and demonstrated that even the apparently non-redundant tasks, such as pressing downward with a single finger to produce a certain force magnitude, involve abundance at the control level, which is used by

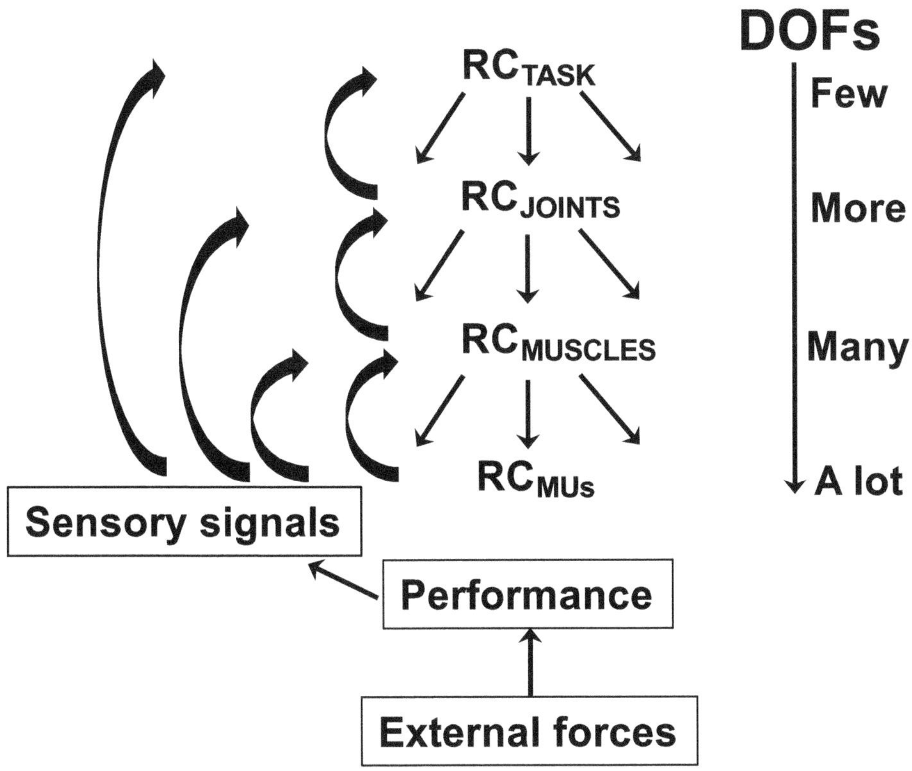

FIGURE C13.1 A schematic illustration of a set of hierarchical levels during the control of an action. At each level, values of spatial referent coordinates (RC) are specified for the effectors. Note the increase in the number of RCs (DOFs) in the top-to-bottom direction. Back-coupling loops both within the CNS and from sensory receptors provide stability of salient performance variables.

the actors to stabilize the salient performance variable, i.e., force magnitude (Ambike et al. 2016; Reschechtko and Latash 2017).

To conclude this section, *abundance is present is all motor tasks, including even the apparently non-redundant ones. Abundance is not a computational nuisance but a powerful and highly useful feature that allows natural movements to be both stable and flexible.*

Is There a Way to Decrease the Number of DOFs?

If *redundancy* is actually *abundance*, then there is no compelling need to reduce the number of DOFs. Consider the hierarchical control scheme with RCs in Figure C13.1, from task-level to individual MU level. At each lower level, uniting variables into groups does not seem to make any of the control processes less redundant; it simply adds another level into the hierarchy. Regularities of co-varied patterns across recorded variables, frequently reported in experiments, reflect a specific stage of skill acquisition – selecting particular, comfortable (or preferred for other reasons including habits) patterns of control variables, e.g., RCs at the joint level reflected in "muscle synergies."

Imagine that one of the transformations illustrated in Figure C13.1 involves mapping of a higher-level one-dimensional RC (RC_{TASK}) on two lower-level RCs for two elements (RC_{EL}),

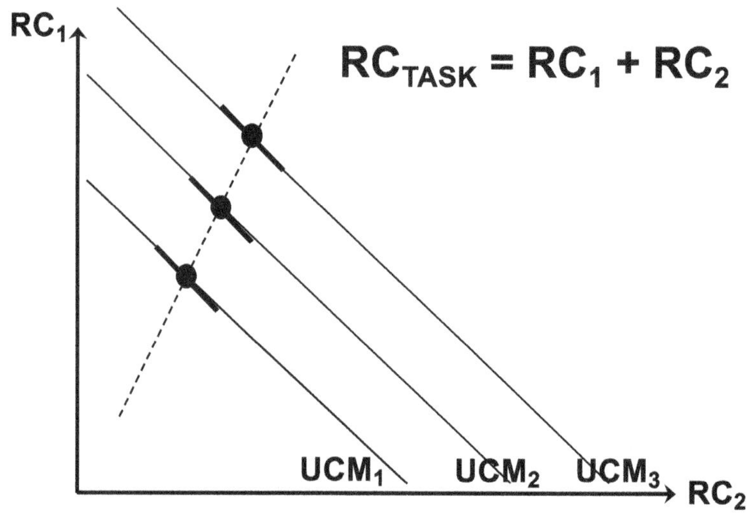

FIGURE C13.2 A task involving two elements coordinated to produce a constant sum of their referent coordinates (RC).
In the two-dimensional space of elemental RCs (RC_1 and RC_2), a constant magnitude of RC_{TASK} corresponds to a slanted solution space (uncontrolled manifold, UCM). Only a range of solutions is typically used (thick lines centered about a preferred solution – black point). Changing the desired magnitude of RC_{TASK} (three magnitudes are illustrated with three UCMs) leads to translation of the preferred solutions and ranges along a line corresponding to a fixed sharing pattern (dashed line).

e.g., during the task of producing a desired level of total force while pressing with two fingers (Figure C13.2). For simplicity, assume a linear transformation: $RC_{TASK} = \Sigma RC_{EL}$. The solution space for this task (the corresponding UCM) is illustrated as the slanted solid line in Figure C13.2. While all points in the UCM are equally able to solve the task, humans typically use only a relatively small range within the UCM (illustrated with the thick line and a dot in its center in Figure C13.2) corresponding to preferred (e.g., comfortable) finger force combinations. With changes in the hand force, the UCM shifts and these preferred patterns are reflected in relatively stable sharing patterns (Li et al. 1998) leading to positive co-variation of individual finger forces. Formally, this pattern qualifies as a synergy in its traditional definition. Note, however, that the number of DOFs at any level of description remains unchanged.

Constraints similar to the one illustrated in Figure C13.2 introduce regularities into changes in recorded variables and allows substituting them with a set of "modes" (e.g., Danion et al. 2003; Krishnamoorthy et al. 2003), but this procedure, by itself, has no effect on the number of variables manipulated by the CNS. This number stays the same at the task level, the ultimate element (e.g., finger) level, the muscle level, and the motor unit level. In some cases, the number of modes is the same as the number of the recorded outputs of the apparent elements (as in the case of finger modes, Danion et al. 2003). In other cases, it is smaller as in studies of multi-muscle systems (Krishnamoorthy et al. 2003; Ivanenko et al. 2004; Ting and Macpherson 2005. Still, the problem of mapping a low-dimensional RC_{TASK} on high-dimensional elemental variables is not solved, it is simply split into two problems. The understanding of traditional synergies as the means of reducing the number of DOFs manipulated by the CNS is a byproduct of the engineering approach where control variables are designed by engineers and typically correspond to the number of elements, e.g., one actuator per joint. This is obviously not true for the hierarchy illustrated in Figure C13.1.

Within the scheme in Figure C13.1, the presence of co-varied changes in variables representing the outputs of elements at a selected level of analysis reflects particular preferences for sharing patterns that define actually observed solutions within the corresponding UCMs. These preferences may reflect anatomical and physiological constraints (e.g., joints have different ranges of motion, muscles have different force capabilities, body segments have different mass-inertial properties, etc.) as well as other, self-imposed, constraints related to such undefined notions as subjective comfort, avoidance of situations perceived as dangerous, habits, etc. They are reflected in specific mappings from task-level RCs to element-level ones. For example, they are likely to lead to observed regularities in the mechanical joint involvement during whole-body actions, which are commonly referred to as "ankle strategy" and "hip strategy" (Horak and Nashner 1986).

To conclude this section, *the concepts of too many DOFs and reduction in the number of DOFs have no relevance to biological movement. The goal of synergies is not to reduce the number of DOFs; only major orthopedic surgery can achieve such an outcome.*

So, What Do Synergies Do?

As it follows from the previous text, synergies have nothing to do with reduction of DOFs. The other major function of synergies, as hypothesized by Bernstein, is to ensure action stability by multi-element kinematic chains. The recent development of the UCM hypothesis (Schöner 1995; reviewed in Latash et al. 2007) follows closely this insight and develops it for any actions, not necessarily involving multi-joint kinematics.

According to the UCM hypothesis, the CNS organizes certain stability properties of a salient performance variable, to which a group of elemental variables contribute. In particular, this leads to certain properties of inter-trial co-variation within the elemental variable space. This does not necessarily mean that the CNS always ensures as high stability as possible because, in certain situations, excessive stability may be counter-productive. For example, when a person gets ready to perform an action associated with a quick change in a performance variable, high stability of that variable acts against the desired change. In other words, stability and agility are competitors in human movements. Two mechanisms of attenuating stability of action have been described. One of them leads to lower stability indices when the actor knows that a quick action may be required, even if no quick action is actually performed (de Freitas et al. 2007; Tillman and Ambike 2017). The other one, termed anticipatory synergy adjustment (ASA) is seen during preparation to an actual quick action; it involves a drop in the stability index 300±100 ms prior to the action initiation (Olafsdottir et al. 2005; Piscitelli et al. 2017).

Several schemes for ensuring task-specific action stability have been suggested including optimal feedback control schemes (Todorov and Jordan 2002; Diedrichsen et al. 2010), which assume computation of cost functions by the CNS over the estimated movement time and, by this feature, belong to the field of engineering and control theory, not to biology. Indeed, no physical (physiological) laws have been suggested that would be able to perform the assumed computations involving future magnitudes of performance variables. Extensive arguments against such approaches to analyze biological movements have been published (Feldman 2015; Latash 2019).

The alternative schemes involve physiological back-coupling loops both from peripheral receptors and within the CNS (Latash et al. 2005; Martin et al. 2009). The obvious advantage of within-the-CNS loops is short time delays that allow facilitating co-variation of elemental variables even during very fast actions. Such schemes are similar to the classical network of Renshaw cells providing quick negative feedback to alpha-motoneurons within a pool that may be viewed as stabilizing the overall output of the pool. There is an obvious drawback of central back-coupling schemes, which is their inability to respond to unpredictable changes in the environment. A number of recent studies have shown that if a task is formulated using visual feedback on a salient variable, turning visual feedback off leads to loss of stability of that variable and its unintentional drift (Ambike et al. 2016; Parsa et al. 2016; Reschechtko and Latash 2017). This happens despite the availability of natural (proprioceptive and tactile) information. Taken together, these studies suggest that both central back-coupling loops and loops using signals from peripheral receptors are likely to work in parallel.

An important advantage of back-coupling schemes is the possibility to modify the gains within such schemes, which can lead to a number of experimentally confirmed phenomena. These include the aforementioned ASAs, facilitating differential sensitivity to sensory signals of different modality, and stabilizing different performance variables produced by the same set of elements (e.g., Zhang et al. 2008). The last feature may be viewed as central for performing a motor task on the background of another ongoing task involving the same set of elements, e.g., kicking a football without falling down.

To conclude this subsection, *synergies ensure controlled, task-specific stability of salient performance variables.*

Synergies in Bernstein's Scheme

Within Bernstein's scheme, the level of synergies rarely represents the task level (the leading level), while it is a major source of background coordination vital for nearly all functional actions. Within our scheme illustrated in Figures C13.1 and C13.2, the level of synergies takes responsibility for mapping between a task-level $RC_{TASK}(t)$ function and $RC_{EL}(t)$ functions to effectors. There are potentially several such mappings involved in any action, which may be distributed throughout the CNS. Possibly, some of such mappings, for example from the muscle level to the motor-unit level, are handled by spinal reflex-like and recurrent-inhibition-like circuits, whereas there is plenty of clinical observations supporting the importance of supraspinal circuitry for ensuring action stability and agility (reviewed in Latash and Huang 2015).

Bernstein associated the level of synergies with such prominent brain structures as the thalamus and globus pallidus. Both structures are involved in the loops through the basal ganglia, while the thalamus is also involved in loops involving the cerebellum. We can conclude that Bernstein's insight on possible neural mechanisms of synergies was surprisingly accurate. Indeed, a number of neurological disorders involving subcortical structures and loops (Parkinson's disease, multi-system atrophy, and multiple sclerosis) lead to a combination of impaired stability of actions and impaired agility reflected in reduced ASAs (Park et al. 2012, 2013; Jo et al. 2017). In contrast, mild cortical stroke results in impaired agility with less pronounced effects on indices of stability (Reisman and Scholz 2003; Jo et al. 2016).

To conclude, *the organization of task-specific stability likely involves neural circuits distributed throughout the CNS, from the spinal cord to subcortical loops involving the cerebellum and the basal ganglia.* This view fits another seminal Bernstein's idea of "brain operators" (cf. "distributed processing units," Houk 2005). But this topic would go too far beyond the scope of this commentary.

References

Ambike S, Mattos D, Zatsiorsky VM, Latash ML (2016) Unsteady steady-states: Central causes of unintentional force drift. *Experimental Brain Research* 234: 3597–3611.

Babinski F (1899) De l'asynergie cerebelleuse. *Revue Neurologique* 7: 806–816.

Danion F, Schöner G, Latash ML, Li S, Scholz JP, Zatsiorsky VM (2003) A force mode hypothesis for finger interaction during multi-finger force production tasks. *Biological Cybernetics* 88: 91–98.

de Freitas SM, Scholz JP, Stehman AJ (2007) Effect of motor planning on use of motor abundance. *Neuroscience Letters* 417: 66–71.

DeWald JP, Pope PS, Given JD, Buchanan TS, Rymer WZ (1995) Abnormal muscle coactivation patterns during isometric torque generation at the elbow and shoulder in hemiparetic subjects. *Brain* 118: 495–510.

Diedrichsen J, Shadmehr R, Ivry RB (2010) The coordination of movement: optimal feedback control and beyond. *Trends in Cognitive Science* 14: 31–39.

Feldman AG (1966) Functional tuning of the nervous system with control of movement or maintenance of a steady posture. II. Controllable parameters of the muscle. *Biophysics* 11: 565–578.

Feldman AG (1986) Once more on the equilibrium-point hypothesis (λ-model) for motor control. *Journal of Motor Behavior* 18: 17–54.

Feldman AG (2015) *Referent Control of Action and Perception: Challenging Conventional Theories in Behavioral Science*. Springer, NY.

Gelfand IM, Latash ML (1998) On the problem of adequate language in movement science. *Motor Control* 2: 306–313.

Horak FB, Nashner LM. (1986) Central programming of postural movements: adaptation to altered support-surface configurations. Journal of Neurophysiology 55: 1369–1381.

Houk JC (2005) Agents of the mind. *Biological Cybernetics* 92: 427–437.

Ivanenko YP, Poppele RE, Lacquaniti F (2004) Five basic muscle activation patterns account for muscle activity during human locomotion. *Journal of Physiology* 556: 267–282.

Jo HJ, Maenza C, Good DC, Huang X, Park J, Sainburg RL, Latash ML (2016) Effects of unilateral stroke on multi-finger synergies and their feed-forward adjustments. *Neuroscience* 319: 194–205.

Jo HJ, Lucassen E, Huang X, Latash ML (2017) Changes in multi-digit synergies and their feed-forward adjustments in multiple sclerosis. *Journal of Motor Behavior* 49: 218–228.

Koshland GF, Gerilovsky L, Hasan Z (1991) Activity of wrist muscles elicited during imposed or voluntary movements about the elbow joint. *Journal of Motor Behavior* 23: 91–100.

Krishnamoorthy V, Goodman SR, Latash ML, Zatsiorsky VM (2003) Muscle synergies during shifts of the center of pressure by standing persons: Identification of muscle modes. *Biological Cybernetics* 89: 152–161.

Latash ML (2008) *Synergy*. Oxford University Press: New York.

Latash ML (2010) Motor synergies and the equilibrium-point hypothesis. *Motor Control* 14: 294–322.

Latash ML (2012) The bliss (not the problem) of motor abundance (not redundancy). *Experimental Brain Research* 217: 1–5.

Latash ML (2016) Towards physics of neural processes and behavior. *Neuroscience and Biobehavioral Reviews* 69: 136–146.

Latash ML (2017) Biological movement and laws of physics. *Motor Control* 21: 327–344.

Latash ML (2019) *Physics of Biological Action and Perception*. Academic Press: New York, NY.

Latash ML, Aruin AS, Shapiro MB (1995) The relation between posture and movement: A study of a simple synergy in a two-joint task. *Human Movement Science* 14: 79–107.

Latash ML, Huang X (2015) Neural control of movement stability: Lessons from studies of neurological patients. *Neuroscience* 301: 39–48.

Latash ML, Scholz JP, Schöner G (2007) Toward a new theory of motor synergies. *Motor Control* 11: 276–308.

Latash ML, Shim JK, Smilga AV, Zatsiorsky V (2005) A central back-coupling hypothesis on the organization of motor synergies: a physical metaphor and a neural model. *Biological Cybernetics* 92: 186–191.

Latash ML, Zatsiorsky VM (2016) *Biomechanics and Motor Control: Defining Central Concepts*. Academic Press: New York, NY.

Li ZM, Latash, ML, Zatsiorsky VM (1998) Force sharing among fingers as a model of the redundancy problem. Experimental Brain Research 119: 276–286.

Martin V, Scholz JP, Schöner G (2009) Redundancy, self-motion, and motor control. *Neural Computations* 21:1371–1414.

Olafsdottir H, Yoshida N, Zatsiorsky VM, Latash ML (2005) Anticipatory covariation of finger forces during self-paced and reaction time force production. *Neuroscience Letters* 381: 92–96.

Park J, Lewis MM, Huang X, Latash ML (2013) Effects of olivo-ponto-cerebellar atrophy (OPCA) on finger interaction and coordination. *Clinical Neurophysiology* 124: 991–998.

Park J, Wu Y-H, Lewis MM, Huang X, Latash ML (2012) Changes in multi-finger interaction and coordination in Parkinson's disease. *Journal of Neurophysiology* 108: 915–924.

Park J, Zatsiorsky VM, Latash ML (2010) Optimality vs. variability: An example of multi-finger redundant tasks. *Experimental Brain Research* 207: 119–132.

Parsa B, O'Shea DJ, Zatsiorsky VM, Latash ML (2016) On the nature of unintentional action: A study of force/moment drifts during multi-finger tasks. *Journal of Neurophysiology* 116: 698–708.

Piscitelli D, Falaki A, Solnik S, Latash ML (2017) Anticipatory postural adjustments and anticipatory synergy adjustments: Preparing to a postural perturbation with predictable and unpredictable direction. *Experimental Brain Research* 235: 713–730.

Profeta VLS, Turvey MT (2018) Bernstein's levels of movement construction: A contemporary perspective. *Human Movement Science* 57: 111–133.

Reisman D, Scholz JP (2003). Aspects of joint coordination are preserved during pointing in persons with post-stroke hemiparesis. *Brain* 126: 2510–2527.

Reschechtko S, Latash ML (2017) Stability of hand force production: I. Hand level control variables and multi-finger synergies. *Journal of Neurophysiology* 118: 3152–3164.

Santello M, Bianchi M, Gabiccini M, Ricciardi E, Salvietti G, et al. (2016) Hand synergies: Integration of robotics and neuroscience for understanding the control of biological and artificial hands. *Physics of Life Reviews* 17: 1–23.

Scholz JP, Schöner G (1999). The uncontrolled manifold concept: Identifying control variables for a functional task. *Experimental Brain Research* 126: 289–306.

Schöner G (1995) Recent developments and problems in human movement science and their conceptual implications. *Ecological Psychology* 8: 291–314.

Tillman M, Ambike S (2017) Cue-induced changes in the stability of finger force-production tasks revealed by the uncontrolled manifold analysis. *Journal of Neurophysiology* 119: 21–32.

Ting LH, Macpherson JM (2005) A limited set of muscle synergies for force control during a postural task. *Journal of Neurophysiology* 93: 609–613.

Ting LH, McKay JL (2007) Neuromechanics of muscle synergies for posture and movement. *Current Opinions in Neurobiology* 17: 622–628.

Todorov E, Jordan MI (2002) Optimal feedback control as a theory of motor coordination. *Nature Neuroscience* 5: 1226–1235.

Tresch MC, Jarc A (2009) The case for and against muscle synergies. *Current Opinions in Neurobiology* 19: 601–607.

Zhang W, Scholz JP, Zatsiorsky VM, Latash ML (2008) What do synergies do? Effects of secondary constraints on multi-digit synergies in accurate force-production tasks. *Journal of Neurophysiology* 99: 500–513.

AUTHOR INDEX

SUBJECT INDEX

For Product Safety Concerns and Information please contact our EU
representative GPSR@taylorandfrancis.com
Taylor & Francis Verlag GmbH, Kaufingerstraße 24, 80331 München, Germany

www.ingramcontent.com/pod-product-compliance
Lightning Source LLC
Chambersburg PA
CBHW080735150926
51501CB00005B/171
* 9 7 8 0 3 6 7 4 1 8 9 0 8 *